Advances in Waste Management

Ajay S. Kalamdhad · Jiwan Singh
Kondusamy Dhamodharan
Editors

Advances in Waste Management

Select Proceedings of Recycle 2016

Springer

Editors
Ajay S. Kalamdhad
Department of Civil Engineering
Indian Institute of Technology Guwahati
Guwahati, Assam
India

Kondusamy Dhamodharan
Department of Civil Engineering
Indian Institute of Technology Guwahati
Guwahati, Assam
India

Jiwan Singh
Department of Environmental Science
Babasaheb Bhimrao Ambedkar University
Lucknow, Uttar Pradesh
India

ISBN 978-981-13-0214-5 ISBN 978-981-13-0215-2 (eBook)
https://doi.org/10.1007/978-981-13-0215-2

Library of Congress Control Number: 2018939948

Printed on acid-free paper

This Springer imprint is published by the registered company Springer Nature Singapore Pte Ltd.
The registered company address is: 152 Beach Road, #21-01/04 Gateway East, Singapore 189721, Singapore

Preface

The amount of solid wastes generated from human, animal, industries, and agricultural activities increases with increasing world's population, consequently causing more pollution problems to the environment. The treatment and disposal of solid wastes, predominantly wastes generated in small, medium, and large urban cities, have become a comparatively serious problem mainly in the developing countries. In economically developing countries, financial, human, and other critical resources generally are very limited. The book presents new concept and strategy of waste management which combines technologies of waste treatment and recycling and emphasizes the benefits to be gained from the use of the recycled products. This book is also focused on wastewater treatment, groundwater treatment using waste materials, application of geographical information system (GIS) for waste management, and photocatalytic treatment of organic pollutants. Waste treatment technologies such as composting, anaerobic digestion, recycling of plastic and agricultural waste in construction are cost-effective and environment-friendly, bring economic returns, and are applicable to most developing countries. This book is intended to be used as a text for the students, scientists, industrialists, and academicians who are working in the field of environmental engineering and sciences. Many institutes/universities worldwide have developed new prospectuses on environment and sustainable development or are offering specialized courses on sustainable waste management, all relating to the subject contents of this book. Environmental professionals and policy makers will find this book a useful reference source for design and operation of solid waste management programs.

The text is arranged into five sections, covering solid waste management (Section A), wastewater and groundwater treatment (Section B), phytoremediation (Section C), aquatic weed management (Section D), and constructed wetland (Section E). This publication has been prepared primarily for two audiences: (1) researchers working in the field of solid waste management and water treatment and (2) professionals convoluted in the management of solid wastes. The information in the publication would also be useful to the students of Environmental Science and Engineering.

Aim and Scope of the Proposed Book

Pollution through solid waste has become one of the serious problems of the current scenario mostly in the developing countries. Urbanization, industrialization, modernization, and population growth are mainly accountable for high increasing rate of solid waste and also a major challenge for environmental sustainable development. Improper waste management can cause serious health problems and other environmental problems. The book presents new technologies of waste management and emphasizes the benefits to be gained from the use of the recycled products. Waste treatment technologies such as composting, anaerobic digestion, recycling of plastic and agricultural waste in construction are cost-effective and environment-friendly, bring economic returns, and are applicable to most developing countries. The aim of the proposed book is to support all those involved in research and teaching related to solid waste management and water treatment study in the leading academic and research organizations around the world. This book will be of great value to postgraduate students, research scientists and professionals in Environmental Science and Engineering, and authorities responsible for the management of solid wastes. Many universities worldwide can develop new prospectuses on sustainable waste management, which are relating to the subject contents of this book. Environmental professionals and policy makers will find this book a useful reference source for design and operation of waste reuse and recycling programs.

Guwahati, India — Ajay S. Kalamdhad
Lucknow, India — Jiwan Singh
Guwahati, India — Kondusamy Dhamodharan

Contents

About the Editors

Dr. Ajay S. Kalamdhad is currently working as an Associate Professor at the Department of Civil Engineering, Indian Institute of Technology Guwahati (IITG). He obtained his Bachelor's, Master's, and Ph.D. in Civil and Environmental Engineering from GEC Jabalpur, VNIT Nagpur, and IIT Roorkee, respectively. Prior to joining IIT Guwahati in 2009, he was an Assistant Professor at VNIT Nagpur and worked on various projects at RRL, Bhopal (Now AMPRI, Bhopal), and NEERI, Nagpur. In addition to his book on rotary drum composting, he has published more than 100 research papers in acclaimed journals and has presented his work in more than 150 national and international conferences/workshops. He is associated with Indian Public Health Engineers, India; International Solid Waste Association, Italy; and National Solid Waste Association of India; and serves as a reviewer for 20 international journals. He is a recipient of ISTE-SGSITS National Award for best research by young teachers of engineering colleges for the year 2012 and IEI Young Engineers Award for 2011–2012 in environmental engineering discipline from Institute of Engineers, India.

Dr. Jiwan Singh is an Assistant Professor at the Department of Environmental Science, BBA University, Lucknow. He obtained his Ph.D. in Civil Engineering from Indian Institute of Technology Guwahati. He has worked as Visiting Scientist at the University of Ulsan, Republic of Korea, and as an Assistant Professor at Kwangwoon University, Republic of Korea. He is currently involved in water and wastewater treatment projects sponsored by Government of India. He has published more than 45 international research papers. His research specialization is in wastewater treatment and solid waste management.

Dr. Kondusamy Dhamodharan is a Research Associate at the Department of Civil Engineering, Indian Institute of Technology Guwahati (IITG). He did his Master's in Energy and Environmental Engineering from VIT University, India, and Ph.D. in Environmental Engineering from IIT Guwahati. His work has received many accolades in both national and international conferences. His research interests include anaerobic digestion, waste to energy, renewable technology, and integrated solid waste management planning. He is presently serving as a reviewer for *Journal of Environmental Chemical Engineering* (Elsevier) and *Bioresource Technology* (Elsevier), 3 *Biotech* (Springer) and *Waste and Biomass Valorization* (Springer).

Earthworm Technology—A Promising Tool for Second Green Revolution

Priyasankar Chaudhuri

Abstract The Green Revolution in India made the country self-sufficient in food grain production. However, indiscriminate use of chemical fertilizers and pesticides during last five decades has made the soil unproductive and caused environmental pollution too. Environmentalists are now planning for a second Green Revolution, where reduced use of chemical fertilizers, pesticides, by recovery of soil health to increase in food grain production through organic farming along with conservation of biodiversity, is in the focus. Present paper deals with adoption of earthworm technology which could be the best choice for second Green Revolution. Earthworm technology is of two types: in-soil and ex-soil. In-soil technology involves mass culture of suitable geophagous and phytogeophagous earthworm species and their direct inoculation in soil with proper organic input. Ex-soil technology, popularly known as vermicomposting, allows quick transformation of organic wastes into plant-nutrient-rich compost through activities of phytophagous earthworm species.

Keywords Earthworm technology · Vermicomposting · *Perionyx excavatus* *Eudrilus eugeniae* · Second Green Revolution

1 Introduction

In order to make the country self-sufficient in food grain production, a package programme comprising of application of chemicals such as fertilizers and pesticides, high-yielding varieties of seeds and extensive irrigation was taken in the mid-sixties (1967–1968) of the last century. Initially, this programme led to dramatic increase in food grain production, which was called Green Revolution (1967–

P. Chaudhuri (✉)
Department of Zoology, Tripura University, Suryamaninagar, Agartala 799022, Tripura, India
e-mail: priya_1956@rediffmail.com

A. S. Kalamdhad et al. (eds.), *Advances in Waste Management*,
https://doi.org/10.1007/978-981-13-0215-2_1

1995). In the first phase, Green Revolution was mainly restricted to Punjab, Haryana and western part of Uttar Pradesh. Later, it spread over to Bihar, West Bengal, Madhya Pradesh and a few other parts of the country. In fact, all those parts that enjoyed the benefit of Green Revolution were plain lands with good physical structure and irrigation facilities. Noticeably, dramatic increase in crop production was noticed up to 1985, and thereafter, growth rate of agricultural production had lost its momentum.

1.1 Weakness of First Green Revolution

Green Revolution made the country self-sufficient in food grain production, yet it had the following drawbacks:

a. Green Revolution was restricted to some specific parts of the country. North-eastern states were totally deprived of its effects.
b. Green Revolution was centred on wheat and paddy that had no impact on production of pulse—the poor man's protein.
c. In the second phase of Green Revolution (1981–1995), agricultural production was not sustainable.
d. Environmental issues were overlooked.

1.2 Cause of Decline in Agricultural Productivity in the Second Phase of First Green Revolution

It is necessary to know the structural components of soil in order to find out the root cause for decline in agricultural productivity in the second phase of Green Revolution. The soil comprises of four components, viz. minerals, organic matter, air and water. Of the total soil volume, about 50% is pore space, 45% mineral materials and 5% organic matter. Under ideal conditions for plant growth, the pore space is divided roughly into 25% of air and 25% water space. The proportions of water and air are subjected to rapid and natural great fluctuations, depending on the extrinsic factors, such as weather conditions and other factors. In fact, intimately mixed conditions of four major soil components encourage interactions within and between them that permits marked variation in the soil environment for the plant growth [1].

Soil is a very active biological laboratory. It harbours a varied population of living organisms like bacteria (more than 1 lakh in 1 g of soil), fungi, actinomycetes, nematodes, earthworms, micro- and macro-arthropods. In the soil ecosystem, activities of soil biota range largely from the physical breakdown of plant residues by earthworms and micro-arthropods to the eventual microbial decomposition of the residues leading to release of several nutrient elements, including nitrogen, phosphorus and sulphur. Humus is one of the most useful products of microbial action.

Good physical health of soils was one of the key factors to boost agricultural productivity in the first phase of Green Revolution. Indiscriminate use of chemical fertilizers and pesticides for the last four decades made the tropical soil unproductive due to burning of soil organic carbon, deterioration of soil physical structure (thus affecting soil porosity and water holding capacity) and crops prone to repeated pest attacks. Agricultural revolution of the pest thus turned into present day's agricultural disasters. Costly chemical inputs lead not only to soil biodiversity crisis but also contamination and pollution of soil, water, air, plants and crops. The damage caused through agrochemical pollution of environment and human health, directly and through the human food chain, is irreparable. In many cases, over ninety per cent of the inorganically produced vegetables, food grains, fruits, milk contain poisonous agrochemical residues harmful and unsuitable for human consumption [2]. As a result, Indian agriculture is now in a crisis due to greying of Green Revolution. So, in the beginning of the twenty-first century, environmental and agricultural scientists were planning for a second Green Revolution for sustainable agriculture (i.e. 'evergreen revolution') where farming would largely be based on rejection or reduced use of chemical fertilizers, recovery of soil health and sustainable increase in food grain production through organic farming and conservation of biodiversity.

2 Steps to Be Taken in Second Green Revolution ('Evergreen Revolution')

2.1 *Adoption of Organic Farming Systems*

Adoption of the following eco-friendly agricultural practices is recommended in order to make the agricultural production sustainable and to keep the environment healthy.

a. Organic manures, viz. worm-worked compost, i.e. vermicompost, ordinary compost, green manures, poultry manures, blood meal, sewage sludge.
b. Bio-fertilizers, viz. *Rhizobium*, *Azospirillum*, *Azotobacter*, *Phosphobacterium*, *VA mycorrhiza*, Blue-green algae, *Azolla*.
c. Crop rotation with intercrop legume for atmospheric nitrogen fixation.
d. Bio-pesticides, viz. neem, tobacco.
e. Biological control: Use of predators, parasitoids and pathogens to suppress pest population.

Organic farming system has been recommended for creating eco-friendly and pollution-free environment, food security and safety for improving public health, ecological balance and micro-environment for growth of soil micro-flora and fauna and aboveground vegetation.

Requirement of manpower is much more in organic farming system unlike machinery-based costly chemical farming practices. So, scope of economic upliftment of poor class farmers is much more in second Green Revolution.

2.2 *Biodiversity Conservation*

Indiscriminate use of chemical fertilizers and pesticides during first Green Revolution led to biodiversity crisis. In fact, only 1% of pesticides act on target organ and the rest through soil bring deleterious effect on soil biodiversity. Overdoses of pesticides cause decline in the population of earthworms, mites, springtails and nitrogen-fixing bacteria [1]. All these organisms are in some way associated with improvement of physico-chemical status of soil. So instead of chemical pesticides, bio-pesticides have been recommended for evergreen revolution.

2.3 *Emphasis on Dry and Tilla Land Farming*

Dry and tilla lands of India were deprived of benefits of the first Green Revolution. Organic farming improves physical structure of the soil through increase in soil aggregation and porosity, thereby increasing the water holding capacity of the soils. Slow release of nutrients from organic manure leads to their proper utilization by plants and sustainable agriculture. It is expected that in the hilly and dry areas of India, crop production could be increased to a large extent following adoption of organic farming.

3 Earthworms—Nature's Best Gift in Organic Farming

Feeding, burrowing and casting activities of earthworms modify the physico-chemical and biological properties of soil, thus supporting aboveground vegetation. In the soil physical properties to be affected include soil aggregation and porosity, while the biological and chemical properties to be modified include nutrient cycling, formation of plant-available nutrients, organic matter dynamics, microbial and faunal activities, decomposition rate [3]. Consequently, plant production above ground may be affected by the activities of earthworms where their three main ecological groups can be recognized: epigeic, endogeic and anecic.

3.1 *Epigeic Species*

Epigeic species are non-burrowing litter and dung-feeding earthworms. By commuting the litter, they modify its physico-chemical characteristics, generally reducing its C/N ratio, making it suitable for further microbial breakdown. They have small-to-medium body size, deep pigmentation, dorsoventrally flattened body, tolerance to disturbance, high metabolic rate, high fecundity, short life cycle and high power of regeneration. Epigeic earthworms form the forest floor community of temperate and tropical countries. They can be utilized for vermicomposting. Examples include *Perionyx excavatus*, *Eisenia fetida* and *Eudrilus eugeniae*.

3.2 *Anecic Species*

The soil inhabiting endogeic and anecic earthworm species are called the 'ecosystem engineers' because of their great role in soil turn over, improvement of water holding capacity, aerobicity and thereby exerting regulatory force in soil functioning. Anecic earthworms live in vertical burrow system that open to the soil surface. They feed on decaying plant litter with some amount of soil and form surface casts. Anecic worms are generally large, moderate to heavily pigmented and intolerant to disturbances. Examples of anecic earthworms are *Drawida grandis*, *Metaphire houlleti*, *Lampito mauritii*. Anecic earthworms show good response to chemical/phytochemical method of extraction from soils [4]. Temperate parts of the world are dominated by both epigeic and anecic earthworms.

3.3 *Endogeic Species*

Polypheretima elongata, *Pontoscolex corethrurus*, *Metaphire posthuma*, etc., are geophagous and form both sub-surface and surface casts. They are non-pigmented or lightly pigmented and form horizontal burrows in top and sub-soils. They establish symbiotic relationship with the gut micro-flora. Endogeic earthworms dominate the tropical parts of the world. Both anecic and endogeic earthworms have long life cycle with the limited power of regeneration. Endogeic earthworm species have diverse effects on soil properties. Compacting species like *P. corethrurus*, *Eutyphoeus gammiei* egest compact globular and tower shaped casts. They increase the proportion of large aggregates along with bulk density of soil. On the other hand, 'decompacting' species such as *L. mauritii* and *M. posthuma* feed on soils along with organic matter and also casts of other compacting species of earthworms egests smaller and fragile aggregates, i.e. granular casts [5]. Thus, the decompacting species of earthworms decrease the amount of large aggregates in soil and also its bulk density. Epigeic species are concerned with humus formation in the surface soil. Both anecic and endogeic species increase overall soil porosity and enhance water infiltration. The three categories of earthworms thus contribute differently to soil fertility and structure and may be manipulated in a given area in a specific agricultural situation.

4 Earthworm Technologies

Earthworm technology involves utilization of suitable species of earthworms to improve soil health and thereby enhancing plant growth. Earthworm technology is of two types: **in-soil technology** and **off-soil technology** [5].

4.1 In-Soil Earthworm Technology

This involves direct management as well as indirect management [5].

Direct Management. In these systems, locally available endogeic and anecic earthworms, following their mass culture, are inoculated directly in the soils of plantation crop with or without proper organic input. Most successful species tested so far for in-soil earthworm technology is *P. corethrurus*. A culture bed of 5 m × 1 m × 20 cm filled with a 3:1 mixture of soil and partly composted sawdust, 200 adult *P. corethrurus* (100 g fresh weight) produced 3355 worms (including immature) within 4 months [6]. In terms of effects, tree seedlings have been shown to be highly responsive to the inoculation of *P. corethrurus* in nursery bags [5]. Following application of selected species of earthworms in the tea gardens at Western ghat, South India, there was significant improvement of overall biological and physico-chemical status of soils over the control plot. In-soil earthworm technology not only resulted in great reduction (30–50%) in the use chemicals, but also led to significant increase (33–80%) in processed tea production when compared with the fertilizer-based management practices [6] and overall long-term sustainability of the soil health conditions [6]. In fact, direct management practice may positively affect plant growth only if a large earthworm biomass (at least equal to a fresh weight value of 35 g m^{-2}) is inoculated initially. Direct management would not be effective in dry lands or high rain-fed zones. Besides these, due to high cost of earthworm production and inoculation, direct management practices can only be realistically applied to high-valued crops like tea, coffee. Considering cost–benefit evaluation and rapid loss of organic carbon, indirect management is preferable to direct management under tropical climatic conditions [6].

Indirect Management. These practices involve mulching, organic matter input, crop rotation, minimum tillage, incorporation of legumes to maintain earthworm populations at a critical level of 30–40 g m^{-2} (fresh biomass), resulting in a significant increase in plant production [6].

In-soil earthworm technology ensures the associated soil biodiversity and long-term sustainability of the soil ecosystem.

4.2 Off-Soil Earthworm Technology: Vermicomposting

Off-soil/ex-soil earthworm technology, popularly known as vermicomposting, allows quick transformation of organic wastes into plant-nutrient-rich compost called vermicompost through synergistic actions of bacteria and epigeic species of earthworms. Details of the process of vermicomposting under Indian conditions have been discussed by Kale [7] and Chaudhuri [8, 9]. The best-known species with potential for waste management under Indian conditions include *P. excavatus*, *E. eugeniae, E. fetida and Eisenia andrei*. There is no significant difference in the quality of compost produced by *Eisenia, Eudrilus and Perionyx*, while *Eudrilus* tops the list of vermicomposting worms when the rate of vermicompost production by this species is considered [10]. In spite of this, most commonly used earthworm species for commercial vermicomposting is *E. fetida* and its closely related *E. andrei*. This is due to the fact that *Eisenia* are generalized feeder, easily handled, a prolific breeder and have a wide range of ecological tolerance. In India, *Perionyx sansibaricus* and *Perionyx pallus* have also been tested for their potential in organic waste degradation [10, 11]. *P. elongata, L. mauritii* and *Dichogaster curgensis* should not be considered as vermicomposting species because they always need soil bedding for their survival.

Management During Vermicomposting. Proper management produces good quality compost within a short period. The key to maximum productivity is to maintain aerobicity in the wastes combined with optimum moisture and temperature conditions. Loss of nutrients through volatilization or leaching becomes minimum during efficient and rapid vermicomposting. The quantity of water to be mixed depends upon the nature of the wastes. Vegetable and fruit wastes do not require much water because such kind of wastes gives out water during the process of decomposition. Agricultural wastes (residues of sugarcane, coconut and betel nut, straw, etc.) mixed with cow dung slurry will hasten up the initial decomposition. Fresh kitchen wastes, poultry and pig wastes contain a large amount of salts and ammonia that killed earthworms. So prior to earthworm inoculation, these should be removed through composting, washing or ageing [12]. High-quality substrates like pig manures, kitchen wastes tend to decompose rapidly even in the absence of earthworms. In fact, greatest gains in the nutrients status can be achieved by adding bulking carbon-rich wastes such as straw, sawdust, paddy husk to enhance the C to N ratio and extend the biological activity, thus maximizing earthworm yields, and the proportion of nutrients to be extracted [12].

During the process of vermicomposting earthworms multiply. The overcrowding will reduce and hamper their growth and reproduction. So, once the population reaches a peak, harvesting of worms is needed immediately. To make the harvesting process easier, vermicomposting tank should be built with placement of porous partition (for horizontal movement of earthworms) inside. Vermicomposting is allowed in one side of the partition. After about one month of vermicomposting, the empty chamber is to be loaded with partly decomposed wastes. This process allows easy movement of worms from the matured compost to new food material.

The fastest processing of organic wastes is achieved by inoculation with nearly fully grown earthworms because of their high rate of feeding activity. Culture beds should be kept free from giant flatworms, termites, rodents and centipedes.

Physical and Chemical Characteristics of Vermicompost: Vermicompost produced from most of the organic wastes is a finely granular homogeneous substance with excellent physical structure, porosity, aeration, drainage and moisture retention capacity [12]. The level of macro- and micro-nutrients is higher than the compost derived from any other method [10]. Remarkable feature of vermicompost is that, during conversion of various organic wastes by earthworms, many of the nutrients are changed into available forms such as nitrate or ammonium nitrogen, exchangeable phosphorous and soluble potassium, calcium and magnesium that are readily taken up by the plants for their growth. So, vermicompost has higher levels of available nutrients than the wastes from which they were formed [13]. Slow but steady release of plant nutrients, improvement of soil physical properties by increasing soil aggregation, enhancement of micronutrient element nutrition in plants through chelation reaction, etc. [14], presence of plant growth substance such as humates, fulvates and large amount of indole acetic acid (IAA), kinetin or gibberellins (plant growth hormone) produced by bacteria within the worm castings were a few among many of its important properties [15]. Another important feature of vermicomposting is that during production of vermicompost, bio-available heavy metals and human pathogens are eliminated and vermicompost possesses much more nitrate nitrogen and humic acid and less C:N ratio than the ordinary compost [16]. Major nutrient contents in the vermicompost derived from various organic wastes are given in Table 1.

Table 1 Major plant nutrients in earthworm-processed wastes

Waste materials	Element content (% dry wt)					Reference
	N	P	K	Ca	Mg	
Cattle solids	2.20	0.40	0.90	1.20	0.25	[3]
Pig solids	2.60	1.70	1.40	3.40	0.55	[3]
Cattle solids + straw	2.50	0.50	2.50	1.55	0.30	[3]
Pig solids + straw	3.00	1.60	2.40	4.40	0.60	[3]
Chick solids + wood shavings	1.80	2.70	2.10	4.80	0.70	[3]
Leaves	0.80	0.34	0.82	0.96	0.53	[25]
Sugarcane	2.67	2.11	0.40	4.08	1.89	[26]
Banana leaf bases	2.50	0.56	3.74	2.36	1.50	[26]
Kitchen wastes	1.81	1.12	0.87	2.95	0.41	[27]
Paper wastes	1.10	0.57	0.29	8.02	0.53	[28]

Vermicompost 'Tea'. Watery extract obtained from vermicompost is popularly called vermicompost 'tea'. Micro-organisms present in vermicompost 'tea' inactivate and suppress the growth of pathogens. Hence, the extract of vermicompost can be used as foliar spray. This liquid organic manure, vermicompost 'tea', can be prepared by taking one kilogram of freshly collected vermicompost with three litres of water and is allowed to soak for two days. The mixture is thoroughly stirred before decanting to use it as spray. This preparation has the water-soluble plant nutrients, mucus secretions of earthworms and also micro-organisms. Liquid organic manures showed positive results with pot experiments [17]. Vermicompost 'tea' also contains plant growth regulators produced by micro-organisms [18].

Vermiwash. Vermiwash is obtained by washing of earthworms in lukewarm water. It can be prepared by using the following method [17].

Earthworms (1/2 kg) when immersed in 500 cc of lukewarm water (37–40 °C) for about 30–40s, coelomic fluid oozes out through their dorsal pores. These heat stressed earthworms are to be immersed again in half a litre of water at room temperature (25–28 °C) to overcome the mild heat shock. Then, the fluid sticking to their bodies mixes with the water. Water in both the trays is mixed which is called vermiwash.

Utility of Vermiwash. The body fluid extract of earthworms has shown the formation of inhibitory zones for some plant pathogens in petri plates at the same time of application [17]. So, it can be used to spray on the seedlings or for soaking seeds before sowing. Excellent results were obtained when leguminous seeds were treated with vermiwash to induce nodulation. It has also been tested as a component of nutrient media in tissue culture and has shown good response to callus growths [17]. A stimulatory effect of root induction in cuttings has also been recorded [17].

4.3 Vermicompost as Plant Growth Medium

Figure 1 summarizes physical, chemical and biological changes brought about by application of vermicompost on the soil. Dramatic effects of vermicompost on growth of plants are absolutely due to improvement of soil physical structures and properties (aggregation, porosity, water holding capacity), presence of plant growth regulators, control of plant pathogens and to a greater extent increase in microbial population and plant-available nutrients [16]. In fact, slow release of nutrients from vermicompost has a sustainable effect on agriculture.

Vermicompost has been shown to promote growth of a wide variety of cereals, pulses, fruit plants, vegetables and ornamental plants [17]. Following application of vermicompost on summer variety wetland paddy at north Bangalore, Kale et al. [19] observed increase in population of beneficial microbes like N-fixing bacteria and mycorrhizae over control plots. In West Tripura, a field trial conducted on upland rice (var. TRC-87-281) using 10 tons vermicompost/hectare and 5 tons

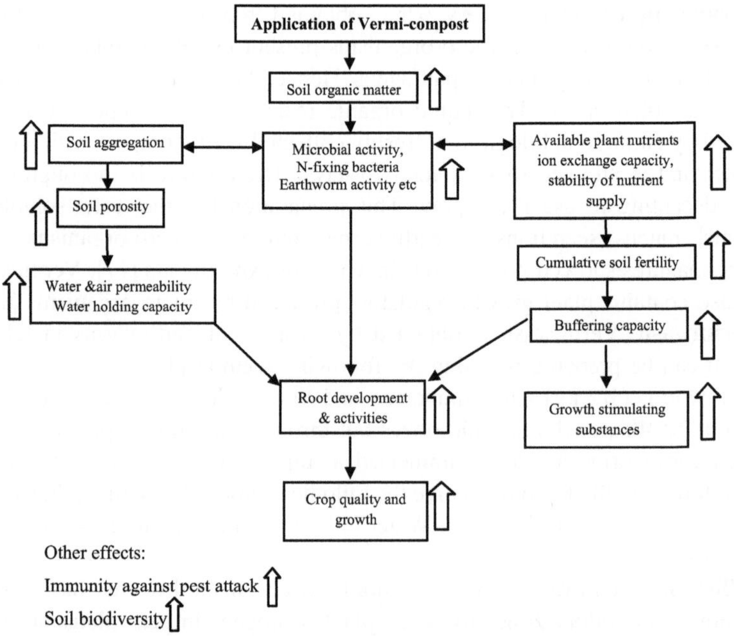

Fig. 1 Physico-chemical changes in the soil brought about by the application of vermicompost

vermicompost along with NPK/hectare led to significantly higher straw and grain yields coupled with better soil aggregation, water use efficiency and uptake of nutrient in vermicompost treated plots than the control and NPK-treated plots [20]. Significant increase in the pineapple yield coupled with the rise in earthworm population was recorded following application of 20 tons vermicompost/hectare [21]. Application of vermicompost reduces the doses of NPK in crops. Vermicompost along with judicious use of chemical fertilizers will not only bring down the cost of cultivation but also present unique opportunities for sustainable agriculture. Opinion of farmers on using vermicompost for different crops is given in Table 2.

Table 2 Opinion of the farmers about using vermicompost for different crops [7, 17]

Crop variety	Opinion	Doses	Crop variety	Opinion	Doses
(A) Cereals		2 t/ac	**(F) Fruits**		2–3 kg/plant
1. Jowar	+++		1. Mango	+++	
2. Rice	++++		2. Banana	++++	
3. Maize	+++		3. Watermelon	++++	
(B) Pulses		2 t/ac	4. Lemon	+++	
1. Garden pea	+++		5. Grapes	++++	
2. Blackgum	+++		6. Jackfruit	+++	
3. Country bean	+++		7. Pomegranate	++++	
(C) Oil seeds		3–5 t/ac	8. Custard apple	++++	
1. Sunflower	+++		**(G) Ornamental plants**		4 t/ac
2. Groundnut	+++		1. Roses	++++	
3. Soya bean	+++		2. Chrysanthemum	++++	
4. Mustard	+++		3. Orchids, vanilla	++++	
(D) Spices		4 t/ac	4. Balsam	+++	
1. Cardamom	+++		5. Marigold	++++	
2. Pepper	++++		6. Lady's lace	++++	
3. Curry leaf plant	++++		7. Tube rose	++++	
4. Turmeric	+++		**(H) Cash crop**		5 t/ac
5. Clove	+++		1. Coffee	+++	
(E) Vegetables		4–6 t/ac	2. Tea	+ +	
1. Cabbage	++++		3. Mulberry	++++	
2. Radish	++++		4. Cotton	+++	
3. Carrot	++++		5. Sugarcane	++++	
4. Potato	++++		6. Beetle leaf	++++	
5. Tomato	++++		**(I) Plantation crop**		5 kg/plant
6. Chillies	++++		1. Coconut	++++	
7. Pumpkin	++++		2. Areca nut	++++	
8. Ribbed gourd	++++		3. Teak	+++	
9. Cucumber	+++				
10. Sweet potato	+++				

++++ (excellent), +++ (very good), ++ (good), + (no difference)

5 Conclusions

The Green Revolution in India is the result of intensive agriculture with the extensive use of chemical fertilizers [10]. Through intensive agricultural practices, farmers could harvest three crops in a year with good irrigation facilities. In the tropical countries, the weather conditions and the soil type do not favour soil carbon conservation unlike in the temperate countries. Indiscriminate use of chemicals (fertilizer and pesticides) increases their leaching due to depletion in organic carbon in tropical soil [10]. Leaching of chemicals encourages the accumulation of salts that changes the pH of the soil. Besides these, the crust formed in soil affects its porosity and water holding capacity so that soil becomes unproductive.

Replenishment organic carbon to the impoverished soils occurs following repeated application of manures in field. High levels of oxidizable organic carbon in soil help in slow release of plant-available nutrients from the manure and thereby checks nutrient loss. Recently, according to Thakur and Sharma [22], the yield, total production, income generation and profit from crops like wheat, maize, peas and rajmash increased significantly under organic farming system as compared to the inorganic farming practices over the years.

Physical, chemical and biological properties of vermicompost influence the growth and yield of crops. Vermicompost, being a very stable granular organic matter, when added to clay soil, loosens the soil to provide aerobicity, thereby increasing porosity. Mucus associated with the earthworm casts being hygroscopic prevents water lodging and improves water holding capacity. Similarly, in the sandy soils, there is a problem of water retention. So the strong mucus-coated aggregates in the vermicompost hold water for a long time [7]. Organic carbon in the worm-worked compost releases nutrients slowly and steadily into the soil ecosystem and thus enables the plants to absorb these nutrients. Vermicompost contains balanced nutrients and provides plant growth promoting factors that are not supplied by the chemical fertilizers. The local soil-dwelling earthworms that get established following application of vermicompost contribute to the structure and turnover of soils [7]. Repeated application of manure to the soil of agricultural land or even wasteland thus leads to its sustained fertility. Sustainability issues could have a cascading impact from a plot level right up to ecosystem or landscape levels, as soil fertility is a critical limiting factor for land use development [23].

Organic farming system may not lead to higher crop production along with income generation in the short run as its returns are of long-term nature. In fact, practice of organic farming system improves health and fertility of soils and thus leads to sustained increase in yield and production. This results in stabilization and high rise in income and sustainable agriculture for food security in the long run [22].

Vermicomposting has gained a momentum due to present days increasing demands on organically grown crops which are free from toxic chemicals, tastier, highly nutritious, healthy, safe and of fresh quality. Being eco-friendly, it is economically viable and a simple technology. This has helped the farming communities of Karnataka (South India) to use it in their own land to improve the soil fertility status and also to make additional income by selling the vermicompost and earthworms produced in excess. Interestingly, this has resulted in improving the standard of living of those farmers [24]. Although farmers, traders and consumers of Karnataka have followed the 'organic way of life', in most of the other states of India, this is not the case, because farmers in those states are at present facing some pertinent constraints and problems, viz. lack of technical know-how of soil and nutrient management, problems in pest and disease control, scarcity of FYM and other organic manures. Thus, there is an immediate need for dissemination of the concept and knowledge of scientific agriculture to create more awareness among farmers for its success and all-round benefits. In fact, whole human society will benefit immensely if the farming communities of our country switch over to extensive production and use of vermicompost [7].

First Green Revolution was restricted to few crops of some specific plain areas with proper irrigation facilities. Application of vermicompost improves physical structure and increases the water holding capacity and nutritional status of the soil. So the vast dry and unutilized tilla lands of India need to come under organic farming. Through genetic engineering, scientists have to develop high-yielding variety of seeds, which would be effective in dryland farming. Crops irrespective of types and nature of land (plain or upland) show very good/excellent response in yield following application of vermicompost. Thus, in second Green Revolution, India's gross agricultural production would seemingly be much more than that of first Green Revolution.

Acknowledgements The author gratefully acknowledges Prof. Durgadas Ghosh, T.U, Dr. Baburam Swami, M.B.B.C for the improvement of language and Mr. Ashish Choudhury, T.U for plagiarism check.

References

1. Brady, N.C.: The Nature and Properties of Soils, p. 13. Macmillan Publishing Co., Inc (1984)
2. Paroda, R.S.: 99 per cent pesticides remain in environment. Tribune (July 19, 2001)
3. Edwards, C.A., Burrows, I.: The potential of earthworm compost as plant growth media. In: Edwards, C.A., Neuhauser, E.F. (eds) Earthworms in Environmental and Waste Management, pp. 211–220. Academic Publishing, The Netherlands (1988)
4. Chaudhuri, P.S., Nanda, D.K., Chaudhuri, D.: Extraction of Octochaetid earthworms, *Eutyphoeus gammiei* using an aqueous extract of *Polygonum hydropiper* Linn, with a comparison of other chemical methods for estimating earthworm population. Philippine J. Sci. **125**, 227–234 (1996)
5. Lavelle, P., Barois, I., Blanchart, E.: Earthworms as a resource in tropical agro ecosystem. Nat. Resour. **34**, 26–41 (1998)
6. Senapati, B.K., Lavelle, P., Giri, S., Pashanasi, B., et al.: In-soil earthworm technologies for tropical earthworms. In: Lavelle, P., Brussaard, L., Hendrix, P. (eds) Earthworm Management in Tropical Agroecosystems, pp. 199–238. CABI Publishing, USA (1999)

7. Kale, R.D.: Earthworm: Cinderella of organic farming. Prism Books Pvt. Ltd., Bangalore (1998)
8. Chaudhuri, P.S.: Vermiculture and vermi-composting as biotechnology for conversion of organic wastes into animal protein and organic fertilizer. Asian J. Microbiol. Biotech. Environ. Sci. **7**, 359–370 (2005)
9. Chaudhuri, P.S.: Vermi-composting as biotechnology for conversion of organic wastes into organic fertilizer and animal protein. In: Singh, S.M. (ed) Earthworms for Solid Waste Management, pp. 75–88. International Book Distributing Co., Lucknow (2007)
10. Kale, R.D.: The use of earthworms: nature's gift for utilization of organic wastes in Asia. In: Edwards, C.A. (ed) Earthworm Ecology, 2nd edn. CRC Press LLC, Florida (2004)
11. Karmegam, N., Daniel, T.: Growth, reproduction biology and life cycle of the vermi-composting earthworm, *Perionyx ceylanensis* Mich. (Oligochaeta: Megascolecidae). Bioresour. Technol. **100**, 4790–4796 (2009)
12. Edwards, C.A.: The use of earthworms in the breakdown and management of organic wastes. In: Edwards, C.A. (ed) Earthworm Ecology, pp. 327–354. CRC press LLC, Florida (1998)
13. Buchanan, M.A., Russell, E., Block, S.D.: Chemical characterization and nitrogen mineralization potentials of vermi-composts derived from differing organic wastes. In: Edwards, C. A., Neuhauser, E.F. (eds) Earthworms in Environmental and West Management, pp. 231–239. SPB Acad. Publ., The Netherlands (1988)
14. Martin, J.P., Focht, D.D.: Biological properties of soils. In: Soils for Management of Organic Wastes and Waste Water, pp. 11–169. ASA, Madison, Wisconsin (1986)
15. Arancon, N.Q., Edwards, C.A.: The utilization of vermicompost in horticulture and agriculture. In: Indo-US Workshop on "Vermitechnology in Human Welfare", Coimbatore (Abstract), p. 23 (2007)
16. Dominguez, J.: State of the art and new perspectives on vermicomposting research. In: Edwards, C.A. (ed) Earthworm Ecology, 2nd edn., pp. 401–424. CRC Press LLC, Florida (2004)
17. Kale, R.D.: Vermicompost—crown jewel of organic farming (A memoir), p. 22. N.D. Kale, Malleswaram, Bangalore (2006)
18. Edwards, C.A., Arancon, N.Q.: The Science of Vermiculture: The use of Earthworms in Organic Waste Management. Indo-US Workshop on "Vermitechnology in Human Welfare", Coimbatore (Abstract), p. 12 (2007)
19. Kale, R.D., Mallesh, B.C., Bano, K., Baggaraj, D.J.: Influence of vermicompost application on the available macronutrients and selected microbial population in a paddy field. Soil Boil. Biochem. **24**, 1317–1320 (1992)
20. Bhattacharjee, G., Chaudhuri, P.S., Datta, M.: Response of paddy (var. TRC-87-251) crop on amendment of the field with different levels of vermi-compost. Asian J. Microbiol. Biotech. Environ. Sci. **3**, 191–196 (2001)
21. Chaudhuri, P.S., Paul, T.K., Dey, A., Datta, M., Dey, S.K.: Effects of rubber leaf litter vermicompost on earthworm population and yield of pineapple (*Ananas comosus*) in West Tripura, India. Int. J. Recycl. Org. Waste Agri. (Springer) **5**, 93–103 (2016)
22. Thakur, D.S., Sharma, K.D.: Organic farming for sustainable agriculture and meeting the challenges of food security in 21st century: an economic analysis. Ind. J. Agri. Econ. **60**(2), 205–219 (2005)
23. Ramakrishnan, P.S.: Sustainable mountain development: The Himalayan tragedy. Curr. Sci. **92**(3), 308–316 (2007)
24. Kale, R.D.: The future of earthworm research. In: The 1st National Symposium on Earthworm Ecology & Environment. Bareilly (Abstract V.), p. 23 (2007)
25. Lofs-Homin, A.: Vermiculture Present Knowledge of the Art of Earthworm Farming. Institute Ecologie Miljovard, Swedish University Agronomic Sciences (1985)
26. Raman, C.J., Romero, G.A.: La Lombricultura una opeions tehnologica. In: Majenoy Disposicio in Final de Residuos So lidos Municipales. Congreso Regional del Sureste, SMISAACY AIDIS, Merida, Mexico, pp. 110–117 (1993)

27. Chaudhuri, P.S., Bhattacharjee, G., Pal, T.K., Dey, S.K.: Chemical characterization of kitchen waste vermi-compost processed by *Perionyx excavatus*. Proc. Zool. Soc. Cal. **54**, 81–83 (2001)
28. Piccone, G., Biasiol, B., Deluca, G., Minelli, L.: Vermicomposting of different organic wastes. In: De Bertoldi, M.P. Ferranti, P.H., Zucconi, F. (eds) Compost: Production, Quality and Use, pp. 818–821. Elsevier, London (1987)

Recycling of Industrial and Municipal Solid Wastes in Cement-Based Applications

Nabajyoti Saikia

Abstract In this communication, the author briefly presented results of experiments, conducted to evaluate the suitabilities of some solid wastes in cement-based applications. The fly and bottom ash from municipal solid waste incineration (MSWI) plants and plastic wastes generated in municipality along with a lead slag were considered. The MSWI fly ash was considered as a raw material in cement clinker production. The MSWI bottom ash as such cannot be considered as replacement of sand in mortar. Treatment of such ash was done to remove impurities like aluminium, sulphate and organics. The replacement of 25 volume per cents of sand by lead slag can improve various properties of cement mortar. Leaching of negligible amounts of toxic elements from waste cement-based composites indicated their environmental suitabilities. The replacement of natural aggregate by plastic waste aggregates over 10 volume per cent deteriorated the strength and other mechanical properties but improved abrasion and toughness behaviours of resulting concrete.

Keywords Waste management · Municipal solid waste · Lead slag
Plastic waste · Cement clinker · Concrete

1 Introduction

For sustainable society, the management and treatment of wastes generated from various industries and municipality are important tasks. Depending on the bulk chemical natures, wastes can be separated into two classes: organic and inorganic wastes. Organic wastes can also be separated into two classes: biodegradable and persistent organics. These wastes ranges from relatively inert, e.g. glass bottles,

N. Saikia (✉)
Department of Chemistry, School of Basic Sciences, Kaziranga University,
Jorhat 785006, Assam, India
e-mail: saikianj@gmail.com

A. S. Kalamdhad et al. (eds.), *Advances in Waste Management*,
https://doi.org/10.1007/978-981-13-0215-2_2

excavated soil, building and demolition wastes, plastics to hazardous waste with high concentrations of heavy metals and toxic organic compounds.

Landfilling or recycling as construction material is the major management option for municipal wastes, more particularly non-biodegradable municipal solid wastes (MSW) [1]. Landfilling, a simple and inexpensive method for disposal of non-recyclable solid wastes, creates many serious problems, such as requirement of landfill sites, groundwater pollution due to the leaching of toxic components from landfill solids and emission of odour. Some alternative methods are therefore necessary to treat municipal wastes.

All types of organic waste materials cannot be used for composting. Biodegradable wastes can only be considered for composting. Thermal treatment or incineration is necessary to treat several municipal wastes. Incineration of MSW has many advantages including a significant reduction in volume (about 70–90%), recovery of energy and complete disinfection. Therefore, incineration of MSW has been adopted in many countries. Non-biodegradable and non-recyclable organic wastes along with some other waste materials can be used for this purpose. Different types of solid residues, including bottom ash and fly ash, are generated due to the incineration of MSW waste [2].

Ferrous and non-ferrous industries are among the most important producers of solid wastes, such as metallurgical slag, wastes generated from air pollution control technology applied in these industries, and various sludges produced from blast furnaces. Majority of these wastes cannot be landfilled due to very high concentrations of toxic constituents.

Although the bulk chemical compositions of the above types of wastes are different, the contents of the predominant components in these wastes usually differ only in a limited range [3]. The predominant part contains four major elements: Si, Al, Ca and Fe. These wastes can therefore be included in the same class of solid wastes. Provided the concentrations of toxic constituents and the physico-chemical characteristics permit, recycling of these ash materials is to be preferred, both in view of environmental considerations (fewer disposals) and of resource management [4].

These ash materials can also be considered for using in cement and building material industries [5–7]. Some such applications are as fuel in cement kilns, as raw material for cement clinker and brick productions, as mineral addition to cement or as granular material in cement mortar and concrete. Therefore, proper evaluation of physico-chemical properties of these waste materials is important for determining their applicability. Moreover, for recycling in cement and concrete-based applications, technical requirements such as strength and durability as well as environmental properties such as leaching of toxic elements from cement-based products must be fulfilled.

In this paper, the author briefly presented some experimental results, those he conducted in last few years to evaluate the suitabilities of some municipal and industrial solid wastes as raw materials in various cement-based products developments. The waste materials considered are fly ash and bottom ash generated from

MSW incineration, industrial slags generated in lead-smelting and plastic wastes from municipal solid waste separation facility.

2 Materials and Methods

2.1 Materials

A municipal solid waste incineration (MSWI) fly ash, collected by bag filters placed at the flue gas cleaning area at around 200–250 °C before wet desulphurization process in a MSWI plant situated in Tokyo metropolitan area was considered as raw materials for cement clinker production. The boiler ash from the steam boiler and the fly ash from the electrofilter (abbreviated as BFA) generated due to the incineration of both sludge and high calorific waste in fluidized bed incinerator was taken to partially substitute the sand fraction of cement mortar. Both types of ash samples were collected together, and the particle sizes of majority of the particles are below 1 mm. For same purpose, a bottom ash (abbreviated as SF) with particle size range of 0.1–2 mm consisted mostly of hard granular materials along with some fibres was considered. A metallurgical slag (abbreviated as SLG) with size range of 0–4 mm generated from a lead blast furnace was also considered as the partial replacement of sand fraction in cement mortar. This slag was very hard granular material and black in colour. Three different types of granular plastic wastes, collected from a post-consumer PET bottles recycling plant (abbreviated as PC, PF, PP), located in Portalegre municipal area, Portugal was also considered as partial replacements of fine and coarse aggregates in concrete productions. The pellet-shaped plastic aggregate (PP) was prepared by heat treatment of shredded waste PET bottle. The details about the various waste materials were presented elsewhere [5, 8, 9].

2.2 Methods

The materials were collected using standard procedures and if necessary stored in airtight container to protect them from any further degradation. Different experimental methods used to characterize various wastes and cement-based products developed from them are presented in Table 1. The details about the methods used in these investigations were presented in some other publications [5, 8–11].

Both gravimetric and inductively coupled plasma (ICP)-based instrumental methods were applied to evaluate the chemical constituents present in various ash and slag samples. For ICP analysis, acid digestion procedure using HNO_3, $HClO_4$ and HF was used to prepare the solutions of various ash samples. The cement-based products were developed using normally used standard procedures. The mineral

Table 1 Characterization methods of various waste samples and their products

Property	Types of constituents/tests	Types of materials	Methods used
Chemical composition	Oxide	All types of waste materials except plastic wastes	Acid digestion, using ICP-AES[a] and ICP-MS[a]
	Minor elements	-do-	-do-
Mineralogical composition	–	-do-	By analysing XRD patterns
Aggregate behaviour	Sieve analysis	All granular waste materials	Using standard C132 method, ASTM
	Potential alkali reactivity	MSWI bottom ash and Lead slag	Using ASTM C289-94 method
	Harmful effect of organics	MSWI bottom ashes	Using ASTM C40-92 method
	Shape index	Plastic waste aggregates and natural aggregates	Using standard European methods
	Loose bulk density	-do-	
	Water absorption	-do-	
Hydration behaviour	Hydrated cement paste	MSWI fly ash-based clinker, MSWI bottom ash and lead slag	Using standard methods; characterization of products: XRD, TG/DTG, SEM analyses
Fresh and hardened properties	Cement mortar and concrete mixes, hardened cement mortar and concrete	MSWI bottom ash, lead slag and plastic aggregate-based composites, along with control	Standard European testing methods
Environmental testing	Waste materials, cement pastes and mortars	Cement clinker, MSW bottom ash, lead slag, cement mortars containing MSWI bottom ash and lead slag	EN 12457-2 test for powdered samples, NEN 7345 test for cement mortar, SPLP test, EPA, USA, for cement clinker

[a]*ICP* inductively coupled plasma; *AES* atomic emission spectrophotometry; *MS* mass spectrophotometry

matters present in ash samples, and their products were detected by X-ray diffraction (XRD) techniques and using standard software. Standard ASTM testing methodologies (C289-94 and C 40-92) were applied to evaluate the potential alkali–silica reactivities and the effects of deleterious organic impurities present in SF and BFA during cement-based applications. The grain size distributions of various granular waste materials were also determined according to ASTM C312 method. The size distributions of various MSWI ash and plastic aggregates were presented in Fig. 1.

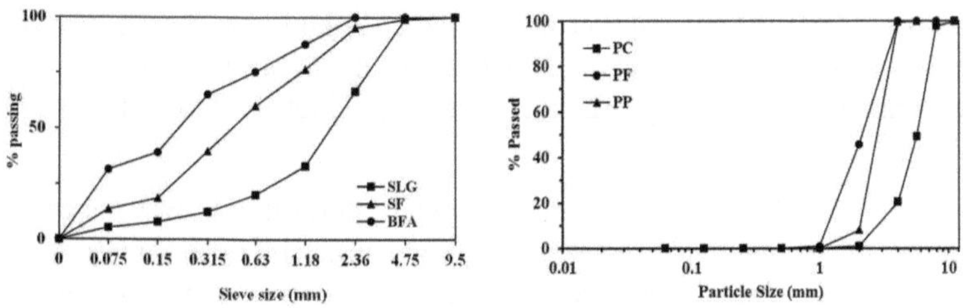

Fig. 1 Particle size distributions of various waste materials used as aggregate in cement mortar

For evaluating various mechanical properties of cement mortars, 4 × 4 × 16 cm^3 mortars were prepared. The preparation, casting, curing and evaluations of various properties of cement mortars were done according to various European standard testing methodologies. Similarly, concretes containing plastic waste as partial replacements of coarse and fine aggregates were prepared and tested according to standard European testing methodologies [9].

The leachabilities of various toxic constituents from wastes and waste cement-based mortars were evaluated according to standard European and Dutch testing methodologies (Table 1). For testing leachabilities of various toxic constituents according to European EN-12457-2 method, finely ground powdered samples were used. Similarly, for Dutch NEN 7345 diffusion leaching test, the 4 × 4 × 4 cm^3 cement mortars were used. The details about these procedures were described in some other publications [8, 10].

3 Results and Discussion

3.1 Use of Municipal Solid Waste Incineration (MSWI) Fly Ash in Cement Clinker Production

Disposal of municipal solid waste incineration (MSWI) ash, particularly fly ash (FA), is a problem as FA contains significant amounts of toxic elements like Zn, Cu, Pb, Cr and As and some highly toxic organic pollutants such as dioxins. Improvements in air pollution control systems, as well as more efficient separation techniques, have further increased the pollution potential of such MSWI fly ash [4]. Therefore, the development of a more reliable and economical technique to treat MSWI fly ash was urgently needed. Considering the following advantages, an investigation was therefore undertaken, where MSWI fly ash was considered as raw material in cement clinker production [5]:

- The major oxide constituents such as CaO, Al_2O_3, SiO_2, Fe_2O_3, present in municipal solid waste incineration (MSWI) fly ash are also the constituents of clinker and its raw materials.

- The use of MSWI fly ash can reduce the CO_2 emission during production of clinker as some part of calcium carbonate in cement raw mix will be replaced by Ca content in MSWI fly ash, the source of which is different from calcium carbonate.

The major problem to apply the investigated MSWI fly ash was the presence of high amounts of alkali chlorides and dioxins [6]. Chlorides can form harmful products during cement hydration and deteriorate the technical properties of solidified products. Dioxins, a class of most toxic organic compounds, are difficult to destroy by applying chemical techniques. However, use of MSWI fly ash in cement kiln can solve these problems because of following advantages:

- High temperature processing of MSWI fly ash leads to the complete destruction of dioxins.
- Metals can be separated by special processes after evaporation or can be fixed in the products. Because of the generation of high pH, the products are also inert towards the leaching of majority of toxic elements.
- It is possible to collect the alkali metal salts and some of the volatile elements by controlling their volatilization behaviours; in this way, it is possible to reduce the corrosion problem during the use of MSWI fly ash in clinker production.

However, considering the presence of high amounts of chlorides in MSWI fly ash, water washing was also considered as a pre-treatment method to reduce the chloride content.

In this investigation, 44–50% MSWI fly ash was used to prepare the cement raw mix with addition of very small amounts of SiO_2 and Fe_2O_3. The amounts of $CaCO_3$ used were also around 50% and lower than the required amounts of $CaCO_3$ that is necessary to prepare ordinary Portland cement (OPC) clinker (more than 70%) since the lime content in MSWI fly ash would provide the additional $CaCO_3$ amounts necessary for clinker formation. The XRD analyses of clinkers produced from various raw mix compositions indicated the formation of various clinker minerals such as belite (Ca_2SiO_4), alite (Ca_3SiO_5), tricalcium aluminate ($Ca_3Al_2O_6$) and tetracalcium aluminoferrite ($4CaO.Al_2O_3.Fe_2O_3$) from 1300 °C with considerable amounts of free lime (>2%) (Fig. 2). However, clinkerization at 1400 °C increased the amounts of various silicate and aluminate minerals in the clinker with reduction of free lime content (<1%) (Fig. 2). Moreover, the quality of cement clinker produced from washed MSWI fly ash was better than unwashed ash-based clinker.

The raw material composition in the best raw mix composition containing unwashed MSWI fly ash was 44.2% MSWI fly ash, 49% $CaCO_3$, 5.5% SiO_2 and 1.3% Fe_2O_3. For washed MSWI fly ash, the best composition of raw mix could be obtained by mixing 45.6% MSWI fly ash with 52% $CaCO_3$, 1.2% SiO_2 and 1.2% Fe_2O_3. The optimized temperature of clinkerization was 1400 °C.

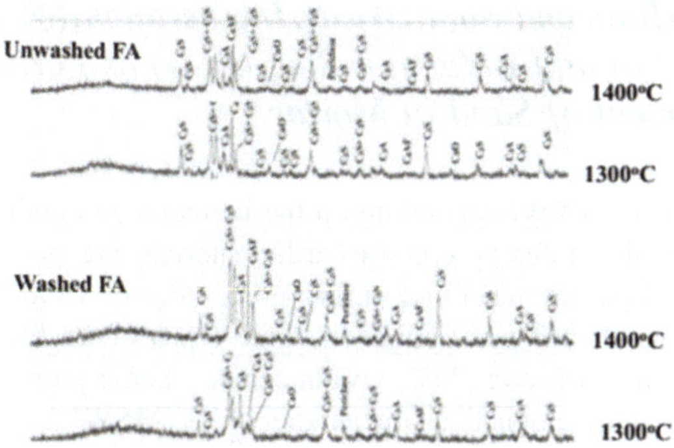

Fig. 2 Cement clinker prepared from unwashed and washed MSWI fly ash

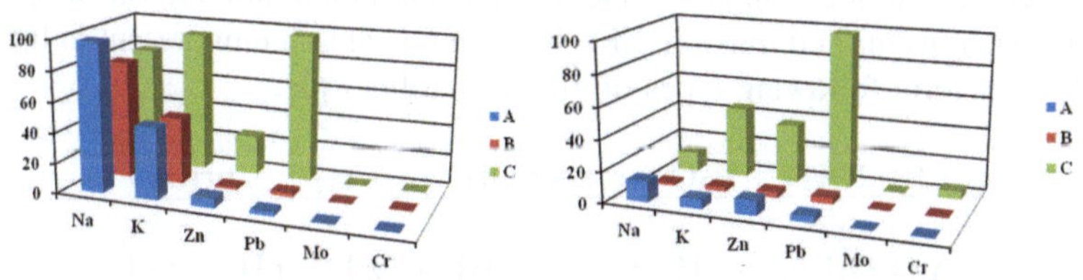

Fig. 3 Concentrations (mg/g) of some elements in cement raw mix (**a**), produced clinker (**b**) and percentage liberated during clinkerization (**c**)

High amounts of alkali metals and some volatile heavy metals such as Pb and Cd (>90%) were emitted into the atmosphere during clinkerization of MSWI fly ash containing raw mix compositions. On the other, almost all amounts of some toxic elements such as Cr, Mo, As, Se, Sn were fixed in the cement clinkers. The emission behaviours of some elements are presented in Fig. 3.

The evaluation of leaching behaviour of some toxic elements suggested that the toxic heavy metals present in clinkers were inert to applied leaching test (Synthetic Precipitation Leaching Procedure, SPLP, US EPA). Finally, the hydration behaviours of cement clinkers obtained from optimized cement raw mixes prepared by using washed and unwashed MSWI fly ash indicated that the reactivity of clinker produced from unwashed MSWI fly ash was considerably higher than that containing washed MSWI fly ash, probably due to the presence of high amounts of chlorides in the former type of cement clinker. Thus, the results obtained from this investigation showed the possibility of using the MSWI fly ash in cement clinker production. Water washing of the ash can be considered as a pre-treatment method of MSWI fly ash as it removes almost all amounts of chlorides.

3.2 *Use of Municipal Solid Waste Incineration (MSWI) Bottom Ash and Industrial Slag (SLG) as Partial Replacement of Sand in Mortar*

In the incineration of MSW, depending on the incineration conditions and type of incinerated materials, different types of ash materials are generated. Similarly, several types of slags are generated during processing of various metals in blast furnace. In a comprehensive investigation, three types of wastes were considered for partial replacement of sand (25%, v/v) in cement mortar preparation [8, 10, 11].

Chemical and mineralogical characterizations of MSWI bottom ash samples indicated the presence of all the major constituents, normally present in various raw materials of cement clinker and concrete [8]. However, major concern regarding the use of these ash materials was the presences of some undesirable components. These include chlorides, sulphates and metallic aluminium in both ash samples and organic materials in sand fraction (SF). The metallic aluminium present in these ash samples can form following types of products at high pH:

$$Al + 2OH^- + H_2O \rightarrow [AlO(OH)_2]^- + H_2\uparrow \quad pH > 7 \tag{1}$$

$$[AlO(OH)_2]^- + H_2O \rightarrow Al(OH)_3 + OH^- \quad pH = 9\text{--}10 \tag{2}$$

$$Al(OH)_3 + OH^- \rightarrow [Al(OH)_4]^- \quad pH > 10 \tag{3}$$

Aluminium hydroxides thus formed can react with soluble sulphates and form expansive ettringite phase at the later stages of hydration (Fig. 4a).

Because of the presence of these constituents, the MSWI ash (BFA and SF) containing mortar cubes showed much lower compressive strengths at all ages (Fig. 5) and could not be used as fine aggregates in cement mortar preparation. Visual observation also showed the craters and spalling on the surface of these mortars, which indicated the effect of metallic aluminium and sulphate during

(a) (b)

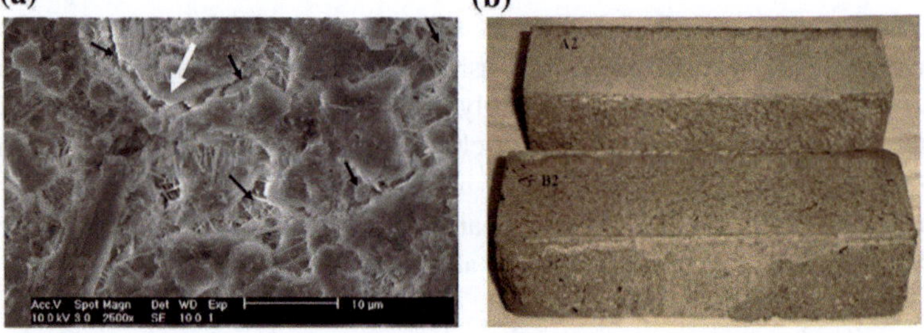

Fig. 4 Formation of the needle-shaped ettringite on the surface of hydrated MSWI bottom ash (BFA) containing cement pastes (**a**) and cracks on the surface of mortar (**b**) (A2 and B2 on the mortar sample represent the mortar containing treated and untreated BFA, respectively)

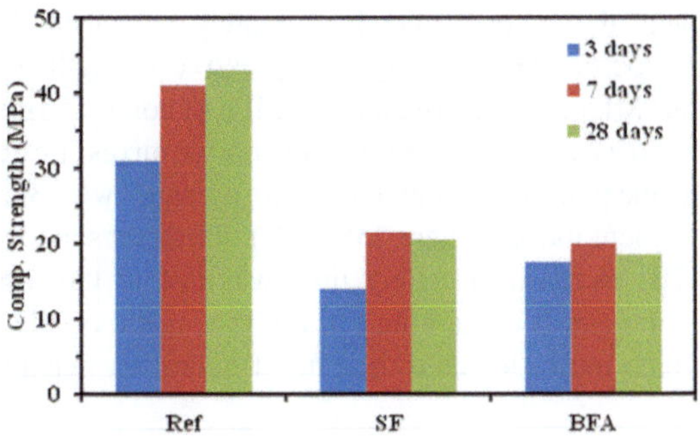

Fig. 5 Compressive strengths of various samples (Ref: reference)

hydration (Fig. 4b). The ash containing mortar bars also showed swelling during the first few hours of curing, indicating the formation of hydrogen gas and cracks in hardened specimens due to delayed ettringite formation.

Presences of harmful organic impurities in SF also adversely affect the compressive strength of the mortars. Thus, removal of the metallic aluminium and sulphate from both ash samples and organic materials from SF was considered to improve the aggregate behaviour of these ash samples.

Accordingly, two separate treatment procedures were adopted to remove deleterious components from these ash aggregates [11]. For SF, heat treatment at about 700 °C was done to remove deleterious organics before or after treatment by a Na_2CO_3 solution. On the other hand, only Na_2CO_3 treatment was done to remove harmful impurities from BFA. The Na_2CO_3 treatment was considered to dissolve sulphate (since Na_2SO_4 is more soluble in water than $CaSO_4$) and to remove metallic aluminium from the ash samples (by increasing pH of the solution):

$$Na_2CO_3\ (aq) + CaSO_4\ (s) \rightarrow Na_2SO_4\ (aq) + CaCO_3\ (s)$$

The proposed treatment methods considerably improved the compressive strengths of the mortar prepared by treated ash samples (Fig. 6).

On the other hand, lead slag (SLG) can be used for partial replacement of the sand fraction in cement mortar [10]. Addition of SLG decreased water requirements

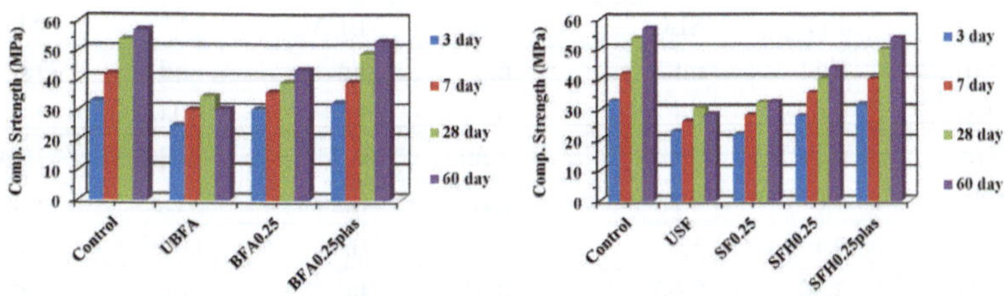

Fig. 6 Compressive strengths of various mortar types (U: untreated; 0.25: 0.25 M $NaCO_3$ solution treated; H0.25: heat treatment before 0.25 M Na_2CO_3 solution treatment; plas: mortar prepared using superplasticizer)

and air contents, which ultimately enhanced the strength and other mechanical properties (e.g. 91 days compressive strength of mortar containing 25% slag was 65 MPa in comparison to 63 MPa of control). However, SLG addition considerably increased the setting times of mortar mixes. For example, the final setting times of cement mortar containing normal sand was 323 min, whereas the same for the cement mortar where 25% (v/v) of natural sand was replaced by SLG was 650 min. The increase in setting time was due to the formation of impermeable layers of compounds containing Pb, Zn, etc., metal on the hydrating cement particles, which temporarily lowered the rate of hydration reaction.

3.2.1 Environmental Behaviour of Cement Mortar

Since all the ash samples contain considerable amounts of various toxic elements such as Pb, Zn, Cd, Cr, As, Se; therefore, environmental evaluations of these wastes and waste-based products were considered to understand their possible effects on environment. Two types of leaching tests were performed to evaluate the environmental suitabilities of these wastes and waste-based products. The EN 12457-2 test was conducted to evaluate leaching behaviour of powdered waste and cement mortars. This test provides information, which can be used to differentiate the nature of waste materials and to adopt future strategies for their applications. The leaching behaviours of some toxic constituents from powdered ash and cement mortar containing such ash samples were presented in Table 2.

Results indicated that the concentrations of some toxic elements such as As, Cd, Cr, Mo, Pb, Se in the leachates of the various powdered ash samples were higher than the specified limit values for those elements. However, the concentrations of majority of the elements except Pb, Cu and Zn in the leachate solutions obtained from powdered waste containing cement mortar are lower than the specified limits of these elements. Long-term leaching information from various waste-based mortars was evaluated by Dutch NEN 7345 test, and some such results for a few

Table 2 Elements leached from the powdered wastes and concrete containing wastes obtained according to EN 12457/2 test

Element[a]	Leaching amount (mg/kg)					
	SLG	SLGC[b]	BFA	BFAC[a]	SF	SFC[a]
As (0.80)	0.11	nd	16.4	nd	nd	nd
Cd (0.03)	0.02	nd	0.17	nd	0.07	nd
Cr (0.50)	0.43	0.34	1.80	0.24	18.5	0.16
Mo (1.5)	1.0	0.25	2.6	0.24	1.8	0.21
Pb (1.8)	49.1	53.3	0.26	0.81	0.32	1.47
Se (0.10)	3.50	0.06	0.31	nd	nd	nd
Zn (2.8)	1.7	7.27	0.83	0.98	0.39	0.29

[a]Values marked off by brackets indicated specific limit of a particular element
[b]Waste containing cement mortar

Table 3 Cumulative leaching amounts of a few toxic elements from cement mortar containing various wastes obtained according to NEN 7345 test

Elements[a]	Cumulative leaching amount (mg/m^2)		
	SLG	BFA	SF
As (27)	nd	nd	nd
Cd (1.1)	nd	nd	nd
Cr (55)	3.3	2.4	11.8
Mo (11)	0.71	5.1	1.2
Pb (60)	9.3	0.30	0.67
Zn (90)	0.79	1.2	2.3

[a]Values (in mg/m^2) marked off by brackets indicated specific limits of a particular element

particular elements are presented in Table 3. Cumulative leaching amounts of almost all toxic elements from the monolithic cement mortar containing waste materials were considerably lower than the specified values for those elements. Therefore, the environmental performance of the materials met the required standards.

3.3 *Plastic Waste as Partial Replacement of Fine and Coarse Aggregate in Concrete*

Huge quantities of plastic-based waste are generated all over the world due to the increasing consumptions of plastics. Plastic waste is now a serious environmental threat to the modern way of living. Because of its economic and ecological advantages, recycling waste plastic to produce new materials such as concrete seems to be one of the best ways to dispose of waste plastics [12]. Therefore, a comprehensive investigation was undertaken to understand the effect of three different types of granular plastic wastes as partial replacement of aggregate in concrete production [9, 13–15]. The experiments were undertaken considering following points:

- The use of fine (PF) and coarse-sized aggregate (PC) generated from same type of plastic waste would help to understand the effect of plastic aggregate size on the behaviour of fresh and hardened states concrete properties;
- The use of flaky (PF and PC) and pellet-shaped (PP) plastic aggregates would help to show the effect of shape of plastic aggregate on the behaviour of fresh and hardened states concrete properties.

Accordingly, several concrete compositions were prepared to understand the fresh and hardened properties. The major findings of these investigations are as follows:

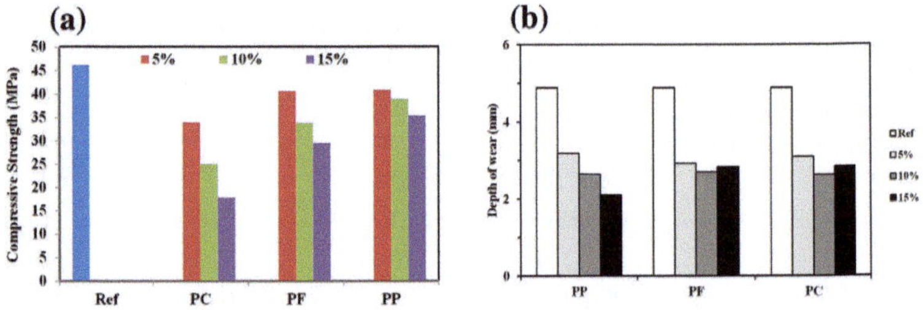

Fig. 7 Compressive strength (**a**) and abrasion behaviour (**b**) of concrete containing various types of waste plastic aggregate

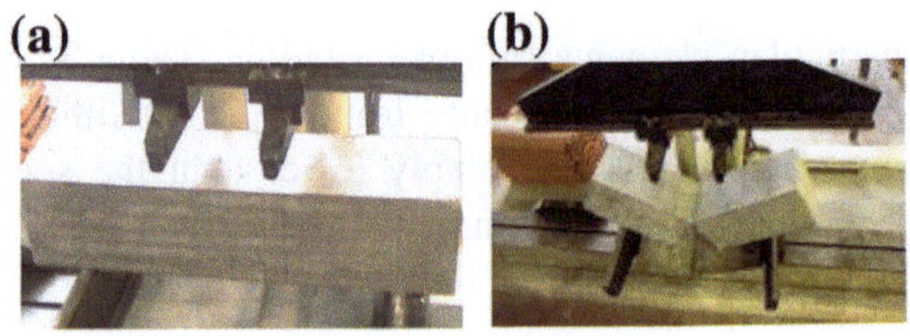

Fig. 8 Cracked concrete specimens after flexural strength test: **a** plastic aggregate-based concrete; **b** reference

- The workability of concrete mixes was greatly dependent on the shape of the plastic aggregates. Spherical and smooth plastic aggregates lowered the W/C ratio and therefore increased the slump value. The angular and non-uniform-shaped plastic aggregates (fine and coarse sized) with sharper edges increased the water requirement of concrete and conversely lowered the slump value of concrete mixes. The density of concrete mixes also decreased with increasing plastic aggregates in concrete.
- The partial substitution of natural aggregates by plastic aggregates deteriorated various mechanical properties of hardened concrete (Fig. 7a). The addition of plastic aggregates also lowered the various durability properties.
- The abrasion resistance of the concrete mixes with various types of plastic aggregate was considerably better than that of the reference concrete (Fig. 7b). The PET-aggregate incorporation improved the toughness behaviour of the resulting concrete too. This behaviour was dependent on PET-aggregate's shape and was maximum for concrete containing coarse, flaky PET-aggregate (PC).
- The elongated and flaky particles of plastic aggregates bridge the two split concrete pieces and therefore prevent concrete from breaking apart after failure during determination of tensile splitting and flexural strengths (Fig. 8).

However, analyses of the results of the various properties of plastic waste-based concrete using various codes and standard specifications, it can be concluded that the quality of plastic waste-based concrete, particularly up to 10% replacement level is good enough for several applications, including structural use [14].

4 Conclusions

The uses of some waste materials in cement-based applications can be a good option for management of these materials. The physico-chemical characterizations of these materials are very important to find the possible application type (i.e. whether it can be used as a raw materials in clinker production or as a granular substitution in concrete) and to find the inherent property in the waste that can improve some special properties of the product.

The MSWI ash generated from municipal waste incineration plant can be used in various cement-based applications. For example, MSWI fly ash can be used for cement clinker production and MSWI bottom ash as a partial replacement of granular fraction in cement mortar and concrete preparations. However, sometimes treatment of these waste materials is necessary to lessen their deleterious effects. For example, water washing can remove the harmful chlorides from MSWI fly ash and therefore improves its suitability in cement clinker production. Similarly, treatments by Na_2CO_3 solution and heat can remove the metallic aluminium, sulphates and deleterious organic components from investigated MSWI bottom ash samples, which is important for their application as granular replacement in cement mortar and concrete productions.

Lead slag as partial replacement of fine aggregate can be a good option for their management if we consider the mechanical and durability performances of resulting cement mortars. However, one concern is the liberation of some toxic elements from lead slag, which also increases the setting time of mortar mixes. Therefore, evaluation of environmental behaviour is very important to assess their environmental suitability. Evaluation of the environmental behaviours of various cement-based products containing MSWI ash and lead slag indicates that these materials can be considered as environmentally inert in cement-based monoliths.

Use of plastic waste as partial replacement (up to 10% replacement level) of natural aggregate in concrete preparation can be a good option. Plastic waste aggregate can improve several important properties such as abrasion behaviour and toughness behaviours of concrete.

Acknowledgements The author gratefully acknowledges the financial supports from FCT, Portugal, the Research Fund of K.U. Leuven, Belgium, and Seikei University, Tokyo.

References

1. Sabbas, T., Polletini, A., Pomi, R., Astrup, T., Hjelmar, O., Mostbauer, P., Cappai, G., Magel, G., Salhofer, S., Speiser, C., Heuss-Assbichler, S., Klein, R., Lechner, P.: Management of municipal solid waste incineration wastes. Waste Manag. **23**, 61–88 (2003)
2. Vandecasteele, C., Wauters, G., Arickx, S., Jaspers, M., Van Gerven, T.: Integrated municipal solid waste treatment using a grate furnace incinerator: the Indaver case. Waste Manag. **27**, 1366–1375 (2007)

3. Theis, T.L., Gardner, K.H.: Environmental assessment of ash disposal. Crit. Rev. Environ. Control **20**, 21–42 (1990)
4. Chandler, A.J., Eighmy, T.T., Hjelmer, O., Kosson, D.S., Sawell, S.E., Vehlow, J., Vander Sloot, H.A., Hartlen, J.: Municipal Solid Waste Incineration Residues, vol. 67, 1st ed., pp. 1–973. Elsevier Science, Amsterdam (1997)
5. Saikia, N., Kato, S., Kojima, T.: Production of cement clinkers from municipal solid waste incineration (MSWI) fly ash. Waste Manag. **27**, 1178–1189 (2007)
6. Polettini, A., Pomi, R., Sirini, P., Testa, F.: Properties of Portland cement stabilized MSWI Fly ashes. J. Hazard. Mater. **88**, 123–138 (2001)
7. Saikia, N., de Brito, J.: Use of industrial waste and municipal solid waste as aggregate, filler or fiber in cement mortar and concrete. In: Maryann, C. W. (ed) Advances in Material Science Research, vol. 3, pp. 65–116, Nova Science Publishers, Inc. 400 Oser Avenue, Suite 1600, New York (2010)
8. Saikia, N., Cornelis, G., Mertens, G., Elsen, J., Van Balen, K., Van Gerven, T., Vandecasteele, C.: Assessment of Pb-slag, MSWI bottom ash and boiler and fly ash for using as a fine aggregate in cement mortar. J. Hazard. Mater. **154**, 766–777 (2008)
9. Saikia, N., de Brito, J.: Mechanical performance and abrasion behaviour of concrete containing recycled polyethylene terephthalate (PET) as a partial substitution of natural aggregate. Constr. Build. Mater. **52**, 236–244 (2014)
10. Saikia, N., Cornelis, G., Cisar, O., Vandecasteele, C., Van Gemert, D., Van Balen, K., Van Gerven, T.: Utilization of Pb-slag as a partial substitution of fine aggregate in cement mortar. J. Mater. Cycles Waste Manage. **14**, 102–112 (2012)
11. Saikia, N., Mertens, G., Van Balen, K., Elsen, J., Van Gerven, T., Vandecasteele, C.: Pre-treatment of municipal solid waste incineration (MSWI) bottom ash for utilisation in cement mortar. In: Construction and Building Materials, vol. 96, pp. 76–85 (2015)
12. Saikia, N., de Brito, J.: Use of plastic waste aggregate in concrete: a review. Constr. Build. Mater. **34**, 385–401 (2012)
13. Ferreira, L., de Brito, J., Saikia, N.: Influence of curing conditions on the mechanical performance of concrete containing recycled plastic aggregate. Constr. Build. Mater. **36**, 196–204 (2012)
14. Saikia, N., de Brito, J.: Waste polyethylene terephthalate as an aggregate in concrete. Mat. Res. **16**, 341–350 (2013)
15. Silva, R.V., de Brito, J., Saikia, N.: Influence of curing conditions on the durability-related performance of concrete made with selected plastic waste aggregates. Cement Concr. Compos. **35**, 23–31 (2013)

Electrochemical Degradation of Reactive Red-195 and Reactive Yellow-81 in Continuous Flow Reactor

Chandrasen F. Rajemahadik and Akash V. Gambhire

Abstract This paper investigates Reactive Red-195 (R.R.) and Reactive Yellow-81 (R.Y.) dyes synthetic solution of 60 mg/L concentration in continuous flow in electrocoagulation (EC) reactor for decolourization. The effects of operational parameters like change in voltage, electrolyte concentrations, inter-electrode distance, initial dye concentration and discharges were examined independently. A 20 V was optimized from range 10–30 V and applied for subsequent experiments. Results indicate with increase in voltage and electrolyte concentration, energy consumption elevates, improving decolourization of both dyes. In addition, as inter-electrode distance increases, colour removal and energy consumptions decrease for R.R. and R.Y. Concentration of dyes influences lower colour removal efficiency, reducing energy consumption at higher concentration. Furthermore, when discharge increases decolourization decreases in both dyes. However, energy consumption reduces with rise in discharge at 4.5 mL/s for R.Y. In addition, noticeable drop in energy consumption was observed in R.R. Finally, effects of operational parameters were successfully elaborated through this investigation.

Keywords Electrochemical · Degradation · Operational parameter
Energy consumption

1 Introduction

Dyeing industries fabricate thousands of dyes [1] and industries consume more than 80,000 tons of reactive dyes each year with potential to pollute [2]. Even when concentrations of dyes are less, they possess serious threats to humans, animals and environment [3]. Mostly, synthetic azo dyes are considered as recalcitrant xeno-

C. F. Rajemahadik (✉) · A. V. Gambhire
Department of Civil Engineering, AMGOI, Kolhapur, India
e-mail: crajemahadik@gmail.com

C. F. Rajemahadik
ACME Consulting and Research, Kolhapur, India

A. S. Kalamdhad et al. (eds.), *Advances in Waste Management*,
https://doi.org/10.1007/978-981-13-0215-2_3

biotic compounds, due to the presence of a N=N bond [4]. They also have groups of aromatic rings that are not easily degraded [5]. Probably, for 1 kg of cotton with 30–60 g of dyestuffs needs 70–150 L of water and 0.6–0.8 kg NaCl [6], and this ascertain a possibility of polluted wastewater discharge. Numerous techniques are employed for the removal of dyes or colour from wastewater such as Fenton [7], photo-catalyst [8], activated carbon [9], nano-materials [10], biological [11] and coagulants and polymers [12]. From the literature studies electrocoagulation (EC) is found to be one of thriving method used for treatment of most industrial liquid waste [13]. EC process proposes some advantages such as less use of chemicals, less sludge, low-cost and compact equipment [14, 15]. The present investigation synthetic reactive red and yellow dye wastewater was treated in continuous flow reactor using EC process. The effects of change in electrical potential, electrolyte concentration, inter-electrode distance, initial dye concentration and flow variation on synthetic dye wastewater were examined.

1.1 Mechanism of EC Process

Electrocoagulation can offer an alternative to traditional chemical coagulation, as metal ions released remove smallest colloidal particles depending on their coagulation capability [16]. EC produces destabilizing agents that neutralize the electric charge of the pollutants present in solution [17]. The reactions occurring in an electrochemical cell involving iron electrodes are as follows:

Anode:

$$Fe_{(s)} \rightarrow Fe^{n+}_{(aq)} + ne^{-} \tag{1}$$

$$Fe^{+n}_{(aq)} + 2OH^{-}_{(aq)} \rightarrow Fe(OH)_{2(s)} \tag{2}$$

In effect of pH of solution, Fe^{3+} may undergo hydration, forming species like Fe $(OH)^{2+}$, $Fe(OH)_2^{+}$, and $Fe(OH)_3$, in an acidic environment. Then solution reaction takes place according to Eq. 3, [18].

$$Fe^{+3}_{(aq)} + H_2O \rightarrow Fe(OH)^{2+}_{(aq)} + 2H^{+}_{(aq)} \tag{3}$$

Apart from Eqs. 1–3, possible anodic reaction for water oxidation is shown below in Eq. 4 and cathode in Eq. 5 [17, 19].

$$2H_2O \rightarrow O_{2(gas)} + 4H^{+} + 4e \tag{4}$$

Cathode:

$$2H_2O + 2e^- \rightarrow H_{2(gas)} + 2OH^-_{(aq)} \quad (5)$$

1.2 *Mechanism for Dye Removal*

The dye degradation depends on interactive mechanisms between dye molecules, hydrolysis [20] and pH of solution [21]. The insoluble metal hydroxide formed, remains in suspension, helps eliminate pollutants from wastewater under mechanism of precipitation and adsorption as shown in Eqs. 7–10 [22]. Furthermore, at low range of pH, precipitation is a key and at pH > 6.5, it is explained as adsorption [21].

Precipitation:

$$\text{Dye} + \text{monomeric Fe} \rightarrow [\text{Dye} - \text{Monomeric Fe}]_{(s)} \quad (6)$$

$$\text{Dye} + \text{polymeric Fe} \rightarrow [\text{Dye} - \text{polymeric Fe}]_{(s)} \quad (7)$$

Adsorption:

$$\text{Dye} + \text{Fe}_n(\text{OH})_{n(s)} \rightarrow [\text{Sludge}] \quad (8)$$

$$[\text{Dye} - \text{polymeric Fe}]_{(s)} + \text{Fe}_n(\text{OH})_{n(s)} \rightarrow [\text{Sludge}] \quad (9)$$

2 Experimental

2.1 *Materials and Methods*

The experiments were performed by mixing Reactive Red-195 (R.R.) and Reactive Yellow-81 (R.Y.) concentrations of 60 mg/l with tap water for 30 min before experimenting. A reactor volume of 300-ml-containing dye solution synthetic was feed with 1 g/L of NaCl electrolyte to adjust strength of ionic solution. A simple continuous flow EC reactor contains iron electrodes with inter-electrode distance (S) of 1 cm and having electrode dimension of 150 mm × 35 mm × 2 mm. The total submerged surface area (A_{sub}) of 45 cm^2 was maintained. The initial discharge of synthetic solution was 105 mL/min (1.75 mL/s). The current was applied through D.C. power supply model FALCON PS 303 D, characterized by range 0–3 A and 0–30 V. The schematic of experimental set-up is as shown in Fig. 1. Electrodes used during experiments were sanded and washed with dilute HCl

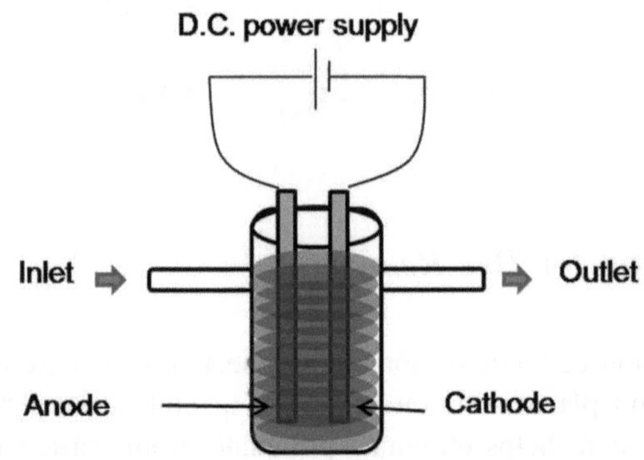

Fig. 1 Experimental laboratory set-up for Reactive Yellow-81 and Reactive Red-195

Table 1 Characterization of synthetic dye wastewater

Characteristics	Units	Red-195	Yellow-81
pH	–	7.69	6.63
Conductivity	mS/cm	2.73	2.78
Absorbance	λ_{max}	542 nm	
Structural formula		($C_{31}H_{19}ClN_7Na_5O_{19}S_6$) 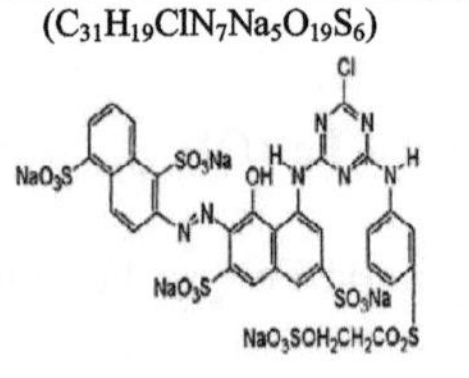 **Molecular Weight:** 1136.32	$C_{52}H_{34}Cl_2N_{18}Na_6O_{20}S_6$ 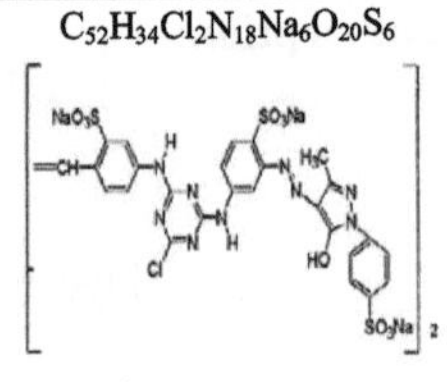 **Molecular Weight:** 1632.18

before each experiment. The pH and conductivity were detected using SPECTRALAB MULTIPARA MP-5 to determine characteristics of synthetic solution. The progress of treatment was monitored by collecting samples of 10 ml in triplicates after discharge of 1 L of wastewater volume. Collected samples were then filtered to eliminate sludge formed during process and were examined for absorbance. The dye absorbance was determined using spectrophotometry (Labtronics LT-290) at λ_{max} of 542 nm for RR-195 and for 351 nm for RY-81, according to Beer–Lambert law, using UV/vis spectrophotometer. Table 1 represents structure of dye and characteristics of synthetic solution specified according to APHA [23]. Chemicals used for experiments such as NaCl, HCl and NaOH were purchased from MERK, and Reactive Red-195 dye was collected from local dyeing industry.

2.2 Calculation

Removal efficiency (η %) was calculated depending in absorbance of colour with respect to electrolysis time of treatment process given in Eq. 10,

$$\eta\,\% = \left[\frac{C - Cx}{C}\right] \times 100 \tag{10}$$

where *C*—initial absorbance, *Cx*—final absorbance at time of test. The electrical energy consumption (*E*), an important parameter in process is calculated using Eq. 11, where *E* is the energy consumption (kWh/m^3), *U* is the applied voltage (V), *I* is the current intensity (A), *t* is the electrocoagulation time (h) and *V* is the volume of the treated wastewater (L).

$$E = \frac{U * I * t}{V} \tag{11}$$

The theoretical amount of electrode dissolved (*C*) was calculated by using Faraday's law as shown in Eq. 12, where *C* (g/L) is metal dissolved in the electrolytic cell, *I* is the current intensity (A), *t* is the electrocoagulation time (h), *M* is the molecular weight of the anode (g/mol), *n* is the chemical equivalence, *F* is the Faraday constant (96,500 C/mol) and *V* is the volume of the treated wastewater (L).

$$C = \frac{I * T * M}{n * F * V} \tag{12}$$

3 Result and Discussion

3.1 Effect of Change in Voltage

Voltage plays a fundamental role in reaction of electrocoagulation process. As voltage increases, current density (CD) increases. CD is defined as current applied per unit surface area of the electrode [24]. Current density also determines coagulant dose, flock production, rate and size of bubbles released from electrode process. [25, 26]. In present study, change in electrical potential in range 10–30 V at constant discharge of 105 mL/min having dye concentration of 60 mg/L and NaCl 1 g/L results are shown in Fig. 2. During investigation, it was observed that as voltage increases colour removal efficiency also increases, increasing energy consumption. The colour removal increased from 88.32 to 99.96% for R.R. and 61.44 to 96.53% for R.Y. at 10–30 V, respectively. Similarly, 1.52–13.57 kWh/m^3 of energy is required in R.R.., whereas RY requires 1.32–14.49 kWh/m^3 for same condition in treatment. After evaluation, R.Y. requires more energy, where lower

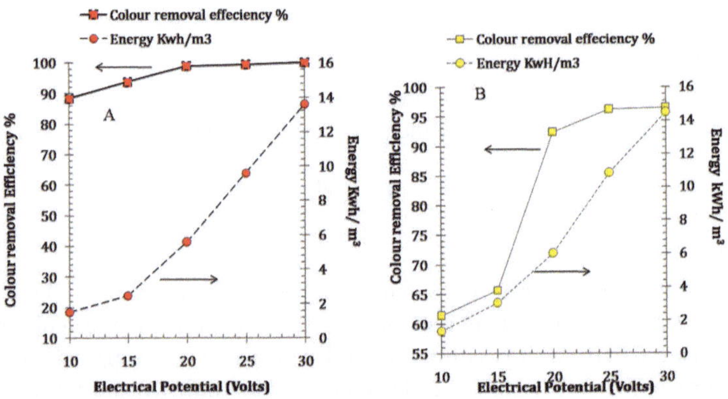

Fig. 2 Effect of change in electrical potential on removal efficiency and energy consumption (voltage range of 10–30 V, RR(A) and RY (B) dye concentration 60 mg/L, NaCl 1 g/L, S = 1 cm, Q = 105 mL/min, iron electrode at anode and cathode and submerged area, A_{sub} = 45 cm^2)

colour removal efficiency was achieved than R.R. dye. From investigation, consequently 20 V was considered as optimum and kept constant for subsequent of experiments.

3.2 *Effect of Change in Electrolyte Concentration*

The slight rise in concentration of electrolyte increases conductivity of solution. Subsequently, conductivity results in passage of excess electrons per unit time improving removal efficiency [24] with increase in current density at constant voltage [21]. Many of electrolytes are available in market such as NaCl, KCl, KI, NH_4Cl, Na_2SO_4, $NaNO_3$, K_2SO_4, $BaCl_2$, NaBr, Na_2SO_3, Na_2CO_3 and KNO_3 [27, 28]. In present study, effect of NaCl concentration in range 0.5–2.0 g/L on colour removal efficiency was investigated on R.R. and R.Y. (Fig. 3). Removal efficiency increased with increased concentration of electrolyte, resulting in reduction of reaction time. The colour removal efficiency increased from 87.95 to 99.63% for R.R. and 66.85 to 97.2% for R.Y. at 0.5 to 2.0 g/L of NaCl concentration, respectively. Similarly, energy consumed was in range of 4.83–7.05 kWh/m^3 for R.R. and 3.48–7.76 kWh/m^3 for R.Y. at NaCl concentration. The increases in NaCl concentration rises colour removal efficiency by 1.0% to R.R. with rise of 1.5 kWh/m^3 and nearly 5.0% rise in colour removal efficiency with rise of 1.8 kWh/m^3 for R.Y. Investigation concludes that as ionic strength increases, it raises colour removal efficiency, elevating energy consumption.

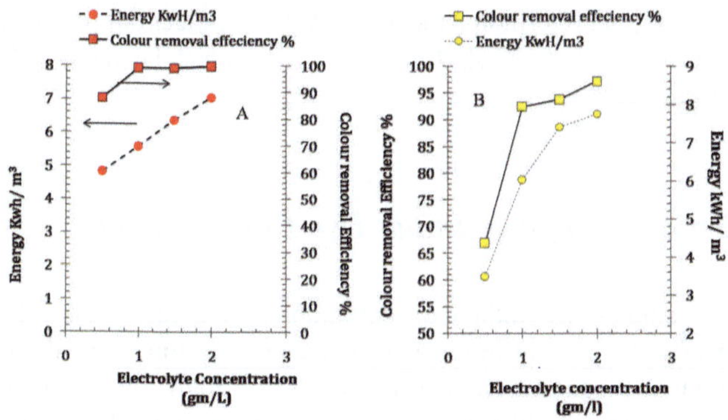

Fig. 3 Effect of electrolyte concentration on removal efficiency and energy consumption (voltage range of 20 V, RR(A) and RY(B) dye concentration 60 mg/L, NaCl 0.5–2.5 g/L, $S = 1$ cm, $Q = 105$ mL/min, iron electrode at anode and cathode and submerged area, $A_{sub} = 45$ cm^2)

3.3 *Effect of Change in Inter-Electrode Distance*

When distance between electrodes increases, IR-drop with resistance reducing contact between ions [25]. In present study, effect of inter-electrode distance in range 1–2.5 cm on colour removal efficiency was investigated (Fig. 4). As the inter-electrode distance increased from 1 to 2.5 cm, colour removal efficiency decreased from 98.91 to 79.15% for R.R. compared to 92.43–27.45% in R.Y. Also, increase in distance energy utilization also decreases from 5.57 to 3.19 kWh/m^3 for R.R., whereas for R.Y. energy reduce from 6.03 to 2.77 kWh/m^3 at same conditions. This may be due to decrease in electrostatic effect, resulting in slower movement of ions, reducing formation of flocs required for coagulating

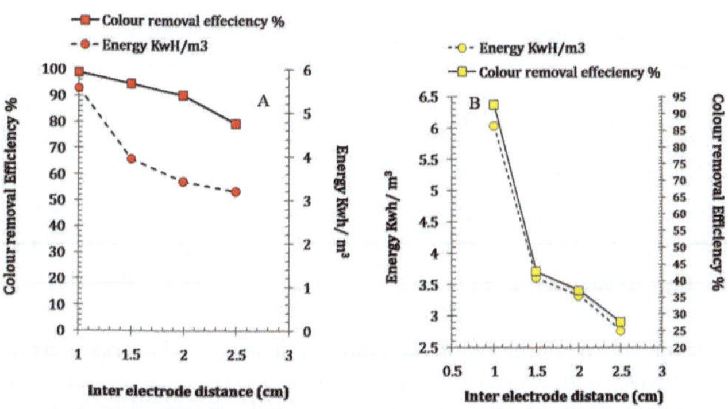

Fig. 4 Effect of inter-electrode distance on removal efficiency and energy consumption (voltage range of 20 V, RR(A) and RY(B) dye concentration 60 mg/L, NaCl 1 g/L, $S = 1$–2.5 cm, $Q = 105$ mL/min, iron electrode at anode and cathode and submerged area, $A_{sub} = 45$ cm^2)

contaminants [29]. Furthermore, the increased inter-electrode distance results in change of current density [30], for that voltage must be increased to achieve required current density [25].

3.4 *Effect of Change in Initial Dye Concentration*

According to studies, as initial concentration of dye increases, colour removal efficiency decreases [25, 31] and increasing electrolysis time [32]. In present study, effect of initial dye concentration in range 60–120 mg/L on colour removal efficiency was investigated (Fig. 5). During investigation of R.R., colour removal efficiency increased till 100 mg/L dye concentration, which was 99.18 and 98.91% for 60 mg/L concentration. Further at 120 mg/L dye concentration, colour removal reduced to 98.6%. As dye concentration increased, energy consumption in R.R. has elevated. The energy consumed ranged from 5.57 to 8.7 kWh/m^3. In R.Y., colour removal decreased as concentration increased from 92.43 to 67.35% at 60 to 120 mg/L, respectively. Consequently, energy consumption dropped down from 7.32 to 5.96 kWh/m^3 as dye concentration increased from 60 to 120 mg/L. These may be due to constant rate of metal hydroxide generated at invariable voltage of 20 V [33]. In EC process, colour is adsorbed onto metallic hydroxide flocs and removed by settling [25]. Consequently, less quantity of contaminants adsorbed with increases in dye concentration resulting in reduction efficiency of process.

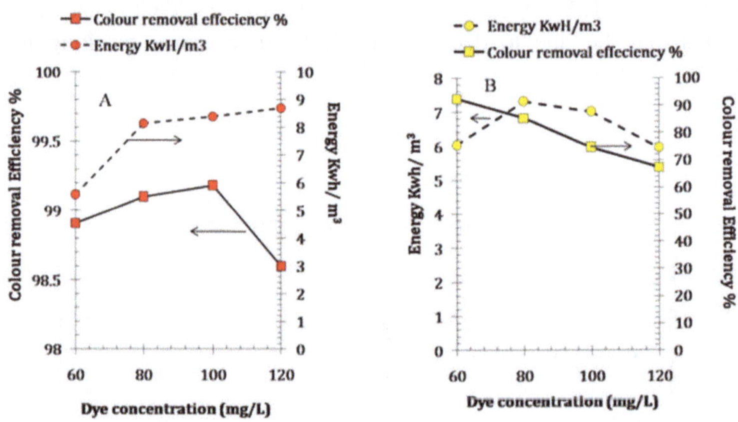

Fig. 5 Effect of initial dye concentration on removal efficiency and energy consumption (voltage range of 20 V, RR(A) and RY(B) dye concentration 60–120 mg/L, NaCl 1 g/L, $S = 1$ cm, $Q = 105$ mL/min, iron electrode at anode and cathode and submerged area, $A_{sub} = 45$ cm^2)

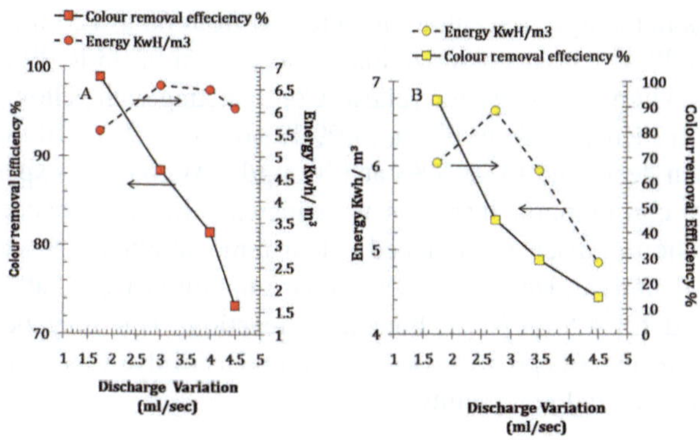

Fig. 6 Effect of discharge on removal efficiency and energy consumption (voltage range of 20 V, RR(A) and RY(B) dye concentration 60 mg/L, NaCl 1 g/L, S = 1 cm, Q = 1.75–4.5 mL/s, iron electrode at anode and cathode and submerged area, A_{sub} = 45 cm^2)

3.5 *Effect of Change in Discharge*

The effect of discharge variation on colour removal efficiency was studied at flow rate ranged between 1.75 and 4.5 mL/s at constant conditions (Fig. 6). The increases in discharge reduce colour removal efficiency, and for RR, removal efficiency reduced from 98.91 to 73.16%, whereas for RY colour removal efficiency reduces from 92.43 to 14.35% at discharge 1.75 to 4.5 mL/s, respectively. In addition, energy consumption increased from 5.57 to 6.60 kWh/m^3 at 3.0 mL/s in RR and reduced at 4.0 and 4.5 mL/s. During these discharges, energy falls from 6.65 to 6.03 kWh/m^3. In R.Y. energy consumption elevated from 6.03 to 6.65 kWh/m^3 at discharge 3.0 mL/s. Further, energy reduced to 4.84 kWh/m^3, 4.5 mL/s with 14.35% colour efficiency.

4 Conclusions

Results show that colour removal efficiency increased from 61.44 to 96.53% for yellow, whereas 88.32–99.07% for red dyes at 10–30 volts. Similarly, energy consumed was 1.269–14.95 and 1.52–13.57 kWh/m^3 for yellow and red, respectively. Reactive red showed higher removal efficiency than yellow on similar conditions with lower count on energy required. Furthermore, electrolyte concentration of 0.5–2.0 g/L was added to both dye wastewaters, to check effect on decolourization. The dose of 1.0 g/L showed 92.43 and 98.91% decolourization, consuming 6.22 and 5.57 kWh/m^3 of energy for yellow and red, respectively, at constant voltage and flow rate. In both cases, colour removal efficiency decreased from 92.43 to 27.45% for yellow and 98.91 to 79.15% with increase inter-electrode distance in range 1–2.5 cm, respectively. Also, energy consumption was reduced.

In addition, as initial dye concentration increased from 60 to 120 mg/L at constant discharge and 20 V. The decolourization reduced from 92.43 to 49.36 consuming 6.22 to 6.42 kWh/m^3 of energy for yellow. Contradicting with yellow, Reactive red decolourization increased from 98.91 to 99.18 for dose 60 to 100 mg/L, respectively, and then decreased to 98.60% at 120 mg/L. At constant experimental condition, energy consumption increases with increase in initial dye concentration. Effect of variation in discharge reduced colour removal efficiency with increase in discharge for both dyes. However, energy consumption increased at 2.75 mL/s and further dropped till 4.5 mL/s for both dyes reaction. This may be due to high mixing or turbulence. EC process used for treatments of dye wastewater may prove advantages to industrial community.

References

1. Janus, M., Kusiak, E., Choina, J., Ziebro, J., Morawski, A.W.: Enhanced adsorption of two azo dyes produced by carbon modification of TiO_2. Desalination **249**, 359–363 (2009)
2. Colindres, P., Yee-Madeira, H., Reguera, E.: Removal of reactive black 5 from aqueous solution by ozone for water reuse in textile dyeing processes. Desalination **258**, 154–158 (2011)
3. Phillips, D.: Environmentally friendly, productive and reliable: priorities for cotton dyes and processes. J. Soc. Dyers Colour. **112**, 183–186 (1996)
4. Rivera, M., Pazos, M., Sanromán, M.Á.: Development of an electrochemical cell for the removal of reactive black 5. Desalination **274**, 39–43 (2011)
5. Idel-aouada, R., Valientea, M., Yaacoubib, A., Tanoutic, B., Lopez-Mesasa, M.: Rapid decolourization and mineralization of the azo dye C.I. Acid Red 14 by heterogeneous Fenton reaction. J. Hazard. Mat. **186**, 745–750 (2011)
6. Akar, S.T., Özcanb, A.S., Akar, T., Özcanb, A., Kaynak, Z.: Biosorption of a reactive textile dye from aqueous solutions utilizing an agro-waste. Desalination **249**, 757–761 (2009)
7. Su, C., Asa, M., Ratanatamskul, C., Lu, M.: Effect of operating parameters on decolorization and COD removal of three reactive dyes by Fenton's reagent using fluidized-bed reactor. Desalination **278**, 211–218 (2011)
8. Subash, B., Krishnakumar, B., Swaminathan, M., Shanthi, M.: Synthesis and characterization of cerium–silver co-doped zinc oxide as a novel sunlight-driven photocatalyst for effective degradation of Reactive Red 120 dye. Mater. Sci. Semicond. Process. **16**, 1070–1078 (2013)
9. Al-Degs, Y.S., El-Barghouthi, M.I., El-Sheikh, A.H., Walker, G.M.: Effect of solution pH, ionic strength, and temperature on adsorption behavior of reactive dyes on activated carbon. Dyes Pigm. **77**, 16–23 (2008)
10. Satapanajarua, T., Chompuchana, C., Suntornchot, P., Pengthamkeerati, P.: Enhancing decolorization of Reactive Black 5 and Reactive Red 198 during nano zerovalent iron treatment. Desalination **266**, 218–230 (2011)
11. Liu, R., Chiu, H.M., Shiau, C.S., Yeh, R.Y.L., Hung, Y.T.: Degradation and sludge production of textile dyes by Fenton and photo-Fenton processes. Dyes Pigm. **73**, 1–6 (2007)
12. Jong, J.D., Shin, W., Choi, J., Choi, S., Kim, M., Han, M., Ha, T., Kime, Y.: Decolorization of reactive dyes using inorganic coagulants and synthetic polymer. Dyes Pigm. **73**, 9–64 (2007)
13. Heidmann, I., Calmano, W.: Removal of Zn(II), Cu(II), Ni(II), Ag(I) and Cr(VI) present in aqueous solutions by aluminum electrocoagulation. J. Hazard. Mat. **152**, 934–941 (2008)

14. Mouedhena, G., Fekia, M., De Petris-Weryb, M., Ayedi, H.: Electrochemical removal of Cr (VI) from aqueous media using iron and aluminum as electrode materials: towards a better understanding of the involved phenomena. J. Hazard. Mater. **168**, 983–991 (2009)
15. Drouichea, N., Ghaffourb, N., Lounicic, H., Mameri, M.: Electrocoagulation of chemical mechanical polishing Wastewater. Desalination **214**, 31–37 (2007)
16. Benhadji, A., Ahmed, M.T., Maachi, R.: Electrocoagulation and effect of cathode materials on the removal of pollutants from tannery wastewater of Rouïba. Desalination **277**, 128–134 (2011)
17. Barrera-Díaza, C.E., Lugo-Lugoa, V., Bilyeu, B.: A review of chemical, electrochemical and biological methods for aqueous Cr(VI) reduction. J. Hazard. Mater. **1**(12), 223–224 (2012)
18. Muftah, H.E., Sulaiman, A.Z., Amal, A.L., Souzan, M.: Assessment of electrocoagulation for the treatment of petroleum refinery wastewater. J. Environ. Manage. **91**, 180–185 (2009)
19. Behbahani, M., Moghaddam, M.R.A., Arami, M.: Techno-economical evaluation of fluoride removal by electrocoagulation process: optimization through response surface methodology. Desalination **271**, 209–218 (2011)
20. Ahmet, G., Mehmet, Y., Cetin, D.: Electrocoagulation of some reactive dyes: a statistical investigation of some electrochemical variables. Waste Manag. **22**(5), 491–499 (2002)
21. Mehmet, K., Orhan, T.C., Mahmut, B.: Treatment of textile wastewaters by electrocoagulation using iron and aluminum electrodes. J. Hazard. Mater. **B100**, 163–178 (2003)
22. Barun, K.N., Sunil, P.: Effects of operational parameters on the removal of brilliant green dye from aqueous solutions by electrocoagulation. Arab. J. Chem. https://doi.org/10.1016/j.arabjc.2013.11.032 (2013)
23. Rice, E.W.: Standard Methods for the Examination of Water and Wastewater. APHA, USA (1998)
24. Murthy, Z.V.P., Parmar, S.: Removal of strontium by electrocoagulation using stainless steel and aluminum electrodes. Desalination **282**, 63–67 (2011)
25. Modirshahla, N., Behnajady, M.A., Kooshaiian, S.: Investigation of the effect of different electrode connections on the removal efficiency of Tartrazine from aqueous solutions by electrocoagulation. Dyes Pigm. **74**, 249–257 (2007)
26. Mollah, M., Schennach, R., Parga, J., Cocke, D.: Electrocoagulation (EC)-science and applications. J. Hazard. Mat. **84**, 29–41 (2001)
27. Jimenez, I.C., Canizares, P., Rodrigo, M.A., Leclerc, J.P., Valentin, G., Lapicque, F.: Effect of the nature of the supporting electrolyte on the treatment of soluble oils by electrocoagulation. Desalination **255**(1–3), 15–20 (2010)
28. Kashefialasl, M., Khosravi, M., Marandi, R., Seyyedi, K.: Treatment of dye solution containing colored index acid yellow 36 by electrocoagulation using iron electrodes. Int. J. Environ. Sci. Technol. **2**(4), 365–371 (2006)
29. Khandegar, V., Anil, K.S.: Electrocoagulation of distillery spentwash for complete organic reduction. Int. J. Chem. Tech. Res. **5**(2), 712–718 (2013)
30. Bouhezila, F., Hariti, M., Lounici, H., Mameri, N.: Treatment of the OUED SMAR town landfill leachate by an electrochemical reactor. Desalination **280**, 347–353 (2011)
31. Adhoum, N., Monser, L., Bellakhal, N., Eddine, J.B.: Treatment of electroplating wastewater containing Cu^{2+}, Zn^{2+} and Cr(VI) by electrocoagulation. J. Hazard. Mat. **112**, 207–213 (2004)
32. Mohammad, M.E., Muttucumaru, S.: Review of pollutants removed by electrocoagulation/flotation processes. J. Environ. Manage. **90**(5), 1663–1679 (2009)
33. Wei-Lung, C., Yen-Hsiang, H.: Electrochemical removal of indium ions from aqueous solution using iron electrodes. J. Hazard. Mater. **172**, 46–53 (2009)

Municipal Solid Waste Collection Time Optimization Using AHP, GMDH and ANN

Debasree Purkayastha, Mrinmoy Majumder and Sumanta Chakrabarti

Abstract Municipal solid waste management is now a global concern and needs to be addressed with proper technological innovations. In this study, a model has been developed using analytical hierarchy process, group method of data handling and artificial neural network to determine optimum time within which the collection of municipal solid waste (MSW) during each visit by each type of waste collection vehicle should be completed; this will economize the collection phase of MSW. A case study was conducted on Agartala Municipality situated in north-eastern part of India. The computed optimum route length in terms of time is 3.84 and 2.75 h for 14 and 8 m^3 garbage compactor, respectively. Adoption of these optimum time estimates will lead to a cost-effective SWM as all the garbage compactors need not have to work in more than one shift, leading to a huge savings in fuel and labour cost.

Keywords Municipal solid waste (MSW) · Collection time · Analytical hierarchy process (AHP) · Group method of data handling (GMDH) · Artificial neural network (ANN)

D. Purkayastha (✉)
Department of Civil Engineering, Indian Institute of Technology Roorkee, Roorkee 247667, Uttarakhand, India
e-mail: debap.dce2015@iitr.ac.in

M. Majumder
Department of Civil Engineering, National Institute of Technology Agartala, P.O.: Former Tripura Engineering College Barjala, Jirania 799046, Tripura (W), India
e-mail: mmajumder15@gmail.com

S. Chakrabarti
Maharaja Bir Bikram University, Agartala 799004, Tripura (W), India
e-mail: imsumanta@gmail.com

A. S. Kalamdhad et al. (eds.), *Advances in Waste Management*,
https://doi.org/10.1007/978-981-13-0215-2_4

1 Introduction

Creation of solid waste is an inevitable part of human activities, especially in urban crowd. The most pressing problem faced by most of the cities around the globe especially in developing countries is municipal solid waste management. In India, the municipal agencies spare 5–25% of the total allocated budget on solid waste management (SWM). And further, almost 85% of this total expenditure in SWM is spent on collection [1]. But unfortunately, high capital investment in the solid waste management sector is not necessarily leading to improvements in the quality of service [2]. Similar observations regarding huge cost involvement in collection phase of SWM have been made by other researchers [3–6] and municipalities as well.

Tripura the third smallest state of India is situated in its north-eastern part. Agartala is the capital of Tripura located in the floodplains of the Haora River. The solid waste management of Agartala city is carried out by Agartala Municipal Corporation (AMC) established in 1871. Municipal area of Agartala is divided into 4 zones and 35 wards (Fig. 1). The overall population of AMC is 4,70,190 with an overall area of 61.718 km^2 as per ward-wise population statement as on August 2013. The total waste generated in AMC per day is about 250 MT. Basically, the waste generated in Agartala is of different types—municipal solid waste, biomedical waste, hazardous waste, construction and demolition waste and e-waste.

AMC carry out collection in two ways—some areas are covered by house-to-house collection which accounts for only 30–40% of total coverage and the other way is through large-sized garbage bins of size 4.5 and 1.1 m^3 placed on the major crossroads of the town. Daily public leave their garbage in these bins

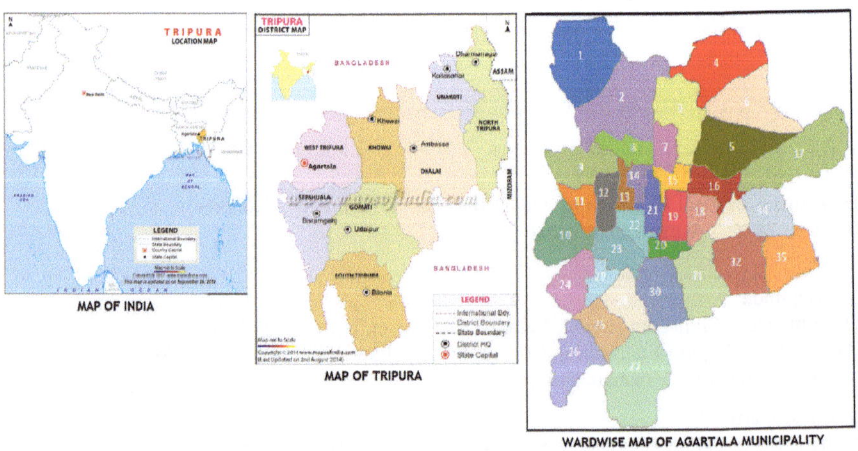

Fig. 1 Ward-wise map of Agartala Municipal Corporation

(a) Garbage compactor (capacity of 8 m3) **(b)** Garbage compactor (capacity of 14 m3)

Fig. 2 Pictures of two types of vehicles used by AMC for waste collection from 1.1 m^3 bin

which are further periodically evacuated by mechanically equipped vehicles of AMC.

AMC has its major disposal ground in Debendrachandra Nagar, located in the northern side of the city. In this study, we are considering collection of waste only from 1.1 m^3 hauled containers to Debendrachandra Nagar disposal site. The collection of MSW from 1.1 m^3 container is carried out by two types of vehicles (Fig. 2)—garbage compactor with capacity of 8 m^3 (10 nos) and 14 m^3 (5 nos).

Several studies have been conducted by various researchers [1, 7–17] to optimize the transportation part of solid waste management using various optimization and heuristics techniques, and many have incorporated GIS also. In this study, an effort has been made to develop a model using analytical hierarchy process (AHP), group method of data handling (GMDH) and artificial neural network (ANN) as tools to determine an optimum time within which the collection of municipal solid waste by each type of vehicle should be completed and a case study has been conducted on Agartala Municipality.

2 Methodology Used

The main objective of this study is to develop a model which can determine the optimum time required by municipal worker to complete waste collection using mechanized garbage compactors. The study used two modelling tools to come up with this model: multi-criteria decision-making (MCDM) and artificial neural network (ANN). Apart from these two techniques, the method used video and motion time measurement survey for field data collection.

2.1 Framework and Data Collection

Initially, two types of survey were carried out: (i) motion time measurement survey—it was conducted to estimate the average speed of the vehicles using stopwatch. The average speed was found to be 25 kmph for the two types of vehicle available for waste collection. (ii) Video survey—this was conducted to estimate the individual time required for carrying out each of the six of activities as described in Table 1 (T_1, T_2, T_3, T_4, T_5, T_6). This survey was done through video captured when the municipality worker started unloading the bins through mechanized garbage compactors and was analysed later in the office. AMC has total 447 numbers of 1.1 m^3 collection bins. Almost 30 videos were collected to extract the time estimates. Further, these videos were analysed and each activity was divided into three types of time estimates: optimistic time, most probable time and pessimistic time. The optimistic time defines the fastest time in which an activity was completed. Most probable time is the time within which the activity was mostly completed, and pessimistic time was considered as the time which was the maximum time taken to complete an activity. All these surveys were carried out for a period of 30 days. And both of these surveys were carried out for the two types of vehicles—garbage compactor of capacity 8 and 14 m^3, collecting wastes from 1.1 m^3 bins.

A survey was also conducted among twenty AMC workers and officials to ascertain the hazards associated with each activity T_1, T_2, T_3, T_4, T_5, T_6.

The number of collection bin that can be emptied per trip (N) was calculated for each type of garbage compactor based on Eq. 1. 'N' was calculated so that the final time estimate can take into account one whole trip for final optimum time computation (Table 2).

$$N = \frac{f.c}{v.u} \tag{1}$$

Table 1 Description of the activities involved in collection work

Activity	Symbol
Driving time from garage to first collection point	T_1
Total time required in loading and unloading the collection bins	T_2
Total time spent driving between the containers	T_3
Time consumed in driving from last collection point to disposal site for unloading collected waste	T_4
Time spent in disposal site	T_5
Time consumed in driving from disposal site to garage	T_6

Table 2 Details regarding various parameters associated with garbage compactor

Compaction factor	Capacity of vehicle (in cum.)	Total volume of waste that can be collected (in cum.)	Volume of each container (in cum.)	Weighted container utilization factor	Number of container that can be emptied per trip	Maximum working hours of labours and drivers
2.5	14	35	1.1	0.85	37	7
	8	20			21	7

where

- f Compaction factor
- c Capacity of vehicle (m^3)
- v Volume of each container (m^3)
- u Weighted container utilization factor (m^3)

2.2 *Weightage Computation Using AHP*

The AHP method requires two steps: (1) selection of criteria and sub-criteria and (2) application of aggregation method.

2.2.1 Selection of Criteria and Sub-criteria

For the present study, the weightage of all the activities (as tabulated in Table 1) needs to be computed. Henceforth, all these activities were considered as sub-criteria. The weight of importance of the sub-criteria needs to be established with respect to some criteria. The weightage associated with each activity was computed based on the criteria 'most probable time estimate' of each activity. Hazard potential was not taken as another criterion because the number of hazards associated with each activity is equal (=2) as can be seen in Table 5, so there is no justification of taking it into consideration. The weightage was computed using analytical hierarchy process (AHP). The hierarchy is shown in Fig. 3.

2.2.2 Application of Aggregation Method

The video surveys from the site were analysed in the office, and each time estimate was calculated for every sub-criteria. Then, each sub-criterion was provided a rank based on the value of most probable time estimate. The ranking was provided in the manner that the sub-criteria under a criterion which was having highest most

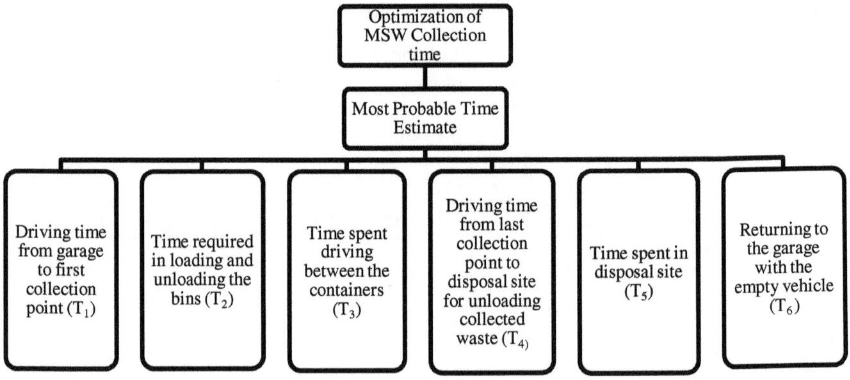

Fig. 3 Hierarchy model for weightage computation using AHP

probable time estimate was ranked 1, the sub-criteria with second highest score was ranked 2 and so on.

A 6 × 6 matrix was developed to find out the weightage of criteria.

$$c = \{n \times n\} \tag{2}$$

where $\{n\} = \{T\} \in R$, where R is the set of real numbers.

Similarly, the sub-criteria are compared with each other based on their importance over each other according to each of the criteria 'n':

$$S = \{f_i \times f_i\} \tag{3}$$

where $\{f_i\} = \{T_1, T_2, T_3, T_4, T_5, T_6\} \in R$, where R is the set of real numbers.

The hierarchy of decision-making is shown in Fig. 1.

In case of AHP, generally the Saaty scale is used which was proposed by Saaty in the year of 1980 [18]. The scale utilized either even or odd number to represent the importance of the criteria and sub-criteria with respect to each other in the pair-wise comparison matrix (PCM). For intermediate importance, the rating in between the evens or odds is utilized. But still, there is a lot of confusion regarding what can be used for representation of high difference of importance and minor difference of importance between two sub-criteria or criteria.

That is why, in the present study, we use the rank of the criteria and sub-criteria based on their magnitude or qualitative ratings and then ration of rank of the criteria/sub-criteria compared and the rank of the other criteria/sub-criteria with which it is being compared was found out (the rank is assigned in such a way that the relationship of the criteria with the decision objective can be reflected). The ratio is then reversed to give the exact difference of importance coherent to decision objective.

The direct use of rank to estimate the importance will ensure uniformity and remove the confusion involving the rating that can be given to depict two different levels of importance that exist between two different criteria or sub-criteria.

2.2.3 Formulation of Most Optimum Collection Time (*T*)

The final '*T*' formula was formulated as:

$$T = (W_{T_1} X\, T_1') + (W_{T_2} X\, T_2') + (W_{T_3} X\, T_3') + (W_{T_4} X\, T_4') + (W_{T_5} X\, T_5') + (W_{T_6} X\, T_6') \quad (4)$$

T_1' Normalized value of T_1
T_2' Normalized value of T_2
T_3' Normalized value of T_3
T_4' Normalized value of T_4
T_5' Normalized value of T_5
T_6' Normalized value of T_6
W_{T_1} Weightage of T_1
W_{T_2} Weightage of T_2
W_{T_3} Weightage of T_3
W_{T_4} Weightage of T_4
W_{T_5} Weightage of T_5
W_{T_6} Weightage of T_6

2.3 *ANN Model Formulation*

To predict S.I., artificial neural network (ANN) was used. GMDH Shell software was used for carrying out the ANN-based prediction. GMDH stands for 'group method of data handling'. The main idea behind GMDH is to develop a feedforward network based on a quadratic node transfer function whose coefficients are obtained using a regression technique Farlow [19]. It was first proposed in 1966 by a Russian cyberneticist, A. G. Ivakhnenko.

The inputs of the model were the random normalized values of T_1, T_2, T_3, T_4, T_5, T_6. The input function of the ANN model is shown in Eq. 4. The output is most optimum collection time (*T*). The model predicted the *T* value for 5000 data.

3 Result and Discussion

3.1 Computation of Time Estimates

The data extracted from the video surveys carried out at various collection bin points were analysed, and the three types of time estimates were found out for two different garbage compactors, viz. 14 and 8 m^3 (Tables 3 and 4).

The survey conducted for hazard potential estimation shows that each activity is associated with equal number of hazard (=2) (Table 5).

Table 3 Time estimates for garbage compactor (with capacity of 14 m^3)

S. No.	Activities/objectives	Optimistic time (in s)	Most likely time (in s)	Pessimistic time (in s)
1	Driving time from garage to first collection point (T_1) = distance/(speed (25 kmph)) (in s)	288	576	864
2a	Time needed to bring the container to lift in the vehicle (in s)	259	444	1665
2b	Time to unload the waste mechanically into the vehicle (in s)	592	999	4440
2c	Time required to load contents from container at pickup location into collection vehicle (uc) (in s)	1036	1443	6105
2d	Time required to put back the container to its original place (in s)	185	333	814
2	Total time required in loading and unloading bins (T_2)	2072	3219	13024
3	Total time spent driving between the containers (T_3)	3105	4645	6350
4	Driving from NP collection point to disposal site for unloading collected waste (s) = distance/(speed (25 kmph)) (T_4)	2061	2560	3042
5	Time spent in disposal site (T_5)	480	600	720
6	Returning to the garage with the empty vehicle = distance/(speed (25 kmph)) (T_6)	994	1000	1200
	Overall length of the trip	9000	12,600	25,200

Table 4 Time estimates for garbage compactor (with capacity of 8 m^3)

S. No.	Activities/objectives	Optimistic time (in s)	Most likely time (in s)	Pessimistic time (in s)
1	Driving time from garage to first collection point (T_1) = distance/(speed (25 kmph)) (in s)	432	792	1152
2a	Time needed to bring the container to lift in the vehicle (in s)	147	252	945
2b	Time to unload the waste mechanically into the vehicle (in s)	336	567	2520
2c	Time required to load contents from container at pickup location into collection vehicle (uc) (in s)	588	819	3465
2d	Time required to put back the container to its original place (in s)	105	189	462
2	Total time required in loading and unloading bins (T_2)	1176	1827	7392
3	Total time spent driving between the containers (T_3)	1298	3461	5916
4	Driving from NP collection point to disposal site for unloading collected waste (s) = distance/(speed (25 kmph)) (T_4)	1140	1440	1740
5	Time spent in disposal site (T_5)	360	480	600
6	Returning to the garage with the empty vehicle = distance/(speed (25 kmph)) (T_6)	994	1000	1200
	Overall length of the trip	5400	9000	18,000

Table 5 Hazard potential associated with each activity

S. No.	Activities/objectives	Hazard potential
1	Driving time from garage to first collection point = distance/(speed (25 kmph)) (T_1)	1. Accident in the roadway
		2. Vehicular failure
2	Total time required in loading and unloading bins (T_2)	1. Health hazard of worker
		2. Vehicular failure
3	Total time spent driving between the containers (T_3)	1. Accident in the roadway
		2. Vehicular failure
4	Driving from NP collection point to disposal site for unloading collected waste (s) = distance/(speed (25 kmph)) (T_4)	1. Accident in the roadway
		2. Vehicular failure
5	Time spent in disposal site (T_5)	1. Health hazard of worker
		2. Vehicular failure
6	Returning to the garage with the empty vehicle = distance/(speed (25 kmph)) (T_6)	1. Accident in the roadway
		2. Vehicular failure

Table 6 Rank of the factors based on most probable time estimate

Activities	Rank
T_1	5
T_2	2
T_3	1
T_4	3
T_5	6
T_6	4

Table 7 Pair-wise comparison matrix (PCM) for factors

	T_1	T_2	T_3	T_4	T_5	T_6
T_1	1.00	2.50	5.00	1.67	0.83	1.25
T_2	0.40	1.00	2.00	0.67	0.33	0.50
T_3	0.20	0.50	1.00	0.33	0.17	0.25
T_4	0.60	1.50	3.00	1.00	0.50	0.75
T_5	1.20	3.00	6.00	2.00	1.00	1.50
T_6	0.80	2.00	4.00	1.33	0.67	1.00

3.2 *Weightage Computation Using AHP*

The rank of the criteria based on the video survey is shown in Table 6. The pair-wise comparison matrix for the sub-criteria is shown in Table 7. The weightages which were finally obtained using analytical hierarchy process (AHP) are tabulated in Table 8.

3.3 *GMDH Model for Optimum Time Computation*

The model predicted by GMDH Shell (Data Science version) software showed the correlation coefficient of the model is 1.00000 which is showing that the predicted model is of high quality. A plot of the model predicted file is shown in Fig. 4.

Table 8 Inverse pair-wise comparison matrix (PCM) for factors

	T_1	T_2	T_3	T_4	T_5	T_6	Geometric mean	Weightage
T_1	1.00	0.40	0.20	0.60	1.20	0.80	0.0064	0.000308645
T_2	2.50	1.00	0.50	1.50	3.00	2.00	0.625	0.030141097
T_3	5.00	2.00	1.00	3.00	6.00	4.00	20	0.964515113
T_4	1.67	0.67	0.33	1.00	2.00	1.33	0.082304527	0.003969198
T_5	0.83	0.33	0.17	0.50	1.00	0.67	0.002572016	0.000124037
T_6	1.25	0.50	0.25	0.75	1.50	1.00	0.01953125	0.000941909
							20.73580779	1.000000

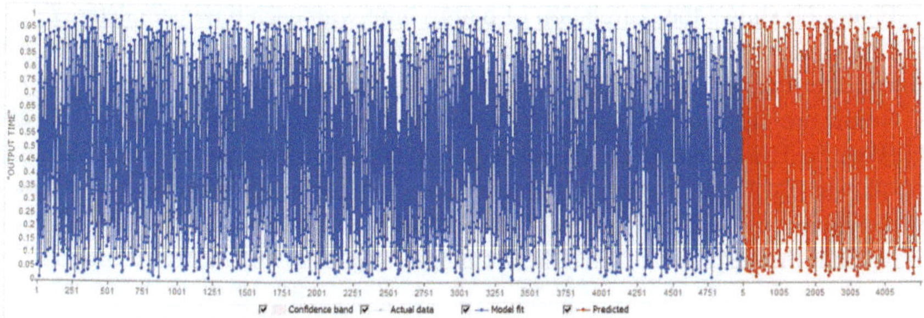

Fig. 4 Plot of model predicted by GMDH shell for output time (*T*)

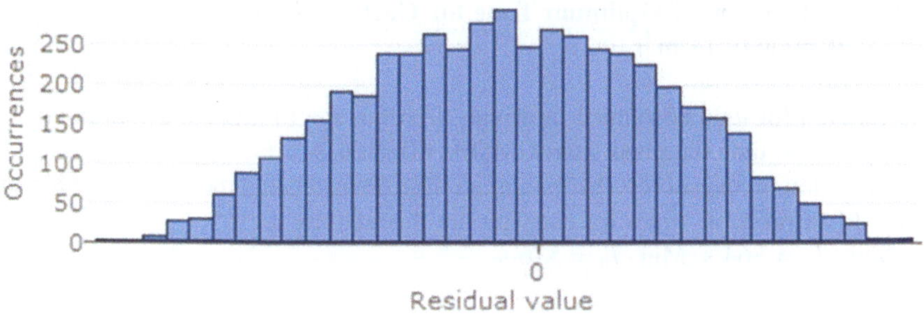

Fig. 5 Residual of model predicted by GMDH Shell for output time (*T*)

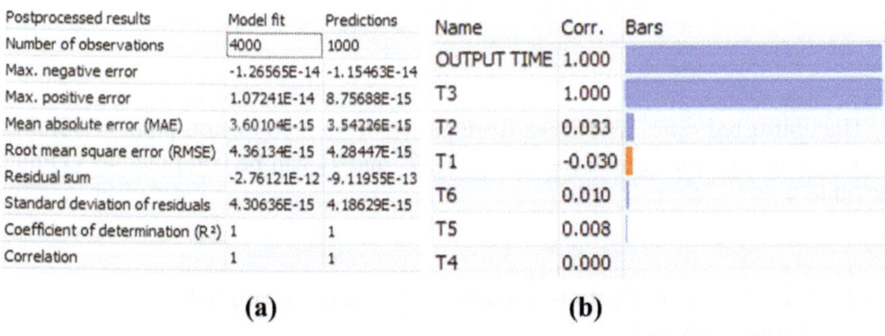

Postprocessed results	Model fit	Predictions
Number of observations	4000	1000
Max. negative error	-1.26565E-14	-1.15463E-14
Max. positive error	1.07241E-14	8.75688E-15
Mean absolute error (MAE)	3.60104E-15	3.54226E-15
Root mean square error (RMSE)	4.36134E-15	4.28447E-15
Residual sum	-2.76121E-12	-9.11955E-13
Standard deviation of residuals	4.30636E-15	4.18629E-15
Coefficient of determination (R²)	1	1
Correlation	1	1

(a)

Name	Corr.	Bars
OUTPUT TIME	1.000	
T3	1.000	
T2	0.033	
T1	-0.030	
T6	0.010	
T5	0.008	
T4	0.000	

(b)

Fig. 6 **a** Accuracy of the global model predicted by GMDH shell for *T*. **b** Correlation of the seven factors with *T*

Figure 5 shows the residual distribution of the data set. The figure shows that the residuals are evenly distributed on both sides of the zero from which it can be said that the net residual of the model is very low. The same can be more clearly depicted from the accuracy data of the model shown in Fig. 6a.

Figure 6b shows the correlation of output time with each time parameter. T_3 has a one-to-one relation with output time T, with highest degree of correlation; following the line are T_2, T_1, T_6, T_5, T_4 in decreasing order of correlation.

Below are the contour maps of T_3 with all other input factors T_2, T_1, T_6, T_5 and T_4 (Fig. 7). The maps clearly depict that T_3 has a monopolistic hold over all other inputs, viz. T_2, T_1, T_6, T_5 and T_4. This behaviour is also clear from the computation of weightage by AHP, in which T_3 alone has a weightage of 0.964515113. This shows that the response of the GMDH model is in line with AHP.

3.4 *Calculation of Optimum Time*

3.4.1 Calculation of Optimum Time for Garbage Compactor (Capacity 14 m^3)

Calculation for total consumed in driving from garage to first collection point, T_1:

Optimized time obtained from GMDH = 0.036016.

According to the survey, the maximum time (Pessimistic time) and the minimum time (Most probable time) for carrying out this activity are:

Max. T_1 = 864 s, Min. T_1 = 576 s

We know,

$$\begin{aligned} \text{Normalised value} &= \frac{x - \text{Min value}}{\text{Max value} - \text{Min value}} \\ \Rightarrow 0.036016 &= \frac{x - 576}{864 - 576} \\ \Rightarrow x &= 586.37\ \text{s} \end{aligned} \tag{5}$$

The optimized time T_1 to drive from garage to first collection point = 586.37 s.

Similarly, optimized time for T_2, T_3, T_4, T_5 and T_6 can be computed as tabulated in Table 9.

3.4.2 Calculation of Optimum Time for Garbage Compactor (Capacity 8 m^3)

Calculations of optimum time for garbage compactor (capacity 8 m^3) are same as the calculations done for garbage compactor (capacity 14 m^3).

The computed optimized times are presented in Table 10.

Total optimal length of trip for 14 and 8 m^3 garbage compactors are 3.84 and 2.73 h to pick up 37 and 21 numbers of 1.1 m^3 bins, respectively.

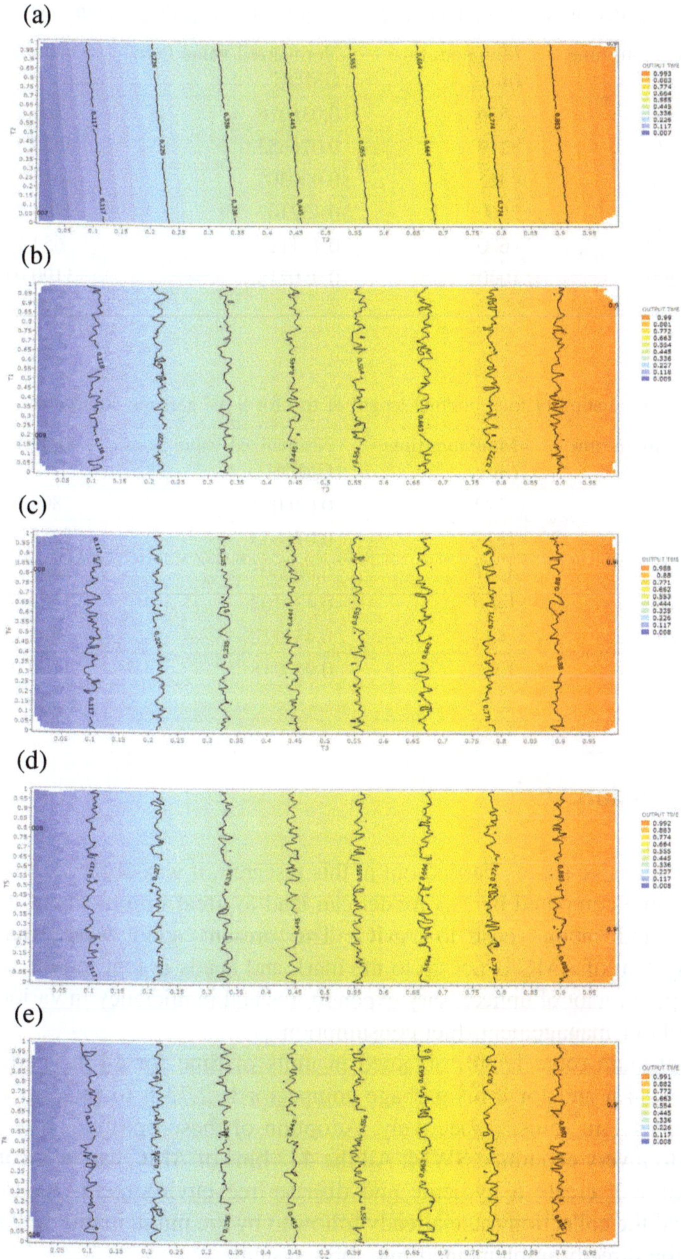

Fig. 7 **a** Contour map of T_3 versus T_2. **b** Contour map of T_3 versus T_1. **c** Contour map of T_3 versus T_6. **d** Contour map of T_3 versus T_5. **e** Contour map of T_3 versus T_4

Table 9 Computation of total optimal length of trip for 14 m^3 garbage compactor

	Maximum time (in s)	Minimum time (in s)	Normalized value from GMDH	Optimized time (in s)
T_1	864	576	0.036016	586.37
T_2	13,024	3219	0.076653	3970.58
T_3	6350	4325	0.000505	4645.86
T_4	3042	2560	0.432102	2768.27
T_5	720	600	0.451112	654.13
T_6	1200	1000	0.950218	1190.04
Total time =				13815.25 (3.84 h)

Table 10 Computation of total optimal length of trip for 8 m^3 garbage compactor

	Maximum time (in s)	Minimum time (in s)	Normalized value from GMDH	Optimized time (in s)
T_1	1152	792	0.036016	804.97
T_2	7392	1827	0.076653	2253.57
T_3	5916	3461	0.000505	3462.24
T_4	1740	1440	0.432102	1569.63
T_5	600	480	0.451112	543.13
T_6	1200	1000	0.950218	1190.04
Total time =				9823.58 (2.73 h)

4 Conclusion

With the extensive survey carried out in this research, it was observed that the total optimum time computed by this model can lead to great improvement in terms of fuel cost and labour cost to AMC. The present solid waste management (SWM) system of AMC is not up to the mark and needs urgent modifications as it gives scope for a lot of unnecessary expenditures and inefficiency in various regards such as labour management, fuel consumption.

The optimum route length obtained in units of time for 14 m^3 garbage compactor is 3.84 h and for 8 m^3 garbage compactor is 2.73 h to pick up 37 and 21 numbers of 1.1 m^3 bins, respectively. Adoption of these optimum time estimates will lead to a very economic SWM. All the 447 bins of AMC can be cleaned every day which will create a hygienic and disease-free environment. As the present model used the collection route already followed by the municipality, this estimated time can be directly implemented into their system.

The future scope of the study includes that based on the optimized collection times an algorithm can be proposed incorporating GIS. If this model can be summed up with an GIS route optimizing algorithm, then the municipalities can directly select the optimized route satisfying this optimization model which will lead them to saving in fuel and labour cost to a great extent.

Acknowledgements The authors want to acknowledge Agartala Municipal Corporation and Urban Development Department, Agartala, Tripura, India, for their consistent help and assistance with all available data and valuable suggestions.

References

1. Ghose, M.K., Dikshit, A.K., Sharma, S.K.: A GIS based transportation model for solid waste disposal—a case study on Asansol municipality. Waste Manage. **26**(11), 1287–1293. ISSN 0956-053X (2006)
2. National Solid Waste Association of India Newsletter.: Economics of Solid Waste Management-Cost Benefit Analysis, 18th issue, March 2010. Retrieved from http://www.nswai.com/images/newsletters/mar2010.pdf (2010)
3. Ghiani, G., Laganà, D., Manni, E., Triki, C.: Capacitated location of collection sites in an urban waste management system. Waste Manage. **32**(7), 1291–1296. ISSN 0956-053X (2012)
4. Ghiani, G., Manni, A., Manni, E., Toraldo, M.: The impact of an efficient collection sites location on the zoning phase in municipal solid waste management. Waste Manage. **34**(11), 1949–1956. ISSN 0956-053X (2014)
5. González-Torre, P.L, Adenso-Díaz, B., Ruiz-Torres, A.: Some comparative factors regarding recycling collection systems in regions of the USA and Europe. J. Environ. Manage. **69**(2), 129–138. ISSN 0301-4797 (2003)
6. Kao, J.-J., Lin, T.-I.: Shortest service location model for planning waste pickup locations. J. Air Waste Manage. Assoc. **52**(5), 585–592. http://dx.doi.org/10.1080/10473289.2002.10470807 (2002)
7. Amponsah, S.K., Salhi, S.: The investigation of a class of capacitated arc routing problems: the collection of garbage in developing countries. Waste Manage. **24**(7), 711–721 (2004)
8. Mourão, M.C., Amado, L.: Heuristic method for a mixed capacitated arc routing problem: a refuse collection application. Eur. J. Oper. Res. **160**(1), 139–153 (2005)
9. Kim, B., Kim, S., Sahoo, S.: Waste collection vehicle routing problem with time windows. Comput. Oper. Res. **33**(12), 3624–3642 (2006)
10. Nuortio, T., Kytöjoki, J., Niska, H., Bräysy, O.: Improved route planning and scheduling of waste collection and transport. Expert Syst. Appl. **30**(2), 223–232 (2006)
11. Simonetto, E.O., Borenstein, D.: A decision support system for the operational planning of solid waste collection. Waste Manage. **27**(10), 1286–1297 (2007)
12. Karadimas, N.V., Papatzelou, K., Loumos, V.G.: Optimal solid waste collection routes identified by the ant colony system algorithm. Waste Manage. Res. **25**(2), 139–147 (2007)
13. Bautista, J., Fernández, E., Pereira, J.: Solving an urban waste collection problem using ants heuristics. Comput. Oper. Res. **35**(9), 3020–3033 (2008)
14. Pop, P.C., Matei, O., Sitar, C.P., Chira, C.: A genetic algorithm for solving the generalized vehicle routing problem. Hybrid Artif. Intell. Syst. **6077**, 119–126 (2010)
15. Kuo, R.J., Zulvia, F.E., Suryadi, K.: Hybrid particle swarm optimization with genetic algorithm for solving capacitated vehicle routing problem with fuzzy demand—a case study on garbage collection system. Appl. Math. Comput. **219**(5), 2574–2588 (2012)
16. Anghinolfi, D., Paolucci, M., Robba, M., Taramasso, A.C.: A dynamic optimization model for solid waste recycling. Waste Manage. **33**(2), 287–296 (2013)
17. Zsigraiova, Z., Semiao, V., Beijoco, F.: Operation costs and pollutant emissions reduction by definition of new collection scheduling and optimization of MSW collection routes using GIS. The case study of Barreiro, Portugal. Waste Manage. **33**(4), 793–806 (2013)
18. Saaty, T.L.: The Analytic Hierarchy Process. McGraw-Hill, New York (1980)
19. Farlow, S.J.: The GMDH algorithm of Ivakhnenko. Am. Stat. **35**(4), 210–215. Available from http://www.jstor.org/stable/2683292 (1981)

Reusing and Recycling of Granite Slurry in Construction Industry

Umang K. Shah, K. Prasanna, K. S. Anandh, R. Annadurai and Kevin Pandya

Abstract The waste generated in the stone processing industry is directly related to the production and processing of various types of stones that have drastically increased every year. Marble industry brings out the amount of waste that has grown nearly three times in last ten years. The processing and polishing waste alone is 20% from mined out reserves. Only insignificant quantities have been utilized, and the rest has been dumped unscrupulously resulting in environmental problems and health issues. The results of the studies, on characteristics of waste generated from stone refrain many factories and reutilize the desolation, for making brick in combination with clay and preparing concrete with replacement of sand/cement has been conferred. Field studies assessed distinctive of waste generated in stone refrain manufactories, the impact of current disposal practices and waste recycling potential. Both physical and chemical characteristics of waste are compared with sand and cement. The environmental concern because of the waste disposal included on ambient air quality was pinpointed at disposal sites, respectively. The results reveal that the waste can be used to replace about 60% of sand and 10% of cement in concrete. Similarly, the waste can replace 40% of clay in clay bricks without compromising the conventional compressive strength.

Keywords Stone processing waste · Concrete · Bricks · Granite slurry
Waste management · Compressive strength

U. K. Shah (✉)
Environmental Engineering at NUS, Singapore, Singapore
e-mail: umang1995@gmail.com

K. Prasanna · K. S. Anandh · R. Annadurai
SRM University, Kattankulathur 603203, Tamil Nadu, India
e-mail: prasanna.env@gmail.com

K. S. Anandh
e-mail: anandh.ks@gmail.com

K. Pandya
IIT Guwahati, Guwahati 781039, Assam, India

A. S. Kalamdhad et al. (eds.), *Advances in Waste Management*,
https://doi.org/10.1007/978-981-13-0215-2_5

1 Introduction

Stone processing units generate a lot of wastes during drilling, cutting and polishing, and cause environmental impacts related to handling and disposal of the wastes. Waste generated in stone processing industry consists of 15% of processing waste and 5% of polishing waste. The associated environmental problems include contamination of water resources, air pollution due to airborne dry powder on windy days, blockage of drainage paths and wastage of valuable natural resources of stones. The objectives of the study are to evaluate the environmental issues in stone processing industries, estimate the quantity and characteristics of waste generated and to assess the impacts of current disposal practices, so as to identify study opportunities for waste minimization and reuse. Mechanization leads to reduction in the waste generation like regularly and accurate analysis, core drilling, non-destructive test, proper training for staff, site-level production plant [1].

The physio-chemical and mineralogical characterization of the sludge from natural granite cutting process showed the need for careful separation of different ingredients. Immense water content (more than half of weight) of the treated sludge and the induced great extent of shrinkage values observed on drying constitute the most critical drawbacks for their direct use [2]. Substituting the high volume of OPC with concrete ternary blend cement is used. One is slow responsive admixture (fly ash, blast furnace slag, etc.), and another one is highly responsive admixture (silica fume, metakaolin, etc.) [3]. The usage of high-volume fly ash concrete is one of the clarifications to eradicate the ecological deterioration being produced by the cement industry [4]. The products in the three selected industries are granite slabs/tiles for flooring and shelves. The size of the slabs and tiles vary with respect to the customer's requirement. About 10% of total water is used by the industry per day during wire dressing, to remove the extra projections of rough blocks and also to get uniform size of block. The waste generated from wire dressing is of small pieces of granite. Next, during block cutting about 40% of total water used by the industry per day is used along with lime and gridiron and a waste is generated in the form of slurry. During the next step of cross-cutting, about 30% of water is used along with lime and the waste generated in the form of slurry. The remaining 20% of water were used for polishing and cleaning the floors.

2 Materials and Methods

An assessment of the stone processing methods starting from extraction of rough blocks from mines to final packing with respect to the production capacity was done by plant walk-through, review of data from the industries and one-to-one discussion with industry personnel of three industries of production capacity in the range of 5000–12,200 m^2/month of granite slabs and tiles. The quantity of waste generated during the processing like wire dressing, block cutting, polishing and cross-stages was estimated by measuring volume of drying pits used for storage of wastes and number and capacity of trucks used to convey the waste disposal site. The production capacity of each industry has been calculated based on the delivery records over a period of one month. The quantity of water used for process was estimated on the size of overhead tank and filling intervals.

Physical characteristics of the waste such as moisture content, bulk density and particle size distribution were determined. The chemical composition was determined in terms of included Fe_2O_3, MnO, Na_2O, MgO, K_2O, Al_2O_3, CaO, SiO_2 expressed as the percentage weight of the dried waste and leachability of heavy metals. Soil characteristics and air quality of the waste disposal area were studied by sampling at three locations for each industry. The soil samples were collected in polythene bags. The physical characteristics such as moisture content, density and permeability were determined. The chemical characteristics for Fe_2O_3, MnO, Na_2O, MgO, K_2O, Al_2O_3, CaO, SiO_2 have been expressed as percentage weight. The results exhibit that the usage of micro-silica and titanium dioxide fillers increased mechanical properties and significantly eradicate all pores. Heterogeneously, the concrete, artificial stone composite achieved in the absence of water pool showed excelling properties that enhance with time [5]. Air quality was assessed using high-volume air samplers regarding suspended particulate matter (SPM) in the ambient air surrounding the waste disposal areas [6–8].

Recycling of potential of the stone processing waste was assessed in terms of replacement of sand or cement in preparing M20 grade concrete and clay in the preparation of bricks. Five concrete cubes of size 15 cm × 15 cm × 15 cm were moulded and tested for compressive strength after a period of 7 days and 28 days for various replacement percentages of sand and cement. The compressive strength of concrete cubes was compared with the standards specified by IS: 456 (2000) for M20 grade concrete. Twenty bricks were moulded manually using rectangular boxes of size 22 cm × 10 cm × 9 cm with different proportions of waste and clay. The dried bricks were burnt at a temperature of 950 °C in a brick kiln for a period of 15 days. The burnt bricks were tested for water absorption, hardness, shape and size, soundness and compressive strength. Water absorption was assessed in terms of weight gain of dried bricks immersed in water for 24 h [9–11]. The pressure required to break the bricks using compression testing machine was taken as compressive strength of the bricks [12–15]. Absence of impression on the surface of brick when scratched with fingernail was taken as an indication of sufficient hardness. All the bricks were closely inspected for their size and shape. The production capacity, water consumption and waste generation in the three selected industries quantified during the study are provided in Table 1.

The quantity of waste is varying with respect to the size and thickness of slabs/tiles produced. The tiles unit of Industry A has an installed capacity of 1,80,000 square metres and is equipped with a processing line from some of the leading manufacturers. The slabs unit of Industry A has an installed capacity of 1,68,000 square metres and is equipped with a processing line from some of the leading manufacturers. The quantity of water used for cutting purposes is 400 and 600 Lpm

Table 1 Quantity of waste generation

S. No.	Industry reference	Water consumption (kL/day)	Waste generated (m^3/day)	Specific water consumption (kL/m^2)	Specific waste generation (m^3/m^2)
1	A	30	177.7	0.074	0.44
2	B	22	40.0	0.071	0.13
3	C	15	42.3	0.090	0.26

for 32 blades and 45 blades, respectively. The water used for polishing purpose is 230 Lpm for both types of blades.

3 Results and Discussion

The products in three selected industries are granite slabs/tiles for flooring and shelves. The size of the slabs and tiles varies with respect to the customer's requirement. About 10% of the total water used by the industry per day during wire dressing removes the extra projections of rough blocks and helps to get uniform size of block. The waste generated from wire dressing removes the extra projections of rough blocks and helps to get uniform size of block [16–18]. The waste generated from wire dressing is in the form of small pieces of granite. During block cutting, about 40% of the total water used by the industry per day is used along with lime and gridiron and a waste is generated in the form of slurry. During the next step of cross-cutting, about 30% of water is used along with lime and the waste is generated in the form of slurry. The remaining 20% of water is used for cleaning the floors and other purposes. The average values of the physical characteristics of waste were compared with the sand and clay properties to examine whether the waste can replace sand and clay [19–21]. The detailed physical characteristics of stone processing waste are given in Table 2.

The chemical components are compared with the oxides of ordinary Portland cement. The parameters like Fe_2O_3, Na_2O, MgO, K_2O, SiO_2 are four times, two times, four times, two times, three times, respectively, greater than the values mentioned in the composition of cement, and other parameters like oxides of CaO are 40 times lesser than, and Al_2O_3 and MnO are equal with the values mentioned in the composition of the cement. Limitations of all the heavy metal concentrations of As, Ba, Cd, Cr, Hg, Pb and Ag are satisfactory as specified by USEPA. The results indicate that there will not be any impact on the leachability of waste.

The physical characteristics of soil in the disposal site in terms of the parameters like moisture content, density and permeability were determined. The moisture content varied from 17 to 21%, and it is termed as too dense and the water absorption from the waste may take time. Density of soil at wet condition varied from 1980 to 2290 kg/m^3, stating that the soil is too dense when compared with normal density of 1800 kg/m^3, because the waste is percolated along the top surface

Table 2 Physical characteristics of stone processing waste

Industry	Sample code	Moisture content (%) Wet	Moisture content (%) Dry	Bulk density (kg/m^3)	Fineness modulus	Effective size (mm)	Coefficient of uniformity	Coefficient of gradation
A	AW1	23.96	1.56	1100	2.02	0.16	1.52	1.32
	AW2	24.32	1.52	1130	2.05	0.17	1.53	1.35
	AW3	22.36	1.61	1110	2.08	0.18	1.56	1.39
B	BW1	21.32	1.62	1120	1.98	0.18	1.60	1.38
	BW2	22.86	1.57	1130	2.03	0.17	1.61	1.37
	BW3	21.68	1.56	1100	2.09	0.16	1.62	1.33
C	CW1	25.36	1.65	1120	2.04	0.18	1.58	1.39
	CW2	24.65	1.63	1110	2.03	0.18	1.60	1.37
	CW3	23.68	1.61	1140	2.07	0.19	1.62	1.41
Average		23.35	1.59	1118	2.04	0.17	1.58	1.37
Comparable properties of sand		25.00	2.50	1200	2.20	0.20	6.00	2.00

Table 3 Chemical characteristics of stone processing waste

Industry	Sample code	Fe_2O_3 (%wt)	MnO (%wt)	Na_2O (%wt)	MgO (%wt)	K_2O (%wt)	Al_2O_3 (%wt)	CaO (%wt)	SiO_2 (%wt)
A	AW1	12.15	0.08	2.07	8.67	2.27	4.81	1.24	63.96
	AW2	11.98	0.07	2.09	8.89	2.29	4.68	1.32	64.06
	AW3	12.26	0.08	2.05	8.56	2.36	4.78	1.48	62.32
B	BW1	11.89	0.09	2.11	8.76	2.32	3.98	1.56	68.09
	BW2	12.05	0.08	2.10	8.78	2.35	4.09	1.62	67.52
	BW3	11.78	0.09	1.98	8.24	2.28	4.12	1.58	66.82
C	CW1	12.12	0.07	2.12	8.92	2.36	4.52	1.76	65.32
	CW2	12.08	0.08	2.11	8.95	2.40	4.58	1.80	64.36
	CW3	11.56	0.08	2.06	8.89	2.38	4.49	1.82	61.28
Average		11.99	0.08	2.08	8.74	2.33	4.45	1.58	64.86
Comparable characteristics of Portland cement		0.55	0.85	0.85	2.15	0.85	5.50	63.50	21.50

of the soil due to the moisture content present in the waste. Permeability of various soil samples was in the range between 1.3×10^{-4} and 2.5×10^{-5} cm/s indicating that it is impervious [22, 23]. The detailed chemical characteristics of stone processing waste are given in Table 3.

Sample collected at 150 m away from each of the disposal site for the same parameters showed the moisture content 13, 12 and 15%, respectively. The density was found to be 1830, 1880 and 1750 kg/m^3, and the coefficient of permeability were 1.5×10^{-3}, 1.3×10^{-2} and 1.9×10^{-3} cm/s in the respective nearby disposal sites. The ambient air quality at three disposal sites of three industries revealed that the suspended particulate matter (SPM) concentrations are 120, 150 and 140 μg/m^3 for three disposal sites. It was found that SPM concentrations are within the regulatory level of 500 μg/m^3, specified by the standards for the industrial zone [24]. It means that there will not be any significant effect in the air due to the waste disposal site.

3.1 Reusing the Waste in Preparation of Concrete

The maximum percentage of replacement of sand was found to be 60% as shown in Fig. 1, when compared with the standard value of 20 N/mm^2 at 28 days curing as recommended by the code. The maximum percentage of replacement of cement by stone processing waste was found to be 10% as shown in Fig. 2, when compared with the standard value of 20 N/mm^2 at twenty-eighth day compressive strength.

3.2 Reusing the Waste in Preparation of Bricks

The burnt bricks are made with stone processing waste sludge and clay in different proportions. The burnt bricks made of different stone processing waste and compositions were tested for water absorption, hardness, shape and size and soundness [25–28]. The results indicate that a maximum of 60% of clay can be replaced. The compressive strength at 40% replacement of clay by waste gives better results when compared with the standard value of 5 N/mm^2. The other combinations of greater than 40% showed values less than 5 N/mm^2. It indicated that a maximum of 40% replacement can be done for bricks. Moreover, the combination of 50 and 60%

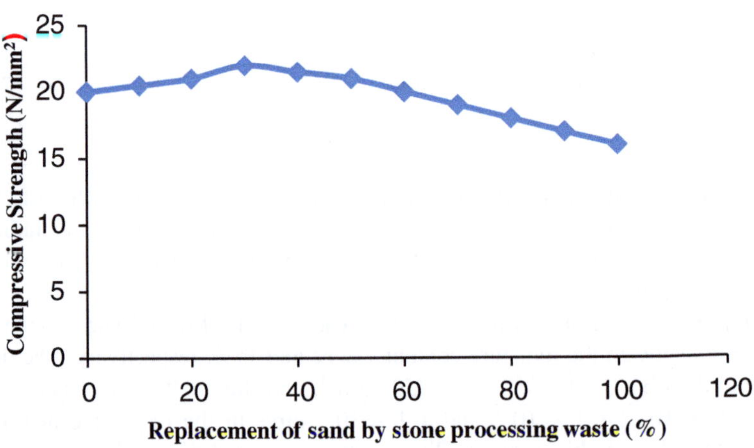

Fig. 1 Variation in 28 days strength of concrete (M20 grade) with replacement of sand by waste

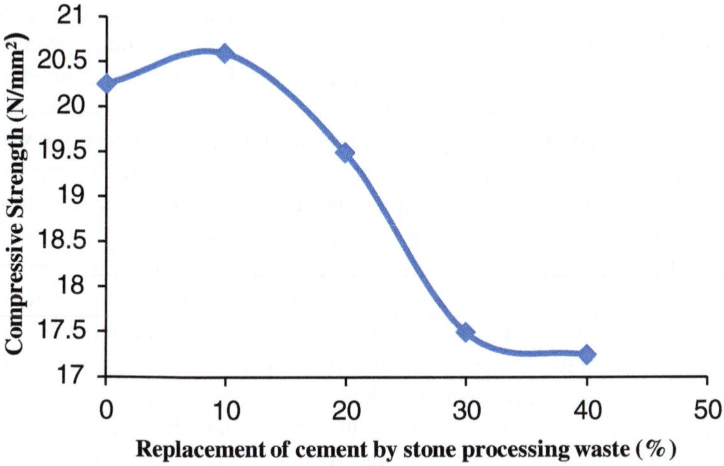

Fig. 2 Variation in 28 days strength of concrete (M20 grade) with replacement of cement by waste

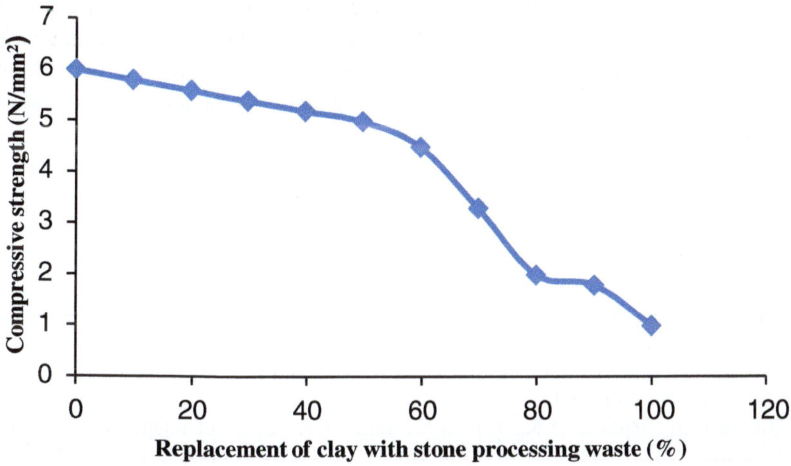

Fig. 3 Compressive strength of bricks with various proportions of replacement of clay with waste

replacements attains the value greater than 3.5 N/mm^2 and that type of bricks can be used for partition walls and other non-load-bearing walls. The variation in the compressive strength is shown in Fig. 3.

4 Conclusion

The physical characteristics of stone processing waste such as moisture content, bulk density and fineness modulus were analyzed, and chemical composition of stone processing waste such as oxides of iron, manganese, sodium, magnesium, potassium, aluminium, calcium and silicon was analyzed as per the standard methods. Heavy metals like arsenic, barium, cadmium, chromium, lead, mercury and silver in the TCLP extract of waste were analyzed using AAS. Characteristics of the soil at the waste disposal site were studied and air quality assessed. Concrete cubes were prepared by replacing sand and cement with stone processing waste. The optimum percentage of replacing sand and cement with stone processing waste was determined by comparison of compressive strength of the concrete cubes with standard values. Bricks were also prepared with different composition of clay and waste, and the suitability of the brick for construction work was assessed by its physical characteristics and compressive strength. The waste can be reused as a replacement of sand and clay with respect to the physical characteristics. The waste can be reused as a replacement of cement with respect to the chemical compositions. The concentrations of heavy metals in the leachate were within the regulatory levels and in the leachate were within the regulatory levels, and thus, there will not be any effect on dumping. The chemical characteristics of soil such as Fe_2O_3, MgO and SiO_2 found about 50, 40 and 30% higher than that of the undisposed site. A maximum replacement of sand would be 60% and a maximum replacement of cement would be 10% which can be achieved in preparation of bricks. Transforming waste into product could be achieved by clustering of industries for setting up a brick kiln or using the existing brick kiln nearby the industries. There is need to create awareness among the public for using bricks made of combination of clay and stone processing waste for construction purposes.

References

1. APHA.: Standard Methods for Examination of Water and Wastewater, 19th edn. American Public Health Association, Washington (1998)
2. Aro Granite Industries Ltd., Hosur.: Cited in www.arotile.com (online reference material) (2005). Accessed on 18 June 2005
3. Kanchan, M., Mullick, A.K., Jain, K.K., Singh, P.K.: Effect of relative levels of mineral admixtures on strength of concrete with ternary cement blend. Int. J. Concr. Struct. Mat. **7**(3), 239–249 (2013)
4. Rashad, A.M., Seleem, H.E.D.H., Shaheen, A.F.: Effect of silica fume and slag on compressive strength and abrasion resistance of HVFA concrete. Int. J. Concr. Struct. Mat. **8** (1), 69–81 (2014)
5. Sarami, N., Mahdavian, L.: Effect of inorganic compound on artificial stones properties. Int. J. Ind. Chem. **6**(3), 213–219 (2015)
6. Das, B.M.: Principles of Geo technical Engineering, pp. 139 168. Thompson Brooks Publishing Group, New Delhi (2004)
7. Wentz, C.A.: Hazardous Waste Management. Mc Graw Hill Inc. Engineering Series, New York (1995)
8. Dijkema, G.P.J., Reuter, M.A., Verhoef, E.V.: A new paradigm for waste management. J. Waste Manag. **20**, 633–638 (2000)
9. Ferreira, J.M.F., Torres, P.M.C., Silva, M.S., Labrincha, J.: Recycling of granite sludges in brick—type and floor tile—type ceramic formulation. J. Euroceram. News (14th Edition) 1–10 (2003)

10. Indian Standards IS: 269.: Code of Practice for Ordinary and Low Heat Portland Cement. Bureau of Indian Standards, New Delhi (1976)
11. Indian Standards IS: 456.: Code of Practice for Plain and Reinforced Cement Concrete. Bureau of Indian Standards, New Delhi (2000)
12. Indian Standards IS: 1077.: Permissible Limits of Structural Parameters of Burnt Bricks. Bureau of Indian Standards, New Delhi (1976)
13. Indian Standards IS: 2368.: Code of Practice for Methods of Test for Aggregates for Concrete. Bureau of Indian Standards, New Delhi (1963)
14. Indian Standards IS: 3495 Part 1-4.: Methods for Testing of Burnt Clay Building Bricks. Bureau of Indian Standards, New Delhi (1976)
15. Indian Standards IS: 5182 Part 4.: Methods for Measurement of Air Pollution. Part 4 Suspended Particulate Matter. Bureau of Indian Standards, New Delhi (1999)
16. Indian Standards IS: 5182 Part 14.: Methods for Measurement of Air Pollution. Part 14 Guidelines for Planning the Sampling of Atmosphere. Bureau of Indian Standards, New Delhi (2000)
17. John, V.M., Zordan, S.E.: Research & development methodology for recycling residues as building materials—a proposal. J. Waste Manag. **21**, 213–219 (2001)
18. Kumar, A., Kulveer, S.: Indian stone industry—an insight. http://www.cdos-india.com/papers (2001). Accessed 15 June 2005
19. Muruganandam, V.: Compressive Test on Bricks. Test Results from Thiagarajar College of Engineering, Madurai (1998)
20. Punmia, B.C.: Soil Mechanics and Foundation, pp. 33–124. Saurabh Publishing Limited, Chennai (1998)
21. Rangwala, S.C.: Engineering Materials, pp. 92–105. Charotar Publishing House, Anand, Gujarat (1982)
22. Pareek, S.: Gainful utilization of marble waste. An effort towards protection of ecology & environment. http://www.cdos-india.com/papers (2001). Accessed 17 June 2005
23. Stone 2000.: Cited in marbleandmore.com (online reference material). Accessed 15 July 2005
24. TIFAC.: Solid waste generation and utilization in calcareous stone industry. Technology Information, Forecasting and Assessment Council. http://www.tifac.org.in (1999). Accessed 18 June 2005
25. Wahlstrom, M., Laine-Ylijoki, J., Maattanen, A., Luotojarvi, Kivekas L.: Environmental quality assurance system for of crushed mineral demolition waste in road constructions. J. Waste Manag. **20**, 225–232 (2000)
26. Prasanna, K., Anandh, K.S.: Analysis on strength of concrete by partial replacement of riversand with sawdust and robosand. ARPN J. Eng. Appl. Sci. **12**(8), 2423–2427 (2017)
27. Prasanna, K., Anandh, K.S., Ravishankar, S.: An experimental study on strengthening of concrete mixed with ground granulated blast furnace slag (GGBS). ARPN J. Eng. Appl. Sci. **12**(8), 2439–2444 (2017)
28. Kandil, A.I., Selim, T.H.: Characteristic of the Marble Industry in Egypt. http://www.aucegypt.edu (2005). Accessed 15 July 2005

Biochemical Changes in *Lantana camara* and *Ipomoea carnea* Growing in the Fly Ash Amended Soil

Shikha Kumari Pandey and Tanushree Bhattacharya

Abstract Proper remediation of fly ash is important as it is harmful for environment and human health due to the presence of metals. This study evaluated the effects of fly ash amended soil on two local species, namely *Lantana camara* and *Ipomoea carnea,* by growing it in different amendments (10%, 20%… up to 100%). Metal fractionation of fly ash showed that Ni, Cd and Mn had higher percentage in the soluble and exchangeable fraction. Plant height, leaf size, number of leaf and biomass of both the plants were substantially increased with the developmental stage. Chlorophyll activity increased with the developmental stage in *L. camara* and *I. carnea.* Peroxidase activity in *L. camara* and in *I. carnea* increased at 120 DAS. Metals accumulation in upper ground part of *I. carnea* and *L. camara* was more as compared to belowground part. It indicates phytoextraction strategy of both the plants. Both the plants performed reasonably well in terms of phytoextraction of metals. Moreover, the biochemical parameters indicated less stress and more tolerance towards metal contamination.

Keywords Fly ash · *Lantana camara* · *Ipomoea carnea* · Sequential extraction Biochemical response · Translocation

1 Introduction

Coal-fired energy generation is the most important source of electricity in the developing countries. Thermal power plant for energy production is going to be the main source of world energy till 2040 as it is growing at an average rate of 1.8% per annum. The adverse side of coal-based electricity production is generation of fly ash (FA) as a by-product. Fly ash is produced due to burning of coal. The FA-contaminated zone is escalating at a stressing scale and has turning into a

S. K. Pandey · T. Bhattacharya (✉)
Department of Civil and Environmental Engineering, Birla Institute of Technology Mesra, Ranchi 835215, Jharkhand, India
e-mail: tbhattacharya@bitmesra.ac.in

A. S. Kalamdhad et al. (eds.), *Advances in Waste Management*,
https://doi.org/10.1007/978-981-13-0215-2_6

worldwide fear. Fly ash is grey colour and very tiny particle having typical diameter of <10 μm and low bulk density with more surface area. According to United States Environmental Protection Agency (USEPA), FA contamination causes cardiovascular and skin ailments, tumour and respiratory disorder. In this way, it is of most extreme significance to look better choices for remediation for the threats of coal FA. According to Central Electricity Authority [1] in India, on an average 48.50% fly ash is utilized in different sectors like cement sector, mine filling (8.26%), roads and embankments (11.65%), low-lying area reclamation (12.73%), tiles and bricks making (6.3%), agriculture (1.74%) and miscellaneous (10.82%). In this manner, there exists a wide extension and a basic need to expand the amount of FA reuse. Particularly there is a wide degree for soil application because of the present consumption rate is very less. In fact, there exist a lot of chances in perspective of the enormous degraded soil existing in the world which should be recovered with proper remediation. Further, thousands of hectares of land are engaged for the storing of FA throughout the world that could be remediated by different plant species [2].

Due to its fugitive emissions, fly ash is usually dumped by preparing slurry in a fly ash pond. These wet storage ash ponds pose a major environmental threat due to release and transport of metals. This can lead to metal contamination in the groundwater, surface water, soil and vegetation. Fly ash amendment with soil has been broadly studied by several researchers [3–5]. Fly ash amendment has some positive effect on soil quality for example; it improves water holding capacity and decreases bulk density that increases pore space. And yet, change in pH and availability of some phytotoxic metals; low nitrogen and phosphorus content, along with metal leaching are the ill effects of fly ash amendments.

Phytomanagement of fly ash is the practice to use vegetation and amendments of soil to minimize the environmental risk caused by fly ash polluted area [6]. The benefit of phytomanagement is to lower the contaminant availability and to prevent ecotoxicological pollution. The success of phytoextraction depends on suitable combination of soil amendments and plant species that tolerate the local condition. Fly ash-contaminated soil surface stabilization is essential to stop wind erosion that reduces hazard of direct consumption by humans and animals. Flowering shrubs such as *Bougainvillea*, *Nerium odorum*, *Rosa damascene*, *Tagetes patula*, *Hibiscus rosa-sinensis* are reported to grow successfully on fly ash amended soils. Some plants are also been reported to stabilize fly ash lagoons and dykes, e.g. *Prosopis juliflora*, *Ziziphus mauritiana*, *Acacia filicifolia*, *Casuarina equisetifolia*, *Dendrocalamus strictus* [7–10]. Fly ash is alkaline in nature, but it has also a little bit amount of Na, Mg, Fe, Ca, K, Ti and P oxides [11]. It involves a few micronutrients which are ideal for plant growth and development. On the opposite side, it has some toxic metal ions like Pb, Cr, V, Hg, Ni, As and Ba [12]. The higher uptake of these heavy metals and metalloids causes crumbling in enzymes activity, damages the membrane, nucleic acids, proteins, chlorophyll and expanded per-oxidation of lipids by the development of free radicals, however at lower uptake it could enhance the growth and productivity of many crops [13]. Foliar application of fly ash enhances the plant morphological properties, metabolism activity rate and photosynthetic activity in *Zea mays* and *Glycine max* [14]. Amendment of Fly ash to soil was accounted to enhance the performance of oil-seed productivity in the species like turnip (*Brassica rapa*), sunflower (*Helianthus* sp.), groundnut (*Arachis hypogaea*), sesame (*Sesamum indicum*) [115] [15]. There are additional reports that 10–25% use of fly ash brought about good development, enhanced photosynthetic activity in *Lactuca sativa* and cultivation of rice [16, 17]. The present study aims to investigate the effect of fly ash amendment on the morphological and biochemical property of *Lantana camara* and *Ipomoea carnea* and to estimate the phytoextraction potential.

2 Material and Methods

2.1 *Samples Collection*

Fly ash was collected from the fly ash dump site of a coal-based thermal power plant of the Patratu thermal power station (PTPS) located at Patratu, Ramgarh, Jharkhand (India). For the pot culture experiment, soil was collected from the garden of BIT Mesra campus from a depth of 20 cm, after scrapping of the surface litters. *L. camara* (Family: Verbenaceae) and *I. carnea* (Family: Convolvulaceae) shoots of the plants were used for plantation. Shoot of the plants was obtained from the plant growing in uncontaminated area.

2.2 *Experimental Set-up*

For the experiment, fly ash was mixed with the garden soil and organic manure in the following percentage: T0 = Control (only garden soil); T1 = 10% fly ash + 90% garden soil; T2 = 20% fly ash + 80% garden soil; T3 = 30% fly ash + 70% garden soil; T4 = 40% fly ash + 60% garden soil; T5 = 50% fly ash + 50% garden soil; T6 = 60% fly ash + 40% garden soil; T7 = 70% fly ash + 30% garden soil; T8 = 80% fly ash + 20% garden soil; T9 = 90% fly ash + 10% garden soil; T10 = 100% fly ash. Total weight of the amended soil in all the pots was 2.5 kg. All the treatments were exposed to the natural weather conditions. Plants were harvested after 60 days of sowing (60 DAS) and after 120 days of sowing (120 DAS).

2.3 *Physicochemical Analysis of Fly Ash Amended Soil*

Soil pH was analysed in a 1:2 soil:solution suspension by the digital pH metre with glass electrode [18]. Electrical conductivity (EC) in soil water solution

(1:2 proportion) was analysed with EC by electrical conductivity metre (HM DIGITAL). Organic carbon (OC) in the fly ash amended soil was analysed by Walkley and Black quick titration technique [19]. Soil-available phosphorus (P) was estimated by Olsen method [20], and available potassium (K) was estimated using N ammonium acetate as extractant and K was quantified in flame photometer [21]. Nitrogen (N) and sulphur (S) were analysed by Elemental analyser (CHNS Analyse, Vario EL). For sequential extraction of metals in fly ash, modified Tessier scheme was followed [22]. Heavy metals (Cd, Cr, Cu, Mn, Ni, Pb) in plants and soil sample were determined by digesting sample in aqua regia (3:1, HNO_3:HCL); then metals were estimated in inductively coupled plasma ray optical emission spectrometry (ICP-OES 2100 DV, Make Perkin Elmer).

2.4 Plant Morphology and Biochemical Parameters

Plant morphological parameters like plant height, size of leaf, biomass and no. of leaves were measured at both the stages of harvesting. For the plant measurement of biomass, plant parts were separated in leaves, roots and stems. Then it was dried for 48 h in hot air oven at 55 °C and weighed with a weight balance. Fully expanded fresh leaves of plants were sampled randomly from each pot for biochemical analysis. For the determination of chlorophyll Arnon 1949, method was followed measured at 645 and 663 nm (systronic spectrophotometer Model-166). The POD activity was determined by guaiacol method at 25 °C studied at 470 nm in the spectrophotometer (systronic spectrophotometer Model-166) [23].

2.5 Phytoextraction Efficiency

Phytoextraction potential is defined as the ratio of metals concentrated in the upper ground parts to that of the total metals ions in soil at final harvesting stage. Phytoextraction potential was used to evaluate efficiency of a plant in metal accumulation and translocation [24]. It depends on metal accumulation in the above-ground part and biomass of the plants. It is calculated as follows:

$$\mathrm{PP}(\%) = \frac{M_{\mathrm{shoot}} \times B_{\mathrm{shoot}}}{M_{\mathrm{soil}} \times B_{\mathrm{soil}}} \times 100$$

where M_{shoot} is the metal (mg/kg) accumulation in above-ground parts of the plant. B_{shoot} represents the plant dry weight (gm). M_{soil} is the metal (mg/kg) concentration in the fly ash amended soil. B_{soil} is the total weight of soil in the pot (gm). It has been calculated for both the plants for every amendment or treatment at 120 DAS.

2.6 *Statistical Analysis*

Standard deviation and ANOVA were calculated using Excel version 10. Origin pro 8.0 was used for plotting graphs.

3 Results and Discussion

3.1 *Physicochemical Properties of Fly Ash Amended Soil*

The physicochemical property of fly ash amended soil for *I. carnea* and *L. camara* with the different treatments is given in Tables 1 and 2, respectively. pH of the amended soil in all the treatments was alkaline in nature. In case of *I. carnea*, pH increased with the fly ash dose. On the other side, pH (7.33) was maximum in T10. There is no substantial change in pH with different sowing stages. The high pH might be due to the neutralization of H^{+} by alkali salts along with solubilization of basic metallic oxides of fly ash in the soil Khan and Khan [25]. Gupta et al. [26] also reported that the increase in soil pH might be a result of precipitation of soluble cations from the fly ash [25, 26]. Electrical conductivity (EC) was found to be of average of 83 ms/m. Low EC suggested that the fly ash samples did not contain high amount of dissolved salts and soluble metals. As, per ANOVA ($P < 0.05$) where $F_{\text{critical}} = 2.97$ for Treatments and sowing stages $F_{\text{critical}} = 4.94$ pH and EC did not show statistically significant variation within different treatments and between harvesting stages in both the plants. Available P and K contents increased with fly ash amendments; the same trend was found in both the plants' set-up. ANOVA ($P < 0.05$) for phosphorous and available K showed statistically significant variation with the plant developmental stage as well as treatment. Nitrogen (%) is less than 1% in every treatment in *L. camara* as well as in *I. carnea*. However, control had more than 1% nitrogen content. Similar kind of findings has been reported in the soil amendments planted with *Brassica chinensis* and *B. parachinensis* by Wong and Wong [27], and *Helianthus annuus* Pandey et al. [28].

3.2 *Chemical Fractionation of Metals in Fly Ash*

Sequential extraction of fly ash results showed that Ni, Cd and Mn had higher percentage in the soluble and exchangeable fraction than the other metals (Fig. 1). The order of metals in the soluble and exchangeable fraction was Cd (25%) > Ni (49%) > Mn (27%) ≈ Pb ≈ Cr ≈ Cu (0%). The trend for carbonate fraction was Cd (38%) > Cu (4%) > Mn (6%) > Ni (3%) > Cr ≈ Fe ≈ Pb (0%). Fe–Mn bound fraction can be released in the environment under anoxic conditions, and its order was Cd (24%) > Mn (11%) > Ni (0%) ≈ Cr (0%) ≈ Pb (0%) ≈ Cu (0%). For

Table 1 Physicochemical property of fly ash amended soil planted with *Ipomoea carnea*

Control	pH (1:2, soil:water)		EC (μs/m)		*N* (%)		Available P (mg/kg)		Available K (mg/kg)	
	60 DAS	120 DAS	60 DAS	120 DAS	60 DAS	120 DAS	60 DAS	120 DAS	60 DAS	120 DAS
T1	7.77 ± 0.01	7.78 ± 0.01	81.6 ± 0.5	84.3 ± 0.5	0.55 ± 0.05	0.53 ± 0.01	127.6 ± 0.5	126.3 ± 0.5	1.57 ± 0.01	1.78 ± 0.01
T2	7.01 ± 0.02	7.16 ± 0.15	82.6 ± 0.5	83 ± 0	0.27 ± 0.05	0.57 ± 0.05	195.6 ± 0.5	197.6 ± 0.5	1.78 ± 0.01	1.89 ± 0.01
T3	7.6 ± 0.1	7.6 ± 0.1	81.6 ± 0.5	85.3 ± 0.5	0.88 ± 0.05	0.88 ± 0	171.6 ± 0.5	178 ± 0	1.26 ± 0.01	1.28 ± 0.01
T4	7.73 ± 0.1	7.73 ± 0.11	81.3 ± 0.5	81.3 ± 0	0.89 ± 0	0.88 ± 0.05	177.6 ± 0.5	178.6 ± 0.5	1.23 ± 0.01	1.25 ± 0.01
T5	7.42 ± 0.02	7.44 ± 0.05	82.3 ± 0.5	83.3 ± 0.5	0.86 ± 0	0.87 ± 0.05	144.6 ± 0.5	147.6 ± 0.5	1.26 ± 0.01	1.58 ± 0.01
T6	6.38 ± 1.7	7.75 ± 0.05	83.3 ± 0.5	82.3 ± 0.5	0.77 ± 0.05	0.87 ± 0	197.6 ± 0.5	198.6 ± 0.5	1.24 ± 0.01	1.28 ± 0.01
T7	7.45 ± 0.05	7.44 ± 0.05	84.3 ± 0.5	81 ± 0	0.82 ± 0.05	0.83 ± 0	176.3 ± 0.5	177.6 ± 0.5	1.28 ± 0.01	1.37 ± 0.01
T8	7.68 ± 0.0	7.67 ± 0.05	82.3 ± 0.5	82 ± 0	0.84 ± 0.05	0.87 ± 0.05	200.6 ± 0.5	204.6 ± 0.5	1.55 ± 0.01	1.58 ± 0.01
T9	7.69 ± 0.0	7.68 ± 0.05	82.3 ± 0.5	88 ± 0	0.55 ± 0.05	0.57 ± 0.01	214.6 ± 0.5	213.6 ± 0.5	1.48 ± 0.01	1.50 ± 0.01
T10	7.03 ± 0.05	7.23 ± 0.2	83.3 ± 0.5	85.3 ± 0.5	0.44 ± 0.01	0.45 ± 0.02	217.6 ± 0.5	212 ± 0.5	1.75 ± 0.01	1.78 ± 0.01
C	7 ± 0	7 ± 0	84.3 ± 0.5	85 ± 0	0.42 ± 0.05	0.46 ± 0.05	124.6 ± 0.5	214 ± 0.5	1.26 ± 0.01	1.29 ± 0.01

T1-10%, T2-20%, T3-30%. T4-40%, T5-50%, T6-60%, T7-70%, T8-80%, T9-90%, T10-100%, C-control, DAS-days after sowing

Table 2 Physicochemical property of fly ash amended soil planted with *Lantana camara*

Sample code	pH (1:2, soil:water)		EC (µs/m)		*N* (%)		Available P (mg/kg)		Available K (mg/kg)	
	60 DAS	120 DAS	60 DAS	120 DAS	60 DAS	120 DAS	60 DAS	120 DAS	60 DAS	120 DAS
T1	6.83 ± 0.46	7.23 ± 0.06	81.3 ± 0.58	83.7 ± 1.53	0.55 ± 0.01	0.55 ± 0.01	121.3 ± 0.58	121.3 ± 0.58	1.97 ± 0.01	1.88 ± 0.01
T2	6.46 ± 0.17	6.40 ± 0.10	82.3 ± 0.58	82.7 ± 0.58	0.58 ± 0.01	0.58 ± 0.01	124.3 ± 0.58	124.3 ± 0.58	1.96 ± 0.01	1.57 ± 0.01
T3	6.50 ± 0.17	6.40 ± 0.10	82.7 ± 0.58	84.3 ± 2.08	0.58 ± 0.01	0.59 ± 0.01	133.7 ± 1.15	133.7 ± 1.15	1.88 ± 0.01	1.78 ± 0.01
T4	6.96 ± 0.06	6.80 ± 0.10	80.3 ± 0.58	82.0 ± 1	0.62 ± 0.01	0.87 ± 0.01	137.3 ± 0.58	137.3 ± 0.58	1.85 ± 0.01	1.84 ± 0.01
T5	6.93 ± 0.12	6.56 ± 0.21	82.3 ± 0.58	83.3 ± 0.58	0.60 ± 0.02	0.87 ± 0.01	137.3 ± 0.58	137.3 ± 0.58	1.83 ± 0.01	1.55 ± 0.01
T6	6.63 ± 0.06	7.13 ± 0.23	83.3 ± 0.58	83.3 ± 0.58	0.58 ± 0.01	0.87 ± 0.01	138.3 ± 0.58	138.3 ± 0.58	1.99 ± 0.01	1.87 ± 0.01
T7	6.43 ± 0.12	7.30 ± 0.26	84.0 ± 1	82.7 ± 2.08	0.55 ± 0.03	0.82 ± 0.01	144.7 ± 0.58	144.7 ± 0.58	1.94 ± 0.01	1.88 ± 0.01
T8	6.20 ± 0.17	6.16 ± 0.29	82.0 ± 1	82.7 ± 0.58	0.53 ± 0.01	0.87 ± 0.01	137.7 ± 0.58	137.7 ± 0.58	1.56 ± 0.01	1.46 ± 0.01
T9	6.93 ± 0.06	7.26 ± 0.12	82.7 ± 0.58	87.0 ± 1	0.61 ± 0.01	0.59 ± 0.01	144.3 ± 0.58	144.3 ± 0.58	1.88 ± 0.01	1.88 ± 0.01
T10	7.33 ± 0.12	7.53 ± 0.32	83.7 ± 0.58	85.0 ± 1	0.59 ± 0.01	0.47 ± 0.01	146.7 ± 0.58	146.7 ± 0.58	1.86 ± 0.01	1.87 ± 0
C	6.93 ± 0.12	6.80 ± 0.10	84.3 ± 0.58	85.3 ± 0.58	1.23 ± 0.06	1.75 ± 0.03	147.7 ± 0.58	147.7 ± 0.58	1.88 ± 0.01	1.85 ± 0.01

T1-10%, T2-20%, T3-30%, T4-40%, T5-50%, T6-60%, T7-70%, T8-80%, T9-90%, T10-100%, C-control, DAS-days after sowing

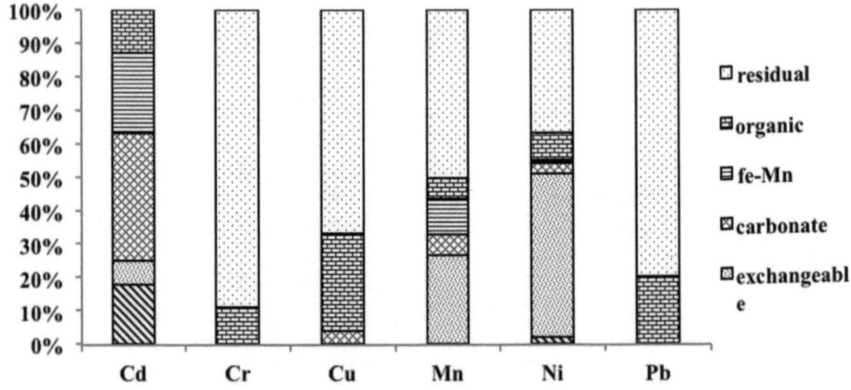

Fig. 1 Sequential extraction of fly ash

organic fraction, it was Cu (29%) > Cu (29%) > Pb (20%) > Cd (13%) > Cr (11%) > Ni (8%) > Mn (6%). Residual fraction is bound to silicate matrix and is the most inert fraction. The trend in this fraction was Cr (89%) > Pb (80%) > Cu (76%) > Mn (51%) > Ni (37%) > Cd (0%). Cr, Pb and Cu were mostly in non-available fractions, and Ni, Mn and Cd were comparatively more in available fractions.

3.3 *Morphological Parameters*

I. carnea and *L. camara* morphological properties such as biomass (gm), plant height (cm), number of leaf and leaf size (cm) were studied at both the harvesting stages (Tables 3 and 4). Plant height, number of leaf, total number of leaves and biomass increased with time. Treatment or amendment-wise variation was also observed in both *I. carnea* and *L. camara*. However, for most of the parameters plants grown in control had higher values than the higher amendments. T1 and T2 amendments showed similar or better response than the control. According to ANOVA ($P < 0.05$) where $F_{\text{critical}} = 2.97$ for treatments and sowing stages $F_{\text{critical}} = 4.94$, all the plant morphological parameters like plant height (cm) ($F = 56.1$ for *L. camara* and $F = 27.09$ for *I. carnea*), leaf size (cm) ($F = 19.1$ for *L. camara* and $F = 195.1$ for *I. carnea*), number of leaf ($F = 91.5$ for *L. camara* and $F = 56.1$ for *I. carnea*) and biomass (gm) ($F = 27.08$ for *L. camara* and $F = 140.87$ for *I. carnea*) showed statistically significant variation only with harvesting stages for *L. camara* and *I. carnea*. Gupta and Sinha [29] have studied *Brassica juncea* with different amendments and stated that only shoot growth was improved but not growth of root by 25% application of fly ash [29]. Finally, it was observed that there was no adverse effect on the plant morphology due to higher exposure of fly ash in both the plants.

Table 3 Morphology of *Ipomoea carnea* growing in fly ash amended soil

Sample code	Plant height (cm)		Leaf size (cm)		Total leaf		Biomass (gm)	
	60 DAS	120 DAS	60 DAS	120 DAS	60 DAS	120 DAS	60 DAS	120 DAS
T1	27.2 ± 0.26	34.6 ± 0.5	4.2 ± 0.1	4.33 ± 0.2	15 ± 0	20 ± 0	8.54 ± 0.31	29.2 ± 0.20
T2	22.6 ± 0.25	44.5 ± 1.32	4.46 ± 0.1	4.56 ± 0.2	13 ± 0	25 ± 0	7.68 ± 0.28	25.1 ± 0.15
T3	26.0 ± 0.11	43.4 ± 0.5	4.23 ± 0.1	4.4 ± 0.1	9 ± 0	23 ± 0	9.21 ± 0.10	23.1 ± 0.20
T4	23.3 ± 0.15	29.6 ± 0.6	5.33 ± 0.1	5.5 ± 0.2	13 ± 0	26 ± 0	5.54 ± 0.28	28.3 ± 0.3
T5	24.0 ± 0.11	27.7 ± 1.0	4.76 ± 0.1	4.73 ± 0.11	11 ± 0	28 ± 0	6.61 ± 0.34	24.2 ± 0.25
T6	22.2 ± 0.15	28.4 ± 1.7	5.8 ± 0.1	5.6 ± 0.36	11 ± 0	27 ± 0	5.41 ± 0.04	26.1 ± 0.15
T7	26.2 ± 0.25	32.1 ± 1.4	5.3 ± 0.2	5.63 ± 0.15	9 ± 0	19 ± 0	4.29 ± 0.06	27.3 ± 0.3
T8	25.0 ± 0.11	32.8 ± 0.5	5.33 ± 0.15	5.5 ± 0.2	6 ± 0	26 ± 0	6.39 ± 0.23	25.3 ± 0.26
T9	22.1 ± 0.20	34.9 ± 0.3	4.6 ± 0.2	4.6 ± 0.2	5 ± 0	25 ± 0	3.14 ± 0.06	23.3 ± 0.3
T10	23.2 ± 0.15	38.8 ± 0.2	4.3 ± 0.2	4.36 ± 0.2	7 ± 0	17 ± 0	10.26 ± 0.25	22.2 ± 0.2
control	26.3 ± 0.28	46.4 ± 0.4	5.4 ± 0.1	5.53 ± 0.25	10 ± 0	20 ± 0	6.73 ± 0.15	30.5 ± 0.60

T1-10%, T2-20%, T3-30%, T4-40%, T5-50%, T6-60%, T7-70%, T8-80%, T9-90%, T10-100%, DAS-days after sowing

Table 4 Morphology of *Lantana camara* growing in fly ash amended soil

Sample code	Plant height (cm)		Leaf size (cm)		Total leaf		Biomass (gm)	
	60 DAS	120 DAS	60 DAS	120 DAS	60 DAS	120 DAS	60 DAS	120 DAS
T1	18.3 ± 0.14	35.3 ± 0.32	2.63 ± 0.15	3.6 ± 0.3	18 ± 0	30 ± 0	12.4 ± 0.45	24.9 ± 0.55
T2	21.2 ± 0.03	40.13 ± 0.15	3.43 ± 0.32	3.83 ± 0.05	12 ± 0	38 ± 0	11.1 ± 0.55	23.4 ± 0.40
T3	19.83 ± 0.14	35.6 ± 0.10	3.66 ± 0.20	4.36 ± 0.32	11 ± 0	31 ± 0	10.4 ± 0.40	21.5 ± 1.26
T4	22.3 ± 0.05	40.6 ± 0.3	2.76 ± 0.15	4.73 ± 0.20	14 ± 0	34 ± 0	13.4 ± 0.20	21.9 ± 0.78
T5	23.6 ± 0.15	32.7 ± 0.15	3.56 ± 0.20	3.6 ± 0.3	18 ± 0	33 ± 0	8.8 ± 0.1	26.3 ± 0.26
T6	18.6 ± 0.23	22.4 ± 0.15	2.56 ± 0.20	3.66 ± 0.11	16 ± 0	36 ± 0	10.46 ± 0.41	24.3 ± 0.40
T7	17.5 ± 0.25	34.4 ± 0.14	2.6 ± 0.2	4.16 ± 0.49	10 ± 0	39 ± 0	12.7 ± 0.17	25.5 ± 0.46
T8	22.5 ± 0.25	45.3 ± 0.3	3.6 ± 0.2	3.83 ± 0.56	13 ± 0	37 ± 0	15.5 ± 0.45	28.2 ± 0.20
T9	19.5 ± 0.3	40.16 ± 0.15	3.73 ± 0.15	4.43 ± 0.15	15 ± 0	32 ± 0	13.43 ± 0.15	23.2 ± 0.2
T10	17.4 ± 0.13	48.3 ± 0.3	2.76 ± 0.15	3.63 ± 0.15	14 ± 0	35 ± 0	12.5 ± 0.30	24.4 ± 0.41
Control	21.2 ± 0.11	54.3 ± 0.3	3.43 ± 0.15	3.76 ± 0.15	12 ± 0	36 ± 0	16.56 ± 0.32	21.4 ± 0.34

T1-10%, T2-20%, T3-30%, T4-40%, T5-50%, T6-60%, T7-70%, T8-80%, T9-90%, T10-100%, C-control, DAS-days after sowing

3.4 Biochemical Parameters

Fly ash amendment effects on chlorophyll content in *I. carnea* and *L. camara* are shown in Fig. 2. Chlorophyll content in *I. carnea* and *L. camara* decreased with increasing amendments and time. However, chlorophyll content in *L. camara* was more than *I. carnea*. At the initial stage (60 DAS), chlorophyll production was high in both the plants and chlorophyll content did not decease with increasing doses of fly ash. After 120 DAS, chlorophyll concentration decreased in *I. carnea* but in *L. camara* it increased with time. During the initial growth stage, metal ions from the fly ash might help for chlorophyll production. After long exposure with the metalliferous soil reduction in chlorophyll in the plants was probably because of replacement of Mg^{2+} ions in chlorophyll molecules by the Cd^{2+}, Cu^{2+}, Ni^{2+}, Zn^{2+} and Pb^{2+}, metal ions. Metals can also suppress chlorophyll activity by inhibiting chlorophyll-synthesizing enzyme activity [30]. ANOVA ($p < 0.05$) for chlorophyll was calculated where $F_{critical} = 2.97$ for treatments and sowing stages $F_{critical} = 4.94$. Chlorophyll concentration varied significantly according to ANOVA ($p < 0.05$) between the harvesting stages but the variation was not significant among the different treatments of fly ash for *L. camara* ($F = 23.7$) as well as *I. carnea* ($F = 7.87$). However, at initial growth stage of plant, chlorophyll content was high also due to increased photosynthetic activity and as the plants matured chlorophyll content decreased in *I. carnea*. Decrease in chlorophyll content is also found in some plants *Sesbania cannabina* and chickpea grown on different fly ash dosed soils [31, 32]. It must be noted that in the present study, plant total chlorophyll was not affected by

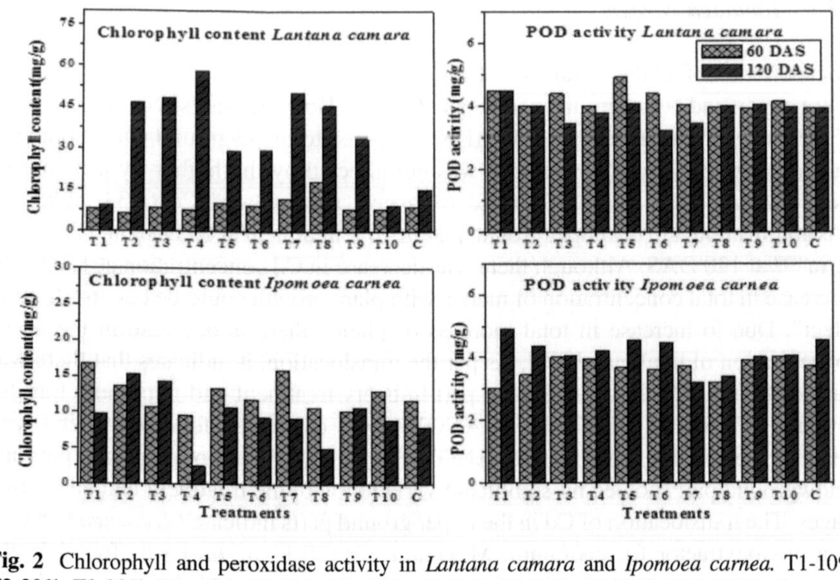

Fig. 2 Chlorophyll and peroxidase activity in *Lantana camara* and *Ipomoea carnea*. T1-10%, T2-20%, T3-30%, T4-40%, T5-50%, T6-60%, T7-70%, T8-80%, T9-90%, T10-100%, C-control, DAS-days after sowing

dose of fly ash in *I. carnea*. Even in 100% fly ash chlorophyll content was not affected. This signified that the plant has good tolerance even when it is exposed to high metal dose and its growth will not be affected. However, in *L. camara* chlorophyll content increased with time but in plants growing on 100% fly ash dose the chlorophyll content was substantially less. The peroxidase activity for both the plants *I. carnea* and *L. camara* is shown in the Fig. 1b, d. In *I. Carnea,* peroxidase activity increased with the different treatments and different harvesting stages. In *L. Camara,* POD activity decreased with the increase in time. Control plant of *L. camara* showed decrease in POD activity at second harvesting stage. With increasing amendments, the metal content increased and hence peroxidase activity also increased. The variation was statistically significant ($P < 0.05$) within different harvesting ($F = 4.96$) stages and not significant among the different treatments of fly ash. POD activity is related to the oxidative reactions due to increase in peroxides and free radicals in the plant cells due to metal stress. In some plants, initiation of POD activity is a normal reaction to uptake of toxic metals. Finally, an increase in POD activity can also be due to plant ageing or senescence. However, no symptoms of toxicity were observed at this study. Similar kind of finding has been reported by Bhattacharya et al. [30] by *Paspalum distichum* grown on sludge-dosed soils. Sharma et al. [32] were reported earlier by Pisum *sativum* L. and Gautam et al. [33] with *Brassica juncea* [30, 32, 33].

3.5 Metal Accumulation

3.5.1 *Ipomoea carnea*

Concentration of all the metals (Cd, Cr, Cu, Mn, Ni, Pb) was analysed in plant tissue in upper part and belowground part (Figs. 3 and 4). Both the plants accumulated most of the metals in the upper ground part as compared to belowground part in both the harvesting stages. However, uptake was not affected by the higher fly ash amendment. Among the six metals studied *I. carnea,* accumulated Cd (4.33 mg/kg) in maximum amount in the upper ground part in 40% treatment at 60 DAS and 2.67 mg/kg in T7 at 120 DAS. Although there was decrease in Cd concentration at 120 DAS. Decrease in total concentration of metals with plant growth could be due to "dilution effect". Due to increase in total biomass of plants, there is decrease in the tissue concentration of total metal [30]. As per the translocation, it indicates that there was translocation of Cd to upper ground part in every treatment and maximum translocation was found at T4. Two-way ANOVA was calculated for every metal with treatments and harvesting. For Cd, ANOVA at ($P < 0.05$) in upper ground part and belowground part showed no significant variation with the treatment and harvesting stages. The translocation of Cd in the upper ground parts indicated *I. carnea* behaved as a phytoextractor for cadmium. Maximum Cr (20.12 mg/kg) was found in the upper ground part after 120 days of sowing at the T1. At 60 DAS Cr concentration in upper ground part was 7 mg/kg in T6 and T10, whereas in belowground part it was

(a)

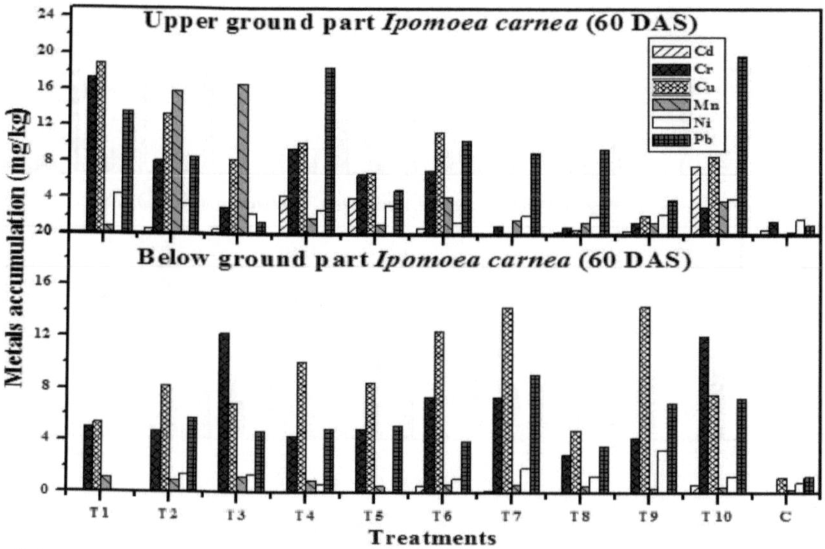

(b)

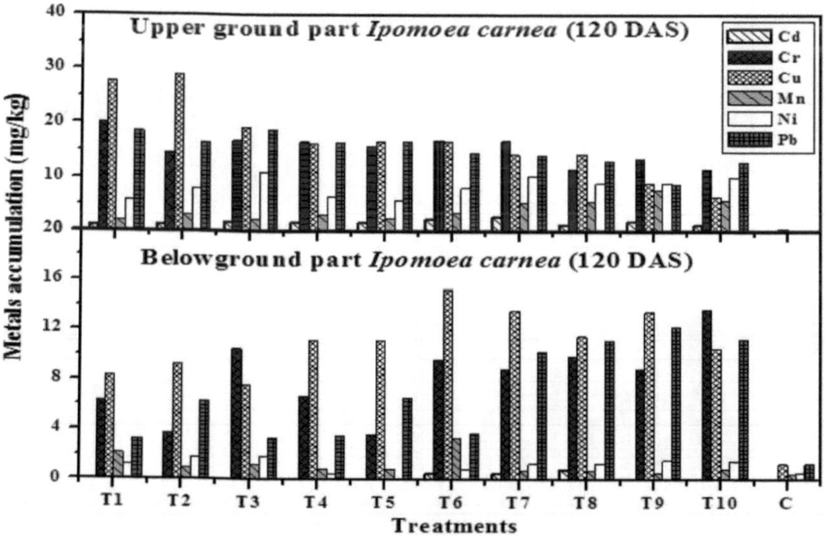

Fig. 3 **a**, **b** Heavy metals uptake by *Ipomoea carnea* at different sowing stages. T1-10%, T2-20%, T3-30%, T4-40%, T5-50%, T6-60%, T7-70%, T8-80%, T9-90%, T10-100%, C-control, DAS-days after sowing

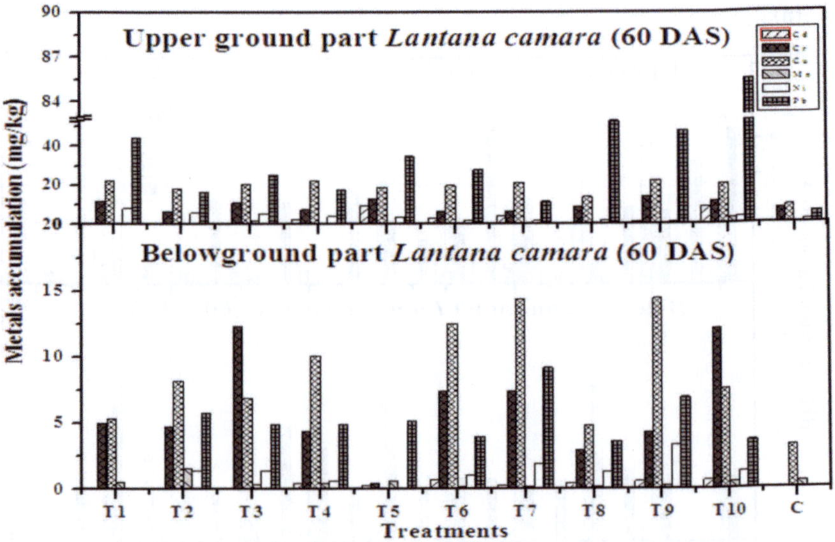

(a)

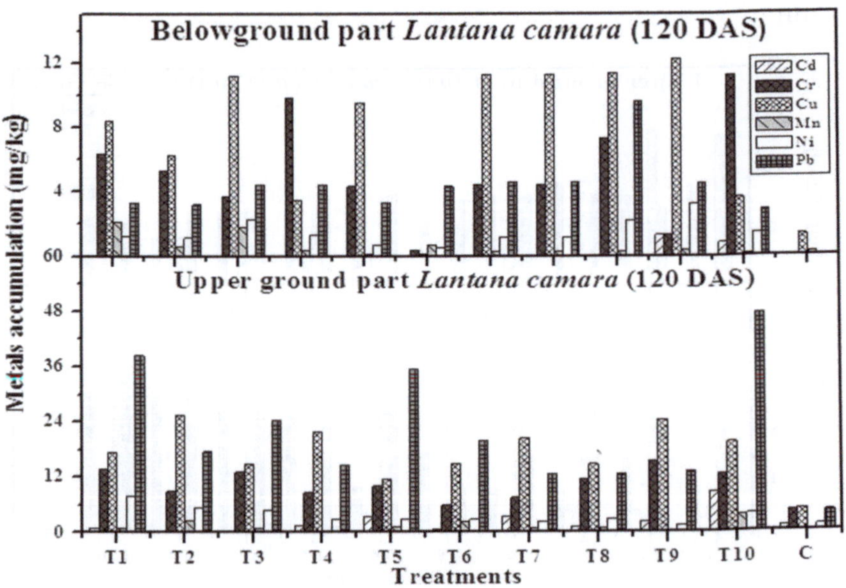

(b)

Fig. 4 **a**, **b** Heavy metals uptake by *Lantana camara* at different sowing stages. T1-10%, T2-20%, T3-30%, T4-40%, T5-50%, T6-60%, T7-70%, T8-80%, T9-90%, T10-100%, C-control DAS-days after sowing more in T3 (12.3 mg/kg). So, at the initial stages Cr was more accumulated in belowground part, and at final stage it was more in upper ground part. According to translocation factor (data not shown) maximum translocation was at T1 and T2. Usually, concentration of Cr found in plant varies widely with the types of tissues and growth stage of plant as the trend for Cr uptake is irregular [34]. At the final stage, Cr was more in upper ground part so the plant behaved similarly like Cd that is as phytoextractor for Cr. Two-way ANOVA ($P < 0.05$) for upper ground part and belowground part was calculated, and it showed statistically significance variation within the treatments in both the stage of plant ($F_{belowground} = 11.5$ and $F_{upper\ ground} = 3.44$). At 60 DAS, Cu accumulation was more in belowground part at T7 (14 mg/kg) than upper ground part (8.95 mg/kg at T1). At the 120 DAS, it was changed as maximum accumulation was in the T2 (29.06 mg/kg) in upper ground part. As per the translocation factor, it showed translocation only in T1, T2 and T3. Cu has important role in the enzymatic activity and also had antagonistic interrelationship with other metals. Cu–Cd interaction was reported by some researchers for both antagonistic and synergistic in the uptake by root [35]. In the present study, Cu and Cd both had good translocation in T1 and T2 suggesting synergistic effect. Similar uptake of Cu was reported in *Cassia siamea* growing on fly ash amended soil (5–30 mg/kg) [34]. ANOVA in upper ground part and belowground part at ($P < 0.05$) was calculated. It showed statistically significant variation only within the treatment in belowground part. Upper ground part showed statistically significant variation within the treatment ($F = 5.80$) and between sowing ($F = 23.3$) also. Mn (16.65 mg/kg) showed more accumulation in T3 in upper ground part at 60 DAS and on the other hand its concentration decreased at 120 DAS to 7.78 mg/kg in T9. The accumulation of metal in belowground part at 120 DAS was less than upper ground part (3.32 mg.kg) at T6. This decrease can also be attributed due to dilution effect or might be due to the translocation in the upper ground parts. Translocation factor showed good translocation of Mn in all the treatments except T1. This indicates *I. carnea* can act as a good phytoextractor for Mn in almost all the amendments. Two-way ANOVA at ($P < 0.05$) in upper ground part and belowground part also suggested that no statistically significant variation in Mn concentration was observed within the treatment or between sowing stage. Reeves [36] reported the range of 20–400 mg/kg of Mn as normal concentration range in plants growing in metalliferous soils [36]. Mn accumulation in *P. juliflora* growing in fly ash amended soil was reported to be 12–32 mg/kg [37]. Ni (10.91 mg/Kg) concentration was highest in T3 at 120 DAS. Ni uptake in the upper ground part was more than belowground part. As per the translocation factor, Ni showed translocation of metals in all the treatments. It had maximum translocation from belowground to upper ground part in T4. According to Kabata and Pendias, Ni toxicity lies in between 10 and 100 mg/kg of dry weight [35]. Ni concentration was within safe limit. Two-way ANOVA for upper ground part and belowground part at ($P < 0.05$) for Ni showed statistically significant variation within the treatments ($F = 3.35$) in belowground part. Pb accumulation was found 18.56 mg/kg at T1 in 120 DAS in upper ground part. Here again the concentration of Pb in upper ground part was more than belowground part. As per the translocation factor, it showed only good translocation in T1, T2, T4 and T9. Study conducted on peanuts by Chaudhari et al. reported the range of Pb concentration in the root and shoot portion from 4.1–6.4 to 1.4–3.7 mg/kg, respectively [37]. Two-way ANOVA for upper ground part and belowground part at ($P < 0.05$) for Pb showed statistically significant variation within the treatments ($F = 4.3$) in belowground part and statistically significant ($F = 5.19$) variation between the sowing stages in upper ground part. At 120 DAS, the overall trend of metal accumulation was in upper ground part was Cu > Pb > Cr > Ni > Mn > Cd and in belowground part was Cu > Cr > Pb > Mn > Ni > Cd.

3.5.2 *Lantana camara*

Among the six metals studied *L. camara* (Fig. 4a, b). accumulated all the metals. Cd was (9.15 mg/kg) in highest amount in the upper ground part in T10, whereas maximum accumulation in belowground part was 0.7 mg/kg at T6. At 120 DAS, upper ground part accumulation decreased to 8 mg/kg at T10 (Fig. 4a, b). But it was more than belowground part. Translocation factor was >1 for all treatment it implies the mobility of Cd in plant upper ground part. It acts as phytoextractor as highest accumulation takes place in plant harvestable part. Two-way ANOVA for upper ground part and belowground part at ($P < 0.05$) showed statistically significant variation with treatment ($F = 7.93$) only in upper ground part. Statistically no significant variation was observed either with treatment or sowing stage. Cr (12.73 mg/kg) accumulation was maximum in the upper ground part at 120 DAS in T9. At 60 DAS, maximum accumulation was 11 mg/kg found in T10. At both the sowing stages, more accumulation was found in upper ground part. For the metalliferous soil, Cu toxicity limit is 5–25 mg/kg as reported by Reeves [36]. Chlorosis is the toxicity symptom of Cu. There was no visible symptom occurred in both the plants. According to translocation factor, it was >1 in all the treatments and maximum was in T6. Two-way ANOVA for upper ground part and belowground part at ($P < 0.05$) was calculated. It showed statistically significant variation within the treatments only in upper ($F = 11.5$) and below ($F = 8.35$)-ground part. No significant variation was found between the sowing stages. Cu (25.4 mg/Kg) was maximum in T2 at 120 DAS in upper ground part. At 60 DAS, Cu accumulation was highest in T1 and T9 (22 mg/kg). *P. juliflora* growing in fly ash amended soil had Cu concentration 17–45 mg/kg [37]. Here Cu is more than toxicity level, but no visibly toxicity symptom occurred. Translocation factor of Cu was >1 for all the treatments, indicating translocation in the upper part. Two-way ANOVA for upper ground part and belowground part at ($p < 0.05$) was calculated. It was statistically significant within treatment in upper ground part ($F = 4.67$) and belowground part ($F = 7.7$). Not significant within the sowing stage. Mn uptake in *L. camara* was more in upper ground part at 60 DAS and 120 DAS. As translocation factor was >1 in all treatments. Maity et al. [37] reported that Mn concentration in naturally growing *B. repens* at fly ash dump site was 7–26 mg/kg [37]. Two-way ANOVA for upper ground part and belowground part at ($p < 0.05$) was calculated. It was statistically significant only within treatment in both the plant part ($F_{\text{upper ground}}$ =

3.77 and $F_{belowground} = 14.4$). Ni (Fig. 4a, b) Accumulation in plant upper ground part was highest 8 mg/kg in T1 at 60 DAS and 7.97 mg/kg in T1 at 120 DAS. Its accumulation in belowground part was less than upper ground part. Translocation factor showed that it was >1 in all treatments. So, *L. camara* was good phytoextractor of Ni. According to Alloway [38], critical concentration of Ni in plants ranges between 10 and 100 mg/kg [38]. In the present study, the Ni concentration was within this limit. Maiti et al. [37] reported Ni concentration in the naturally growing vegetation on fly ash dumpsite *A. aspera* was between 1.2 and 58 mg/kg [37]. Two-way ANOVA for upper ground part and belowground part ($p < 0.05$) was calculated. It showed statistically significant variation within the treatments in belowground part ($F = 11.17$) and upper ground part ($F = 32.9$). Pb highest accumulation was found in the upper ground part at 120 DAS 47.32 mg/kg in T10. And total uptake of metal was more in upper ground part as compared to the belowground part. Translocation factor for Pb showed that it was >1 in all the treatment. Accumulation of Pb in *cynodon* growing in normal soil was reported as 12.4 mg/kg [38]. According to ANOVA in upper ground part, the variation in Pb concentration was statistically significant for both treatment ($F = 5.24$) and harvesting stages ($F = 3.88$). The overall metals accumulation at the final harvesting stage in *L. camara* followed the order Cu > Cr > Pb > Ni > Mn > Cd for belowground part and Pb > Cu > Cr > Ni > Mn > Cd for upper ground part. Both the plants accumulated Cu, Pb and Cr more than the other metals. Cd uptake was least for both the plants. Both the plants translocated most of the metals in the upper ground part or harvestable part. So, with proper harvesting strategies, metals can be scavenged from the soil amendments or fly ash.

3.6 *Metals in the Fly Ash Amendment Soil*

Metal concentration in the fly ash amended soil planted with *I. carnea* and *L. camara* showed decrease in metal concentration with time (Fig. 5a, b). The decrease of Cd was least as it was not taken up by plants. The average percentage decrease of metals in soil amendments planted with *L. camara* was in the following order Ni (38%) > Mn (34%) > Cd (27%) > Cr (21%) > Pb (11%) > Cu (10%). These average values in the parenthesis are based on the decrease in metal content with respect to the initial metal concentration. The order of decrease of metals for *I. carnea* planted soil was as follows: Cd (45%) > Ni (41%) > Mn (24%) > Cu (23%) > Pb (20%) > Cr (18%).

3.7 *Phytoextraction Efficiency*

Phytoextraction potential for *I. carnea* and *L. camara* is shown in Fig. 6. *I. carnea* showed good phytoextraction efficiency for all the metals (Cd, Cr, Cu, Mn, Ni, Pb)

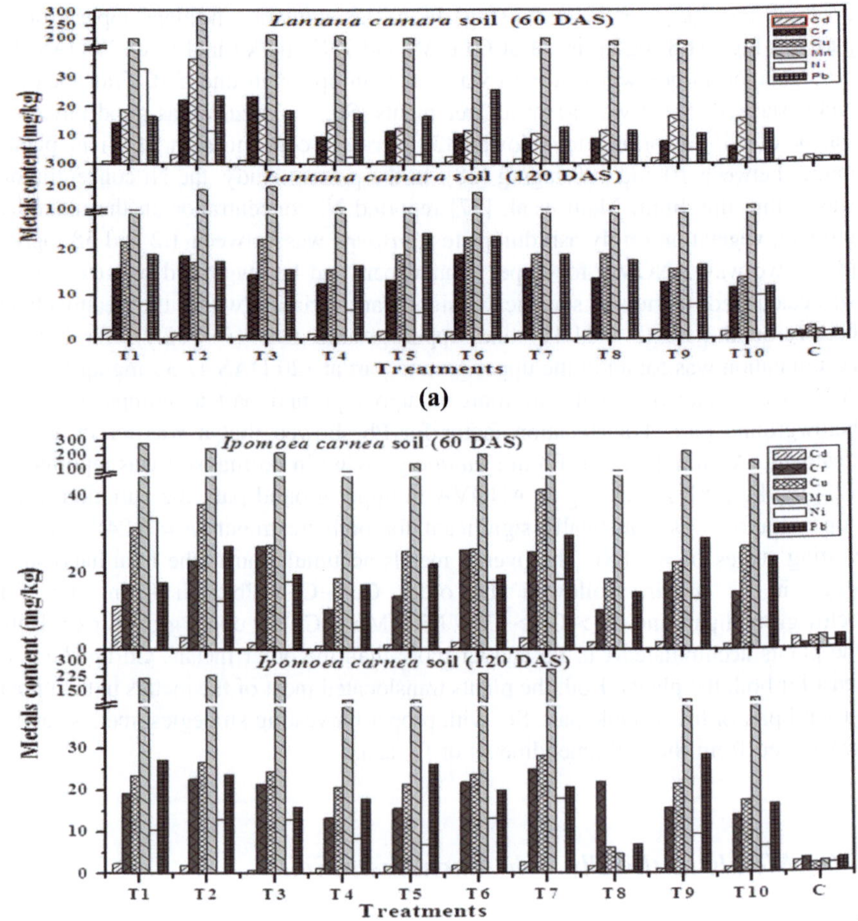

(b)

Fig. 5 **a**, **b** Variation of metal concentration in soil planted with *Lantana camara* and *Ipomoea carnea*. T1-10%, T2-20%, T3-30%, T4-40%, T5-50%, T6-60%, T7-70%, T8-80%, T9-90%, T10-100, DAS-days after sowing

and for all the treatments, whereas *L. camara* showed good phytoextraction for Pb at efficiency T10. It was reported in the literature that perfect species for phytoextraction must have good biomass as metal accumulation can lead to decrease in biomass. In the present study, higher metal exposures did not affect the biomass of the plants. It is also evident from Fig. 6 that the phytoextraction potential increased with the higher dose of fly ash. This signifies gain that as the biomass increased, the metal uptake increased and hence the phytoextraction efficiency. However, *L. camara* showed better phytoextraction potential for all the metals.

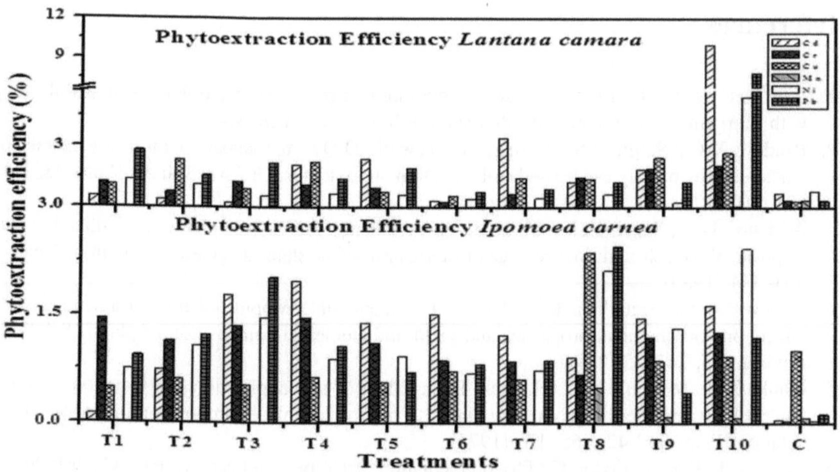

Fig. 6 Phytoextraction *efficiency* of *Lantana camara* and *Ipomoea carnea*. T1-10%, T2-20%, T3-30%, T4-40%, T5-50%, T6-60%, T7-70%, T8-80%, T9-90%, T10-100%, C-control

4 Conclusion

Overall behaviour of *I. carnea* and *L. camara* showed good phytoextraction capability for all the metals (Cd, Cr, Cu, Mn, Ni and Pb), as the metals were accumulated in the harvestable part of the plants. However, *L. camara* showed better phytoextraction potential. Morphological parameters like plant height (cm), leaf size (cm), number of leaf and biomass (gm) increased with time in both the plants. No stress reflected by the estimation of the morphological parameters. Likewise, chlorophyll production was high in both the plants, and chlorophyll content did not decease with increasing doses of fly ash. This indicates growth parameters and the biochemical parameters of both the plants were not affected by metal stress. So, these plants can be used successfully to remediate fly ash-contaminated soils with proper harvesting strategy.

Acknowledgements The first author is thankful to Jawaharlal Nehru Memorial Fund for Doctoral Studies (Ref: SU-1/291/2016-17/83) for providing research fellowship. The authors are thankful to Mr. Ashwini Kumar of Central Instrumentation Facility, Birla Institute of Technology for analysis of samples in ICP-OES.

References

1. Ram, L.C., Mast, E.: Fly ash for soil amelioration: a review on the influence of ash blending with inorganic and organic amendments. Earth Sci. Rev. **128**, 52–74 (2014)
2. Pandey, V.C., Singh, J.S., Kumar, A., Tewari, D.D.: Accumulation of heavy metals by chickpea grown in fly ash treated soil: effect on antioxidants. CLEAN Soil Air Water **38**, 111–628 (2010)
3. Adriano, D.C., Page, A.L., Elseewi, A.A., Chang, A.C., Straugham, I.: Utilization and disposal of fly ash and coal residues in terrestrial ecosystem: a review. J. Environ. Qual. **9**, 333–344 (1980)
4. Elseewi, A.A., Straughan, I.R., Page, A.L.: Sequential cropping of fly ash amended soils: effects on soil chemical properties and yield and elemental composition of plants. Sci. Total Environ. **15**, 247–259 (1980)
5. Plank, C.O., Martens, D.C., Hallock, D.L.: Effect of soil application of fly ash on chemical composition and yield of corn (*Zea mays* L.) and on chemical composition of displaced soil solution. Plant Soil **42**, 465–476 (1975)
6. Barcelo, J., Poschenrieder, C.: Phytoremediation: principles and perspectives. Contrib. Sci. **2**, 333–344 (2003)
7. Jambhulkar, H.P., Juwarkar, A.A.: Assessment of bioaccumulation of heavy metals by different plant species grown on fly ash dump. Ecotoxicol. Environ. Saf. **72**, 1122–1128 (2009)
8. Kapur, S.L., Shyam, A.K., Soni, R.: Greener management practices ash mound reclamation. TERI Inf. Dig. Energ. Environ. **1**(4), 559–567 (2002)
9. Zevenbergem, C., Bradley, J.P., Shyam, A.K., Jenner, H.A., Platenburg, R.J.P.M.: Sustainable ash pond development in India—a resource for forestry and agri-culture. Waste Mater. **1**, 533–540 (2000)
10. Adholeya, A., Bhatia, N.P., Kanwar, S., Kumar, S.: Fly ash source and substrate for growth of sustainable agro-forestry system. In: Proceeding of Regional Workshop Cum Symposium on Fly Ash Disposal and Utilization. Kota Thermal Power Station, RSEB, Kota, Rajasthan, India (1998)
11. Aitken, R.L., Campbell, D.J., Bell, L.C.: Properties of Australian fly ashes relevant to their agronomic utilization. Aust. J. Soil Res. **22**(4), 44–353 (1984)
12. Robab, M.I., Hisamuddin, Azam, T.: Impact of fly ash on vegetative growth and photosynthetic pigment concentrations of *Solanum nigrum* L. Nanobiotechnica universale **1** (2): 133–138 (2010)
13. Gupta, A.K., Dwivedi, S., Sinha, S., Tripathi, R.D., Rai, U.N., Singh, S.N.: Metal accumulation and growth performance of *Phaseolus vulgaris* grown in fly ash amended soil. Bioresour. Technol. **98**, 3404–3407 (2007)
14. Mishra, L.C., Shukla, K.N.: Effects of fly-ash deposition on growth, metabolism and dry matter production of maize and soybean. Environ. Pollut. **42**, 1–13 (1986)
15. Basu, M., Pande, M., Bhadoria, P.B.S., Mahapatra, S.C.: Potential fly-ash utilization in agriculture: a global review. Prog. Nat. Sci. **19**, 1173–1186 (2009)
16. Srivastava, K., Farooqui, A., Kulshreshtha, K., Ahmed, K.J.: Effect of flyash amended soil on growth of *Lactuca sativa* L. J. Environ. Biol. **16**(2), 93–96 (1995)
17. Dwivedi, S., Tripathi, R.D., Srivastava, S., Mishra, S., Shukla, M.K., Tiwari, K.K.: Growth performance and biochemical responses of three rice (*Oryza sativa* L.) cultivars grown in fly ash-amended soil. Chemosphere **67**, 140–151 (2007)
18. Jackson, M.L.: Soil Chemical Analysis. Prentice-Hall of India, New Delhi (1968)
19. Allison, F.A.: Soil Organic Matter and Its Role in Crop Production. Elsevier, Amsterdam (1973)
20. Olson, R.A., Rhodes, M.B., Dreier, A.F.: Available phosphorus status of Nebraska soils in relation to series classification, time of sampling and method of measurement. Agron. J. **46**, 175–180 (1954)

21. Hanway, R.H., Heidel, H.: Soil analysis method as used in Iowa State College soil testing laboratory. Iowa Agron. **57**, 1–31 (1952)
22. Tessier, A., Campbell, P.G.C., Blsson, M.: Sequential extraction procedure for the speciation particulate trace metals. Anal. Chem. **51**(7), 844–851 (1979)
23. Lagrimini, L.M.: Wound-induced deposition of polyphenols in transgenic plants overexpressing peroxidase. Plant Physiol. **96**, 577–583 (1991)
24. Sun, L.N., Yang, X.B., Wang, W.Q., Ma, L., Chen, S.: Spatial distribution of Cd and Cu in soils in Shenyang Zhangshi Irrigation Area (SZIA), China. J. Zhejiang Univ. Sci. B **9**(3), 271–278 (2008)
25. Khan, M.R., Khan, M.W.: The effect of fly ash on plant growth and yield of tomato. Environ. Pollut. **92**, 105–111 (1996)
26. Gupta, A.K., Dwivedi, S., Sinha, S., Tripathi, R.D., Rai, U.N., Singh, S.N.: Metal accumulation and growth performance of *Phaseolus vulgaris* grown in fly ash amended soil. Bioresour. Technol. **98**, 3404–3407 (2007)
27. Wong, M.H., Wong, J.W.C.: Germination and seedling growth of vegetable crops in fly ash amended soils. Agric. Ecosyst. Environ. **26**, 23–25 (1989)
28. Pandey, V., Misra, J., Singh, S.N., Singh, S., Yunus, M., Ahmad, K.J.: Growth response of *Helianthus annuus* L. grown on fly ash amended soil. Indian. J. Environ. Biol. **15**, 117–125 (1994)
29. Sinha, S., Gupta, A.K.: Translocation of metals from flyash amended soil in the plant of *Sesbania cannabina* L. Ritz: effect on antioxidants. Chemosphere **61**, 1204–1214 (2005)
30. Bhattacharya, T., Chakaraborty, S., Banerjee, D.K.: Heavy metal uptake and its effect on macronutrients, chlorophyll, protein, and peroxidase activity of *Paspalum distichum* grown on sludge-dosed soils'. Environ. Monit. Assess. **169**(1), 15–26 (2009)
31. Pandey, V.C., Singh, J.S., Kumar, A., Tewari, D.D.: Accumulation of heavy metals by chickpea grown in fly ash treated soil: effect on antioxidants. CLEAN Soil Air Water **38**, 1116–1289 (2010)
32. Sharma, S., Sharma, P.P.: Bioaccumulation of heavy metals in *Pisum sativum* L. growing in fly ash amended Mehrotra soil. J. Am. Sci. **6**(6), 43–50 (2010)
33. Gautam, S., Singh, A., Singh, J., Shikha: Effect of ffiyash amended soil on growth and yield of Indian mustard (*Brassica juncea*). Adv. Biores. **3**(4), 39–45 (2012)
34. Tripathi, R.D., Vajpayee, P., Singh, N., Rai, U.N., Kumar, A., Ali, M.B.: Efficacy of various amendments for amelioration of fly ash toxicity: growth performance and metal composition of *Cassia siamea* lamk. Chemosphere **54**, 1581–1588 (2004)
35. Kabata-Pendias, A., Pendias, H.: Trace Elements in Soils and Plants. CRC Press, Boca Raton, FL (2001)
36. Reeves, R.D., Baker, A.J.M.: Metal-accumulating plants. In: Raskin, I., Ensley, B.D. (eds.) Phytoremediation of Toxic Metals. Using Plants to Clean Up the Environment, pp. 193–229. Wiley, New York (2000)
37. Maiti, S.K., Jaiswal, S.: Bioaccumulation and translocation of metals in the natural vegetation growing on fly ash lagoons: a field study from Santaldih thermal power. Environ. Monit. Assess. **136**, 355–370 (2008)
38. Alloway, B.J.: Heavy Metals in Soils, p. 339. Blackie, Glasgow, UK (1990)

Designing and Utilizing of the Solar Water Heater for Digestion of Lignocellulosic Biomass

Dawit Gudeta Gunjo, Pinakeswar Mahanta and P. S. Robi

Abstract Conversion of woody biomass, animal waste such as cattle dung, chicken litters, pig manure, municipal solid, and agricultural wastes to methane gas under oxygen-free environment and favorable temperature range is called anaerobic digestion. But achieving the required temperature range under normal condition is difficult which necessitates the development of a heating system for better digestion process. The endeavor of the present investigation was developing 1.8 m^2 solar collectors intended to heat biogas digester and investigating its performance. The solar collector was tested experimentally and numerically for heating water before implementing in actual model. A three-dimensional computational fluid dynamics model was developed by taking a single straight tube attached with an absorber plate at the bottom to predict outlet water and absorber plate temperature. The developed model predicted the outlet water temperature and absorber plate temperature with a reasonable accuracy. Moreover, 53 °C of outlet water temperature and 44% daily average efficiency were achieved at a flow rate of 0.021 kg/s. As the obtained temperature was found to be good enough for anaerobic digestion, the developed solar collector could be regarded as an alternate option.

Keywords Anaerobic digestion · Thermal pretreatment · Instantaneous efficiency Lignocellulosic biomass · CFD

D. G. Gunjo · P. Mahanta · P. S. Robi (✉)
Department of Mechanical Engineering, Indian Institute of Technology Guwahati (IITG), Guwahati 781039, Assam, India
e-mail: psr@iitg.ernet.in

D. G. Gunjo
e-mail: dawit@iitg.ernet.in; dawitphd@gmail.com

P. Mahanta
e-mail: pinak@iitg.ernet.in

A. S. Kalamdhad et al. (eds.), *Advances in Waste Management*,
https://doi.org/10.1007/978-981-13-0215-2_7

1 Introduction

Lignocellulose is the biomass existing freely in large quantities on the earth surface. This can be exploited as useful sources of renewable energy. It is composed of carbohydrate polymer (cellulose and hemicellulose) and aromatic polymer (lignin). Since the carbohydrate polymers are rich in sugar, they can be used as source of fermentable sugar. Lignin is the aromatic polymer which is formed during biosynthetic processes. Its function is to act as a protective layer for the plant walls. Conversion of lignocellulose (woody biomass), animal waste, municipal solid, and agricultural wastes, etc., to methane gas under oxygen-free environment is called anaerobic digestion. The process involves four stages, namely hydrolysis, acidogenesis, acetogenesis, and methanogenesis, and these processes are temperature-dependent. But achieving the required temperature under normal condition is difficult. The motivation behind the present research was to design and test a solar heating device for achieving the temperature required for efficient digestion of lignocellulosic biomass.

Among the various solar heating collectors, the most common device which converts solar energy to heat is flat plate collector (FPC) [1]. This collector can deliver moderate temperature up to 100 °C. Properly designed FPC should have high thermal performance, low cost, low maintenance, suitable for selected fluid, low weight, long effective life, and breakage resistance of the glazing because of thermal expansion and other causes [1, 2]. According to Ayompe and Duffy [3], a properly designed solar water heating system can achieve annual average efficiency of 35–40%. Whiller [4] suggested the use of plastic covers to gain the advantage of low weight and good resistance to shock and also the influence of cover material on thermal efficiency. Bilgen and Bakeka [5] developed a solar water heater and estimated its thermal efficiency and economics. Zambolin and Del Col [6] compared the performance of evacuated and flat plate collector experimentally. The result indicated higher performance for evacuated collector compared to flat plate collector under the same conditions.

Presently due to the rapid growth of computational works, analytical models are often preferred owing to their simplicity and practicality. Sekhar et al. [7] estimated the top heat loss coefficient from single glazing solar collector theoretically and experimentally. The result indicated high thermal performance at lower top heat loss coefficient. Selmi et al. [8] employed computational fluid dynamics (CFD) to estimate the working fluid temperature and found closely matching result. Gunjo et al. [9] carried out both experimental and CFD study on flat plate collector thermal performance. The developed model predicted the outlet water and absorber plate temperature with reasonable accuracy. Martinopoulos et al. [10] designed a solar collector and used CFD for predicting various temperatures and found good agreement with the experimental result. Further, the effects of operating parameters such as flow rate, temperature, solar insolation were also covered. Facão [11] carried out both experimental and numerical study to optimize the flow rates in the riser and header FPC based on temperature distribution. The result revealed uniform temperature in both riser tube and absorber plate when the flow rates were less than 0.0167 kg/s.

Literature reveals several attempts by researchers to enhance heat transfer and estimate thermal performance of FPC. But majority of the research employed either simulations or experimental methods. Studies involving a combined approach were a few. The endeavor of this study is developing, and performance evaluation of straight tube solar water heating system intended to use as a heat source during biogas production process. The solar water heating system was isolated and tested experimentally and numerically. The numerical model was carried out using Fluent 14.5 software to predict the outlet water temperature and absorber plate temperature. Further, the predicted numerical value was validated with experimental results.

2 Material and Methods

2.1 Element of Flat Plate Solar Collector

- Absorber plate is a thin metallic plate made up of copper or aluminum sheet metal. It is painted with black color to increase absorptivity.
- Transparent cover is the top surface of the solar collector that covers the absorber plate. It is made up of transparent glass or plastic sheet to allow penetration of solar beams to heat the absorber plate.
- Insulating materials, having very low thermal conductivity, are installed in the bottom and sides of the collector for reducing heat loss.
- Heat transfer medium is the circulating fluid that flows through the solar collector to transfer the heat from the absorber plate to the utilization system.
- Casing is usually made up of plastic, metal, or wood. The glass front cover must be sealed at the back and sides for reducing heat losses. Further, the casing also protects the entire system from dirt, insects, and humidity. The model of a FPC indicating various components is depicted in Fig. 1.

2.2 Experimental Setup

The schematic and photograph of the developed experimental setup are depicted in Fig. 2a, b. It comprises of reservoir tank, FPC, connecting pipe, thermocouple, and pyranometer. The temperature of the glass cover, absorber plate, and working fluid were measured by thermocouples. The flow rate of the working fluid was adjusted by rotameter. The solar insolation on the surface of the collector was measured by the help of pyranometer. When water is pumped through the riser tube of the collector, it absorbs heat. The heated water from the FPC enters to the storage tank to heat the

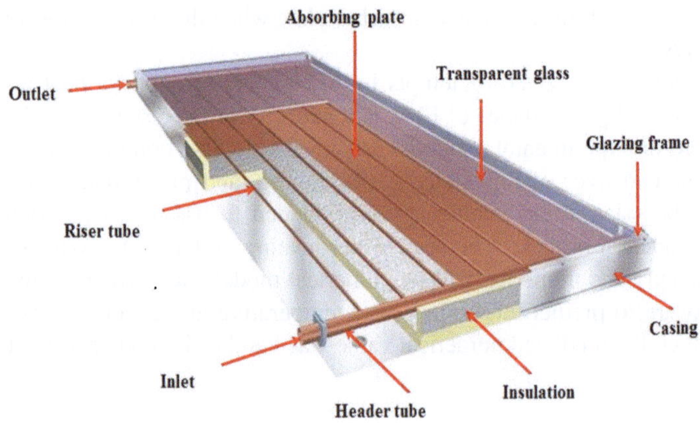

Fig. 1 Various parts of flat plate collector

biomass through heat exchanger. Experimental setup was placed at 26.7°N latitude and 91.7°E longitude.

2.3 *Numerical Model*

The study was carried out to predict the outlet water and absorber plate temperature using computational fluid dynamics (CFD) and validating the result with the measured data. An absorber plate with a single riser tube attached from the top was considered as the flow domain. The performance of a solar water heating system depends on the flow distribution in the riser tubes. Facão [11] carried out both experimental and numerical study on FPC to investigate temperature distribution on the absorber plate and flow distribution in the riser tube. The result revealed that low Reynolds number (<1000) exhibited uniform temperature on the absorber plate and flow distribution in the riser tube. Based on this report, the present study considered a laminar flow mass flow rates of 0.021 kg s^{-1}. Flow in each riser tube was one-tenth of the header (the ratio of the total flow rate through the header pipe to the number of riser tubes). Figure 3 illustrates the model geometry and meshed part of the straight riser tube. Analysis was made based on the following assumptions:

- Flow through the riser tube is laminar and steady.
- The thermal and physical properties of water, tube, and plate are constant.
- Working fluid is incompressible and continuous.

For modeling, the absorber plate length, width, tube diameter, and thickness were taken as 2, 0.1, 0.0125, and 0.007 m, respectively, which was same as the dimension of the experimental setup. The continuity, momentum, and energy

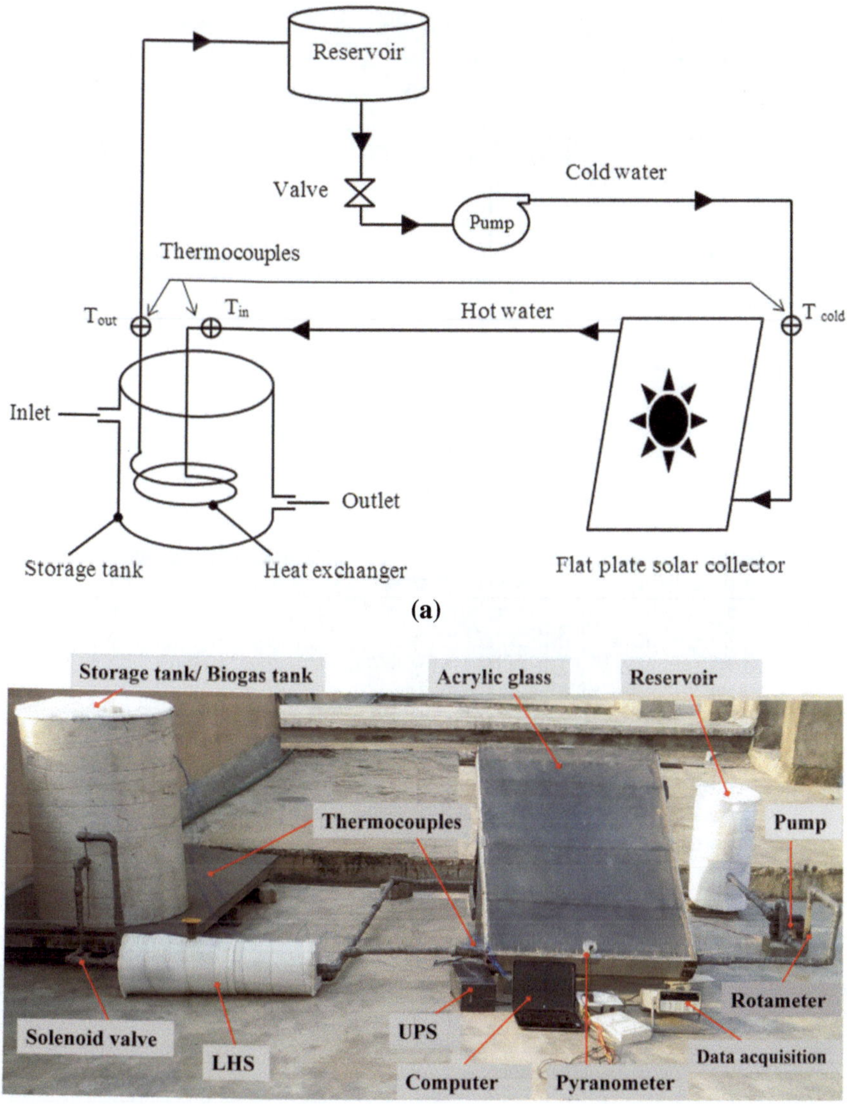

Fig. 2 **a** Schematic diagram of **a** completed setup, **b** photograph of the experimental setup

equations were solved to estimate the velocity of the working fluid and temperature of absorber plate. The fluid zones and the corresponding faces convective heat transfer were solved by coupling the energy and momentum equations. The SIMPLE scheme was employed for coupling momentum and energy equations. The second-order upwind is used for discretization energy and momentum equations and first-order upwind for the pressure. The grid independency test was

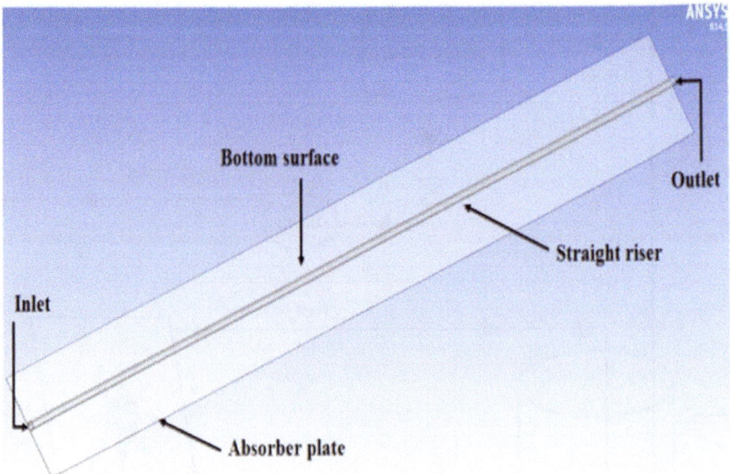

(a)

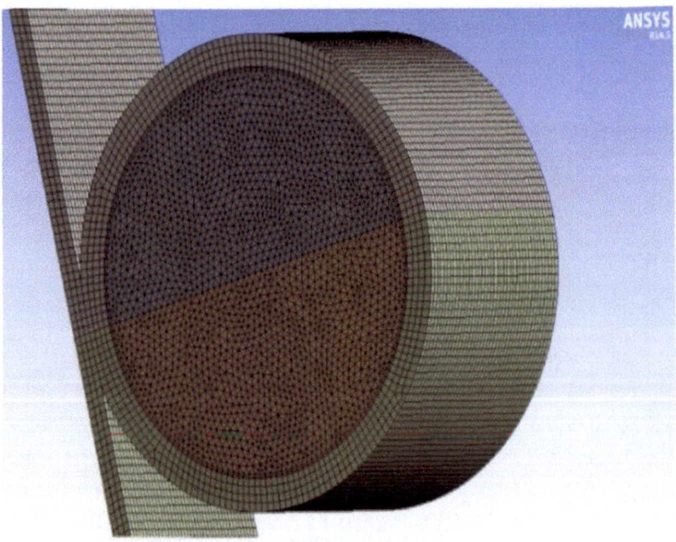

(b)

Fig. 3 **a** Model geometry of straight riser tube and **b** magnified meshed part

carried to check the quality of mesh and its effect on the solution. The solutions converged for the values of residuals below 1×10^{-6} in the computational domain. Further grid refinement resulted in an error less than 0.6%. The number of elements for the computational domain was 400,000 which consists water, water tube, and absorber plate.

The top surface of the FPC was receiving constant heat flux (solar insolation). Convective wall boundary was set at the bottom surface of the FPC. The convective heat transfer coefficient at the bottom surface obtained by the relationship [12]:

$$h_b = 2.8 + 3V_w \tag{1}$$

The heat energy absorbed by the solar collector is determined by the relationship:

$$\dot{Q}_i = A_c(\alpha\tau)I \tag{2}$$

The thermal efficiency of the FPC (η) is determined by the expression [13]:

$$\eta = \frac{Q}{IA_c} \tag{3}$$

The heat absorbed by the water ($\dot{Q}$) is calculated by the expression:

$$\dot{Q} = \dot{m}C_p(To - Ti) \tag{4}$$

The instantaneous solar collector efficiency in terms of inlet and outlet water temperature is calculated using Eq. (5) which was obtained by substituting Eq. (4) into Eq. (3).

$$\eta = \frac{\dot{m}C_p(To - Ti)}{IA_c} \tag{5}$$

3 Result and Discussions

3.1 *Input Parameters*

The simulation was carried out by taking the average values of measured data within one-hour interval as input parameters. The measured experimental data such as inlet water velocity (v), inlet water temperature (Ti), ambient temperature (Ta), and solar radiation (I) were applied as boundary condition for this simulation. Figure 4 illustrates the hourly variation of the measured Ti, Ta, and I for mass flow rate of 0.021 kg s^{-1}.

Figure 4 shows that the average solar radiation (I) increases from 8:00 h and reaches the maximum value at 12:00 h. Also, the water inlet temperature variation was nearly 4 °C for all flow rates. The present study provided insulation to the cold water storage tank to lower the inlet water temperature rise, thus enhanced the efficiency of the collector. Experiments carried out beyond 16:00 h, and results decrease in both inlet water temperature (Ti) and solar radiation (I).

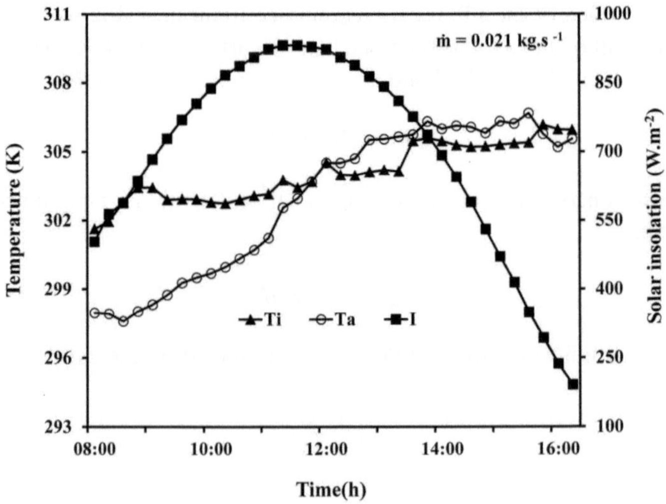

Fig. 4 Measured experimental result of *Ti*, *Ta*, *I* versus time

3.2 *Simulated Result of Outlet Water and Absorber Plate Temperature*

Validation was carried out by comparing the simulated results with the experimental values. The CFD simulation results in the variations of water outlet temperatures along straight riser tube, at the outlet pipe, and on the absorber plate at 11:00 h for the mass flow rate of 0.021 kg s^{-1} depicted in Fig. 5. The variations of water temperature along straight tube length at 11:00 h shown in Fig. 5a indicate higher water outlet temperatures at the upper portion of the collector compared to the lower portion. Figure 5b illustrates the variation of temperature across the outlet tube diameter at 11:00 h. Almost half of the tube side is brazed with the hot absorber plate; the temperature near the brazed surface of the outlet tube was higher than that of the center tube. Since it takes more time for the heat to reach the center of the tube by convection, the water temperature at the center of the tube was 14 K lower than at the surface. The possible mechanism to explain this scenario could be as follows. The heat transfer taking place by a sequence of radiation and conduction through the tube thickness follows convective heat transfer inside the tube.

Figure 5c presents the variation of absorber plate temperature at 10:00 h. Since heat is absorbed by the cold fluid flowing through the riser pipe which is in contact with the absorber at the mid-width section, the temperature variation across the absorber plate length indicates a difference of 42 K between the inlet and outlet side of the absorber plate.

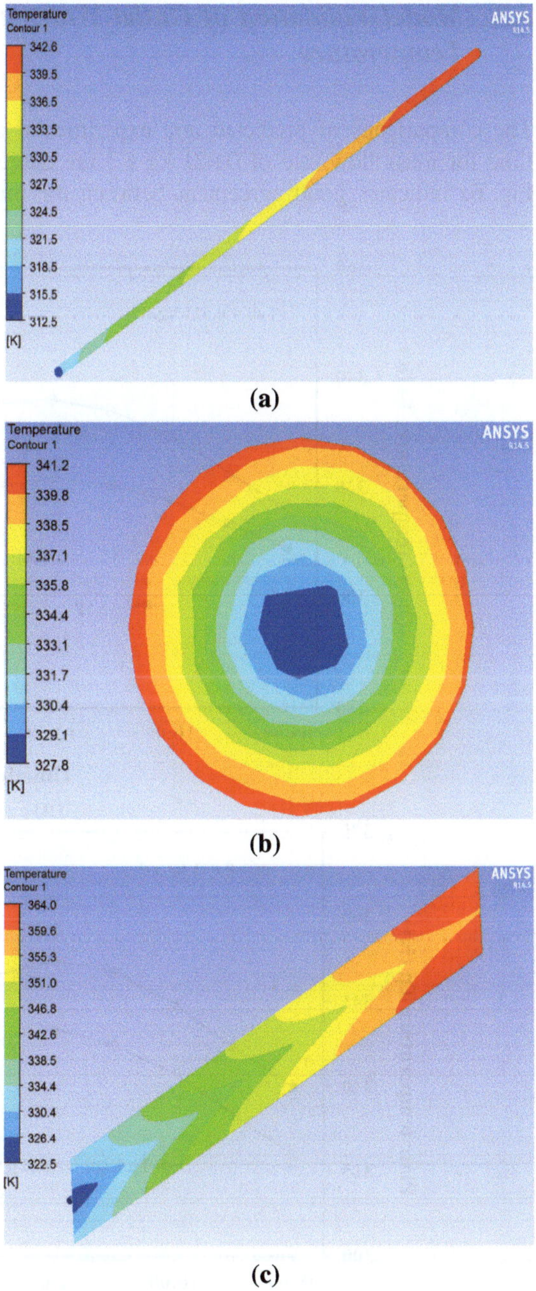

Fig. 5 Variation of water temperature **a** along the riser pipe and **b** at the outlet pipe at 11.00 h, **c** variation of plate temperature at 11.00 h

3.3 *Model Validation of Outlet Water and Absorber Plate Temperature*

The comparisons of predicted and experimental outlet water temperature versus time for mass flow rate of 0.021 kg s^{-1} are depicted in Fig. 6a. The result from Fig. 6a indicates good agreement between the predicted and experimental outlet

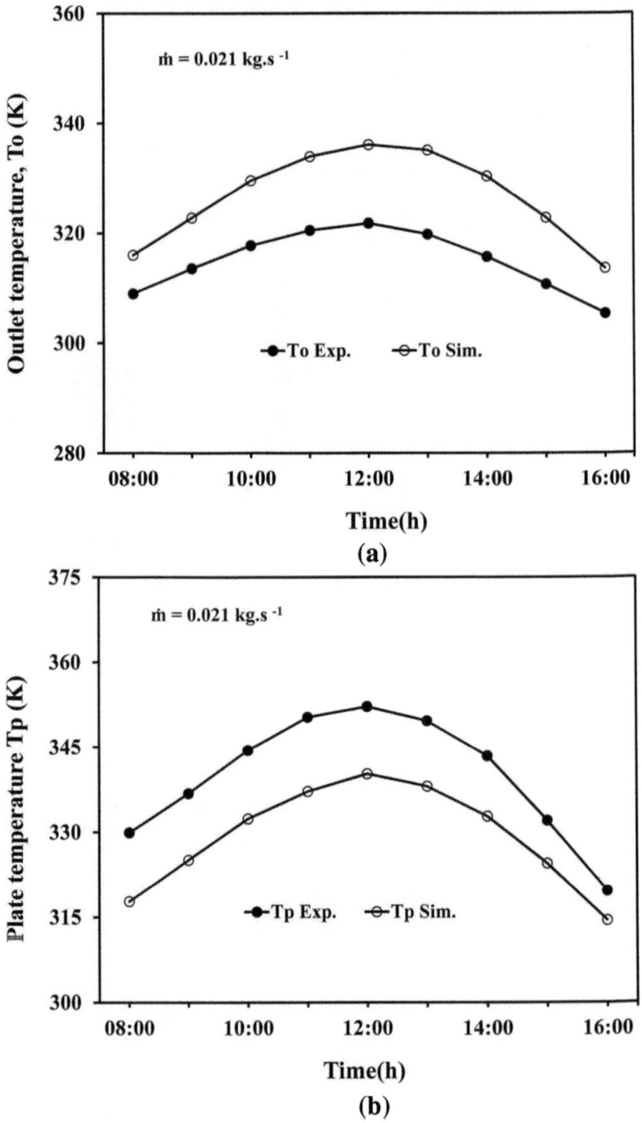

Fig. 6 Plot of predicted and experimental variation of **a** outlet water temperature and **b** plate temperature versus time

water temperature. The maximum deviation in error between experiment and simulation was ±6.2%. It is also found that the experimental outlet water temperature increment was lower as compared to simulated result for higher solar radiation, and this leads to higher deviation in error values.

This was expected since during the test, the experimental result was influenced by climatic conditions, making it tough to predict coefficient of convective heat transfer throughout the day. For the computational cases, convective heat transfer coefficient was taken as constant for the entire duration. Hence, the error values at peak solar insolation were higher.

The variations of predicted and experimental absorber plate temperature with time for mass flow rate of 0.021 kg s^{-1} are depicted in Fig. 6b. It is evident from the figure that the developed heat transfer model could predict the absorber plate temperature for the above-mentioned mass flow rate with deviation in error of ±4.2%. The experimental plate temperature was higher compared to simulated plate temperature. Since the simulation was based on the assumption of a perfect contact between the absorber plate and the tube, better heat transfer is obtained between the plate and the tube during simulation compared to the actual case which resulted in lower simulated absorber plate temperature.

3.4 Experimental Findings

The variation of outlet water temperature versus time for clear and cloudy sky weather conditions at flow rate of 0.021 kg/s is depicted in Fig. 7a. It is observed from Fig. 7a that the outlet temperature of the solar collector was higher for clear sky compared to cloudy sky condition. The maximum outlet temperature of the working fluid for clear and cloudy sky conditions was 53 and 34 °C, respectively.

Variation of instantaneous efficiency, outlet temperature, inlet temperature, ambient temperature, and solar radiation of the measured experimental data with time is shown in Fig. 7b. It was found that the maximum thermal efficiency was 50% and the daily average thermal efficiency was 44%. The experiment was performed September 2015.

3.5 Parametric Study

The parametric study was performed to observe the effects of working parameters on thermal efficiency of the FPC. The working parameters are: (i) inlet water temperature, (ii) ambient temperature, (iii) wind speed, and (iv) transmissivity coefficient of the glass cover.

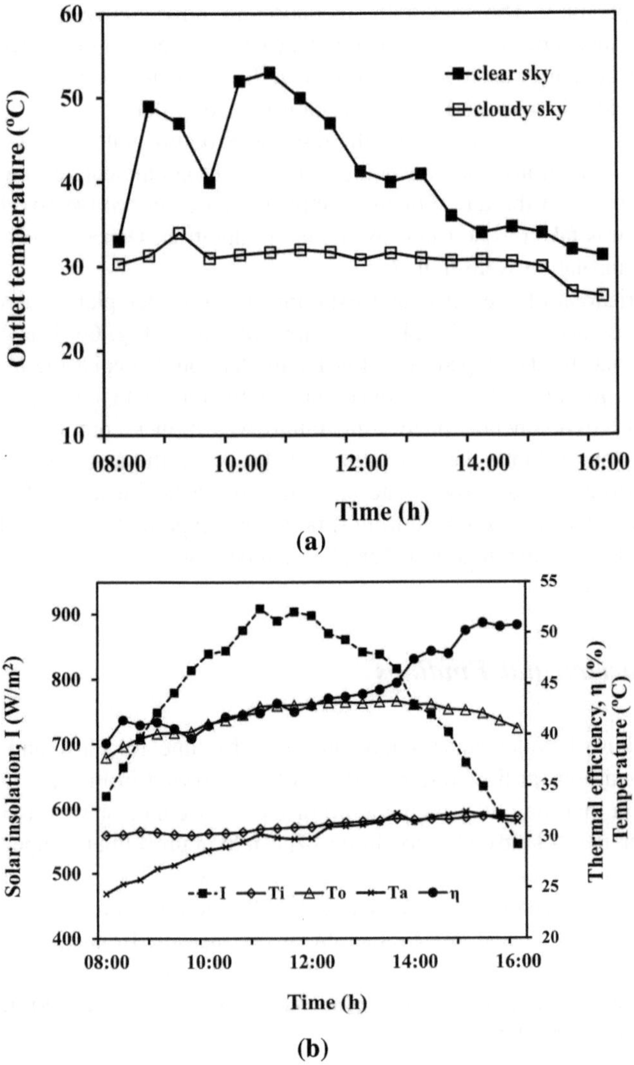

Fig. 7 Plot of **a** water temperature and **b** I, Ta, Ti, To, and η versus time

3.5.1 Effect of Water Inlet Temperature and Solar Intensity on Thermal Efficiency

The variation of thermal efficiency versus inlet water temperature for different solar radiation value is depicted in Fig. 8a. It is evident from Fig. 8a that for a constant solar radiation value, the thermal efficiency is higher for lower inlet water temperature. The efficiency of the collector is higher for higher value of solar intensity and lower value of inlet water temperature.

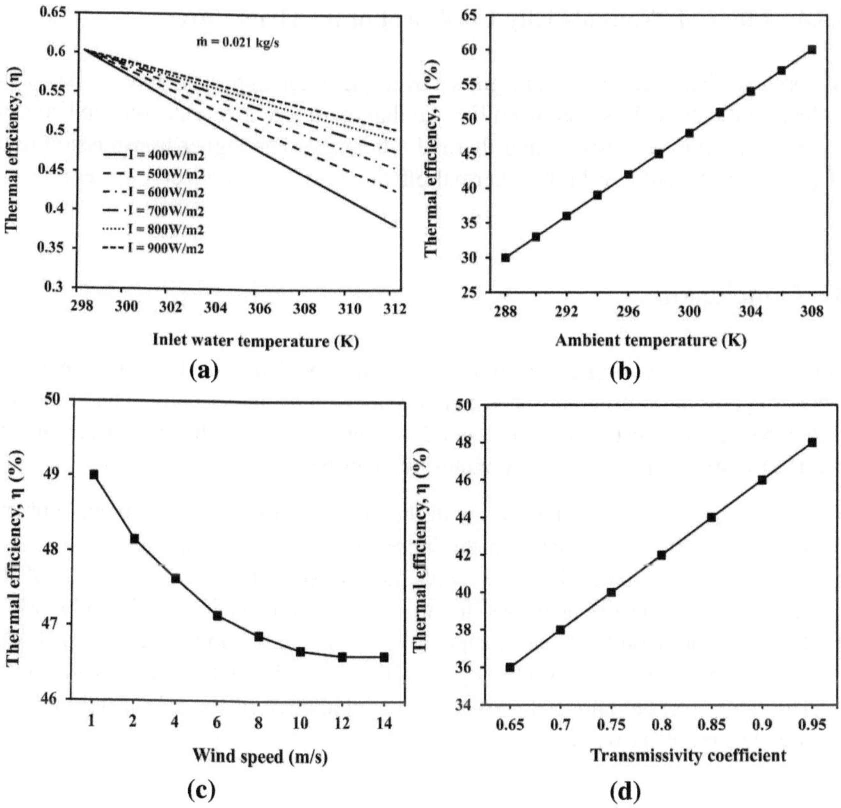

Fig. 8 Variation of η versus **a** Ti, **b** Ta, **c** V_w, and **d** $\tau\alpha$

3.5.2 Effect of Ambient Temperature on Thermal Efficiency

The thermal efficiency of the solar collector depends on ambient temperature. The variation of thermal efficiency versus ambient temperature is illustrated in Fig. 8b. It is evident from Fig. 8b that as the ambient temperature increases, the system heat loss by conduction, convection, and radiation decreases resulting in an increase in the thermal efficiency.

3.5.3 Effect of Wind Speed on Thermal Efficiency

The variation of thermal efficiency versus wind speed is depicted in Fig. 8c. As the wind speed increases, the convective heat losses from the collector increases, thereby reducing the thermal efficiency. Hence, covering with glazing material is mandatory to reduce overall heat loss.

3.5.4 Effect of Transmissivity Coefficient of the Glass Cover

An optical characteristic of the glass cover plays an important role in flat plate collector efficiency. It is seen from Fig. 8d that there is a direct relationship between the transmissivity coefficient and thermal efficiency. The higher transmission ratio of glass cover results in higher thermal efficiency of solar collector system.

4 Conclusion

In the present study, proper design procedure and performance investigation of flat plate collector are presented experimentally as well as numerically. The numerical and experimental result indicated good agreement. From the experimental and parametric study, the following conclusions can be drawn:

- The thermal efficiency increases with the increase of solar radiation and ambient temperature but decreases with the increase of ambient temperature.
- The thermal efficiency increases with the increase of the transmissivity of the glass by reducing the heat loss due to radiation, convection, and conduction.
- The maximum outlet water temperature at the mass flow rate of 0.021 kg/s obtained for clear and cloudy sky conditions was 53 and 34 °C, respectively.
- The daily average and pick instantaneous efficiency of the fabricated collector was 44 and 50%, respectively.

References

1. Kalogirou, S.A.: Solar thermal collectors and applications. Prog. Energy Combust. Sci. **30**, 231–295 (2004)
2. Ramsey, J.W., Borzoni, J.T., Holland, T.H.: Development of Flat-Plate Solar Collectors for the Heating and Cooling of Buildings (1975)
3. Ayompe, L.M., Duffy, A.: Analysis of the thermal performance of a solar water heating system with flat plate collectors in a temperate climate. Appl. Therm. Eng. **58**, 447–454 (2013)
4. Whillier, A.: Plastic covers for solar collectors. Sol. Energy **7**, 148–151 (1963)
5. Bilgen, E., Bakeka, B.J.D.: Solar collector systems to provide hot air in rural applications. Renew. Energ. **33**, 1461–1468 (2008)
6. Zambolin, E., Del Col, D.: Experimental analysis of thermal performance of flat plate and evacuated tube solar collectors in stationary standard and daily conditions. Sol. Energy **84**, 1382–1396 (2010)
7. Sekhar, Y.R., Sharma, K.V., Rao, M.B.: Evaluation of heat loss coefficients in solar flat plate collectors. ARPN J. Eng. Appl. Sci. **4**, 15–19 (2009)
8. Selmi, M., Al-Khawaja, M.J., Marafia, A.: Validation of CFD simulation for flat plate solar energy collector. Renew. Energ. **33**, 383–387 (2008)
9. Gunjo, D.G., Mahanta, P., Robi, P.S.: CFD and experimental investigation of flat plate solar water heating system under steady state condition. Renew. Energ. **106**, 24–36 (2017)

10. Martinopoulos, G., Missirlis, D., Tsilingiridis, G., Yakinthos, K., Kyriakis, N.: CFD modeling of a polymer solar collector. Renew. Energ. **35**, 1499–1508 (2010)
11. Facão, J.: Optimization of flow distribution in flat plate solar thermal collectors with riser and header arrangements. Sol. Energy **120**, 104–112 (2015)
12. MacAdams, W.H., McAdams, W.H.: Heat Transmission. McGraw-Hill, New York (1954)
13. Fatigun, A.T., Adesakin, G.E., Gwani, M.: Experimental determination of the effect of tube spacing on the performance of a flat plate solar collector. Int. J. Environ. Sci. **3**, 363–370 (2012)

Utilization of Water Treatment Plant (WTP) Sludge for Pretreatment of Dye Wastewater Using Coagulation/ Flocculation

Y. Shiva Shankar, Kumar Ankur, Prashant Bhushan and Devendra Mohan

Abstract Water treatment plants generate large amounts of sludge during the coagulation and flocculation of raw water that poses a challenge in the disposal. Potential savings in coagulant dosage could be achieved through the recovery of coagulant from sludge or reusing the sludge during the treatment. Current research focuses on the reuse of WTP sludge for the treatment of dyeing wastewater. The study area, i.e. Varanasi, is famous for its Banarasi Saree industries, most of which are small to medium scale in magnitude. Dyes being used have been polluting nearby water bodies such as Ganga, Varuna and Assi. At present, there is an immediate need for the development of low-cost methods to treat the coloured wastewater as commonly adopted methods are costly that cannot provide an economical treatment option for such small- and medium-scale industries. Therefore, use of low-cost materials like WTP sludge for pretreatment could be a feasible option for achieving the desired results after final treatment. The study was conducted using coagulation and flocculation by WTP sludge for the treatment of Acid Red 94, Acid Yellow 1, Direct Green 26 and Reactive Blue 21 dyes because these

Y. Shiva Shankar
Department of Civil Engineering, Ujjain Engineering College, Ujjain, India
e-mail: shivjuet@gmail.com

K. Ankur
Bharat Heavy Electricals Limited, New Delhi, India
e-mail: ankurmec04@gmail.com

P. Bhushan
Department of Ophthalmology, Institute of Medical Sciences, Banaras Hindu University, Varanasi, India
e-mail: dr.prashantbhushan@gmail.com

D. Mohan (✉)
Department of Civil Engineering, Indian Institute of Technology (Banaras Hindu University), Varanasi, India
e-mail: devendra.civ@itbhu.ac.in

A. S. Kalamdhad et al. (eds.), *Advances in Waste Management*,
https://doi.org/10.1007/978-981-13-0215-2_8

were frequently used in dyeing units in and around Varanasi city. The maximum removal for Acid Red 94, Acid Yellow 1, Direct Green 26 and Reactive Blue 21 dyes was found to be at the sieve size of 90–125 μm, and the corresponding values were 41.5, 27, 43.5 and 26.2%, respectively.

Keywords Coagulation/flocculation · Water treatment plant (WTP) sludge
Dyes · Pretreatment · Colour removal

Abbreviations

WTP sludge	Water treatment plant sludge
AR 94	Acid Red 94
RB 21	Reactive Blue 21
DG 26	Direct Green 26
AY 1	Acid Yellow 1

1 Introduction

Dyes are coloured chemicals, consisting of ionizing and aromatic organic compounds having the affinity towards the substrate they are applied. Synthetic dyes have been increasingly used in the textile and dyeing industries because of their ease of application and cost-effectiveness in synthesis, high stability to light, temperature, detergent and microbial attack. More than 10,000 chemically different dyes are being manufactured. The world dyestuff and dye intermediates production are estimated to be around 7×10^8 kg per annum [1]. Synthetic colouring agents are harmful to human beings and animals because of their toxicity carcinogenicity, mutagenicity, teratogenicity and other such characteristics that adversely affect the human health [2–4]. US International Trade Commission classifies dyes into 12 types. Out of these reactive dyes, acid dyes and direct dyes are most commonly used, which have been found to be present in most of wastewater's originating from the textile industries [5].

Several physical, chemical and biological methods are available for the treatments of dye-house effluents. Guendy et al. [6] have carried out the studies for Acid Red dye removal using combination of coagulation and adsorption processes. Alum and ferric chloride were used as coagulants, and granular-activated carbon (GAC) was used as an adsorbent. The results revealed that application of coagulation before adsorption was effective for colour removal which could be further enhanced by decreasing the dye concentration. Daud et al. [7] have examined the decolourization of Acid Red 1 (AR 1) dye using Fe(III) oxide immobilized on Montmorillonite K10 (Fe–MK10) catalyst in the presence of hydrogen peroxide in a batch process. The results indicated that by using 5.0 g/L catalyst dosage of 0.14 wt% Fe–MK10 at pH 2.5 and 16 mM of hydrogen peroxide, 99% of 50 mg/L of AR 1 was found to be decolourized within 150 min.

Daneshvar et al. [8] have conducted experiments using photo-catalysis with zinc oxide instead of titanium dioxide for degradation of Acid Red 14 (AR 14) dye in the presence of UV light. The results indicated that UV/ZnO process could be efficiently used to degrade the AR 14. German et al. [9] have carried out studies on aerobic biodegradation of an azo dye Acid Red 151 (AR 151) in a sequencing batch biofilter packed with a porous volcanic rock piece. Results suggest a colour removal up to 99% for an initial concentration of 50 mg/L of AR 151.

Bahadir et al. [10] have investigated the electrochemical decolourization of textile dye and removal of metal ions using binary mixtures of Acid Blue 29, Reactive Red 2, Acid Red 97, applying iron anodes and Na_2SO_4 electrolyte in a batch electrochemical reactor. Depending on electrochemical reaction conditions, the percentages of decolourization and metal ion removal were observed to lie between 70.6–96.7 and 64.9–100% for textile dye decolourization and metal iron removal, respectively.

Furlan et al. [11] have studied the removal of reactive dyes like Reactive Black 5 (RB 5) and Reactive Orange 16 (RO 16) through a combined treatment with coagulation and adsorption. Activated carbon derived from coconut shells was used as the adsorbent and aluminium chloride, as the coagulant. The removal efficiencies were found to be approximately 90 and 84% for RB 5 and RO 16, respectively.

From the above-mentioned studies, it could be concluded that application of processes like adsorption, photo-catalysis and biofiltration has proved to be successful in the treatment of coloured effluents and the removal was found to be up to 100%. But, application of these methods may not provide an economically feasible solution for small- and medium-scale industries due to limitations such as higher costs and problems in maintaining the operational parameters for the treatment. Therefore, the present study was carried out using coagulation/flocculation technique with WTP sludge as a coagulant. WTP sludge was used due to its low cost and availability to solve the problems mentioned above. The work examines the efficiency of WTP sludge for dye removal, on various classes of dyes commonly used in dyeing process (such as acid dye, reactive dye and direct dye).

1.1 WTP Sludge

Coagulation/flocculation has been one of the most commonly used methods due to its simplicity in application, cost-effectiveness and feasibility to use as a main treatment or pretreatment [12]. The major drawback in application of coagulation and flocculation is the management of generated sludge. Generated sludge is being managed through the direct discharge to water bodies, discharge to sanitary sewers, lagooning, sand bed drying, vacuum filtration, filter pressing, centrifugation, freezing, heating and pressurizing, polyelectrolyte-aided thickening and land disposal. Discharged WTP sludge is mostly rich in pathogens and heavy metals that pollute the surroundings [13–15]. This sludge consists of nearly about 40–50% of unused alum, and therefore, reusing the sludge through the recovery of coagulant, recycling for utilization as a construction material and reusing the processed sludge in wastewater treatment have been observed to be as feasible options as studied by various researchers.

Structural applications

Applications of WTP sludge as building materials have been studied extensively. Various studies have been undertaken for manufacturing the bricks using WTP sludge as a partial replacement in combination with rice husk and fly ash [16–18]. The lightweight bricks manufactured from WTP sludge and rice husk by controlling the optimum sintering temperature and amendment ratios have proved to be a feasible option [16]. Hegazy et al. [18] have studied the substitution of brick clay by water treatment plant sludge incorporated with rice husk ash. Studies were undertaken with three different series of sludge to rice husk ash proportions (25, 50 and 75% of total weight). The results revealed that WTP could be replaced up to 75%, by operating at the temperature commonly reached in brick kiln. Properties of the produced bricks were found to be superior to the conventional clay bricks. Fungaro and Silva [17] have concluded that massive bricks built with the composition of 60% soil, 12% cement, 8% coal fly ash and 20% WTP sludge showed satisfactory results.

Wolff et al. [19] studied the properties of bricks manufactured using waste materials (WTP sludge, dregs, grits and lime mud) from the pulp industry and granite fines. Manufactured bricks using such materials showed higher mechanical resistance along with providing an ecologically feasible solution for managing the wastes. Gastaldini et al. [13] investigated the possibility of using the processed WTP sludge as a mineral addition. WTP sludge was substituted for ordinary Portland cement in concentrations ranging from 5 to 30% in three different water/binder (w/b) ratios (0.35, 0.50 and 0.65). Results proved that use of WTP sludge ash improves the strength of concrete mixes as compared to the concrete produced with rice husk ash or with silica fume.

Environmental applications

Recovery of coagulants from water works sludge using methods such as acidification, alkalization, ion exchange and membrane separation has proved to be beneficial in sludge handling reducing the operating costs of water treatment plant. Studies were carried out for recovery of coagulant using sulphuric acid and hydrochloric acid. The results suggested that maintenance of lower pH has potential for recovery between 70 and 90%. In the case of recovery using alkalization process, the studies were done with sodium hydroxide and calcium hydroxide and the results showed that removal up to 90% was possible at pH of 12 using sodium hydroxide. Effectiveness of these processes could be enhanced by applying membrane, but fouling of the membranes because of particulate matter posed limitation in recovery [14, 20–22]. The studies have also confirmed the presence of soluble substances along with the recovered coagulant that affect the quality of water if used for water treatment. Hence, application of the recovered coagulant was suggested for wastewater treatment [14, 20, 22].

Ishikawa et al. [21] have studied the application of recovered alum using sulphuric acid for the treatment of landfill wastewaters and wastewaters from sewage treatment plants. The results have shown that removal of chemical oxygen demand, total nitrogen and total phosphorous with the recovered coagulant was higher as compared to commercial aluminium sulphate or polyaluminium chloride. Li et al. [23] have investigated the advantage of alum sludge for dewatering the sewage sludge. The results proved that alum sludge from water treatment plant contained residual polyaluminium chloride and relatively large amounts of inorganic matters that improved the sewage sludge dewatering.

Chu [24] has studied extensively dye removal from textile industry wastewater using recycled alum sludge (RAS) by coagulation process. RAS was found to be good material for removing hydrophobic dyes from wastewater, as well as reducing the fresh alum dosage. Results have shown to decrease the fresh alum dosage by one-third fraction.

Moghaddam et al. [25] examined the application of coagulation and flocculation processes using WTP sludge and optimization through response surface methodology for ferric chloride sludge (FCS) for removal of Acid Red 119 (AR 119) from aqueous solutions. Dye removal of 96.53% was observed, and it was found that reusing FCS as a low-cost material in coagulation/flocculation process in wastewater treatment plants can offer certain advantages such as high efficiency for AR 119 dye removal and economic savings on overall treatment plant operation costs.

1.2 Study Area

Varanasi has been known for long as a centre of trade in silk and related products. There is production of Banarasi Sarees (women's fabric) in the sector, in and around Varanasi city. Sant Ravidas Nagar (known earlier as Bhadohi) has been famous internationally for carpet manufacturing industries. Most of these units are operating as small-scale units. All such industrial units generate a large amount of wastewater, containing different types of colouring agents (dyestuff). Since most of these units are in the unorganized sector, they are unable to afford treatment units for such wastewaters. Experimental study was conducted for developing a low-cost industrial waste water treatment option related to garment and carpet manufacturing industries in the region, which commonly use different classes of dyes for colouring the fabrics.

Considering the factors mentioned above, the present study was done for the colour removal of dye solutions prepared with Acid Red 94 (AR 94), Reactive Blue 21 (RB 21), Direct Green 26 and Acid Yellow 1 (AY1), by coagulation/flocculation using WTP sludge as coagulant.

2 Materials and Methods

2.1 *Dyes Used*

Dyes used in the study are Acid Red 94 (AR 94), Reactive Blue 21 (RB 21), Direct Green 26 and Acid Yellow 1 (AY1), and these dyes were selected as they were being used in dyeing units, representing various classes of dyes. The properties of the dyes were shown in Table 1. Acid dyes belong to the class of dye that is mostly applied using an acidic solution. These are water-soluble anionic dyes that are applied to fibres such as silk, wool, nylon and modified acrylic fibres from near neutral to acidic dye baths. Attachment to the fibre is attributed to salt formation between anionic groups in the dyes and cationic groups present in the fibre. In textiles, acid dyes are effective on protein fibres, i.e. animal hair, fibres like wool, alpaca and mohair. These are also effective on silk and nylon [26]. Reactive dyes are coloured compounds with side chains that react to form covalent bonds with the substrate. These can be used to provide very fast (resistant) dyeing of cellulose, etc. Direct dyes are acidic dyes with large molecules that bind to cellulose fibres directly [27].

2.2 *WTP Sludge*

The sludge used in study was obtained from water treatment plant located at Bhelupur in Varanasi city in which polyaluminium chloride is being used as a coagulant. Figure 1 shows the XRD of WTP sludge. According to the observed peaks in XRD, the following compounds were detected: crednerite, copper manganese iron oxide, potassium aluminium oxide, manganese borate hydroxide, aluminium manganese, lead aluminium fluoride, sodium 4-aminosalicylate dehydrate, lead aluminium fluoride and aluminium fluoride. XRD of alum has indicated the presence of metals. Concentrations of metals were analysed using atomic absorption spectrophotometer for WTP sludge processed at different sieve sizes

Table 1 Properties of the dyes [29–32]

Property	Acid Red 94	Acid Yellow 1	Reactive Blue 21	Direct Green 26
C.I. number	45,440	10,316	21	34,045
Molecular formula	$C_{20}H_2Cl_4I_4Na_2O_5$	$C_{10}H_6N_2O_8SNa_2$	$C_{18}H_{15}N_7O_{18}S_{5.5}Na$	$C_{50}H_{33}N_{12}Na_5O_{18}S_4$
Reported wavelength of maximum absorption (λ_{max}) (nm)	548–549	427	624	410
Colour	Red	Yellow	Blue	Green

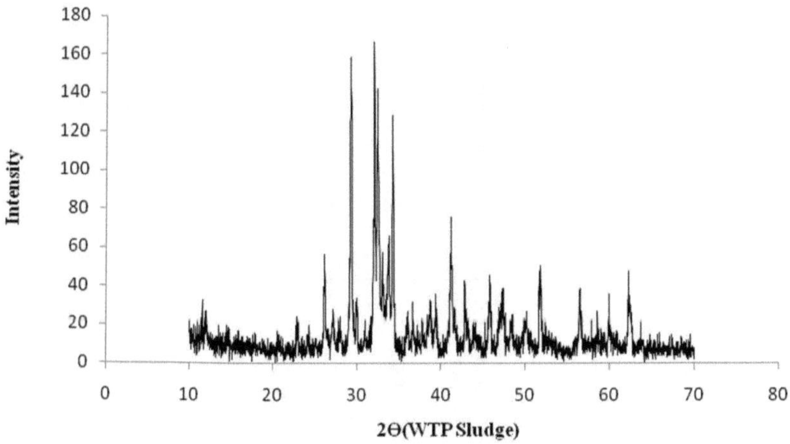

Fig. 1 XRD of WTP sludge

(manufacturer: ECIL, model: AAS4141). The results of the analysis are shown in Table 2. The presence of heavy metals in WTP sludge indicates the adverse impacts of its disposal on the environment.

2.3 Preparation of Dye Solution

Dye solution was prepared by adding 5 grams of dye to 500 mL of the distilled water. After mixing properly, 1 mL of this solution diluted to one litre of distilled water gives a concentration of 10 mg/L of dye.

2.4 Preparation of WTP Sludge Solution

WTP sludge collected from local WTP was subjected to activities like drying at 103 °C for 48 h; followed by grinding, performing sieve analysis and then collecting dry sludge samples at various sieve size range from higher than 75 μm to less than 355 μm. Fifty grams of the sludge samples of various sieve size were taken and dissolved in 500 mL of distilled water. To ensure thorough mixing, these were agitated on a reciprocating shaker for 2 h. After settlement of settleable parts, the supernatant was decanted and used as a coagulant solution for removal of the colour. One mL of this solution will be equivalent to the active ingredient extracted from 100 mg/L of the WTP sludge.

Table 2 Atomic absorption spectrophotometer analysis of WTP sludge

Sample	Heavy metals in WTP sludge (in PPM)					
	Cu (copper) λ_{max}: 324.5 nm (ppm)	Zn (zinc) λ_{max}: 213.5 nm (ppm)	Cd (cadmium) λ_{max}: 228.2 nm (ppm)	Fe (iron) λ_{max}: 247.9 nm (ppm)	Pb (lead) λ_{max}: 216.9 nm (ppm)	Mn (manganese) λ_{max}: 279.3 nm (ppm)
Sample 1 (sieve size: 75–90 μm)	4.75	0.241	2.681	1.560	3.881	2.681
Sample 2 (sieve size: 90–125 μm)	3.17	0	2.11	0.843	2.847	2.204
Sample 3 (sieve size: 125–355 μm)	1.58	0	1.935	0.727	2.15	1.960

2.5 Performing the Jar Test

Jar test was performed using a six glass beakers of one litre capacity each. A stirring device was used that mixed the contents of each jar simultaneously with a uniform power input. Each jar was filled up to one litre mark, and turbidity, alkalinity and pH values of contents of each jar were predetermined. After coagulant dosing, the contents in jar were mixed rapidly for one minute to ensure complete dispersion of the chemicals and subsequently these were mixed slowly for 15–20 min to facilitate the formation of flocs. Then, the contents were allowed to settle for about 30 min or until the clarification occurred (BIS 2001 [28]). Absorbance was measured for further clarified solution (obtained using a centrifuge) by UV–Vis spectrophotometer (Systronics, Model Number–66). Concentrations of the dye solutions were kept constant, i.e. 20, 40, 80 and 50 mg/L for AR 94, AY 1, DG 26 and RB 21, respectively. Concentrations of the dyes were evaluated with respect to the maximum removal at constant alum dosage and varying dye concentration.

2.6 Calculating Percentage Removal

After gravitational settling and centrifuging, the absorbance of the dye solution was determined using spectrophotometer and then corresponding concentrations were determined from respective calibration curves. Percentage removal was calculated using the following formula:

$$\text{Percentage removal} = [(C_{\text{i}} - C_{\text{O}})/C_{\text{O}}] \times 100$$

where C_{i} = initial concentration and C_{O} = final concentration.

3 Results and Discussion

3.1 Acid Red 94 (AR 94)

At optimum conditions for the maximum removal at a dye concentration of 20 mg/L, percentage removal for Acid Red 94 (AR 94) was determined with WTP sludge at sieve sizes ranging between 355 and 75 μm. The maximum percentage removal values were determined with sieve size ranging from 75–90, 90–125 and 125–355 μm. Percentage removal values observed were 34.2% at a dosage of 1300 mg/L, 41.5% at a dosage of 700 mg/L and 23.4% at a dosage of 600 mg/L of WTP sludge in sieve size range of 75–90, 90–125 and 125–355 μm, respectively. The results are shown graphically in Fig. 2a–c. After attaining the maximum removal, it was observed that there was an increase in turbidity and decrease in

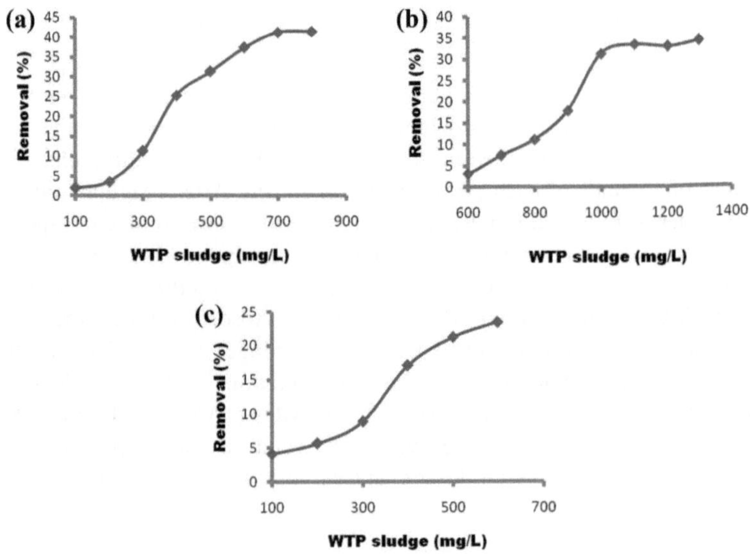

Fig. 2 Optimum coagulant dosage at dye concentration (AR 94) of 20 mg/L and sieve size ranges of **a** 75–90 μm, **b** 90–125 μm and **c** 125–355 μm

percentage removal. Thus, the maximum removal was found to be 41.5% at dosage of 700 mg/L for sieve size range of 90–125 μm.

3.2 *Acid Yellow 1 (AY 1)*

At optimum conditions for the maximum removal at a dye concentration of 40 mg/L, percentage removal for the dye Acid Yellow 1 (AY 1) was determined with WTP sludge at sieve sizes ranging between 355 and 75 μm. Values of the maximum percentage removal with WTP sludge were determined with sieve size ranges of 75–90, 90–125 and 125–355 μm. The values of the maximum removal found were 19.2% at dosage of 1400 mg/L, 27% at a dosage of 1000 mg/L and 10.5% at a dosage of 600 mg/L of WTP sludge in sieve size ranges of 75–90, 90–125 and 125–355 μm, respectively. The results are shown graphically in Fig. 3a–c. After reaching the maximum value of removal, it was observed that there was an increase in turbidity and decrease in the removal. Maximum removal was observed to be 27% at a dosage of 1000 mg/l in sieve size range of 90–125 μm.

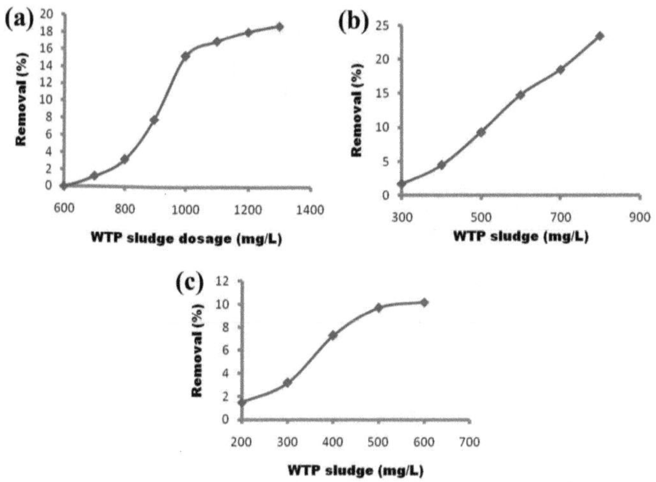

Fig. 3 Optimum coagulant dosage at dye concentration (AY 1) of 40 mg/L and sieve size ranges of **a** 75–90 μm, **b** 90–125 μm and **c** 125–355 μm

3.3 Direct Green 26 (DG 26)

At optimum conditions for the maximum removal at a dye concentration of 80 mg/L, percentage removal for Direct Green (DG 26) was determined with WTP sludge at sieve sizes ranging between 355 and 75 μm. The maximum removal with WTP sludge was determined with sieve size ranges of 75–90, 90–125 and 125–355 μm. Percentage removal reached the maximum up to 22.8% at dosage of 1800 mg/L, 26.2% at a dosage of 1200 mg/L and 17.2% at a dosage of 800 mg/L of WTP sludge in sieve size ranges of 75–90, 90–125 and 125–355 μm, respectively. The results are shown graphically in Fig. 4a–c. After reaching the maximum removal, it was observed that there was an increase in turbidity and decrease in percentage removal. The maximum removal was found to be 26.2% at a dosage of 1200 mg/L for sieve size range of 90–125 μm.

3.4 Reactive Blue 21 (RB 21)

At optimum conditions for the maximum removal at a dye concentration of 50 mg/L, percentage removal for Reactive Blue (RB 21) was determined with WTP sludge at sieve sizes ranging between 355 and 75 μm. The maximum values of percentage removal with WTP sludge were determined with sieve size ranging from 75–90, 90–125 and 125–355 μm. Percentage values of removal were observed up to 39% at dosage of 1300 mg/L, 45.3% at a dosage of 1000 mg/L and 25% at a dosage of 800 mg/L of WTP sludge for sieve size range of 75–90, 90–125 and 125–355 μm,

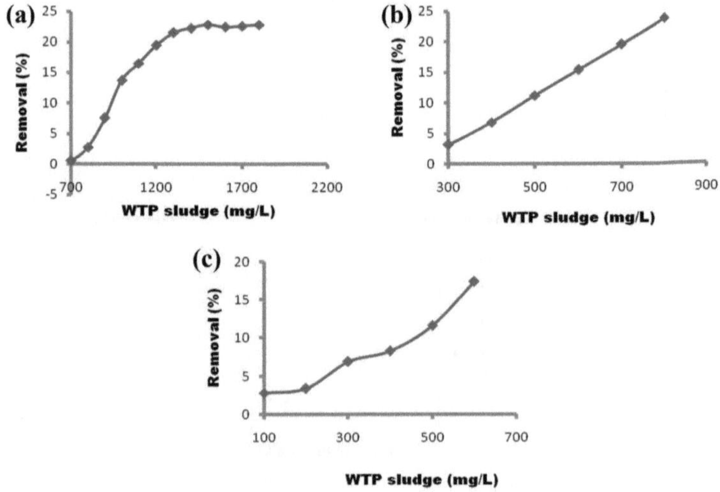

Fig. 4 Optimum coagulant dosage at dye concentration (DG 26) of 80 mg/L and sieve size ranges of **a** 75–90 μm, **b** 90–125 μm and **c** 125–355 μm

respectively. The results are shown graphically in Fig. 5a–c. After attaining the maximum removal, it was observed that there was an increase in turbidity and decrease in percentage removal. The maximum removal was found to be 43.5% at a dosage of 1000 mg/L for sieve size range of 90–125 μm.

3.5 *Discussion*

The dye removal was observed due to the mechanisms such as adsorption and charge neutralization followed by the sweep coagulation. The WTP sludge contained certain recovered coagulants which were responsible for the colour removal. The results indicated that amounts of the coagulant recovered into solution from processed WTP sludge varied with the sieve sizes. Efficiency of WTP sludge at sieve size 75–90 μm was found to be minimum even at higher dosages. The maximum removal was observed with WTP sludge at sieve size 90–125 μm even with the minimum dosage. Addition of WTP sludge increased the colloidal concentration at this sieve size, and this enhanced the removal. Further increase in WTP sludge dosage might have lead to restabilization of the colloids, which was reflected in results (decreased removal) for solution at sieve size 125–355 μm.

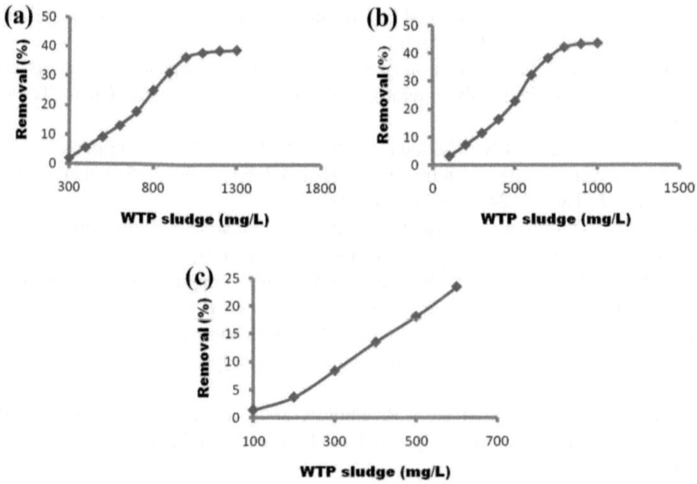

Fig. 5 Optimum coagulant dosage at dye concentration (RB 21) of 50 mg/L and sieve size ranges of **a** 75–90 μm, **b** 90–125 μm and **c** 125–355 μm

4 Conclusions

The experimental studies carried out have proved that coagulation and flocculation using WTP sludge can be effectively used for pretreatment of the effluents from small- and medium-scale industries (garment and carpet manufacturing), situated in and around Varanasi city. It was also observed that the best results were obtained with WTP sludge at sieve size range of 90–125 μm. The maximum values of removal for Acid Red 94, Acid Yellow 1, Direct Green 26 and Reactive Blue 21 were observed for the corresponding dye concentrations of 20, 40, 80 and 50 mg/L. All of these were observed to be at a sieve size range of 90–125 μm, and the values of the corresponding removal were found to be 41.5, 27, 26.2 and 43.5%, respectively. WTP sludge proved to be a more effective coagulant for the treatment of Acid Red 94 and Reactive Blue 21 in comparison with the other two dyes. For further enhancement in colour removal and potential savings in treatment cost, WTP sludge can be used in combination with another suitable treatment method for reducing the coagulant dosage.

References

1. Fatih, D., Sengul, K.: Removal of basic red 46 dye from aqueous solution by pine tree leaves. Chem. Eng. J. **170**, 67–74 (2011)
2. Abdou, R.F., Abdel-Aal, S.A., Waly, E.A., Khalil, Z.H.: Effect of cyanine dyes analogues in breaking dormancy and their mutagenic effects in garlic (*Allium sativum* L.), Ishs acta

horticulture: 208. In: VI International Symposium on Medicinal and Aromatic Plants (2001). http://www.actahort.org/books/208/208_17.html
3. Birhanli, A., Ozmen, M.: Evaluation of the toxicity and teratogenenicity of six commercial textile dyes using the frog embryo teratogenesis assay xenopus. Drug Chem. Toxicol. **1**, 51–65 (2005)
4. Marks, T.A., Gupta, B.N., Ledoux, T.A., Staples, R.E.: Teratogenic evaluation of 2-nitro-P-phenylenediamine, 4-nitro-O-phenylenediamine, and 2, 5-toluenediamine sulfate in the mouse. J. Teratol. **24**, 253–265 (1981)
5. Uygur, A.: An Overview of Oxidative and Photooxidative Decolorisation Treatments of Textile Waste Waters. J. Soc. Dyers Col. **113**, 211–217 (1997)
6. Guendy, H.R.: Treatment and reuse of wastewater in the textile industry by means of coagulation and adsorption techniques. J. Appl. Sci. Res. **6**, 964–972 (2010)
7. Daud, N.K., Ahmad, M.A., Hameed, B.H.: Decolorization of acid red 1 dye solution by fenton-like process using Fe–montmorillonite K10 catalyst. Chem. Eng. J. **165**, 111–116 (2010)
8. Daneshvar, N., Salari, D., Khataee, A.R.: Photocatalytic degradation of azo dye acid red 14 in water on Zno as an alternative catalyst to Tio_2. J. Photochem. Photobiol. A: Chem. **162**, 317–322 (2004)
9. German, B., Maribel, Q., Gloria, M.: Aerobic degradation of the azo dye acid red 151 in a sequencing batch biofilter. Bioresour. Technol. **92**, 143–149 (2004)
10. Bahadir, K.K., Kahraman, A., Cihan, G., Ayla, O.: Electrochemical decolourization of textile dyes and removal of metal ions from textile dye and metal ion binary mixtures. Chem. Eng. J. **173**, 677–688 (2011)
11. Furlan, F.R., Silva, L.G.D.M.D., Morgado, A.F., de Souza, A.A.U., de Souza, S.M.A.G.U.: Removal of reactive dyes from aqueous solutions using combined coagulation/flocculation and adsorption on activated carbon. Resour. Conserv. Recycl. **54**, 283–290 (2010)
12. Verma, A.K., Dash, R.R., Bhunia, P.: A review on chemical coagulation/flocculation technologies for removal of colour from textile wastewaters. J. Environ. Manag. **93**, 154–168 (2012)
13. Gastaldini, A.L.G., Hengen, M.F., Gastaldini, M.C.C., Amaral, F.D., Antolini, M.B., Coletto, T.: The use of water treatment plant sludge ash as a mineral addition. Constr. Build. Mater. **94**, 513–520 (2015)
14. King, P.H., Chen, B.H.H, Weeks, R.K.: Recovery and Reuse of Coagulants from Treatment of Water and Wastewater, Bulletin 77. Virginia Water Resources Research Center, Virginia Polytechnic Institute and State University (1975). https://vtechworks.lib.vt.edu/bitstream/handle/10919/25734/WRRC_Bull_77.pdf?sequence=1
15. Lee, J., Choi, S., Thiruvenkatachari, R., Shim, W., Moon, H.: Evaluation of the performance of adsorption and coagulation processes for the maximum removal of reactive dyes. Dyes Pigm. **69**, 196–203 (2006)
16. Chiang, K.Y., Chou, P.H., Chien, J.L., Wu, C.C.: Novel lightweight building bricks manufactured from water treatment plant sludge and agricultural waste. J. Residuals Sci. Technol. **6**(4), 185–191 (2009)
17. Fungaro, D.A., Silva, M.V.: Utilization of water treatment plant sludge and coal fly ash in brick manufacturing. Am. J. Environ. Prot. **2**(5), 83–88 (2014)
18. Hegazy, B.E.E., Fouad, H.A., Hassanain, A.M.: Brick manufacturing from water treatment sludge and rice husk ash. Aus. J. Basic. Appl. Sci. **6**(3), 453–461 (2012)
19. Wolff, E., Schwabe, W.K., Conceicao, S.V.: Utilization of water treatment plant sludge in structural ceramics. J. Cleaner Prod. **96**, 282–289 (2015)
20. Evuti, A.M., Lawal, M.: Recovery of coagulants from water works sludge: a review. Adv. Appl. Sci. Res. **2**(6), 410–417 (2011)
21. Ishikawa, S., Ueda, N., Okumura, Y., Iida, Y., Baba, K.: Recovery of coagulant from water supply plant sludge and its effect on clarification. J. Mater. Cycles Waste Manag. **9**(2), 167–172 (2007)

22. Joshi, S., Shrivastava, K.: Recovery of alum coagulant from water treatment plant sludge: a greener approach for water purification. Int. J. Adv. Comput. Res. **1**(2), 101–103 (2011)
23. Li, J., Liu, L., Liu, J., Ma, T., Yan, A., Ni, Y.: Effect of adding alum sludge from water treatment plant on sewage sludge dewatering. J. Environ. Chem. Eng. **4**(1), 746–752 (2016)
24. Chu, W.: Dye removal from textile dye wastewater using recycled alum sludge. Water Res. **35**, 3147–3152 (2001)
25. Moghaddam, S.S., Moghaddam, M.R.A., Arami, M.: Coagulation/flocculation process for dye removal using sludge from water treatment plant: optimization through response surface methodology. J. Hazard. Mater. **175**, 651–657 (2010)
26. Ozer, A., Akkaya, G., Turabik, M.: The removal of acid red 274 from wastewater: combined biosorption and biocoagulation with *Spirogyra rhizopus*. Dyes Pigm. **71**, 83–89 (2006)
27. Kiernan, J.A.: Classification and naming of dyes, stains and fluorochromes. Biotech. Histochem. **76**, 261–278 (2001)
28. Bureau of Indian Standards (BIS). IS 3025 (Part 50), Methods of Sampling and Tests (Physical and Chemical) for Water and Waste Water, Jar Test (2001)
29. Aldrich Chemical Catalogue, Aldrich Chemical Company, Milwaukee, W.I., U.S.A. Details Available Online at http://www.sigmaaldrich.com/india.html (1992)
30. Gurr, E.: Synthetic Dyes in Biology: Medicine and Chemistry. Academic Press, London, England (1971)
31. SCCP (Scientific Committee on Consumer Products), Opinion on Acid Yellow 1 (2008). http://ec.europa.eu/health/ph_risk/committees/04_sccp/docs/sccp_o_155.pdf
32. World Dye Variety. http://www.worlddyevariety.com/direct-dyes/direct-green-26.html

Comparison of Plant Tolerance Towards Air Pollution of Rural, Urban and Mine Sites of Jharkhand: A Biochemical Approach to Identify Air Pollutant Sink

Shuchita Jain, Tanushree Bhattacharya and Sukalyan Chakraborty

Abstract Air pollution is one of the most threatening challenges faced by humans. Rapid industrialization coupled with increase in vehicular emissions is responsible for degradation of air quality. Plants can be effectively used as bio-indicators for air quality. Air pollution tolerance index (APTI) is a parameter to assess the tolerance or sensitivity of a plant towards pollution. Ten common plant species have been selected from rural, urban and mine sites of Jharkhand. ANOVA showed no significant variation in APTI between different sites for monsoon and winter season at 0.5 level of significance. *t*-test showed significant seasonal variation in biochemical parameters of different species of a particular site. Anticipated performance index of the plants has also been calculated to assess the overall performance of a plant in a particular region. *Azadirachta indica* and *Mangifera indica* were found to be the good performers amongst the selected plants and *Artocarpus heterophyllus* to be the poor performer.

Keywords APTI · API · Biochemical parameters · Bio-monitor ANOVA

S. Jain (✉) · T. Bhattacharya · S. Chakraborty
Department of Civil and Environment Engineering, Birla Institute of Technology Mesra, Ranchi 835215, Jharkhand, India
e-mail: shuchi_567@yahoo.co.in

T. Bhattacharya
e-mail: tbhattacharya@bitmesra.ac.in

S. Chakraborty
e-mail: sukalyanchakraborty@bitmesra.ac.in

A. S. Kalamdhad et al. (eds.), *Advances in Waste Management*,
https://doi.org/10.1007/978-981-13-0215-2_9

1 Introduction

Air pollution can be defined as the introduction of chemicals, particulate matter and other biological matter into the atmosphere in such quantities and for such time duration that it causes harm and discomfort to humans, animals, plants and the surrounding environment [1]. Plants are an integral part of the ecosystem and are most likely to be affected by air pollution. Plants that are exposed to air pollution for an extended period of time show slow growth rate, and its physiology is affected thereby affecting the entire metabolism. Plants act as an effective indicator of air pollution. They absorb the gaseous air pollutants and act as natural cleansing agents of the environment. Bio-monitoring of air pollution by vegetation is a cost-effective method. The susceptibility of plants to pollution varies from species to species and can be measured by its biochemical parameters [2]. Air pollution tolerance index (APTI) and anticipated performance index (API) are two indexes which are useful to assess pollution tolerance of plants. The vegetation in urban areas has gained significance as it is an important indicator of air quality. Depending on the resistance of particular species to air pollutants, the level of pollution can be regulated by planting those specific plants in the affected area. It is possible that plants may store and absorb the pollutants and later release them. These processes include chemical transformation, accumulation and incorporation in the metabolic system [3]. Air pollution tolerance index (APTI) and anticipated performance index (API) methodology can be applied in any place in the world as they are based on biochemical parameters and universally relevant biological and socio-economic characters. Selected tree species are evaluated for biological and socio-economic parameters with various biochemical factors. These parameters are subjected to a grading scale to determine the anticipated performance index (API) of tree species [4]. This type of study will be essentially helpful for plantation purposes in urban and rural development programmes and green belt development. Plants with high APTI values indicate more tolerance towards pollution and hence can serve as sink of pollutants. Finally, plants also serve as bio-monitors of air quality which can be reflected from the APTI value of selected plants, without installing expensive monitoring equipment.

2 Materials and Methods

2.1 Sampling Sites

Jharkhand being a forest rich region has abundance of plant diversity. So far, air pollution tolerance index of plants of this region has not been reported. Three sites, namely Birla Institute of Technology campus in Mesra (rural site with lots of forest cover), Doranda region in Ranchi (urban site with high traffic density and one of the

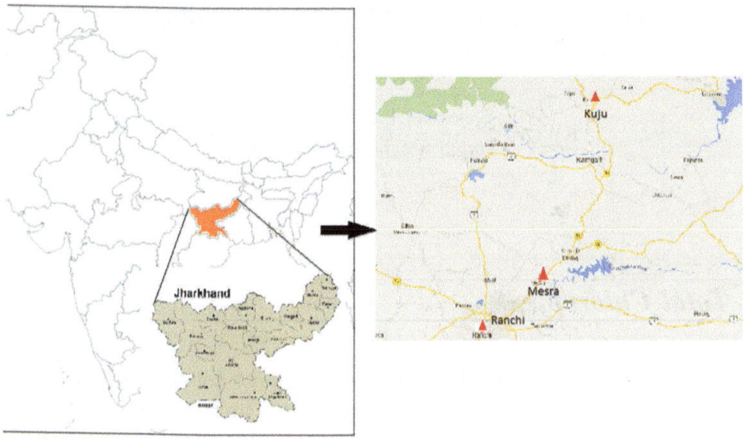

Fig. 1 Location of selected sites

main commercial hub of the city) and Kuju region (coal mine site in Hazaribagh) have been taken for the study. The biochemical parameters studied were ascorbic acid, relative water content, pH, total chlorophyll. Reducing sugar and protein have also been studied for the winter season to conform the results (Fig. 1).

Leaves of ten common plant species were collected from all the three sites in early morning hours. The mature leaf samples of the most common plants in all the three regions such as *Mangifera indica* (mango), *Ficus benghalensis* (bargad), *Dalbergia sissoo* (sissoo), *Shorea robusta* (sal), *Azadirachta indica* (neem), *Ficus religiosa* (peepal), *Bauhinia* sp. (kachnar), *Cassia fistula* (amaltas), *Neolamarckia cadamba* (kadam) and *Artocarpus heterophyllus* (jackfruit) were taken. *D. sissoo* and *N. cadamba* were unavailable at the mine site, and the former was unavailable in the winter season due to leaf shedding.

Samples were immediately kept in zipper packets to avoid moisture loss. The collected samples were analysed for ascorbic acid, chlorophyll, relative water content and leaf extract pH. The mean values of different parameters were used for computing the index.

The various biochemical parameters are analysed by the following procedure given below.

2.2 *Relative Water Content*

Relative water content (RWC) was determined by following the method of Sen and Bhandari [5]. Individual leaves of different plant species were cleaned and weighed immediately. They were dipped into water in a beaker overnight. Next day, the leaves were blotted dry and reweighed before being dried at 80 °C for 24 h and reweighed. The following formula was used to calculate the relative water content

$$\text{RWC} = \frac{\text{Fresh weight} - \text{Dry weight}}{\text{Turgid weight} - \text{Dry weight}} * 100.$$

2.3 Total Chlorophyll

Total chlorophyll was estimated by the method given by Arnon [6]. For analysis, 1-g-fresh leaves material was grounded to paste in a mortar and then extracted with 20 ml of 80% acetone. The liquid portion was decanted into another test tube and centrifuged at 2500 rpm for 5 min. The supernatant was then collected, and the absorbance taken at 645 and 663 nm using a spectrophotometer. Calculations were done using the formula below

$$\text{Chlorophyll a} = \frac{(12.7 * D663 - 2.69 * D645) * V}{1000 * W}\ \text{mg/g}$$

$$\text{Chlorophyll b} = \frac{(22.9 * D645 - 4.68 * D663) * V}{1000 * W}\ \text{mg/g}$$

$$\text{Total Chlorophyll} = \frac{(20.2 * D645 + 8.02 * D663) * V}{1000 * W}\ \text{mg/g}$$

where

D_x Absorbance of the extract at the wavelength x nm.
V Total volume of the chlorophyll solution (ml).
W Weight of the leaf extracted (g).

2.4 Leaf Extract pH

1 g of leaf material was ground to paste and dissolved in 50 ml of deionized water, and leaf extract pH was measured by using calibrated digital pH metre [7].

2.5 Ascorbic Acid Content

The ascorbic acid content (mg/g) was determined using the modified calorimetric 2,6-dichlorophenol indophenol method given by Sadasivam [8]. 1-g sample is extracted in 4% oxalic acid and made up to a known volume (20 ml) and centrifuged at 4500 rpm for 5 min. 5 ml of this supernatant is pipette out, and 10-ml oxalic acid is added and titrated against the dye.

$$\text{Ascorbic Acid} = \frac{0.5\ \text{mg} * V_2 * 20\ \text{ml}}{V_1\ \text{ml} * 5\ \text{ml} * \text{wt of the sample}}\ \text{mg/g}$$

where

V_1 Volume of dye titrated against the working standard.
V_2 Volume of dye titrated against the sample.

2.6 Reducing Sugar Estimation

The reducing sugar content of the samples was determined by dinitrosalicylic acid method as developed by Miller [9]. In this method, the sugar was first extracted from 100 mg of the sample with hot 80% ethanol twice (5 ml each time). The contents were cooled, and the intensity of the dark red colour was recorded at 510 nm.

2.7 Protein Estimation

"Lowry's method" was used for protein determination [10]. 500 mg of leaf sample was extracted with 10 ml of 0.2M phosphate buffer (pH—7.0). The extract was centrifuged and the supernatant collected. Folin–ciocalteu reagent was used for colour development. Blue colour is developed, and the readings are taken at 660 nm.

2.8 Statistical Analysis

t-test was done in order to find the seasonal variation of the biochemical parameters of each species. ANOVA was done using Excel Analysis ToolPak version 15.

Quality control was achieved by using analytical grade chemicals and by estimating the parameters in triplicates. Double distilled water was used for preparation of the reagents.

2.9 Air Pollution Tolerance Index (APTI)

APTI of a species was determined by using the formula established by Singh and Rao [11].

$$\mathrm{APTI} = \frac{A(T+P)+R}{10}$$

where

A Ascorbic acid content (mg/g).
T Total chlorophyll (mg/g).
P Leaf extract pH.
R Relative water content (%).

2.10 Anticipated Performance Index

By combining the resultant APTI values with some significant biological and socio-economic characters (plant habit, canopy structure, type of plant, economic value and laminar structure), the API was calculated for different species. Based on these characteristics, different grades (+ or −) are allotted to plants. Different plants are scored according to their grades. The criteria used for calculating the API of various plant species are done by referring to the method given by Prajapati [3]. The percentage scoring based on which API grades are calculated is given by the formula

$$\%\ \mathrm{Scoring} = \frac{\mathrm{Total\ plus}}{16} \times 100$$

Maximum number of grades that can be allotted to a plant = 16.

3 Results and Discussions

Ambient air quality characteristics for all the three regions are different. The average (24 h) SO_2 concentration for rural, mine and urban site are 2.27, 11.11 and 37.71 μg/m^3, respectively. The average (24 h) NO_x concentration values for rural, mine and urban site are 13.86, 23.44 and 68.7 μg/m^3, respectively. $PM_{2.5}$ has an average (24 h) value of 7.67, 178.25 and 21.25 μg/m^3 for rural, mine and urban site, respectively, and PM_{10} has an average (24 h) value of 10.56, 230.75 and 37.71 μg/m^3 for rural, mine and urban site, respectively. The NAAQS values for industrial, residential and rural sites for SO_2 is 50 μg/m^3, NO_x is 40 μg/m^3, PM_{10} is 60 μg/m^3 and $PM_{2.5}$ is 40 μg/m^3. The values for NO_x for urban region are beyond the permissible limit. PM_{10} and $PM_{2.5}$ values for mine site are very high and exceed the permissible limit.

3.1 Relative Water Content

Moisture content for all the plants have shown higher values in the monsoon season and lower in the winter season (refer Figs. 2, 3 and 4). This can be explained by the high water content in the soil and lower rate of transpiration from leaves in monsoon season. High water content in plants facilitates the plant to maintain its physiological balance under high-stress conditions [12]. The values range from 70 to 97% amongst the various species. Higher values of water content indicate more tolerant plants as they are more resistant to extreme conditions [13].

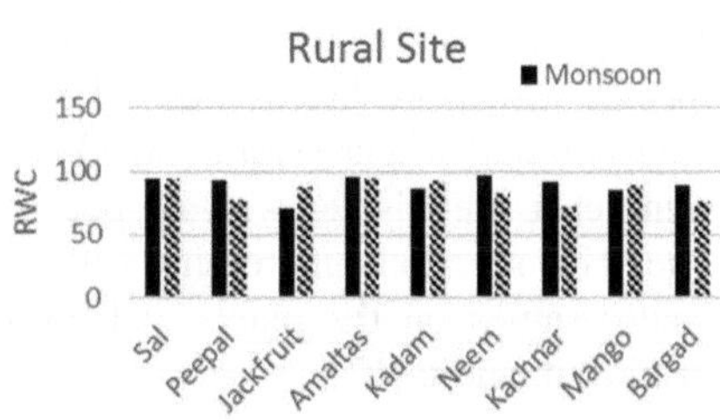

Fig. 2 Rural site RWC values

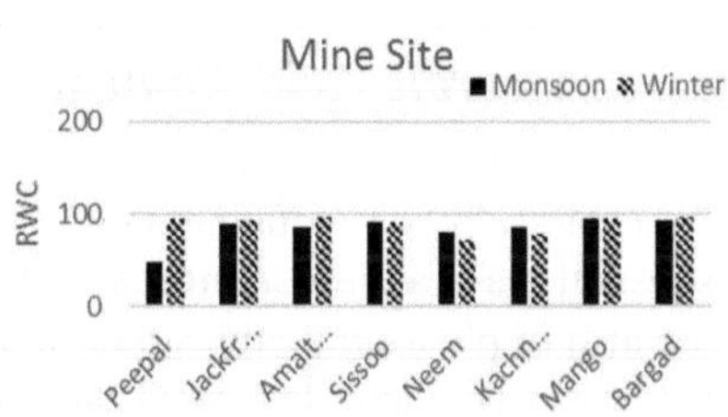

Fig. 3 Mine site RWC values

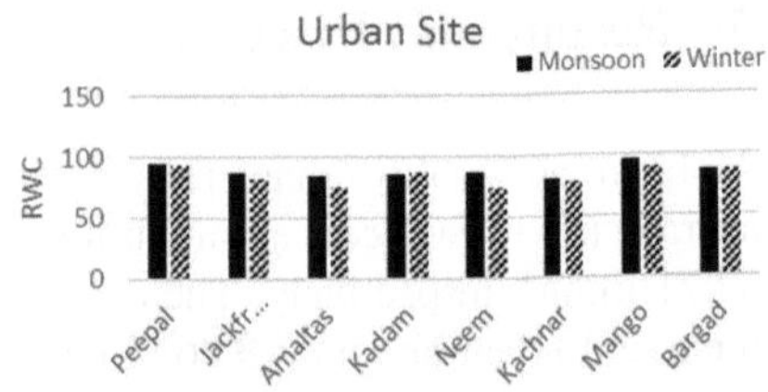

Fig. 4 Urban site RWC values

For the rural site, maximum values were shown by *A. indica* (95.2%) and *C. fistula* (97.03%) while for the mine site *M. indica* (95.28%) and *F. benghalensis* (93.18%) showed the highest values. *M. indica* (96.13%) and *F. religiosa* (95.52%) showed the highest values in urban site exhibiting higher tolerance.

3.2 Leaf Extract pH

Leaf extract pH values for all the sites increased in the winter season (refer Figs. 5, 6 and 7), the slightest increase indicated by *M. indica* (5.74–6.16) and *A. indica* (6.95–7.04) for rural site; *M. indica* (5.96–6.61) and *A. indica* (6.44–6.98) in mine site, whereas in the urban site, the slightest increase was exhibited by *M. indica* (5.72–6.71) and *A. indica* (6.55–7.03). The values indicated increasing pH in the winter season owing to the fact that in winter season there are stable atmospheric conditions which inhibits dispersion and makes more SPM settle on the leaves. During the monsoon season, the same SPM is washed off lowering the pH value [14]. The increase in pH is supported by the findings of Das [15] for *A. indica* (5.3–6.6) and *M. indica* (4.6–5.3). Scholz and Reck [16] reported that the higher decline in pH indicates sensitive species. Hence, minimum decline in pH indicates tolerant species which are listed above. Leaf extract pH showed significant variation in monsoon and winter season. ANOVA ($F_{\text{critical}} = 2.99$, $F = 7.91$) results indicated significant site-wise as well as species-wise variation at 0.05 level of significance. The pH is seen to be increasing with increased pollution. The suspended particulate matter settles on the plants and consequently dissolves in the cell sap thereby making it acidic [17].

3.3 Ascorbic Acid Content

Ascorbic acid is a reducing agent which is activated as soon as the plant is under stress. It plays an important role in the cell wall synthesis, defence mechanism and the also in photosynthetic carbon fixation. The ascorbic acid is the most important factor in calculating the APTI due to its pivotal role in defence and detoxification mechanism and thus used as a multiplication factor in the formula [14].

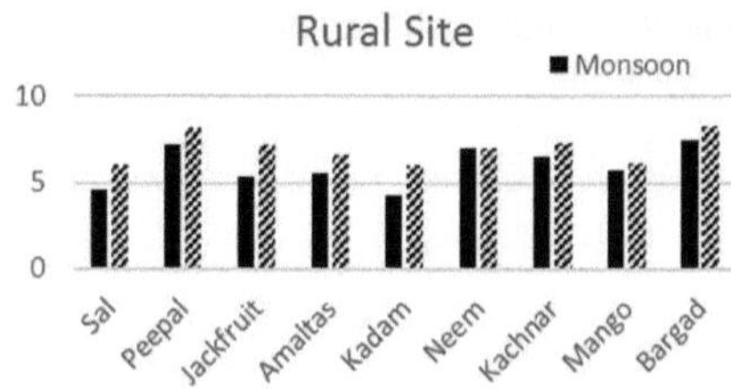

Fig. 5 Leaf extract pH for rural site

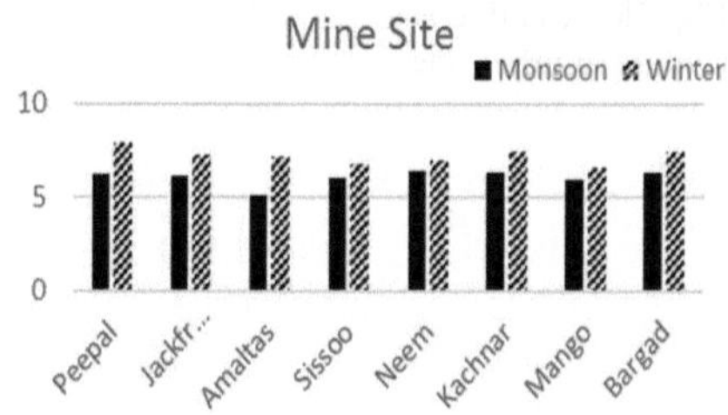

Fig. 6 Leaf extract pH for mine site

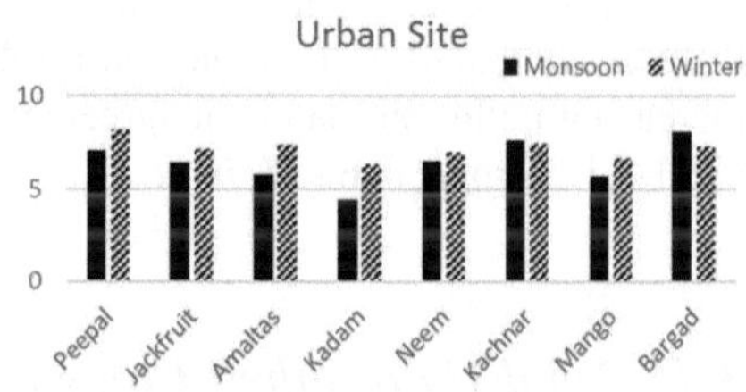

Fig. 7 Leaf extract pH for urban site

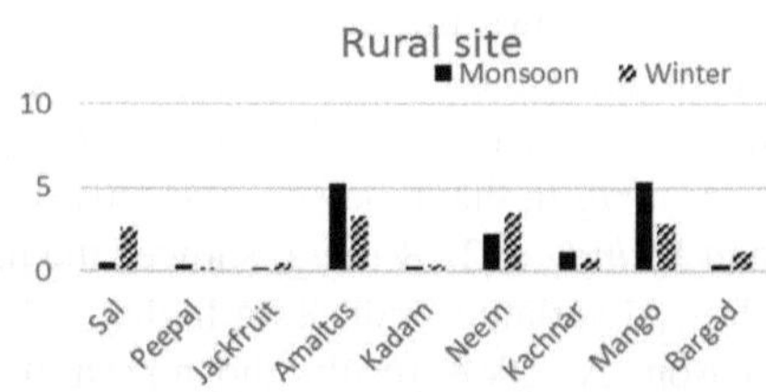

Fig. 8 Ascorbic acid content in rural site

The highest ascorbic acid values have been shown by *C. Fistula*, *A. indica* and *M. indica* in all the three sites (Figs. 8, 9 and 10). As pollution load is the lowest in the rural site, the ascorbic acid values have also shown lower values—5.27, 2.23, 5.33 mg/g, respectively. The maximum values for these plants in the urban site being 21.48, 7.48 and 9.71 mg/g, respectively. The mine site values have shown lower values indicating that these plants are not able to perform very well in that area owing to the very heavy pollution load with the values being 3.91, 2.25 and 2.85 mg/g, respectively. Ascorbic acid shows a positive correlation with SO_2 and NO_x. The production of reactive oxygen species (ROS) under pollution stress increases the production of ascorbic acid, which in turn prevents plants from oxidative stress by

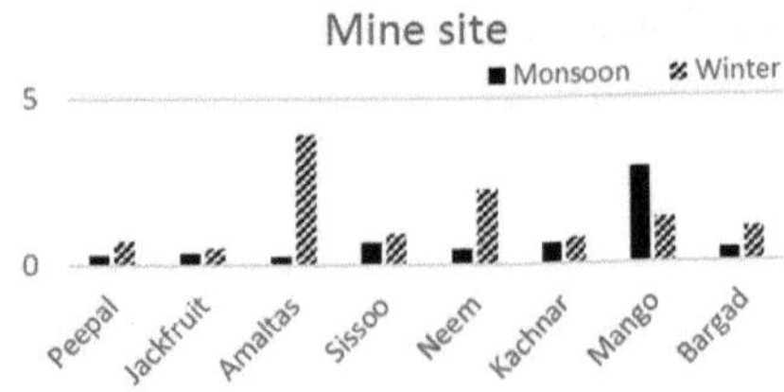

Fig. 9 Ascorbic acid content in mine site

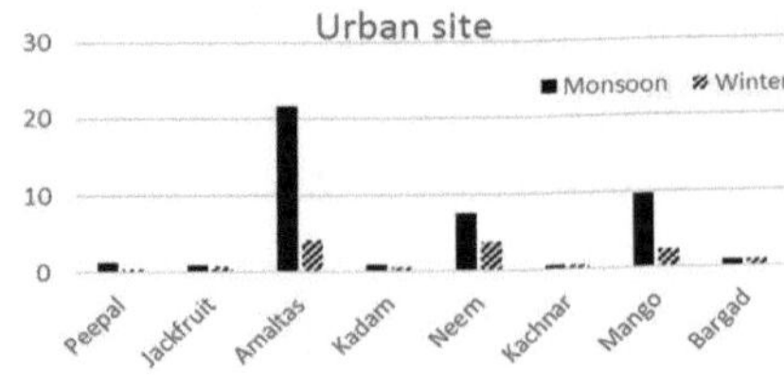

Fig. 10 Ascorbic acid content in urban site

scavenging the ROS [18]. This might be one of the possible reasons of increased ascorbic acid levels in urban site as the SO_2 and NO_x concentration are much higher there as compared to the mine and rural site. Increasing ascorbic acid content with increasing pollution load is in agreement with the findings of [16, 19] for *A. indica* (5.91–11.33 mg/g) and *M. indica* (0.69–7.5 mg/g).

3.4 Total Chlorophyll Content

Table 1 shows the chlorophyll values in both monsoon and winter season. The maximum values have been shown by *C. fistula* and *A. indica* in all the rural and mine sites with the values being 4.16 and 3.51 mg/g in the rural site, 1.21 and 5.89 mg/g in the mine site, respectively. In the urban site *Bauhinia* sp. (2.71 mg/g) and *M. indica* (2.78 mg/g) showed the maximum values. *t*-test for seasonal variation showed variation indicating higher values in monsoon season and lower in winter season. ANOVA results did not show any significant variation species-wise and site-wise for monsoon and winter season. The lower values in winter season are due to the fact that dust is accumulated on the moist leaf surface during winter thereby inhibiting chlorophyll synthesis. Higher total chlorophyll values for monsoon season is supported by the findings of [3, 16, 20] for *M. indica* (1.49–2.41 mg/g), *C. fistula* (3.87 mg/g) and *A. indica* (6.8 mg/g) Also, the chlorophyll content was seen to be decreasing for the mine and urban site. This is because of the fact that chloroplast is the attacked by the air pollutants thereby decreasing its synthesis [21]. Higher chlorophyll values are an indicator of tolerance in plants [22].

Table 1 Total chlorophyll content (mg/g) for the species in all sites

Common name	Rural site		Mine site		Urban site	
	Monsoon	Winter	Monsoon	Winter	Monsoon	Winter
Sal	1.09	1.09	NA	0.67	0.79	NA
Peepal	2.126	1.34	2.41	1.79	1.35	1.63
Jackfruit	2.3	1.48	1.14	1.72	1.93	1.52
Amaltas	4.156	1.22	1.21	1.58	2.42	1.91
Sissoo	0.86	NA	1.37	1.22	2.62	NA
Kadam	1.56	1.70	NA	NA	2.22	1.35
Neem	3.39	1.00	5.89	1.85	1.38	1.02
Kachnar	1.93	1.34	1.61	1.45	2.71	1.77
Mango	2.53	1.10	2.99	1.54	2.78	1.63
Bargad	2.94	1.44	2.13	1.28	1.21	1.35

3.5 *Reducing Sugar and Protein Content*

Reducing sugar is a source of energy for plants and is indicative of the physiological activity of plants [21]. Studies reveal that reducing sugar is accumulated in plants due to increase in pollution stress [23]. The values increased with increasing pollution load for *F. religiosa*, *C. fistula*, *F. benghalensis*, *N. cadamba*, *S. robusta and M. indica* (refer Figs. 11 and 12) in mine and urban site as compared to the rural site. Increased amounts in polluted sites in this study suggest that the plant is under stress. ANOVA ($F_{\text{critical}} = 3.88$, $F = 5.10$) results for reducing sugar show significant variation site-wise. Protein shows increasing and decreasing trends when plants are under stress, and it depends upon the inherent capacity of the species [22]. In this study, the protein content showed increasing trend with increasing pollution load. One possible reason to this can be proline accumulation. Proline is accumulated when plants are subjected to stress. Due to this, protein degradation is slowed down [24]. Higher protein and reducing content in polluted site are also reported by Bakiyaraj and Ayyappan [15].

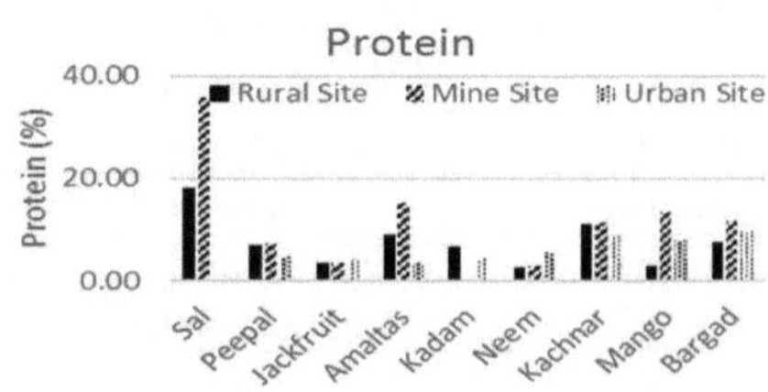

Fig. 11 Protein content (mg/g) in winter season

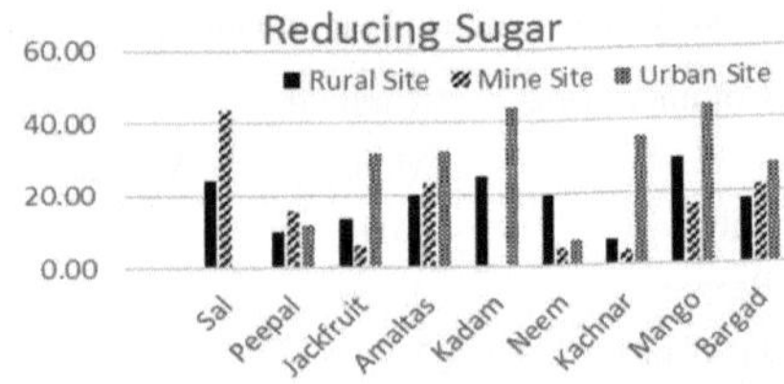

Fig. 12 Reducing sugar (mg/g) in winter season

3.6 *Air Pollution Tolerance Index*

t-test showed significant seasonal variation in the APTI values as shown in Table 7. The values are seen to be increasing with the increase in the pollution load. The APTI value is highest for *C. fistula* and *M. indica* and lowest for *A. heterophyllus* and *Bauhinia* sp. The values for the APTI have been shown in Table 2. Higher APTI values with increase in pollution are in agreement with the researches of [19, 22, 25].

3.7 *Anticipated Performance Index*

API is a broader term which takes into consideration not only the biochemical parameters but also the socio-economic parameters giving a holistic perspective to it. Various plant parameters such as the leaf size, canopy structure also regulate the plant capacity to abate pollution. Greater leaf surface area expedites dust interception by direct impaction, and rough texture facilitates the dust particles to be captured by the leaves making the plant more efficient to function as a pollutant sink. According to the average APTI values of winter and monsoon season, the API values for the plants have been calculated for the urban and mine site. Table 6 shows which plant performs better which may be used for plantation purpose for

Table 2 APTI values of plants at various sites

Common name	Scientific name	Rural site	Mine site	Urban site
Sal	*Shorea robusta*	10.49	10.86	9.46
Peepal	*Ficus religiosa*	11.69	10.15	6.97
Jackfruit	*Artocarpus heterophyllus*	8.47	9.04	9.42
Amaltas	*Cassia fistula*	12.55	18.79	10.25
Sissoo	*Dalbergia sissoo*	9.50	9.05	9.74
Kadam	*Neolamarckia cadamba*	8.58	9.16	NA
Neem	*Azadirachta indica*	11.56	12.66	8.80
Kachnar	*Bauhinia* sp.	9.03	8.49	8.84
Mango	*Mangifera indica*	11.95	13.50	11.31
Bargad	*Ficus benghalensis*	8.79	9.59	10.12

developing an effective green belt around these highly polluted regions [22] (Tables 3, 4 and 5).

Table 6 shows the plant performance at various sites. *M. indica* is seen to be the best performer amongst the selected plant species. *A. indica*, *F. religiosa* and *N. cadamba* are good performers. It can be seen from the result that the same plant is performing differently in different sites which is due to the pollution characteristics of the area. As the mine site is very heavily polluted, the only good performer seen here is *M. indica.* More plant species can be assessed for this area apart from the selected ones for better performance.

From the study, it is evident that the overall performance of a plant does not depend on the biochemical parameters alone but the socio-economic parameters also play an important role. For example, *C. fistula* has the highest APTI values for both the regions but due to its socio-economic parameters, it has been categorized as a poor, very poor and moderate and performer in rural, mine and urban site, respectively. From Table 6, it can be seen that for the rural site *M. indica* and *F. religiosa* are the best performers and for the mine and urban site *M. indica* is the best performer and is recommended for plantation.

3.8 Statistical Analysis

Seasonal variation for the parameters has been statistically evaluated by the *t*-test (refer Table 7). All the parameters have shown significant variation. The parameter-wise seasonal variation has been explained earlier for all the parameters. APTI values have also shown significant variation. The APTI has increased for almost all the plants in the mine site and the rural site. The urban site values have shown decreasing trend from monsoon to winter season. *M. indica*, *Bauhinia* sp. and *C. fistula* exhibited higher values in the monsoon season than in winter season. All the other plant species showed higher values in the winter season owing to increased pollution concentration due to stable atmospheric conditions.

4 Conclusions

This study concluded that for any region APTI and API can prove to be very effective for development of green belt and mitigating the effect of pollution depending on the pollution characteristics of the area. APTI value was highest for *C. fistula* and *M. indica* and lowest for *A. heterophyllus* and *Bauhinia* sp. API assessment showed *M. indica* as the best performer and *A. indica*, *F. religiosa* and *N. cadamba* as good performers in all the three sites. The above species are recommended for social forestry, mine reclamation projects and urban planning as these have high APTI values as well as good socio-economic uses. *C. fistula* has high APTI value but due to its poor socio-economic parameters (type of plant,

Table 3 Evaluation of plant species based on APTI and socio-economic parameters at the rural site

Scientific name	APTI	Plant habit	Canopy structure	Type of plant	Size	Texture	Hardiness	Economic value	Total	% scoring
Ficus religiosa	+	++	+	+	+	+	+	++	10	62.5
Artocarpus heterophyllus	–	+	+	+	+	–	–	++	6	37.5
Cassia fistula	++	+	+	–	+	–	+	+	7	43.75
Dalbergia sissoo	+	++	++	–	–	–	–	+	6	37.5
Ficus benghalensis	–	++	+	+	++	–	+	+	8	50
Mangifera indica	+	++	++	+	+	+	+	++	11	68.75
Bauhinia sp.	+	+	+	+	+	+	+	++	9	56.25
Azadirachta indica	+	++	++	+	–	–	+	++	9	56.25
Shorea robusta	+	++	+	–	+	–	–	++	7	43.75
Neolamarckia cadamba	–	++	++	+	++	–	–	++	9	56.25

Table 4 Evaluation of plant species based on APTI and socio-economic parameters at the mine site

Scientific name	APTI	Plant habit	Canopy structure	Type of plant	Size	Texture	Hardiness	Economic value	Total plus	% scoring
Ficus religiosa	–	++	+	+	+	+	+	++	9	56.25
Artocarpus heterophyllus	+	+	+	+	+	–	–	++	7	43.75
Cassia fistula	+	+	+	–	+	–	+	+	6	37.5
Dalbergia sissoo	+	++	++	–	–	–	–	+	6	37.5
Ficus benghalensis	+	++	+	+	++	–	+	+	9	56.25
Mangifera indica	+	++	++	+	+	+	+	++	11	68.75
Bauhinia sp.	–	+	+	+	+	+	+	++	8	50
Azadirachta indica	–	++	++	+	–	–	+	++	8	50
Shorea robusta	+	++	+	–	+	–	–	++	7	43.75

Table 5 Evaluation of plant species based on APTI and socio-economic parameters at the urban site

Scientific name	APTI	Plant habit	Canopy structure	Type of plant	Size	Texture	Hardiness	Economic value	Total plus	% scoring
Ficus religiosa	+	++	+	+	+	+	+	++	10	62.5
Artocarpus heterophyllus	+	+	+	+	+	–	–	++	7	43.75
Cassia fistula	++++	+	+	–	+	–	+	+	9	56.25
Dalbergia sissoo	+	++	++	–	–	–	–	+	6	37.5
Ficus benghalensis	+	++	+	+	++	–	+	+	9	56.25
Mangifera indica	++	++	++	+	+	+	+	++	12	75
Neolamarckia cadamba	+	++	++	+	++	–	–	++	10	62.5
Bauhinia sp.	–	+	+	+	+	+	+	++	8	50
Azadirachta indica	++	++	++	+	–	–	+	++	10	62.5
Shorea robusta	+	++	+	–	+	–	–	++	7	43.75

Table 6 Anticipated performance index of studied plant species

Scientific name	API grade			Assessment category		
	Rural site	Mine site	Urban site	Rural site	Mine site	Urban site
Shorea robusta	2	2	2	Poor	Poor	Poor
Ficus religiosa	4	3	4	Good	Moderate	Good
Artocarpus heterophyllus	1	2	2	Very poor	Poor	Poor
Cassia fistula	2	1	3	Poor	Very poor	Moderate
Dalbergia sissoo	1	1	1	Very poor	Very poor	Very poor
Azadirachta indica	3	3	4	Moderate	Moderate	Good
Bauhinia sp.	3	3	2	Moderate	Moderate	Poor
Mangifera indica	4	4	5	Good	Good	Very good
Ficus benghalensis	2	3	3	Poor	Moderate	Moderate
Neolamarckia cadamba	3	–	4	Moderate	–	Good

Table 7 *t*-test for seasonal variation of the parameters

Biochemical parameter	Rural site		Mine site		Urban site	
	t	*t* Stat	*t*	*t* Stat	*t*	*t* Stat
Ascorbic acid	2.119905	−0.08876	2.13145	−1.43897	2.228139	1.371672
Total chlorophyll	2.22813885	2.979756	2.306004	1.562794	2.1788128	1.69529083
RWC	2.20098516	−0.75955	2.2281388	−1.19961	2.1447866	0.85303289
pH	2.10981558	−2.25714	2.1314495	−4.80641	2.1447866	−2.196829
APTI	2.109815578	−0.41303	2.2009851	−1.57878	2.22813885	1.490776685

texture, economic use), the API is low for this species. Species with good API values provide good canopy cover and rough leaf texture for dust capturing, better habitat for wildlife. This method can prove to be an effective strategy for air pollution management without any expensive equipments.

References

1. Bernatzky, A.: Tree ecology and preservation. In: Development in Agricultural and Management Forest Ecology, vol. 2. Elsevier Scientific Publishing Co., New York (1978)
2. Singh, S.K., Rao, D.N., Agrawal, M., Pandey, J., Narayan, D.: Air pollution tolerance index of plants. J. Environ. Manage. **32**, 45–55 (1991)
3. Prajapati, S.K., Tripathi, B.D.: Anticipated performance index of some tree species considered for green belt development in and around an urban area: a case study of Varanasi city, India. J. Environ. Manag. **88**, 1343–1349 (2007)
4. Dhankhar, R., Mor, V., Lilly, S., Chopra, K., Khokhar, A.: Evaluation of anticipated performance index of some tree species of Rohtak city, Haryana, India. Int. J. Recent Sci. Res. **6**(3), 2890–2896 (2015)
5. Sen, D.N., Bhandari, M.C.: Ecological and water relation to two *Citrullus* spp. In: Althawadi, A.M. (ed.) Indian Arid Zone, Environmental Physiology and Ecology of Plants, pp. 203–228 (1978)
6. Arnon, D.I.: Copper enzymes in isolated chloroplast polyphenol oxidase in *Beta vulgaris*. Plant Physiol. **24**, 1–15 (1949)
7. Liu, Y.J., Ding, H.: Variation in air pollution tolerance index of plants near a steel factory: implications for landscape-plant species selection for industrial areas. WSEAS Trans. Environ. Dev. **4**, 24–32 (2008)
8. Sadasivam, S., Manickam, A.: Biochemical Methods, 2nd edn. New Age International (P) Limited, New Delhi (1996)
9. Miller, G.L.: Use of DNS reagent for the determination of glucose. Anal. Chem. **31**, 426–428 (1972)
10. Lowry, O.H., Rosebrough, N.J., Farr, A.L., Randall, R.J.: Protein measurement with the Folin phenol reagent. J. Biol. Chem, 293–265 (1951)
11. Singh, S.K., Rao, D.N.: Evaluation of plants for their tolerance to air pollution. Proceeding of Symposium on Air Pollution Control, vol. 1, pp. 218–224. Indian Association of Air Pollution Control, New Delhi (1983)
12. Agarwal, S., Tiwari, S.L.: Susceptibility level of few plants on the basis of air pollution tolerance index. Indian Forester, 319–322 (1997)
13. Keller, T.: The electrical conductivity of Norway spruce needle diffusate as affected by air pollutants. Tree Physiol. **1**, 85–94 (1986)
14. Keller and Schwager: Air pollution and ascorbic acid. Eur. J. For. Pathol. **7**, 338–350 (1977)
15. Bakiyaraj, R., Ayyappan, D.: Air pollution tolerance index of some terrestrial plants around an industrial area. Int. J. Mod. Res. Rev. **2**(1), 1–7 (2014)
16. Das, S., Prasad, P.: Seasonal variation of air pollution tolerance indices and selection of plant species for industrial areas of Rourkela. IJEP **30**(12), 978–988 (2010)
17. Scholz, F., Reck, S.: Effect of acids on forest trees as measured by titration in vitro, inheritance of buffering capacity in Piceaabies. Water Air Soil Pollut. **8**, 41–45 (1977)
18. Bhattacharya, T., Kriplani, L., Chakraborty, S.: Seasonal variation in air pollution tolerance index of various plant species of Baroda city. Univ. J. Environ. Res. Technol. **3**(2), 199–208 (2013)
19. Lakshmi, P.S., Sravanti, K.L., Srinivas, N.: Air pollution tolerance index of various plant species growing in industrial areas. Ecoscan **2**, 203–206 (2008)

20. Rai, P.K., Panda, L.S.: Roadside plants as bio indicators of air pollution in an industrial region, Rourkela, India. Int. J. Adv. Res. Technol. **4**(1) (2015). ISSN 2278-7763
21. Tripathi, A.K., Gautam, M.: Biochemical parameters of plants as indicators of air pollution. J. Environ. Biol. **28**(1), 127–132 (2007)
22. Mondal, D., Gupta, S., Datta, J.K.: Anticipated performance index of some tree species considered for green belt development in an urban area. Int. Res. J. Plant Sci. **2**(4), 099–106 (2011). ISSN: 2141-5447
23. Prado, F.E., Boero, C., Gallarodo, M., Gonzalez, J.A.: Effect of NaCl on germination, growth and soluble sugar content in *Chenopodium quinoa* wild seeds. Bot. Bull. Acad. Sin. **41**, 27–34 (2000)
24. Thomas, H.: Accumulation and consumption of solutes in swards of Loliumperenne during drought and after rewatering. New Phytol. **118**, 35–48 (1991)
25. Agbaire, P.O., Akporhonor, E.E.: The effects of air pollution on plants around the vicinity of the delta steel company, Ovwian-Aladja, Delta State, Nigeria. IOSR J. Environ. Sci. Toxicol. Food Technol. **8**(7), 61–65 (2014). e-ISSN: 2319-2402, p-ISSN: 2319-2399

Trend Analysis of Long-Term MSW Leachate Characteristics

Tamru Tesseme and Sumedha Chakma

Abstract The landfill leachate characteristics are found varied significantly with stages of solid waste degradation and continued even after closure. The objective of this paper is to notice the trend of long-term leachate characteristics using time series analysis methods. The historical observed leachate characteristics data were taken from the literature and split into three groups based on their biodegradable and dissolvable nature. The developed trend showed higher biodegradable and dissolvable leachate characteristics depend on the quantity of biodegradable matter presence. The non-biodegradable and highly dissolvable leachate characteristics had a constant trend. The non-biodegradable and non-dissolvable leachate characteristics were found increasing trend in the analysis. The developed long-term trend showed time series analysis estimation was an encouraging method for the prediction of long-term leachate characteristics without further laboratory investigation and thus would help to communicate to the general public, the field professionals, and the policy makers about the future status of the landfill.

Keywords Landfill · Leachate · Biodegradable · Dissolvable · Time series
Trend analysis

1 Introduction

Solid waste management nowadays becomes a growing global alarm due to its dynamic generation rates and variation of quantities and compositions with time. Even though recently different types of integrated solid waste management options are implemented worldwide, landfilling remains the best disposal option for

---

T. Tesseme (✉) · S. Chakma
Department of Civil Engineering, Indian Institute of Technology (IIT) Delhi,
Hauz Khas, New Delhi 110016, India
e-mail: tamruiit@gmail.com; tessemetamru@gmail.com

S. Chakma
e-mail: chakma@civil.iitd.ac.in

A. S. Kalamdhad et al. (eds.), *Advances in Waste Management*,
https://doi.org/10.1007/978-981-13-0215-2_10

municipal solid waste compared to the other alternatives [1, 2]. The reasons that make landfilling the best attractive option correlated with others are its economic advantages, and other alternatives are also leftover 10–20% residue which can be sent again to the landfill. In India, the population of the New Delhi has expected to grow about 40% in the year of 2020 and thus can lead the generation of huge amount of municipal solid waste (MSW). The developing countries like India thus facing serious problems of managing its MSW due to urbanization and increase in population. In India, every year about 50 million tons of MSW has been collected by the civic authorities; however, 90% of the total collected MSW is dumped in low-lying areas at the outskirts [3]. In landfill, a complex sequence of physical, chemical, and biological reactions occurs [4, 5]. The major effects that take place from the reactions in and around the dumping sites are the generation of leachate, emissions of methane, greenhouse effects, ozone depletion, differential settlement, and water pollution [6–8]. The leachate pollutant is generated, while the external percolated water eroded and flashed organic, inorganic, and heavy metal pollutants from the landfill [9, 10]. Landfilling technologies need to improve from open dumping to highly engineered bioreactor landfill that may reduce the potential for contamination while enhancing the economic benefits [11–13]. However, quantifying the strength of leachate with respect to time is crucial in all developed landfill technologies to treat the leachate and safely dispose into the natural water bodies [14, 15].

Quantifying the present and future leachate characteristics strength and its impacts on the adjacent waters resources needs frequently monitoring of the landfill and recording leachate pollutant for longer periods. To carry out a long-term monitoring and quantification, need huge resources which may be difficult to afford by several countries. Besides this, adopting of one landfill leachate characteristics to the other landfills is often problematic. Several studies on the leachate characteristics have been performed during the last 2–3 decades in the test lysimeter [16, 17] and pilot scale model [18, 19] and full-scale landfills [20] revealed that the leachate characteristics were widely varied among different landfills. The leading factors that make leachate characteristics to be varied among landfills are solid waste type and composition, rainfall intensity, ambiance temperature, pH, moisture level and stage of solid waste degradation and ages of the landfill [20]. Moreover, the leachate characteristics even in the same landfills also varied over a period of 30–40 years according to stages of municipal solid waste stabilization [21, 22]. As a result of this, a simple method that would be accurately estimated and quantify the strength of leachate with time is the need of the present system [23, 24]. A simple analytical computational technique would be essential to estimate the future leachate characteristics with respect of times based on the historical observed leachate characteristics during the resources are limited and also no satisfactory alternative approaches [25].

Therefore, the objective of this study is to estimate the future long-term leachate characteristics trend using time series multiplicative decomposition model based on historical observed data. The computed trend analyses data for selected leachate parameters revealed that the time series multiplicative decomposition model was an encouraging forecasting method for long-term monitoring and controlling of the leachate characteristics without carried out intensive field leachate examination and thus saves the amount for future investigation.

2 Materials and Methods

The leachate characteristics used in this particular study were obtained from unlined landfill site at Toronto, Canada [26]. The leachate sample was taken from leachate collection pond and analyzed for period of six years from 2004 to 2009 with a frequency of twelve times a year. A total of 72 historical observed leachate characteristics were under-considered in the study to estimate the future long-term trends of leachate characteristics. The historical observed leachate characteristics were split into three groups according to their degradation and dissolvable behaviors. In the first group those highly biodegradable and dissolvable leachate characteristics such as BOD_5 and COD were kept because these elements are highly degradation and dissolvable natures. In the second group, those non-biodegradable and rapidly dissolvable characteristics, namely chloride and iron, were considered, and in the third group, those elements neither degradation nor dissolvable behaviors were also undertaken. The classified leachate characteristics were plotted with the leachate concentration versus age (months) to estimate the future long-term trend. In the study, leachate characteristics analyzing period (month) was considered as the age of the landfill by referring the first month of 2004 as the starting period of the time series analyzing data. Then after doing this, the landfill leachate characteristics were analyzed per recommendation of the time series decomposition method using Minitab 16 software [27]. The time series decomposition can perform using one of the options such as multiplicative model and the additive model. The multiplicative model is estimated using Eq. (1), and similarly, the additive model is computed using Eq. (2). According to Worrall and Burt [28], time series multiplicative decomposition technique is a simple and rapid method for development of a wide range of parameters for estimating the future long-term data based on the historical observation. In addition to this, the multiplicative decomposition technique has also 5–10% better estimation accuracy compared with the additive model. Therefore, in this particular study, the time series trend analysis was performed using the multiplicative decomposition model.

$$y_t = T_t \times S_e \times e_t; \quad t = 1\ldots, n \tag{1}$$

$$y_t = T_t + S_t + e_t; \quad t = 1\ldots, n \tag{2}$$

where t is time; y_t is historical observed data; T_t is trend component; S_t is seasonal component; and e_t is errors with respect to times.

The accuracy of the time series analysis results was computed using statistical error estimation techniques such as mean absolute percentage error (MAPE), mean absolute deviation (MAD), and mean squared deviation (MSD) which have been mentioned in Eqs. (3)–(5). The smaller error estimated values indicated a better fitting result [29].

$$\text{MAPE} = \frac{\sum \left|\left(y_t - y_t^1\right)/y_t\right|}{n} \times 100 \quad (y_t \neq 0) \tag{3}$$

$$\text{MAD} = \frac{\sum_{t=1}^{n} \left|y_t - y_t^1\right|}{n} \tag{4}$$

$$\text{MSD} = \frac{\sum_{t=1}^{n} \left(y_t - y_t^1\right)^2}{n} \tag{5}$$

where y_t *is* the actual value; y_t^1 *is* the forecast value; and n is the number of forecasts

3 Results and Discussion

The historical observed leachate characteristics data were analyzed by time series multiplicative decomposition model from 2004 to 2009, and the long-term leachate characteristics were estimated for next 48 future months. The developed long-term trend of the leachate characteristics was presented below in to three subdivision groups based on their degradation and dissolvable behaviors.

3.1 Highly Biodegradable and Dissolvable Leachate Characteristics

Under highly biodegradable and dissolvable group, the concentrations of COD, BOD_5, and their corresponding ratio of BOD_5/COD values were kept and their trend analysis was also performed by time series multiplicative decomposition method. The values of COD, BOD_5, and the ratio BOD_5/ COD are presented in Figs. 1, 2, and 3. The values of COD were observed in the range of 500–2000 mg/L, and the values of BOD_5 were indicated between 0 and 500 mg/L. The result showed that the magnitude of the observed and the forecasted long-term trend did not vary significantly. However, leachate characteristics in younger landfills noticed that the values of COD and BOD_5 concentrations would be initially low when the organic fraction of waste is decomposed, flashed, and then reached to their peaks followed by decreasing trends while the biodegradables organic fraction of wastes depleted from the landfill via washout and degradation.

The small magnitude of both values in the trend developed shown in Figs. 1 and 2 demonstrated that the landfills were in maturation stages which depleted its organic fraction by biological degradation and washing out. Besides to this, in the estimated future long-term trend observed data there were promote decline the values of COD and BOD_5 in the future due to occurrence of higher stabilization processes. Nevertheless, the values younger refuse leachate characteristics having low pH (5–6), high BOD_5 (4000–13000 mg/L) and COD (10,000–60,000 mg/L) values. Correspondingly, the trend of the BOD_5/COD (Fig. 3) showed initial increasing trend followed by decreasing trend associated with the decomposition and depletion of dissolved biodegradable organic matters.

3.2 Non-biodegradable and Highly Dissolvable Leachate Characteristics

It has been well documented in the literature the characteristics of leachate taken at the bottom of the landfill widely varied with respected times [30–32]. It has been common to observe the concentration profiles of biodegradable organics in leachate such as BOD_5 and COD that initially increasing to reach a peak value and followed by reducing trend with respect to the waste stabilization stages. However, there are some particular landfill leachates characteristics which are poorly biodegradable but highly dissolution behaviors such as chloride which is tend to persist in the leachate for several years. It can observe from Fig. 4 the concentration of chloride in the leachate were more or less constant in the historical observed as well as in estimated future long-term trend due to the chloride is poorly biodegradable but highly dissolution element.

The concentration of chloride was observed remain constant (Fig. 4) due to the concentration of chloride in the landfill is independent with respect of waste biodegradation stages but its concentration in the leachate depend only on the quantity of moisture percolated and hydraulic conductivity of the landfills.

Thus, the non-degradable and conservative behavior of chloride in the landfill makes the chloride characteristics to use commonly for tracer studies. Moreover, as Figs. 4 and 5 showed that the concentrations of chloride and iron in the landfill do not show significant variation with respect to the landfill age. It was due to the landfill in maturation stages and thus may show no significant difference in chloride and iron concentrations during the acidogenic and methanogenic phases. However, the estimated chloride values (Fig. 4) could be increased in little magnitude due to washing out of chloride in the landfill by external percolated water.

3.3 Non-biodegradable and Non-dissolvable Leachate Characteristics

In the non-biodegradable and non-dissolvable leachate characteristics groups those leachate characteristics their concentration which do not generated with the landfill waste degradation and dissolvable but they are the indicators of the landfill ecology status. In this group, the values of pH and conductivity were analyzed to estimate the long-term trends as shown in Figs. 6 and 7, and thus the values of pH and conductivity may use to evaluate the overall status of the landfill stabilization stages.

The value of the pH is shown in Fig. 6. The pH trend was gradually increased with the age of landfill leachate due to the depletion organic matters (measured as COD, BOD_5, and DOC) occurred in the landfill. Besides this, the values of the landfill pH widely varied at the different stages of the municipal solid waste stabilization which is influenced by different microbial actions and their by-product. The values of pH in the historical observed data of the landfill were detected in range of 6–8 (Fig. 6) and thus showed the landfill maturation stages. This gradual increment of pH values in the landfill demonstrates the depletion of organic matter via anaerobic degradation processes in maturation stages. The time series estimated the long-term trend values as depicted in Fig. 6 and observed that the pH values again further increased which was associated with the depletion of acid by production of methane gas and other pollutants. However at the maturation stages of the landfill stabilization due to the organic fraction of solid waste consumed the landfill cases to increase pH values, and reduced the generation of heavy metals. The trend of the conductivity follows the inverse of pH trend, which has higher values during early stabilization stages and gradually declines values during the maturation stage as shown in Fig. 7.

Table 1 summarized the long-term trend leachate characteristics estimation equations which were generated using Figs. 1, 2, 3, 4, 5, 6, 7, 8, and 9. The estimated equations have large errors due to the landfill leachate characteristics by nature varied widely with time.

Table 1 Long-term trend estimation statistical error values

Parameters	MAPE	MAD	MSD	Equations
COD (mg/L)	49	399	476,845	$y_t = 916 + 2.80^{*}\ t$
BOD_5 (mg/L)	108	197	176,467	$y_t = 157.3 + 2.26^{*}\ t$
BOD_5/COD	56.89	0.09	0.02	$y_t = 0.1366 + 0.00174^{*}\ t$
Chloride (mg/L)	27.5	159	39995.8	$y_t = 701 - 0.450624^{*}\ t$
Iron (mg/L)	120.824	7.841	781.103	$y_t = 6.83 + 0.03318^{*}\ t$
pH	2.30017	0.16706	0.06168	$y_t = 6.9942 + 0.00650^{*}\ t$
Conductivity	40	3233	41337313	$y_t = 9979 - 69.9799^{*}\ t$
Temperature (°C)	38.94	5.9962	58.8959	$y_t = 10.94 + 0.142^{*}\ t$
Precipitation (mm)	28.196	13.567	732.314	$y_t = 68.50 - 0.016331^{*}\ t$

Fig. 1 Long-term trend analysis for COD

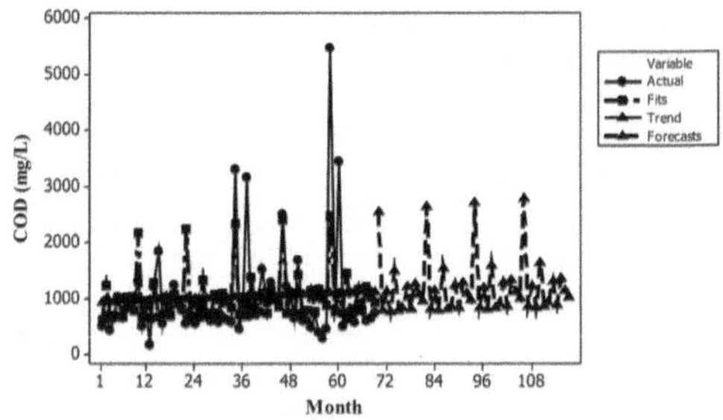

Fig. 2 Long-term trend analysis for BOD_5

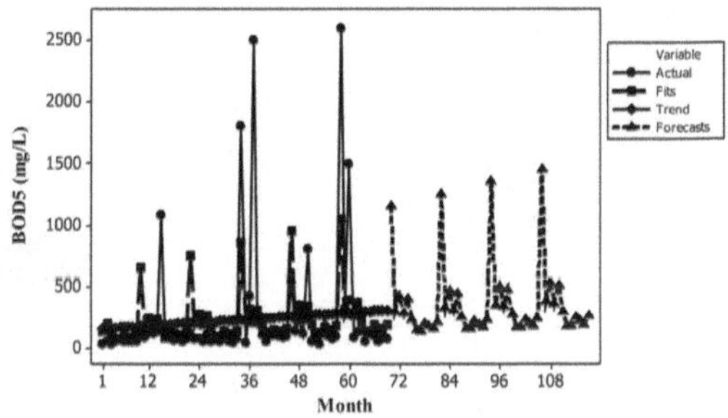

Fig. 3 Long-term trend analysis for BOD_5/COD

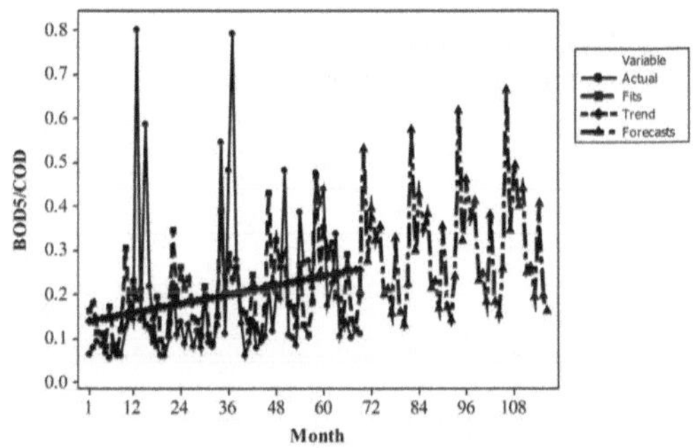

Fig. 4 Long-term trend analysis for chloride

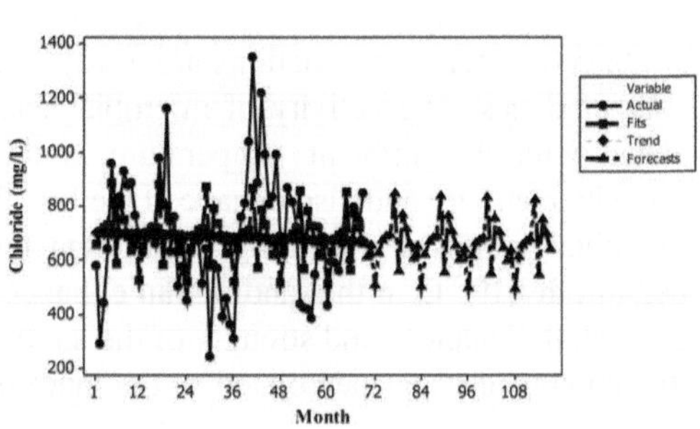

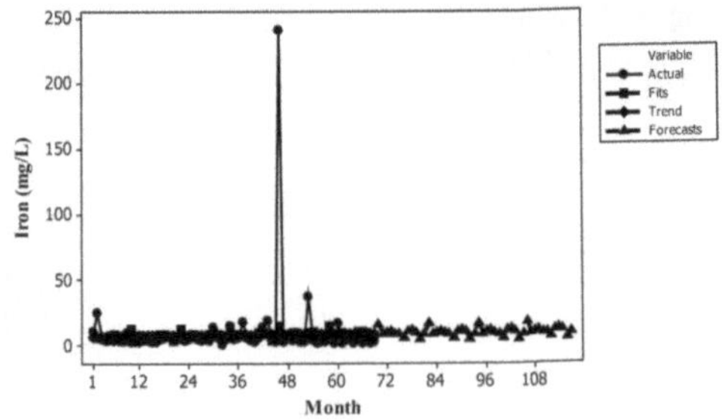

Fig. 5 Long-term trend analysis for iron

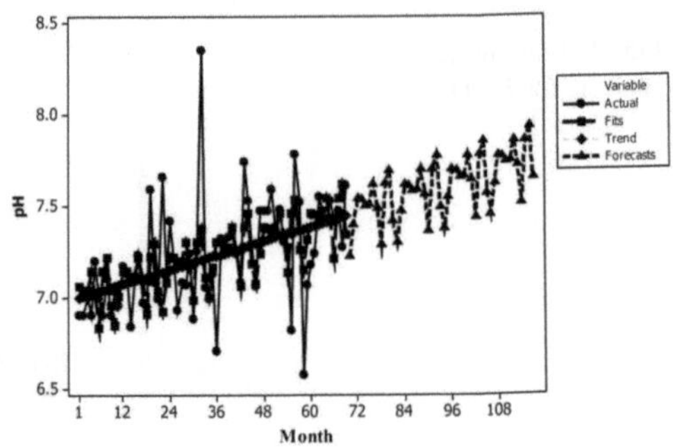

Fig. 6 Long-term trend analysis for pH

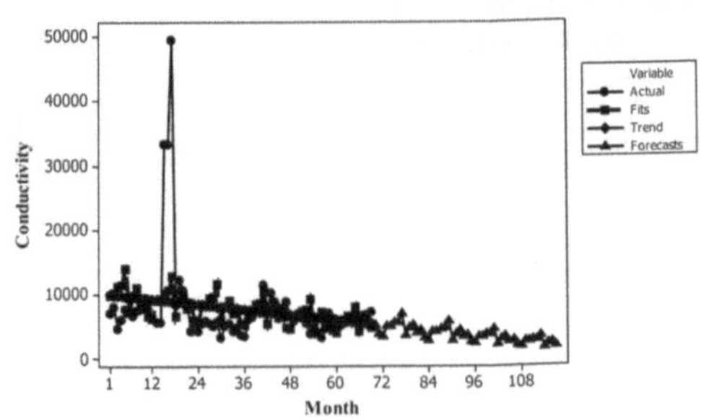

Fig. 7 Long-term trend analysis for conductivity

3.4 Seasonal Variation of the Landfills

Ambient temperature variations are always significantly influenced on the leachate characteristics. The activity of microbial community in the landfill directly associated with the ambient temperature, and thus, the temperature increased the microbial activity and also enhanced the landfill stabilization. Even though different microbial groups are working at different temperature ranges, increment of temperature by 10 °C in the landfill can enhance stabilization processes by double. In general, the quantity and strength of the landfill leachate are directly associated with the amount of water percolated to the landfill. It has been shown in Figs. 8 and 9

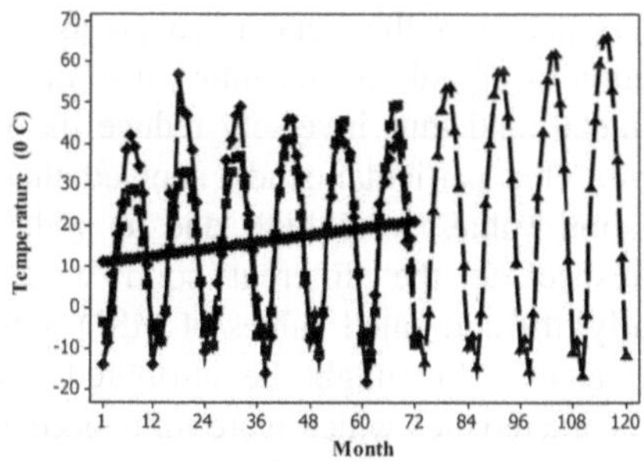

Fig. 8 Long-term trend analysis for temperature

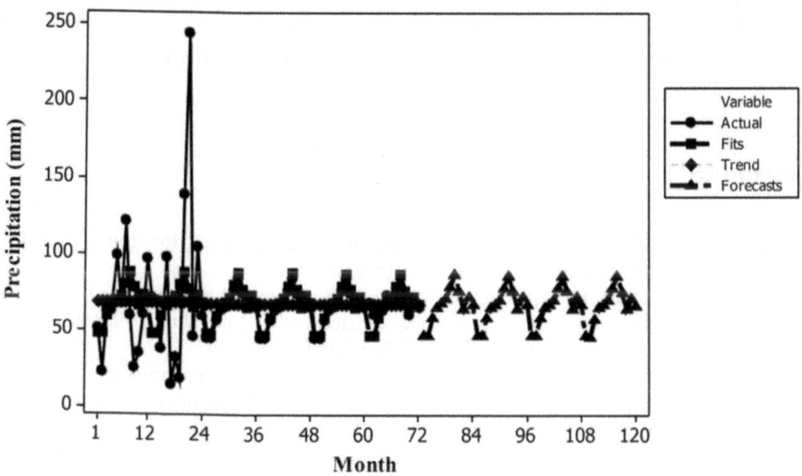

Fig. 9 Long-term trend analysis for precipitation

that the values of temperature and precipitations were moving in waveform that repeated itself with specific interval of times based on the seasons of the years.

4 Conclusions

In this study, the time series analysis by multiplicative decomposition model was performed to estimate the future long-term trend of leachate characteristics. The developed long-term trend results demonstrated that highly biodegradable organic matters were depleted through degradation and dissolution and thus may be the cause for decline of the long-term concentrations of COD and BOD_5 values. However, the chloride concentration profile in the developed long-term trend confirmed that the concentration of chloride was independent on the biodegradation rates of organic matters and thus the trend is found as remained constant. The degradation and depletion of organic matter may cause increment of pH, decrease dissolution of heavy metal, and thus inversely reduce the values of conductivity trend in the long term. This particular study showed that the developed trend statistical error estimation values were high due to wide variation of leachate characteristics with respect to the different solid waste stabilization stages. Furthermore, particularly, the computed values of MSD were very high compared with the other error values. This might be attributed due to highly dynamic behaviors of leachate characteristics which were influenced by the different factors such as moisture percolation, landfill pH, and ambiance temperature. The developed long-term trend analysis revealed that estimation of the long-term trend behaviors of leachate characteristics using time series analysis multiplicative decomposition model was useful when little knowledge is available on the underlying data generating process or when there is no sufficient resources also available to examine intensively the long-term leachate characteristics and quantifications.

References

1. Chakma, S., Mathur, S.: Modelling gas generation for landfill. J. Environ. Technol., 1–8 (2016). https://doi.org/10.1080/09593330.2016.1231226
2. Remmas, N., Melidis, P., Katsioupi, E., Ntougias, S.: Effects of high organic load on amoA and nirS gene diversity of an intermittently aerated and fed membrane bioreactor treating landfill leachate. Biores. Technol. **220**, 557–565 (2016)
3. Chakma, S., Mathur, S.: Leachate: a threat to groundwater contamination at Delhi. In: International Conference on Water Resource Management: Challenges and Opportunities in the 21st Century, pp. 47–48. Silchar, Assam (2007)
4. Ahmed, F.N., Lan, C.Q.: Treatment of landfill leachate using membrane bioreactors: a review. Desalination **287**, 41–45 (2012)
5. Kurniawan, T.A., Lo, W.H., Chan, G.Y.: Physico-chemical treatments for removal of recalcitrant contaminants from landfill leachate. J. Hazard. Mater. **129**(1), 80–100 (2006)
6. Chakma S., Jha N., Gupta V., Bisht P.S.: Leachate and its characteristics from Open MSW Dumps. In: 2nd International Conference on Solid waste Management and Exhibition, 8–10 Nov, Kolkata, India (2011)
7. Chakma, S., Singh, V.K.: Bio-Reactor Landfill Potential in Allahabad. In: ENSURE 2012: Environmentally Sustainable Urban Ecosystems, 24–26 Feb 2012 at IIT Guwahati, Assam, India (2012)
8. Chakma, S., Mathur, S.: Post closure long-term settlement for MSW landfills. J. Hazard. Toxic Radioactive Waste ASCE J. **17**(2), 81–88 (2013)
9. Xie, B., Xiong, S., Liang, S., Hu, C., Zhang, X., Lu, J.: Performance and bacterial compositions of aged refuse reactors treating mature landfill leachate. Biores. Technol. **103** (1), 71–77 (2012)
10. Jun, D., Yongsheng, Z., Weihong, Z., Mei, H.: Laboratory study on sequenced permeable reactive barrier remediation for landfill leachate-contaminated groundwater. J. Hazard. Mater. **161**(1), 224–230 (2009)
11. Warith, M.: Bioreactor landfills: experimental and field results. Waste Manag. **22**(1), 7–17 (2002)
12. Benson, C.H., Barlaz, M.A., Lane, D.T., Rawe, J.M.: Practice review of five bioreactor/recirculation landfills. Waste Manag. **27**(1), 13–29 (2007)

13. Jain, P., Townsend, T.G., Tolaymat, T.M.: Steady-state design of vertical wells for liquids addition at bioreactor landfills. Waste Manag. **30**(11), 2022–2029 (2010)
14. Wang, C.C., Lee, P.H., Kumar, M., Huang, Y.T., Sung, S., Lin, J.G.: Simultaneous partial nitrification, anaerobic ammonium oxidation and denitrification (SNAD) in a full-scale landfill-leachate treatment plant. J. Hazard. Mater. **175**(1), 622–628 (2010)
15. Tamru, A.T., Chakma, S.: Mathematical modelling for landfill leachate pollution index error estimation. Discovery **41**(189), 123–129 (2015)
16. Yang, Y., Yue, B., Yang, Y., Huang, Q.: Influence of semi-aerobic and anaerobic landfill operation with leachate recirculation on stabilization processes. Waste Manage. Res. **30**(3), 255–265 (2011)
17. Bilgili, M.S., Demir, A., Varank, G.: Effect of leachate recirculation and aeration on volatile fatty acid concentrations in aerobic and anaerobic landfill leachate. Waste Manage. Res. **30** (2), 161–170 (2011)
18. Huang, F.S., Hung, J.M., Lu, C.J.: Enhanced leachate recirculation and stabilization in a pilot landfill bioreactor in Taiwan. Waste Manage. Res. **30**(8), 849–858 (2012)
19. Pohland, F.G.: Sanitary Landfill Stabilization with Leachate Recycle and Residual Treatment. Environmental Protection Agency, Washington, D.C. (1973)
20. Reinhart, D.R.: Full-scale experiences with leachate recirculating landfills: case studies. Waste Manage. Res. **14**(4), 347–365 (1996)
21. Kulikowska, D., Klimiuk, E.: The effect of landfill age on municipal leachate composition. Biores. Technol. **99**(13), 5981–5985 (2008)
22. Tamru, T., Chakma, S.: Effects of landfilled MSW stabilization stages on composition of landfill leachate: a review. In: National Conference on Recent Advancement in Civil & Environmental Engineering, RACEE-2015, 28th–29th November, 2015, pp. 112–115. Bhiwani, Haryana, India: BRCM College of Engineering and Technology (2015b)
23. Gönüllü, M.T.: Analytical modelling of organic contaminants in leachate. Waste Manage. Res. **12**(2), 141–150 (1994)
24. Lee, K.K., Suk, H., Choi, S.I., Lee, C.H., Chung, S.Y.: Numerical evaluation of landfill stabilization by leachate circulation. J. Environ. Eng. **127**(6), 555–563 (2001)
25. Raveh, A., Avnimelech, Y.: Leaching of pollutants from sanitary landfill models. J. (Water Pollut. Control Fed.) 2705–2716 (1979)
26. Lee, A.H.: Prediction of Pollutant Leaching from Landfill. Division of Engineering and Science, Department of Civil Engineering, Curtin University of Technology (2010)
27. Minitab: Minitab User's Guide 2: Data Analysis and Quality Tools. Windows® 95, Windows® 98, and Windows NT™, USA (2000)
28. Worrall, F., Burt, T.: Decomposition of river water nitrate time-series comparing agricultural and urban signals. Sci. Total Environ. **210**(211), 153–162 (1998)
29. Armstrong, J., Collopy, F., Yokum, T.: Decomposition by causal forces: a procedure for forecasting complex time series. Int. J. Forecast. **21**(1), 25–36 (2004)
30. Kjeldsen, P., Barlaz, M.A., Rooker, A.P., Baun, A., Ledin, A., Christensen, T.H.: Present and long-term composition of MSW landfill leachate: a review. Crit. Rev. Environ. Sci. Technol. **32**(4), 297–336 (2002)
31. Chakma, S., Mathur, S.: Estimation of primary and mechanical compression in MSW landfills. J. Hazard. Toxic Radioactive Waste ASCE J. **16**(4), 298–303 (2012)
32. Chakma, S.: Ground water and air pollution from open dumps MSW in India. In: 23rd Indian Engineering Congress, Warangal, Andhra Pradesh, 12–13 Dec, pp. 65–67 (2008)

Experimental Investigation for the Feasible Applications of Processed Recyclable Plastic Waste in Construction Sector

Kaushal Kishor Sahoo, Mohanish Gupta, Ravi Sahu, Kshitij Mudgal and Y. Shiva Shankar

Abstract Plastics have become an integral part of modern lifestyle due to its availability in various forms leading to increased generation of plastic waste. The major problem with waste plastics on disposal is due to their non-biodegradable nature that affects the environment in longer run. Recycling of plastics is the best practice to reduce the detrimental effects on environment. Options for the recycling are converted into raw material for the industry, fuel production, and utilization as construction material. Utilization of recycled plastics in construction sector provides ecological and economical benefits, due to increasing demand for building materials. The present work investigates the feasible applications for the utilization of processed recyclable waste plastic materials in construction sector. Materials such as plastic coarse aggregate (PCA) and plastic fine aggregate (PFA) were processed from the plastic bottles generated in the university campus. The application of these materials in areas such as concrete production, block manufacturing, pavement construction, and soil stabilization has been studied. The results were compared with natural fine aggregate (NFA) and natural coarse aggregate (NCA). Experimental results for PCA suggest that physical and mechanical properties proved to be in conformance with conventional NCA, for applications such as concrete production and pavement construction. The properties of PFA for the replacement of natural fine aggregate in pavement construction and soil stabiliza-

K. K. Sahoo · M. Gupta · R. Sahu · K. Mudgal · Y. S. Shankar (✉)
Department of Civil Engineering, Jaypee University of Engineering and Technology, Guna, India
e-mail: shivjuet@gmail.com

K. K. Sahoo
e-mail: sahook.k.0000@gmail.com

M. Gupta
e-mail: mohanish10@gmail.com

R. Sahu
e-mail: ravisahu1266@gmail.com

K. Mudgal
e-mail: kshitij.mudgal01@gmail.com

A. S. Kalamdhad et al. (eds.), *Advances in Waste Management*,
https://doi.org/10.1007/978-981-13-0215-2_11

tion has been studied through Marshall stability test (as 100% replacement to NFA) and CBR value (as 10, 20, and 30% replacement of NFA). The results of aggregate properties indicated the conformance of PFA with natural sand. The results of CBR test indicate that PFA has proved to enhance the CBR value of black cotton up to 20% replacement. Blocks manufactured with PFA (as 100% replacement to NFA) have shown better properties for application in framed structure. Marshall stability test results indicated that presence of PFA has enhanced the load resistance of the specimen, the results of air voids, voids filled with bitumen, and flow values indicate that addition of PFA enhances the durability, shear strength, and fatigue resistance improving the life of pavement.

Keywords Plastic coarse aggregate · Plastic fine aggregate · Waste plastics Concrete production · Soil stabilization · Construction sector

Abbreviations

PP	Polypropylene
PS	Polystyrene
PET	Polyethylene tetraphthalate
HDPE	High-density polyethylene
LDPE	Low-density polyethylene
PVC	Polyvinyl chloride
PCA	Plastic coarse aggregate
PFA	Plastic fine aggregate
NFA	Natural fine aggregate
NCA	Natural coarse aggregate
CBR	California bearing ratio

1 Introduction

Plastics have become integral part of the life due to its availability in various forms such as bottles, packing material, containers, shopping and garbage covers, household products. Commonly used plastics for various applications are categorized as thermoplastics and thermoset plastics. Thermoplastics constitute 80% of total market utilization in various forms such as polypropylene (PP), polystyrene (PS), high-density polyethylene (HDPE), low-density polyethylene (LDPE), polyethylene tetraphthalate (PET). In addition to the above construction waste in the form polyvinyl chloride (PVC) has also been increasing causing problems in the management. Majority of these forms are being recovered through rag pickers, especially PET bottles due to their resale value. However, due to their existence in mixed form along with the remaining fractions of the solid waste, most of these waste plastics remain in the environment [1, 2].

Increased utilization of plastics has been posing problems to the environment in terms of the volume of the waste generated. Civic bodies have been facing problems in the collection, transportation, and disposal of municipal solid waste due to rapid changes in the waste composition, urbanization, and increasing per capita waste generation rates. The waste plastics due to their non-biodegradable nature have a negative impact on environment in long run. The commonly practiced methods for management of solid waste in India are open dumping and landfilling. On disposing the plastics in combination of municipal solid waste, plastics pollute the soil and in certain cases they percolate through the soil reaching the groundwater table [3–5].

Recycling the waste plastics is a feasible solution to reduce the adverse impact on environment. In many of the urban areas plastic recycling units are increasing, these units process the wastes into usable form after sorting, separation, and treatment. Studies have suggested that diversion of waste streams such as plastics, wood, papers offers a feasible solution to increased space requirements for landfills. The possible options for recycling the solid wastes are conversion/processing the wastes for replacement of virgin materials in manufacturing plastic products, mechanical processing of plastics into pellets, flakes or powders for reuse in modified form, and energy recovery. The quality of the processed materials has been major issue for the replacement of virgin materials. Various thermochemical methods exist to enhance properties of the waste plastics but the limitation of these methods has been increasing in the cost of production. One of the beneficial applications of plastic waste has been through the energy recovery for utilizing the higher calorific value. Technologies such as incineration, gasification, and pyrolysis could be used for deriving the valuable fuel as a replacement to the conventional sources. Pyrolysis and gasification have been proved to be suitable techniques for energy recovery as the burning of plastics releases a variety of poisonous chemicals into the air, including dioxins, one of the most toxic substances. Beneficial utilization and recycling of waste would be feasible only through segregation at the source. Intermixing of the various components of solid waste and recovery of the components post-disposal reduces the efficiency of these processes [2, 4–9].

1.1 Applications in Construction Sector

Increasing demand for construction materials and depleting resources have exerted pressure on researchers to study the alternatives for conventional materials. Materials such as cement, sand, and coarse aggregate exert higher ecological footprint in terms of resource requirements. Utilization of processed waste plastics as building materials has been projected to be a feasible option. Various researchers have studied the application of waste plastics in form of fibers, aggregate, powder, flakes for utilization in concrete and bitumen. The preparation of aggregates from plastic wastes was done by cutting and grinding the wastes followed by washing and sieving to obtain the required fractions. Some researchers have adopted the modifications through heating and processing the waste through addition of sand, etc., to improve the quality of the material [5, 10].

Bajracharya et al. [3] investigated the mechanical behavior of mixed plastic solid waste under different loading conditions for utilization as construction material using coupon and full-scale specimens. From coupon test, the strength values were found to be 14.8, 19.8, 20, 5.6 MPa in tension, compression, flexure, and shear, respectively, while the modulus of elasticity is 0.91, 1.03, 0.72 GPa in tension, compression, and flexure, respectively. The results indicated that strength properties of mixed plastic solid waste were comparable to softwood structural timber indicating the feasibility for reuse.

1.1.1 Pavements

Indian road congress has suggested that the addition of modifiers such as plastics/ rubber improves the properties of bitumen. The addition of modifier provides a suitable solution to varying temperature conditions responsible for early development of distress like rutting, cracking, bleeding, shoving, and potholing of bituminous surfaces [11]. Plastics in form of carry bags, disposable cups, and PET bottles collected from garbage dumps have proved to be a feasible construction material. Increasing demand for construction materials and depleting resources have exerted pressure on researchers to study the alternatives for conventional materials. Materials such as cement, sand, and coarse aggregate exert higher ecological footprint in terms of resource requirements. Utilization of processed waste plastics as building materials has been projected to be a feasible option. Various researchers have studied the application of waste plastics in form of fibers, aggregate, powder, flakes, etc., for utilization in concrete and bitumen. The preparation of aggregates from plastic wastes was done by cutting and grinding the wastes followed by washing and sieving to obtain the required fractions. Some researchers have adopted the modifications through heating and processing the waste through addition of sand, etc., to improve the quality of the material [5, 10].

Vasudevan et al. [12] proved that softened waste plastics form a film-like structure over the aggregate, when it is sprayed over the hot aggregate at 160 °C. The formed plastic-coated aggregate when mixed with bitumen showed improved binding property and poor wetting property. Roads constructed using this method in various parts of Tamil Nadu (India) has proved to offer better performance as compared to conventional pavements. The studies have proved that the addition of shredded wastes to hot bitumen has improved the temperature susceptibility of bitumen by forming an oily coat over the aggregate [13].

Ahmadinia et al. [14] examined the effect of waste plastic bottles (Polyethylene Terephthalate [PET]) on the engineering properties of stone mastic asphalt (SMA) mixture. Studies were carried out at various percentages of PET (0, 2, 4, 6, 8, and 10%). The results show that the addition of PET has a significant positive effect on the properties of SMA.

Performance of bitumen with PVC waste was studied by Köfteci et al. and Behl et al. [15, 16]. Köfteci et al. [16] investigated the performance of bitumen modified by three groups of polyvinyl chloride (PVC)-based waste plastics (window, blinds, and cable wastes). Experimental investigations were carried out at 1, 3, 5% replacement with modifier. The results proved that the performance of the mixtures varied with temperature, PVC (from window and blinds wastes) has shown improved performance at high temperatures, whereas PVC (from cable wastes) has shown better performance at low temperatures. Behl et al. [15] have studied the properties of bitumen with addition of chemically modified waste PVC up to a level of 3 and 5% of bitumen. The results indicate that strength and stability of the mix increased with incorporation of PVC pipe waste.

Ahmed et al. [1] investigated the efficiency of recycled gypsum from waste plasterboard and waste plastic trays for the ground improvement. Test results proved that increasing recycled gypsum content has significant effect on compressive strength as compared to tensile strength, whereas the addition of waste plastic strips to samples treated with recycled gypsum enhanced both splitting tensile and compressive strengths.

1.1.2 Concrete and Mortar

Ruiz-Herrero et al. [9] studied the effect of PE and PVC waste that obtained from crushing of electrical copper and aluminum cables, on properties of mortar and concrete. The studied mortar and concrete materials had low density and increased porosity due to the presence of plastic aggregates. However, the results indicated the poor mechanical properties due to the poor bonding between the materials. The results proved that use of non-reusable plastic waste could be used in civil engineering applications providing mortars and concretes with low mechanical properties but increased thermal insulation properties.

Azhdarpour et al. [17] investigated the effect of plastic waste addition as fine sand replacement on the engineering properties of concrete. The results of laboratory tests showed that the compressive, tensile, and flexural strength of samples increased, when 5–10% of the concrete fine aggregates were replaced by the same percentage of PET fragments. Corinaldesi et al. [18] investigated the properties of plasters, where fine sand was replaced up to 100% by waste PET particles, wood particles, and pulverized glass fiber reinforced plastic (GFRP) waste, respectively. The results proved that the plasters possessed low conductivity values reducing the energy consumption in construction.

Yang et al. [19] examined the effect of incorporating recycled modified PP plastic particles on the workability and mechanical behavior of self-compacting lightweight concrete (SCLC). Four replacement levels (10, 15, 20, and 30%) of sand by plastic on volume basis were introduced. The compressive strength, splitting tensile strength, and flexural tensile strength have increased with the replacement level up to 15%. Ghernouti et al. [20] exhaustively investigated the fresh and hardened properties of self-compacting concrete containing plastic bag waste fibers. Fourteen mixtures of concrete with 0.40 of water/cement ratio were studied by varying the length of fibers (2, 4, and 6 cm) with different levels of incorporation (1, 3, 5, and 7 kg/m^3). The results of these mixtures were compared with concrete prepared using 1 kg/m^3 of PP fibers and another without fiber as reference. The obtained results proved that incorporation of plastic bag waste fibers in concrete delayed the location of micro-cracks.

Sommerhuber et al. [21] examined the properties of wood–plastic composites prepared with recycled wood and plastics (HDPE) from packaging waste. The results proved that at lower wood content (30%) the properties of processed material were found to be identical to wood plastic composites made from virgin materials. Yin et al. [22] studied the applications of macro-plastic fibers in reinforced concrete. The results proved that the macro-plastic fibers decreased the workability of the fresh concrete, but effectively control plastic shrinkage cracking with no significant effect on compressive and flexural strengths. Safi et al. [23] have examined the properties of self-compacting mortars prepared with plastic waste as a partial replacement to fine aggregate. The studies proved that replacement up to 50% was acceptable for the preparation of lightweight mortars.

1.2 Objectives of the Research

Utilization of waste plastics under various circumstances as attempted by researchers proved to be beneficial either up to a partial replacement or in a specific application. The present work has attempted to process the waste plastic bottles generated in the university campus (Jaypee University of Engineering and Technology, Guna) identical to virgin materials (coarse and fine fractions) for understanding the applications in construction sector. The followings were the objectives of present research:

- To investigate the properties of processed plastic coarse aggregate (PCA) and plastic fine aggregate (PFA) for their utilization as aggregate in concrete preparation and pavement construction.
- To study the mechanical properties of blocks prepared with PFA as substitute for fine sand.
- To understand the application of PFA (with 10, 20, and 30% replacement by volume) in soil stabilization through California bearing ratio (CBR unsoaked) test.
- To understand the effect of PFA as a replacement to NFA on load resistance of specimen through Marshall stability test.

2 Materials and Methods

2.1 Methodology

Figure 1 explains the procedure for processing the waste plastics into suitable aggregates (PCA and PFA). The raw materials (i.e., waste plastics) were collected from the campus; the collected bottles were predominantly soft drink bottles, i.e., PET type followed by PP. The materials were processed in the following sequence shredding, heating to its melting point under controlled conditions, collection of melted fluid in mold, crushing the solidified mass to obtain PCA. The PCA was further processed in Los Angeles abrasion testing machine (at 700 revolutions) to obtain PFA.

2.2 Materials Used

2.2.1 Cement

Portland pozzolana cement (PPC) confirming to IS 1489—1991 (Manufacturer: Jaypee cement) has been used for the study [24].

2.2.2 Fine Aggregate

Fine aggregates in conformance with Zone II as per IS: 383—1970 were adopted in the study [25].

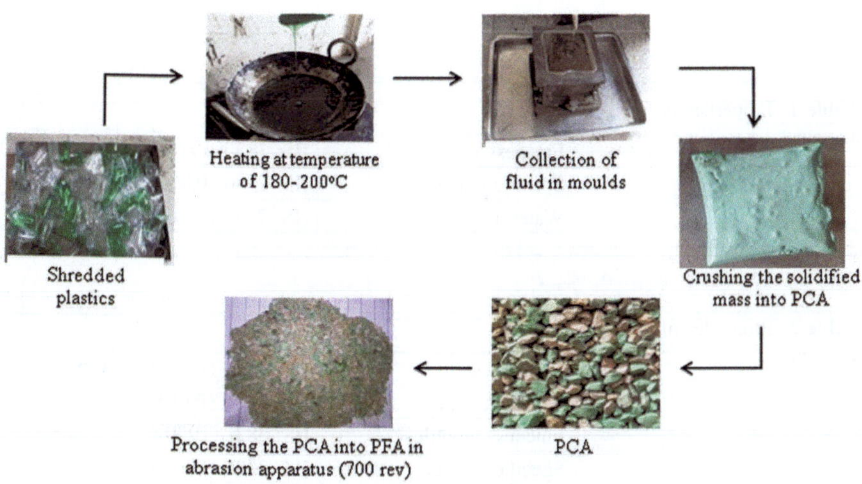

Fig. 1 Methodology adopted for processing the waste plastics

2.2.3 Processed Materials (PCA and PFA)

The obtained PCA was observed to be passing through 12.5 mm and retained on 10 mm, PFA used was in accordance with Zone II sand as per IS: 383—1970.

2.3 Testing Program

The processed materials (PFA and PCA) were investigated for understanding the feasible applications in the construction sector such as block manufacturing, soil stabilization, and pavement construction. Processed materials (both PFA and PCA) were studied with respect to physical and mechanical properties of aggregate [25, 26]. Tables 1 and 2, show the parameters investigated and the standard procedures adopted in the study.

2.3.1 Blocks

Blocks were prepared in size (200 × 200 × 450) mm. The proportions for preparing the blocks were maintained in conformance with the blocks manufactured in the campus for construction purpose. Blocks were prepared with 1:2:3 proportions of cement, stone dust, and coarse aggregate, respectively. In the present study, stone dust has been replaced with PFA (100%); water to cement ratio was maintained as 0.5 for preparing the blocks. Coarse aggregates used in the study were in the size range passing through BIS sieve 10 mm and retained on 4.75 mm [27] sieves were considered. The prepared blocks were tested for their compressive strength and were compared with the standard blocks used for construction in the campus.

Table 1 Properties of fine aggregate

Property	IS code
Fineness modulus	[IS: 383—1970]
Specific gravity	[IS: 2386 (Part-III)—1963]
Water absorption	[IS: 2386 (Part-III)—1963]

Table 2 Properties of coarse aggregate

Property	IS code
Impact test	[IS: 2386 (Part-IV)—1963
Abrasion value	[IS: 2386 (Part-IV)—1963]
Fineness modulus	[IS: 383—1970]
Specific gravity	[IS: 2386 (Part-III)—1963]
Water absorption	[IS: 2386 (Part-III)—1963]

2.3.2 Marshall Stability

Marshall stability test was performed in the present study with a primary objective of understanding influence of PFA (as 100% replacement to NFA) on the load resistance, flow value, air voids, and voids filled with bitumen. The specimen was prepared by maintaining varying binder content (4, 5 and 6%). The influence of PFA on the binder content and NCA has been studied [11, 28].

2.3.3 California Bearing Ratio Test (Unsoaked)

The CBR test is meant for the evaluation of subgrade strength of pavements. The test essentially measures the soil resistance to penetration prior to reaching its ultimate shearing value. Optimum moisture content was maintained as 7% for all the specimens after evaluation on trial specimens at varying percentages. The present study was undertaken on black cotton soil, PFA was added up to 30% (by volume) of soil. Specimens were prepared at 10, 20, and 30% replacement levels, respectively, and tested as per the standard procedure [29].

3 Results and Discussion

3.1 Aggregate Properties

The properties of PCA compared with NCA were shown in Table 3. Properties of PCA were evaluated to assess their suitability for application as pavement material and concrete production. The results indicated the conformance with the standard values. Mechanical properties such as crushing value, impact value, and abrasion value were compared with required properties of NCA for application in wearing course of the pavement; the results suggested the suitability for utilization in wearing course. Physical properties such as fineness modulus, specific gravity, and water absorption have shown the conformance with standard values in accordance with IS: 383—1970 [25, 26].

Table 3 Physical and mechanical properties of aggregates [25]

Test	PCA	NCA
Crushing value (%)	27.2	<30
Impact value (%)	23.9	<30
Abrasion value (%)	33	30–40
Fineness modulus	7.04	5.5–8
Specific gravity	1.22	2.6
Water absorption	Negligible	1–2%
Bulk density (g/cm^3)	0.8	1.65

Table 4 Comparison of PFA with NFA

Test	PFA	NFA
Fineness modulus	2.59	2.6
Specific gravity	1.53	2.5
Bulk density (g/cm^3)	0.72	1.3

The properties of PFA compared with NFA were shown in Table 4. The obtained values were in conformance with desirable properties as per IS: 383—1970.

Discussion

- The properties of PCA and PFA were found to be identical to the virgin materials commonly used in concrete production.
- The specific gravity of PCA was found to be 1.22 (as compared to 2.6 for NCA), indicating the suitability of the aggregate in production of lightweight concrete. The low density of the concrete provides relative savings in construction cost and reduced structural load on its components.
- The water absorption of PCA was found to be negligible. The concrete produced with materials such as PCA shows lesser inclination to segregation maintaining the homogeneity of mix providing better workability.
- PCA has shown the better resistance at varying load conditions (gradual and sudden) and wear. Indicating the utilization as pavement material due to its resistance to disintegration, compressive stresses, and moving loads.
- The properties of PFA proved to be feasible for application in preparation of lightweight mortars.

3.2 *Concrete*

Previously, the authors have studied the properties of PCA in concrete production. PCA was replaced up to 20% of NCA for preparing M20 grade concrete. Compressive strength of specimens was tested after 28 days. Results were shown in Fig. 2. Compressive strength has been evaluated to indicate the overall quality of concrete. The results indicated that at 10% replacement, maximum compressive strength of 30.4 MPa was observed; which was 14.3% higher than the target mean strength. Up to 20% replacement, concrete has shown better performance that decreased with further replacement [30].

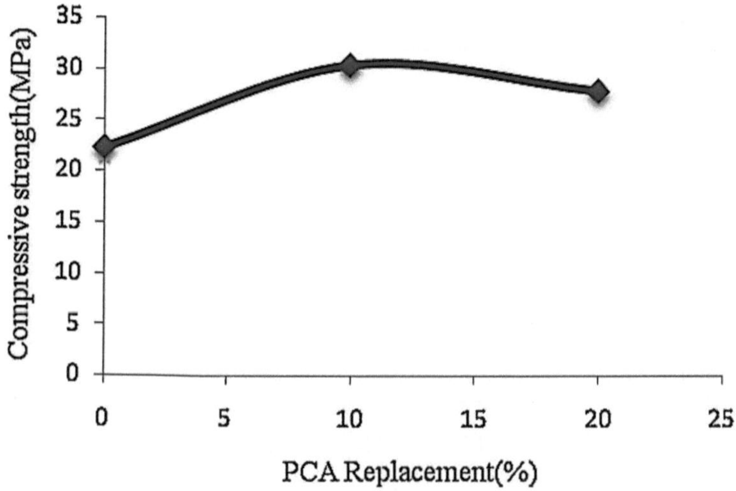

Fig. 2 Compressive strength of concrete with partial replacement of PCA

3.3 Blocks

The results of compressive strength of blocks prepared with 100% PFA were compared with the standards related to IS: 2185 (Part-I)—2005 [31]. The compressive strength of blocks was observed to be 4.73 MPa, which was higher than the blocks being used in the campus for construction (i.e., 4.23 MPa). The results indicate that the crushing strength was found to be 12% higher than the blocks being utilized for construction, and the weight of blocks was also reduced by same percentage. The obtained values were much higher than the crushing strength values for standard blocks as prescribed in IS: 2185 (Part-)—2005 (i.e., 3.5 MPa).

3.4 Soil Stabilization

CBR values were studied at penetration values of 2.5 and 5 mm on specimens replaced with 10, 20, and 30% PFA. The values of CBR at 5 mm penetration were found to be much higher as compared to 2.5 mm after verification. Hence the values 5 mm penetration were considered as CBR values. With increasing the percentage replacement of PFA, the results indicate the enhancement of subgrade strength at partial replacement (up to 20%). Maximum value was observed at 10% replacement, but however specimens have shown marginal increase up to 30% replacement. Results could be improved further by addition of coarser fraction with lesser number of revolutions in abrasion testing machine while processing PFA.

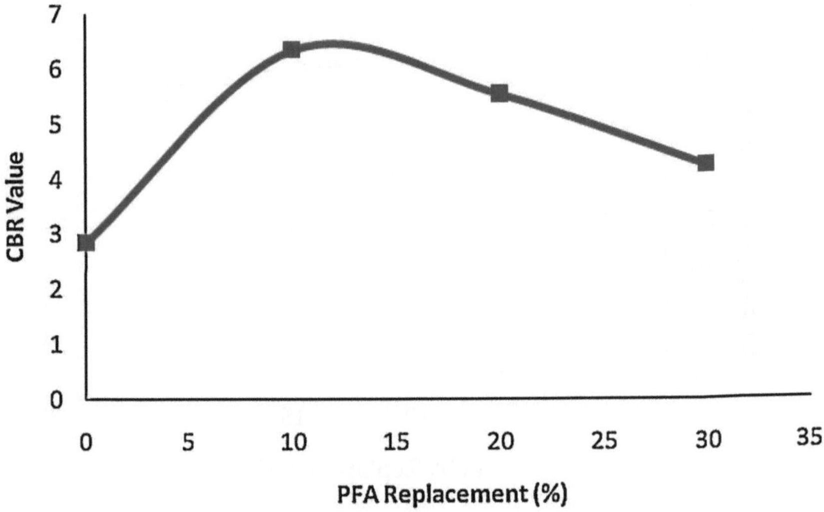

Fig. 3 CBR value at penetration 5 mm

However, the test indicated the suitability of PFA for application in soil stabilization for enhancing the strength of subgrade layer for black cotton soil (Fig. 3).

3.5 Pavements

Marshall stability test was performed with an objective of assessing the load resistance, flow value, air voids, and voids filled with bitumen, for the specimen prepared with varying binder content (4, 5, and 6%) and the fine fraction was totally replaced with PFA (100%) in preparation of specimen. The results obtained were compared with normal specimen prepared with aggregate (coarse and fine), filler, and binder as shown in Table 5. Optimum binder content has been obtained at 5% replacement. The presence of PFA enhanced the load resistance above 400% as compared to normal specimen. Flow value of the specimen containing PFA was observed to be much lesser than normal specimen, indicating the adequate workability preventing segregation. Lower air voids as compared to normal specimen prevent the percolation of water through the pavement reducing the deformation due to rutting. The results indicate the suitability of PFA to be used in conjunction with binder (bitumen) enhancing the properties of the bitumen. Hence the result proves that addition of PFA increases the durability, shear strength, and fatigue resistance enhancing the life of pavement.

Table 5 Marshall stability value

S. No.	Type	Bitumen content (%)	Flow value (mm)	Stability value (kg)
1	Normal	4	2.4	407.72
2		5	3.9	534
3		6	6.1	443.83
4	Plastic (PFA)	4	1.2	1734.85
5		5	1.8	2364.66
6		6	2.5	1397.91

4 Conclusion

The present work has investigated the feasible options for utilization of processed recyclable plastics in construction sector. The followings were the conclusions drawn from the study

- The processed materials (PCA and PFA) proved to be in conformance with the properties of virgin materials suggesting the suitability for application in construction sector.
- Mechanical properties of PCA have proved to be effective for utilization as aggregate in wearing course of pavement and production of lightweight concrete.
- Blocks manufactured with PFA possessed properties such as lightweight and adequate crushing value as compared to the standard blocks currently being used in the campus for construction.
- The result of the Marshall stability test has proved that the specimen containing PFA as replacement to NFA provided better properties, suggesting the compatibility with the binder for enhancing the life of the pavement.
- PFA has proved to enhance the CBR value of black cotton soil effectively up to 20% replacement.

4.1 Suggestions

The present work has broadly examined the feasible applications of recycled plastic waste in the construction sector, the following aspects could be considered for further research:

- Development of modified processing technique under controlled conditions for capturing the higher calorific value volatile gas produced during the melting of plastics provides multiple benefits.

- The present work has examined the suitable applications of the recyclable plastics only; the work could be extended to mixed plastics stream recovered from the solid waste.
- The effect of atmospheric agencies on structural components incorporating these processed materials has to be further investigated for a wider application.
- The study could be extended further by processing PFA through controlling the number of revolutions (300–700 revolutions), for obtaining well-graded material.

References

1. Ahmed, A., Ugai, K., Kamei, T.: Investigation of recycled gypsum in conjunction with waste plastic trays for ground improvement. Constr. Build. Mater. **25**, 208–217 (2011)
2. CPCB: Assessment of plastic waste and its management at airports and railway stations in Delhi, Central Pollution Control Board. http://cpcb.nic.in/upload/newitems/newitem_155_final_rite_report.pdf (2009)
3. Bajracharya, R.M., Manalo, A.C., Karunasena, W., Lau, K.: Characterisation of recycled mixed plastic solid wastes: coupon and full-scale investigation. Waste. Manag. **48**, 72–80 (2016)
4. Gug, J., Cacciola, D., Sobkowicz, M.J.: Processing and properties of a solid energy fuel from municipal solid waste (MSW) and recycled plastics. Waste Manag. **35**, 283–292 (2015)
5. Saikia, N., Brito, J.: Use of plastic waste as aggregate in cement mortar and concrete preparation: a review. Constr. Build. Mater. **34**, 385–401 (2012)
6. Al-Salem, S.M., Lettieri, P., Baeyens, J.: Recycling and recovery routes of plastic solid waste (PSW): a review. Waste Manag. **29**, 2625–2643 (2009)
7. Fletcher, B.L., Mackay, M.E.: A Model of plastics recycling: does recycling reduce the amount of waste? Resour. Conserv. Recycl. **17**, 141–151 (1996)
8. Rigamontia, L., Grossoa, M., Møllerb, J., Martinez Sanchez, V., Magnania, S., Christensen, T.H.: Environmental evaluation of plastic waste management scenarios. Resour. Conserv. Recycl. **85**, 42–53 (2014)
9. Ruiz-Herrero, J.L., Nieto, D.V., López-Gil, A., Arranz, A., Fernández, A., Lorenzana, A., Merino, S., De Saja, J.A., Rodríguez-Pérez, M.A.: Mechanical and thermal performance of concrete and mortar cellular materials containing plastic waste. Constr. Build. Mater. **104**, 298–310 (2016)
10. Sharma, R., Bansal, P.P.: Use of different forms of waste plastic in concrete—a review. J. Clean. Prod. **112**, 473–482 (2016)
11. Indian Road Congress (IRC): Guidelines on use of modified bitumen in road construction, IRC SP: 53 (2010)
12. Vasudevan, R., Sekar, A.R.C., Sundarakannan, B., Velkennedy, R.: A technique to dispose waste plastics in an ecofriendly way—application in construction of flexible pavements. Constr. Build. Mater. **28**, 311–320 (2012)
13. García-Morales, M., Partal, P., Navarro, F.J., Gallegos, C.: Effect of waste polymer addition on the rheology of modified bitumen. Fuel. **85**, 936–943 (2006)
14. Ahmadinia, E., Zargar, M., Karim, M.R., Abdelaziz, M., Shafigh, P.: Using waste plastic bottles as additive for stone mastic asphalt. Mater. Des. **32**, 4844–4849 (2011)
15. Behl, A., Sharma, G., Kumar, G.: A sustainable approach: utilization of waste PVC in asphalting of roads. Constr. Build. Mater. **54**, 113–117 (2014)

16. Köfteci, S., Ahmedzade, P., Kultayev, B.: Performance evaluation of bitumen modified by various types of waste plastics. Constr. Build. Mater. **73**, 592–602 (2014)
17. Azhdarpour, A.M., Nikoudel, M.R., Taheri, M.: The effect of using polyethylene terephthalate particles on physical and strength-related properties of concrete; a laboratory evaluation. Constr. Build. Mater. **109**, 55–62 (2016)
18. Corinaldesi, V., Donnini, J., Nardinocchi, A.: Lightweight plasters containing plastic waste for sustainable and energy-efficient building. Constr. Build. Mater. **94**, 337–345 (2015)
19. Yang, S., Yue, X., Liu, X., Tong, Y.: Properties of self-compacting lightweight concrete containing recycled plastic particles. Constr. Build. Mater. **84**, 444–453 (2015)
20. Ghernouti, Y., Rabehi, B., Bouziani, T., Ghezraoui, H., Makhloufi, A.: Fresh and hardened properties of self-compacting concrete containing plastic bag waste fibers (WFSCC). Constr. Build. Mater. **82**, 89–100 (2015)
21. Sommerhuber, P.F., Welling, J., Krause, A.: Substitution potentials of recycled hdpe and wood particles from post-consumer packaging waste in wood–plastic composites. Waste Manage. **46**, 76–85 (2015)
22. Yin, S., Tuladhar, R., Shi, F., Combe, M., Collister, T., Sivakugan, N.: Use of macro plastic fibres in concrete: a review. Constr. Build. Mater. **93**, 180–188 (2015)
23. Safi, B., Saidi, M., Aboutaleb, D., Mallem, M.: The use of plastic waste as fine aggregate in the self-compacting mortars: effect on physical and mechanical properties. Constr. Build. Mater. **43**, 436–442 (2013)
24. Bureau of Indian Standards: Specification for Portland Pozzolana Cement, IS 1489 Part 1, New Delhi (1991)
25. Bureau of Indian Standards: Specification for Coarse and Fine Aggregates from Natural Source of Concrete, IS: 383, New Delhi (1970)
26. Bureau of Indian Standards: Methods of Test for Aggregates for Concrete, IS: 2386, New Delhi (1963)
27. Bureau of Indian Standards: Specifications for Test Sieves, IS: 460, New Delhi (1985)
28. American Society for Testing and Materials (ASTM): Standard Test Method for Marshall Stability and Flow of Asphalt Mixtures. ASTM D6927 (2015)
29. Bureau of Indian Standards: Methods of Test for Soil, IS: 2720, New Delhi (1980)
30. Sharma, R., Sahu, K.K., Gupta, M., Bashani, K.: Recycled plastic as coarse aggregate for concrete. International UKIERI concrete congress on concrete research driving profit and sustainability, pp. 1070–1074 (2015)
31. Bureau of Indian Standards: Concrete Masonry Units—Specification, Hollow and Concrete Solid Concrete Blocks, IS: 2185 (Part 1), New Delhi (2005)

Estimation of Landfill Gas Using Genetic Programming

Kshitija S. Tikhe, Basavraj S. Balapgol and Sandip T. Mali

Abstract Large amount of greenhouse gases (CO_2 and CH_4) are generated through the disposal of municipal solid waste in landfill. Methane is an increasing concern of greenhouse gas. Complex processes taking place within the landfill leads to formation of gas which has to be managed. In order to manage this landfill gas (LFG), it is necessary to estimate it on daily basis. Conventional hard computing techniques as well as modern soft computing techniques have been used to model LFG estimation. The present study uses soft computing method of linear genetic programming (LGP), to estimate the landfill gas emission for Pune city (India). Data from the simulated laboratory-scale landfill have been used, and the temporal models are developed. Landfill gas is estimated using previous values of the gas recorded. The performance of the models has been analyzed using correlation coefficient (r) and root mean square error (RMSE). It is found that the model results are in good agreement with the actual values.

Keywords Landfill gas · Genetic programming · Pune city · Municipal solid waste

K. S. Tikhe (✉) · B. S. Balapgol
Department of Civil Engineering, D.Y. Patil College of Engineering,
Akurdi, Pradhikaran, Pune 411044, Maharashtra, India
e-mail: kshitija24@gmail.com

B. S. Balapgol
e-mail: b_basavraj@yahoo.com

S. T. Mali
Department of Civil Engineering, Pimpri Chinchwad College of Engineering,
Nigdi, Pradhikaran, Pune 411044, Maharashtra, India
e-mail: sandipmali05@gmail.com

A. S. Kalamdhad et al. (eds.), *Advances in Waste Management*,
https://doi.org/10.1007/978-981-13-0215-2_12

1 Introduction

Anaerobic digestion is the well-established bioprocessing technology to produce highly energetic biogas. Mathematical modeling of the anaerobic digestion is very complex [1]. Landfill gas (LFG) modeling is the practice of forecasting gas generation and recovery, based upon present and past waste disposal histories. LFG modeling is mainly carried out for regulatory and the non-regulatory purposes.

Studies have been carried out by different researchers for understanding and evaluating the gas generation and biodegradation process. In these studies, predictions have been carried out using empirical functions [2–4]. Model results reveal some drawbacks in prediction when empirical functions have been used. This leads to the requirement of another modeling technique which would overcome these drawbacks, and hence, the researchers have used the soft computing tools [5]. The two well-known soft computing tools, namely artificial neural network and fuzzy logic, were used for estimating the gas generation. Along with these tools, a relatively new tool 'linear genetic programming (LGP)' is also used for evaluating some environmental issues [6].

The present work consists of the development of LGP models to estimate the LFG. Multiple linear regression (MLR) is used as a benchmarking tool, and the results of LGP are compared with MLR.

2 Landfill Gas Generation Process

The landfill gas generation process starts with the breaking down of the organic material into simpler material in an aerobic and anaerobic condition. The gas generation process is divided into five stages as shown in Fig. 1. In stage one, the organic matter present in the waste gets decomposed in the presence of oxygen and moisture, leading to the formation of the gases CO_2 and H_2O. In the second stage, the oxygen present in the waste goes on reducing and the facultative bacteria start taking part in further degradation process. These bacteria hydrolyze the organic acids, proteins, and fatty acids to sugars. In stage three, acetogenic bacteria present in the waste further ferment the acids generated in the previous stage. The next stage (stage four) is the most important phase in gas generation which is known as the methane generation stage where methanogenic bacteria convert the acids formed in the initial stages into CH_4 and CO_2. In the last stage (stage five), i.e., maturation stage, the gas generation process gradually stops and some portion of CH_4 generated in the earlier stage gets oxidized into CO_2 and H_2O [7].

2.1 Landfill Gas Estimation Models

Figure 1 shows the process which takes place at the landfill site or in the simulated reactor by which CH_4 is generated with the passage of time. In order to estimate the generated gas, some of the hard computing methods like default methodology, first-order decay, triangular method, and modified triangular methods are used [8]. In default methodology, the model developed is based on the mass balance concept, while in the first-order decay method, it is assumed that the total gas generation process takes place in two phases. In the first phase, gas generation rate goes on increasing till a peak is reached, and then, it starts reducing till the material gets

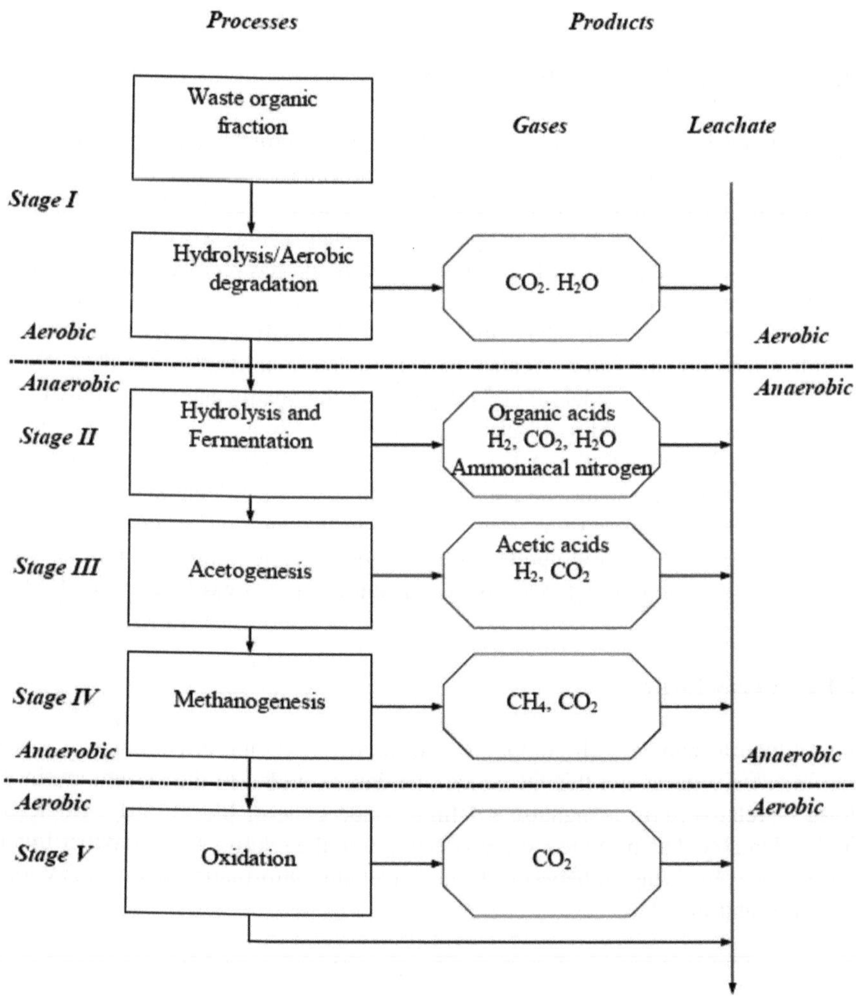

Fig. 1 Stages of waste degradation and gas generation (*Source* Abushammala et al. [7])

stabilized. In the triangular and modified triangular method, the amount of landfill gas generated corresponds to the area of the triangle for a particular period of time [8].

Hard computing models or conventional computing models need considerable amount of time for their computation as well as they follow a particular algorithm during these estimation processes. Also, the accuracy given by hard computing models is comparatively less. Hence, in this circumstance, soft computing techniques prove beneficial over hard computing techniques [9].

Soft computing techniques can work with limited dataset as well as they can work with ambiguous and noisy data. Moreover, the time required for soft computation is comparatively less than conventional hard computing techniques [9]. Following are some of the soft computing methods applied for LFG estimation.

2.1.1 Artificial Neural Network (ANN)

Artificial neural network (ANN) is biologically inspired parallel computational models. They consist of simple highly interconnected processing elements which process the input similar to human brain. ANN acquires, represents, and computes mapping from one multivariate space to another using self-learning. This is regarded as an intelligent, cost-effective approach and has received much attention in environmental engineering. ANN is a universal approximator which does not need any prior knowledge about the system to be analyzed. ANN has ability of adaptive learning and can recognize the patterns but is tolerant of the data. As the performance of ANN models entirely depends on the quality and the quantity of the data, they are considered as black box. Deciding ANN architecture is a complex process which demands the skill of the programmer as there is no fixed rule to decide the data division and the number of neurons in the hidden layer. Sometimes, ANN models may suffer from overfitting due to the addition of too many hidden neurons or executing large number of iterations. Despite many advantages, some researchers have found that ANN is insufficient to predict extreme events [10].

2.1.2 Fuzzy Logic

Fuzzy logic is another soft computing tool used for solving environmental issues. As its name suggests, in this technique, the data provided to the system are in the form of relative term as against a definite value or term like 0 or 1—Boolean's logic. Here, the data provided to the system are in the relative term of either true or false with value lying in between 0 and 1. Detailed information about fuzzy logic can be found in [11].

2.1.3 Genetic Programming

Genetic programming is a relatively new evalutionary based data-driven tool. It is similar to genetic algorithm (GA) which follows the Darwinian principle of the survival of the fittest and which obtains the solution to the problem through the process of crossover, mutation, and reproduction. Computer program or an equation is obtained as a solution of GP, whereas set of numbers are obtained in GA [10]. As a result, GP can be suitably used as a regression tool. The evaluation process of GP starts with computer programs constructed from the dataset. Then, the genetic operations are performed using function and terminal set. Function set comprises of the operator to be used such as addition, subtraction, logarithm, square root. Terminal set consists of values such as inputs, constants, temporary variables on which function set operates. From the population so generated, four programs are randomly selected and a tournament is conducted. Performance of each program is measured by GP, and the two programs are selected on the basis of performance. The two winner programs are copied by GP algorithm and transformed into two new programs via crossover and mutation as per fitness. In an iterative manner, offsprings are created till the specified number of offsprings is created in generation and also till the specified number of generations is produced. The solution of the problem is obtained as the resulting offspring at the end of the process in the form of either a computer program or an equation. The generated equation or program can be directly applied to unseen data to obtain required predictions. Details about genetic programming can be found in [12]. To the best of authors' knowledge, LGP has not been used for LFG estimation for the study area (Fig. 2).

3 Study Area and Data

Data used for the analysis have been obtained from a simulated laboratory-scale model of a landfill to study the effect of the operational factors and the leachate recirculation on the amount of the gas generation. The solid waste used for the study is collected from the landfill site of Uruli Devachi which is situated on Pune Saswad Road in Pune (Maharashtra, India). Five reactors simulating different landfill conditions are designed. These reactors are divided into two groups: Group 1 consists of three reactors (R1, R2, and R3) which are fed with the mixed waste, while the second group consisted of two reactors (R4 and R5) which have only organic waste. Reactor R1 simulates open dump without leachate recirculation,

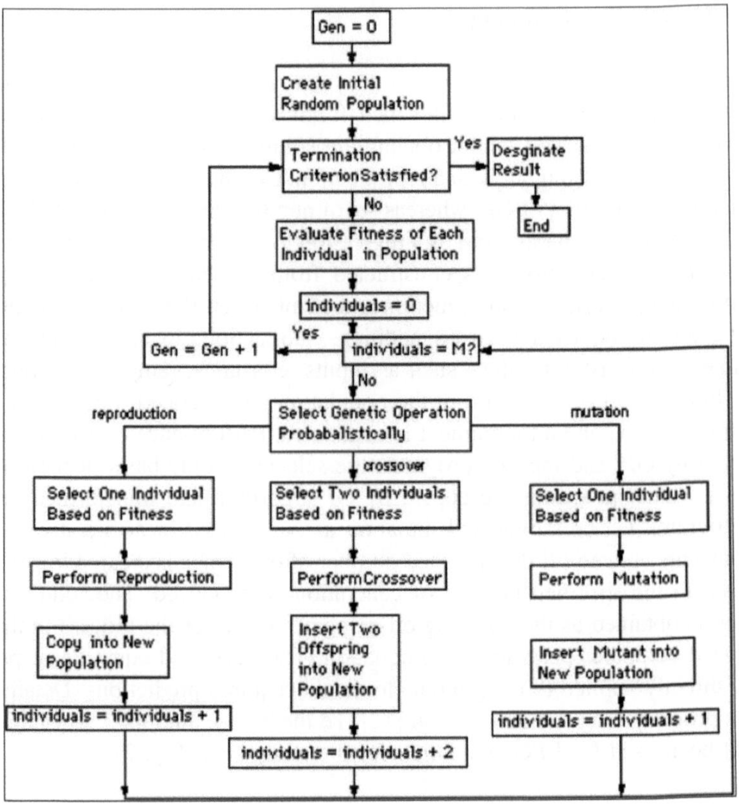

Fig. 2 Flowchart of genetic programming (*Source* Koza [12])

whereas the leachate was recirculated in Reactor R2. In case of Reactor R3, gravel and microbial culture have been added to the waste to enhance its functionality. Along with this, leachate has been also added after a definite period of time. For Reactor R4, only leachate has been recirculated, whereas in Reactor R5, all the treatments similar to Reactor 3 are given. The experimental measure includes daily biogas measurements. The measurements are taken for nine months from February 2010 to October 2010. Different features and configuration of bioreactors are specified in Table 1.

Table 1 Configuration and features of bioreactor (*Source* Mali [13])

S. No.	Parameters	Group first (mixed MSW)			Group second (compostable MSW)	
		R1	R2	R3	R4	R5
1	Type of bioreactor	Open dump	Anaerobic bioreactor landfill	Anaerobic bioreactor landfill	Anaerobic bioreactor landfill	Anaerobic bioreactor landfill
2	Quantity of wet waste, kg	32	26.5	30	30	35
3	Moisture content, %	16.78	16.78	16.78	22.84	22.84
4	Quantity of dry waste, kg	26.30	22.05	24.97	23.15	27
5	Size of shredded MSW, mm	30–70	30–70	30–70	30–70	30–70
6	Quantity of anaerobic digested sludge added at starting, L	No	16	10	14.5	10
7	Gravel mixing at starting, kg	No	No	10	No	10
8	Soil addition, kg	6	10	10	10	10
9	Density, kg m^{-3}	538.24	517	708.21	566.57	779.03
10	Aeration at starting (first 14 days)	No	No	Yes	No	Yes
11	Addition of bioculture (150 g) and EM solution (600 ml)	No	No	Yes	No	Yes
12	Weight reduction after aeration process (after 14 days)	–	–	2.4 kg (9.6%)[a]	–	1.95 kg (7.22%)[a]
13	Leachate recirculation	Not	Yes	Yes	Yes	Yes
14	Frequency of leachate recirculation (10–13.5%, v/v)	–	2	2–4	2	2–4
15	Leachate recirculation variation during process	–	Yes	Yes	No	Yes
16	Maintain pH of leachate during leachate recirculation	–	Yes	Yes	Yes	Yes
17	Adding of anaerobic digested sludge intermediately during leachate recirculation	–	No	Yes	No	Yes

[a]Indicates the percentage of weight reduced with respect to the dry weight of waste after aeration

Table 2 Optimum number of input parameters

S. No.	Reactor	Optimum number of input parameters
1	R1	2
2	R2	4
3	R3	2
4	R4	2
5	R5	6

4 Model Development

Temporal models are developed using Multiple Linear Regression (MLR) and Linear Genetic Programming (LGP). In order to decide the optimum number of inputs for the temporal model, initially regression models are developed with the addition of inputs one by one till six in order to study the weekly variation of the LFG. The correlation of the model output with the observed values is calculated. Optimum numbers of inputs are chosen as the inputs for which the highest value of the correlation coefficient is observed. Optimum number of inputs for all the models is listed in Table 2. MLR and LGP models are developed and analyzed using the optimum number of inputs as suggested by correlation analysis.

MLR models are developed with an objective of establishing the relationship between the input and output. MLR models are developed using excel tool of MS Office version 10. Confidence level was set as 95%. Similarly, linear genetic programming (LGP) models are developed for each of the reactors using software of Discipulus by setting the initial parameters as population size = 500, crossover frequency = 50%, and mutation frequency = 95%.

In order to build the model, the available data have been checked for its continuity and gaps are filled using linear interpolation method. The data division of the model for MLR and LGP is set to 70% (training data) and 30% (testing data).

5 Results and Discussion

5.1 General

Landfill gas generation involves different physical and chemical processes. Numbers of factors influence the gas generation process. Integrating all the influential factors and developing a gas estimation model are a challenge. In the present study, gas generation data as obtained from five reactors have been used to estimate the amount of LFG generation. Multiple linear regression (MLR) which is a well-established tool has been used as a benchmarking, and the feasibility of LGP is assessed for forecasting of LFG. Model's performance is assessed using '*r*' and 'RMSE' as the evaluators, and the results are displayed in Table 3.

Table 3 Performance evaluation

Reactor	Correlation coefficient (*r*)		Root mean square error (RMSE)	
	MLR	LGP	MLR	LGP
R1	0.7448	0.7874	0.9543	0.4040
R2	0.9143	0.9324	15.2719	7.2674
R3	0.8727	0.9159	6.7169	5.3071
R4	0.9064	0.9060	1.4756	1.2272
R5	0.9142	0.9546	9.6305	4.5042

5.2 Results

The first model is developed for Reactor R1 which has two previous days of LFG values as their input while LFG of the same day as the output. The LGP estimated LFG reasonably well with $C_{LGP} = 0.7874$ as compared to MLR ($C_{MLR} = 0.7448$). Figure 3 shows LFG estimation by both the tools. It can be seen that GP has attempted to map almost all the peaks where MLR has failed.

The second model is developed for Reactor R2 with four previous days of LFG values as input while LFG of the same day as the output. In this model too, LGP has performed better than regression with $C_{LGP} = 0.9324$ while $C_{MLR} = 0.9143$. Here, GP has mapped the peak reasonably well, but MLR has failed to do so which can be seen from Fig. 4.

The next model developed is for Reactor R3 with two previous days of LFG values as inputs and same day value as output. Both MLR and GP models estimated LFG reasonably well with LGP slightly better ($C_{MLR} = 0.8727$, $C_{LGP} = 0.9159$). Figure 5 represents the graphs showing the estimation of LFG for Reactor R3.

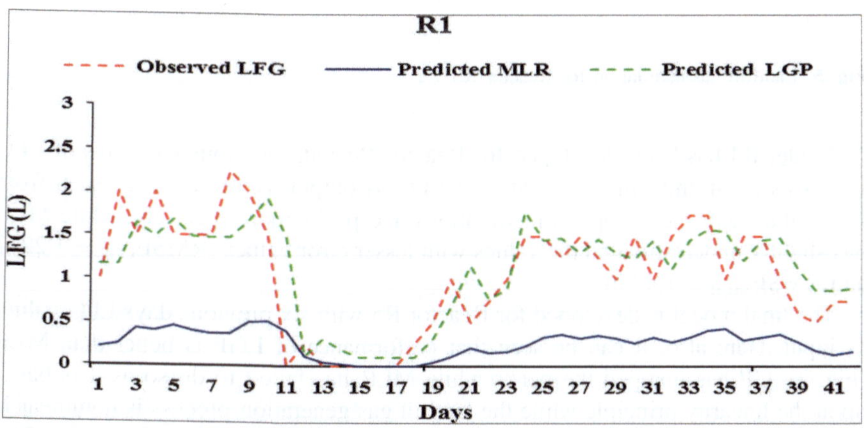

Fig. 3 Landfill gas estimation for Reactor R1

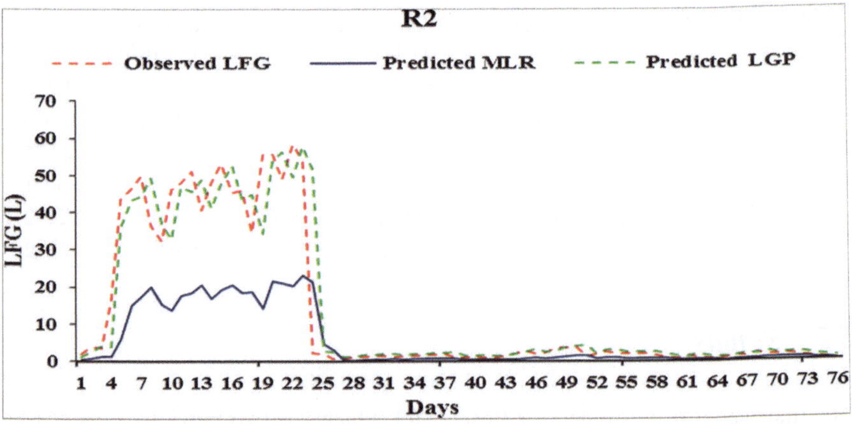

Fig. 4 Landfill gas estimation for Reactor R2

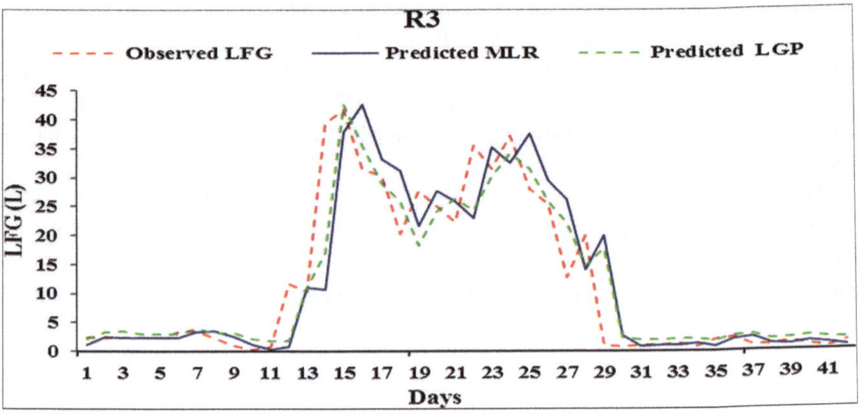

Fig. 5 Landfill gas estimation for Reactor R3

Model R4 has been developed for Reactor R4 with previous two days of LFG values as input and same day value of LFG as output. Here, it can be seen from Fig. 6 that LGP has mapped the variations and peaks more precisely, while MLR has slightly underestimated the values with lesser error values ($RMSE_{LGP} = 1.2272$ and $RMSE_{MLR} = 1.4756$).

The final model is developed for Reactor R5 with six previous days LFG values as input. Here also, it can be seen that performance of LGP is better than MLR (Fig. 7). GP has mapped the peaks while MLR has failed to do so as it is based upon the linearity principle while the landfill gas generation process is nonlinear in nature. The performance of the LGP is better than MLR ($C_{LGP} = 0.9546$, $C_{MLR} = 0.9142$).

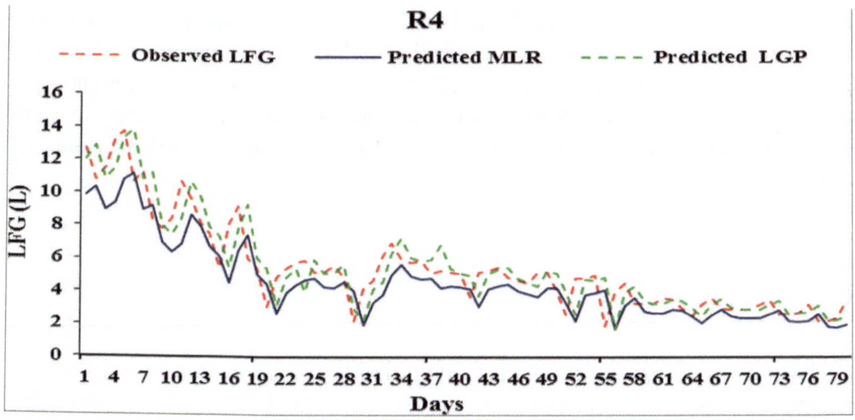

Fig. 6 Landfill gas estimation for Reactor R4

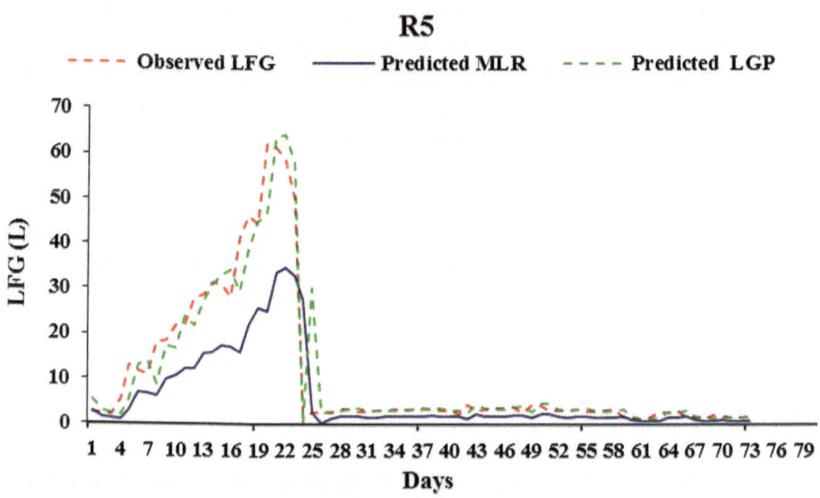

Fig. 7 Landfill gas estimation for Reactor R5

5.3 *Discussion*

It can be seen from all the models that LGP performed consistently better than MLR. It is able to trace the nonlinearity in the LFG generation process unlike MLR which is based on linearity principle. For R1 model, the correlation between the observed LFG and estimated LFG was less. The reactor was simulating an open dump without leachate recirculation and addition of sludge as in case of Reactor R3 and Reactor R5 whose corresponding models show good correlation. LGP has tried to map the variations in output due to variations in operating conditions of the reactor. Thus, it can be indicated that LGP works well as it can modify itself based upon the data type, volume of data, and population function tree for ultimately getting an optimal and physically interpretable model. However, a common phenomenon is observed in all the LGP models, and some MLR models that, the values were estimated nearly accurate but there is a positive lag between measured and estimated one.

6 Conclusions

In the present study, linear genetic programming has been used to estimate LFG using a dataset from simulated laboratory-scale landfill experimentation. Five models representing five reactors are developed. The results of the models are compared with the well-established tool of multiple linear regression.

From all the models, it can be stated that LGP performed better as compared to MLR. GP could map the peaks and variations which are there in the observed LFG values. In general models achieved acceptable statistical measures in terms of correlation coefficient and root mean square error. Although GP performed well, it showed a positive lag in its estimation which is required to be studied as a future scope. Moreover, performance of GP is mainly dependent on the quality and quantity of the data. In this analysis, the phenomenon of leachate recirculation is considered while preparing the models, but the stage-wise analysis of the LFG has not been considered. In all, it can be concluded from the results that GP can be used as one of the soft computing tools for estimation and prediction of LFG, provided all the limitations mentioned above are taken into account.

References

1. Adak, A., Mazumder, D., Bandyopadhyay, P.: Simulation of a process design model for anaerobic digestion of municipal solid wastes. Int. J. Civil Env. Eng. **3**(3), 177–182 (2011)
2. Ubavin, D., Madous, N., Milovanovic, D., Stege, G.A., Leatherwood, C.: Preliminary estimate of methane production at Belgrade MSW landfill "Vinca". In: The 6th PSU-UNS International Conference on Engineering and Technology, pp. T7–2.5 (2013)
3. Abushamala, M.F.M., Basri, N.E.A., Younges, M.K.: Landfill Methane oxidation: predictive model development. J. Appl. Sci. **15**(2), 283–288 (2015)
4. Abdallah, M., Warith, M., Narbaitz, R., Petriu, E., Kennedy, K.: Combining fuzzy logic and neural networks in modeling landfill gas production. World Acad. Sci. Eng. Technol **5**, 445–451 (2011)
5. Rada, E.C., Ragazzi, M., Stefani, P., Schiavon, M., Torretta, V.: Modelling the potential biogas productivity range from a MSW landfill for its sustainable exploitation. Sustanaility, 482–495 (2015) (MDPI Publications, Switzerland)

6. Tikhe, S.S., Khare, K.C., Londhe, S.N.: Forecasting criteria air pollutants using data driven approaches: an Indian case study. IOSR J. Env. Sci. Toxicol. Food Technol. **3**(5), 01–08 (2013)
7. Abushammala, M.F.M., Basri, N.E.A., Kadhum, A.A.H.: Review on landfill gas emission to the atmosphere. Eur. J. Sci. Res. **30**(3), 427–436 (2009)
8. Mali, S.T., Khare, K.C., Biradar, A.H.: Estimation of methane gas emission from municipal solid waste landfill: Pune city, India. EVRJ **6**, 1–11 (2012)
9. Londhe, S.N., Dixit, P.R.: Genetic programming: a novel computing approach in modeling water flows, pp. 1080–1089. Intech Publication (2012)
10. Bose, N.K., Liang, P.: Neural network fundamentals with graphs, algorithms and applications. Tata McGraw Hill Publishing Company Limited, New Delhi, India (1996)
11. Jang, J.S.R., Gulley, N.: Fuzzy logic toolbox for use with Matlab. The Math Work Inc., Massachusettes, USA (1995)
12. Koza, J.R.: Genetic Programming. MIT Press (1992)
13. Mali, S.T.: Anaerobic bioreactor landfill for bio-stabilization and green renewable energy generation from municipal solid waste: a case study. Ph.D. Thesis, University of Pune, India (2013)

Integrated Water Hyacinth Control and Waste Management Plan: A Case of Futala Lake, Nagpur

Nishant Joshi, R. K. Tomar, Madhuri Kumari and Shweta Khatri

Abstract Water hyacinth belongs to family Pontederiaceae and is a very remarkable and rapidly growing angiospermic plant, generally found in the freshwater bodies. Cost-effective and efficient control of water hyacinth growth and managing its resources is a challenge in most places where the problem has occurred. The present study identifies the effluents from upstream of surrounding area, immersion of idols of Gods during festivals and dumping of domestic garbage as the principal reasons for the growth of water hyacinth in Futala Lake, Nagpur, and proposes an integrated management plan to control the water hyacinth growth and eliminate the municipal solid waste being dumped in the lake at its source. Various studies have determined that besides having pernicious and inimical characteristics, water hyacinth is a remarkable source of biomass and hence makes good organic compost. The study compares the effectiveness of the mechanical, environmental, chemical and biological water hyacinth control methods in conjunction with other waste-to-compost and waste-to-energy technologies in terms of volume reduction, ease of operations, restrictions and conservations on type of waste processing, quality of compost, odour control, ease of organizing human resources, financial feasibility and uses the technology-specific emission factors in conjunction with the Intergovernmental Panel on Climate Change methodologies to determine the annual greenhouse gas (CO_2 eq.) emissions from them. The proposed integrated plan uses the inferences obtained from comparing the technical, financial

N. Joshi (✉) · R. K. Tomar · M. Kumari
Department of Civil Engineering, Amity University, Noida 201313, Uttar Pradesh, India
e-mail: ntj1304@gmail.com

R. K. Tomar
e-mail: brktomar@amity.edu

M. Kumari
e-mail: madhurikumari@gmail.com

S. Khatri
Faculty of Planning, CEPT University, Ahmedabad 380009, India
e-mail: shweta.khatri.mplan14@cept.ac.in

A. S. Kalamdhad et al. (eds.), *Advances in Waste Management*,
https://doi.org/10.1007/978-981-13-0215-2_13

and operational feasibility of applying different water hyacinth control methods in Futala Lake to make one integrated water hyacinth control and management plan and explains the stepwise procedure to eliminate the water hyacinth problem and control the wild dumps of municipal solid wastes near the lake by using waste-to-compost machines working on the principle of thermophilic decomposition. Thus, with respect to the analysis of the efficiency and effectiveness of the proposed management and control plan for Futala Lake, the study proposes that similar plans could be used to check the growth of water hyacinth in any freshwater body.

Keywords Water hyacinth · Pollution · Thermophillic decomposition Integrated management · Compatibility

1 Introduction

Water hyacinth (*Eichhornia crassipes*) is a floating plant found in many parts of the world, mostly in the freshwater bodies [1]. It usually grows in the freshwater bodies which have high nutrient content and has lavender-coloured flowers and round leather-like leaves which are attached on spongy stalks. Due to its rapid reproductive rate and intricate root structure, it normally forms thick and intertwining mats [2]. It grows like a floating mat on the freshwater bodies which doubles its size within one to two and a half weeks [3]. A water hyacinth mat of 1-acre area can weigh up to 200 tonnes (wet). It was found first in the Amazon Basin of Brazil but has then been introduced to many tropical and subtropical regions around the globe [4]. It has been recorded that by 1977, 56 countries had reported water hyacinth as a noxious weed [5–7]. The tropical and subtropical regions where there are higher rates of deforestation, agricultural runoff and inadequate wastewater treatment have recorded a rapid growth of water hyacinth. The spread of the mat on the water surface degrades the water quality by blocking the air–water interface and reducing the dissolved oxygen levels in the water. This not only blocks the access of aquatic plants and animals to oxygen and other nutrients but also adversely affects the terrestrial animals which depend on the water bodies for shelter and nesting [8]. The strongest determinant factors for the growth and reproduction of water hyacinth are temperature and nutrient contents [9]. But in the areas where temperature and nutrients are sufficient for faster regeneration and spread of water hyacinth, salinity has been determined as a limiting factor [10]. Even though *E. crassipes* is excluded from the colder regions because of the temperature limitations, it does show the signs of degree of freeze tolerance. The sections of the plant which are killed by the moderate-freeze events can potentially regrow from the underwater stem tips which are secluded from freezing by water [4]. It has been studied that water hyacinth is bigoted of salty conditions [5]. The regeneration potential of water hyacinth and its growth is inversely proportional to salinity [11]. Hence, apart from the freshwater bodies, it is only found in the low salinity coastal lagoon surroundings in the rainy season. Water hyacinth can reproduce both sexually and asexually [2]. Both the modes are equally important to the growth and development of the invasive species [12]. Explosive growth rates which double the population every 1–3 weeks are usually found in the places having high nutrient content during early spring to late fall [2, 13] during which it is capable of yielding an abundance of seeds [14]. Under such suitable conditions, water hyacinth has an average growth rate of 17.5 metric tonnes per hectare per day [15].

Water quality is affected in an adverse manner by the widespread and intense water hyacinth mats [16], which also blocks waterways [17]. In case of Futala Lake, which is a tourist and recreation spot in Nagpur, the widespread of water hyacinth has made boating and other recreational activities difficult to be conducted. Depletion of dissolved oxygen and fish kills are the direct impacts of plant respiration and decay of biomass. The blockage of irrigation channels, choking off of direction-finding routes, clogging of rice paddies, loss of fishing spots, amplification in breeding habitat to disease-transmitting mosquitoes are some of the major and common documented harmful economic impacts of its growth and spread [18]. There have been several researches towards searching ways of determining the value of otherwise invasive and bothersome weed, water hyacinth. There have been a number of uses of water hyacinth, determined by some researches in particular areas. But, these solutions are site-respective. The uses comprise of utilization of the weed as a phytoremediation agent [19], as compost [20], for power alcohol and biogas production [21, 22], as a biosorbent for toxic metals [23], and for animal fodder/fish feed [24]. Other uses which are being studied are its use as pulp material for making greaseproof paper [25] and manufacturing fibreboards for a diversity of end uses like bituminized board for cheaper roofing material and indoor partitioning. Processing of water hyacinth-based medication has been reported in India to treat certain diseases [26, 27]. In a developing country like India, where the skilled manpower and financial resources available for such purposes are limited, we need a method which can be applied in all alike cases. Thus, this study compares all the four methods used for water hyacinth control and management and then based on the technical, operational and financial feasibility proposes an integrated plan for Futala Lake and all alike cases.

2 Control Methods of Water Hyacinth

Mechanical, chemical, environmental and biological are the four methods which have been developed to control and eliminate the growth of water hyacinth [28]. The first step towards the effective control and management of water hyacinth is to identify the plants which fall under two categories: vascular plants and algae. Similar to familiar garden plants, the vascular plants have built-in transportation systems for water and nutrients [29]. These can be further divided into submergent, emergent, marginal and floating species.

Submergent plants are the ones which are attached to the bottom; the ones that are rooted in shallow water with the body extending above the surface are called emergent and the ones that are unrooted and can drift with the currents of wind and water are called floating species. The marginal species are found along the shoreline in shallow waters or heavily saturated soils. Hence, the control method which is chosen should be with respect to the regional climatic and topographical conditions and the type of weed.

Nagpur has a hot and humid climate. Municipal solid waste from the surrounding restaurants and residential area is dumped in the lake around the year. The sewage disposal in the lake has decreased over the past few years, but it is still there in some proportions. These factors aid the growth of water hyacinth in the lake which is then catalysed by the immersion of idols during the festive seasons. Thus, the technical, operational and financial feasibilities of the below-stated methods were studied to choose one or to integrate them and implement a water hyacinth control and municipal solid waste management plan.

2.1 Mechanical Method

Various machines such as draglines, boats fitted with mesh rakes, booms, mechanical harvesters, bulldozers and conveyors are used to mechanically control and check the water hyacinth spread [30]. Large floating weeds are removed by the harvesters, and the small plants are controlled by cutting, especially in their early growing season. These methods do not require the introduction of any kind of chemicals or bacteria in the environment, and hence, the operations are easier to handle [31]. But, this method tends to become labour-intensive and expensive.

This method can also be used for the submergent and emergent species, but regrowth may occur soon after the shredding, which would make it necessary to cut several times during a growing season. Moreover, it is also important to remove the fragments or cuttings from water because the species like hydrilla can even grow from the fragments floating on the water surface. If these fragments flow down to other parts of the water body, then they could also grow in the regions which were previously uninfested. Table 1 details the advantages and disadvantages of the mechanical method for water hyacinth control.

2.2 Chemical Control

Chelated copper compounds are widely used as herbicides for aquatic weed control as they are slightly less toxic to fishes and have been approved for use in drinking water [39]. Organic herbicides like 2,4-D (2,4-dichlorophenoxyacetic acid), diquat, endothall, glyphosate fluridone and others are widely available for aquatic weed control [40]. But using herbicides is never a permanent solution [41]. The seeds and

Table 1 Advantages and disadvantages of mechanical control method

Advantages of mechanical control	Disadvantages of mechanical control
In cases where the formation of dense mats on the surface have made the water body inaccessible [32]	In such cases, it requires the use of specialized equipment and hence it is expensive [36]
Once the water body is made accessible, then cheaper control methods like biological and chemical control can be used [33]	Seeds/pieces/parts of plants which are left floating on the surface are able to form new plants within a short span of time [37]
Plants can directly be removed and hence eutrophic conditions which may arise due to large volumes of decaying matter within the water body can be easily avoided [34]	Use of specialized equipment increases the operation and maintenance cost [38]
The removed plants or parts of plants can be productively used as biomass or as its substitute [35]	

tubers of aquatic weeds are not killed by the herbicides and hence could sprout within a few days if the conditions are favourable for their germination [42].

The herbicides are mostly applied to actively growing weeds in the spring after the temperatures are in the 60–70 °F range [43]. The topographical and climatic conditions of Futala Lake aid the germination and hence during summers, the dissolved oxygen levels decrease. This might be catalyzed if chemical dozing in the lake is not monitored properly [44, 45] and the consequences of which can also result in a fish kill. Generally, one-fourth of the water body is treated at a time, and the next section is treated after one to two weeks [46]. Surfactants and adjuvants are commercially available compounds which help the herbicides disperse evenly and also enable them to penetrate leaves in a better way [47]. However, the herbicides and compounds should not be applied on cloudy days as it increases the chances of oxygen depletion [48]. Table 2 details the advantages and disadvantages of chemical method for water hyacinth control.

2.3 Environmental Control

Reducing the amount of shallow water by reshaping the shoreline in order to remove the long gradual slopes and thereby reduce the shoreline vegetation. Sunlight directly reaches to the plants growing in shallow waters and hence catalyses the plant growth [52]. In such cases, reducing the amount of water available to the plants to allow a major fraction of vegetation to dry out can also be effective. Another method is to limit the sunlight so that plants cannot make adequate food. There are dyes and related chemicals which are commercially available, which reduce the penetration of light into water.

But, this technique is less effective when plants start growing closer to the surface [29]. Fertilizing the pond to enhance the growth of phytoplankton is also a

Table 2 Advantages and disadvantages of chemical control method

Advantages of chemical control	Disadvantages of chemical control
Impact on high and low-density water hyacinth infestations could be achieved within 6 weeks of application [49]	Chemicals are expensive to procure and store
Weeds are completely killed by the chemicals, which reduces the chances of regrowth as happens in manual and mechanical control methods	Decay of large volumes of weed creates anaerobic conditions which result in fish kills. The dose of herbicides can also impact some of the non-target organisms [50]. (The impacts are site-respective. As no such experiment was performed on the lake, the impacts are not quantified)
	Excessive use of chemicals may also have an adverse impact on the water quality, making it unfit for domestic use [51]

Table 3 Advantages and disadvantages of environmental control

Advantages of environmental control	Disadvantages of environmental control
The control measures and solutions are natural and hence sustainable in comparison to the manual, mechanical and chemical methods	Designing the control method is very difficult and needs to be continuously monitored and corrected in the initial phase
Helps in growth and development of many invertebrate species and fishes in the lake or pond	Does not yield satisfactory results when the plants have started growing closer to the water surface

way to limit light penetration. This would also be helpful as it would provide food for many invertebrate species which would then be eaten by fishes. There is a high phosphorous content in the granular and liquid fertilizers. These are available in various formulations or can also be formulated as per need. Granular fertilizers are less effective than liquid fertilizers as it is difficult to apply those [53]. Table 3 details the advantages and disadvantages of environmental control method.

2.4 Biological Control

The introduction of natural enemies to remove the weed's competitive advantage and check its growth rate to a level at par with the natural vegetation is the major objective of the biological control method [54]. Stocking white amur or triploid grass carp is a common method adopted to biologically limit the growth of water hyacinth [55]. Young grass carp have been observed to eat 50–300% of their body weight every day [56]. But, these fish would also consume other plants above or under the water surface and hence if they are introduced in large numbers, then they might even consume the entire pond or lake vegetation. In some cases, insects are

Table 4 Advantages and disadvantages of biological control method

Advantages of biological control	Disadvantages of biological control
The most sustainable and environment-friendly method of all the water hyacinth control methods	The results are achieved at the slowest rate in comparison to other methods [58]
Does not require long-term maintenance [57]	Occasional outbreaks and lower level of infestation do remain when only biological control method is adopted [59]

also used for biological control. As alligatorweed flea beetle is often used to control alligatorweed, there are other insects available to control water hyacinth, water-lettuce and other weed species. Table 4 details the advantages and disadvantages of biological control method.

2.5 *Integrated Control Method—Integrating the Advantages to Minimize the Shortcomings*

Each of the above four methods has its own advantages and disadvantages. In the cases of a catchment-wide scale or a lake on which a number of commercial activities are dependent, where levels of water hyacinth infestations vary, where there are various permutations of ecological and economic considerations possible which engage and number of stakeholders, integrated control method has been observed to be the most appropriate option [60]. The relative cost-effectiveness of different methods under the prevalent conditions and social, environmental and economic benefits associated with each method and each combination of methods must be assessed before drafting the integrated control method [61].

Thus, in the case of water hyacinth control in Futala Lake, Nagpur, which is one of the major recreation facilities in the city, where water hyacinth growth is observed at periodic intervals and direct dumping of municipal solid wastes in the lake is a nuisance and a cause of water hyacinth growth in the lake, the hyacinth control plan should be such that it serves both the purposes in a better manner and with minimum investment (considering long-term solutions) and maximum ease of operations. Hence, the following methodology outlines the water hyacinth control and waste management plan with respect to Futala Lake. Although it is subjected to Futala Lake, the integrated control and management plan could be applied in any other case wherein there are human establishments near the lake or river or similar commercial activities are dependent on the subject water body.

3 Methodology

The proposed methodology is a combination of the mechanical method for water hyacinth control and waste-to-compost technology for using the biomass available from the aquatic weed to generate useful by-product. The following sections detail the stepwise method to be followed for installing the integrated water hyacinth control and waste management plan.

3.1 *Classification of Land Use Near Futala Lake, Nagpur*

The Futala Lake is a closed water body located in Nagpur city in Maharashtra (21° 8′ 44′′N and 79° 03′ 48′′E). It is not potable and hence is used for irrigation, recreational and commercial fisheries purposes only. As it does not have self-cleansing capacity, eutrophic conditions have developed due to the accumulation of nutrients which flow into the lake from multiple polluting sources (mainly from the residential area around the lake). The lake is rainwater impoundment with and area of 26.3 hectors is about 6–7 m deep during monsoon. Figure 1a shows the classification of land use near Futala Lake, Nagpur, and Fig. 1b shows its satellite image.

Due to the release of sewage, direct dumping of municipal solid wastes and immersion of idols during festive season, almost half of the lake is now covered

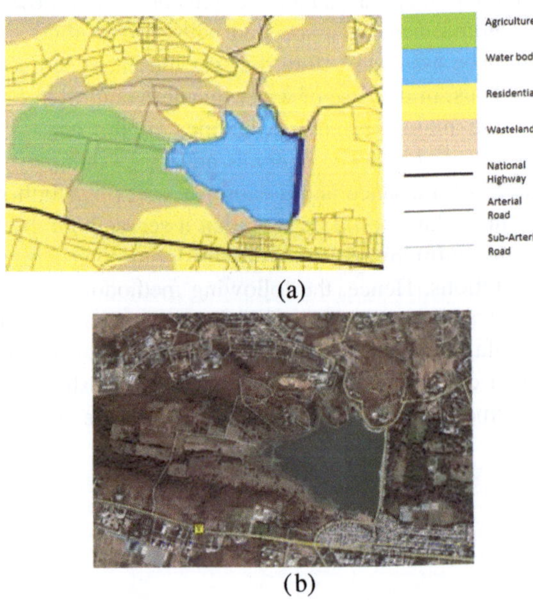

Fig. 1 **a** Classification of land use near Futala Lake, Nagpur; **b** satellite image of Futala Lake, Nagpur

with aquatic weeds. Initially the effects of eutrophication were seen towards west, but now they are clearly evident on the northern and southern sides of the lake.

3.2 *Determination of Quantity of Water Hyacinth to Be Treated*

The second step towards implementing the integrated waste management plan is to determine the quantity of water hyacinth to be treated and the period after which it would have to be treated again. Several studies have revealed that a healthy acre of water hyacinth could weigh somewhere around 200 tonnes (wet).

Now, as the surface area of the lake is approximately 65 acres and almost half of the lake is covered with the spread of water hyacinth, the spread of the aquatic weed is approximately on 30–32 acres of the lake. Hence, the total weight of water hyacinth in the lake is about 4000–4500 tonnes (wet). Hence, the minimum quantity of water hyacinth to be treated everyday should be 10 tonnes (wet, minimum).

3.3 *Determination of Quantity of Municipal Solid Waste to Be Treated*

As Futala Lake is now used as a recreational and picnic spot, there are about 25 restaurants or eatery spots near the lake. The composition of their municipal solid waste is as in shown in Fig. 2.

Every restaurant produces about 50–55 kg of food waste each day during weekdays and 100–110 kg of food waste each day during weekends. Hence, the recreational facilities near Futala Lake produce 1–1.5 tonnes of treatable organic waste every day.

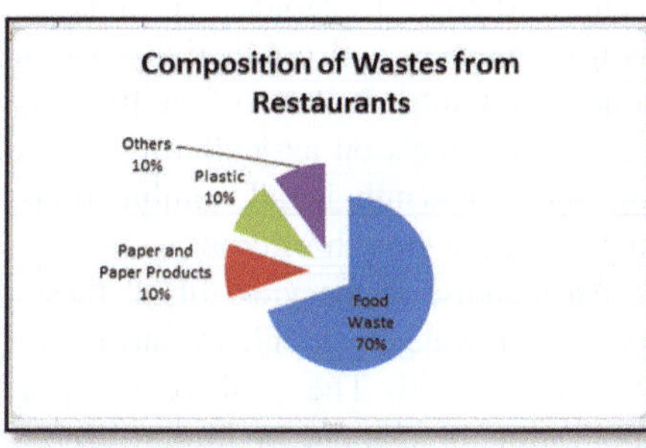

Fig. 2 Composition of wastes from restaurants near Futala Lake, Nagpur

3.4 Installation and Monitoring of Integrated Water Hyacinth and Waste Management Plan

From the above determinations, it is evident that the average volume of water hyacinth to be treated every day is 10 tonnes (wet) and average municipal solid waste to be treated each day should be 1–1.5 tonnes. Now, the following steps have to be followed to implement the integrated water hyacinth control and waste management plan:

- Study the sources of waste generation and spread of water hyacinth cover.
- Procure water hyacinth weed removal equipment capable to remove 8–10 tonnes (wet) of aquatic weeds per day.
- Procure an automatically operated waste-to-compost machine (which works on the principle of thermophilic decomposition).
- Arrange vehicles sufficient to mobilize the harvested weeds and collected municipal solid waste to a nearby site (10 tonnes of aquatic weeds (wet) and 1.5 tonnes of wastes).
- Draft management resources, complete safety requirements and brief the stakeholders about the tangible benefits from the process.
- Establish continuous monitoring and reporting practices.
- Monitor the process and the improvisations in the lake's water quality.

4 Results and Discussion

4.1 Quality of Compost

Amongst the alternatives available for exploiting the uses of the aquatic weeds, the use of its decayed tissues in order to provide nutrients for crops is a simpler option to implement in comparison to the available alternatives [62–66]. The use of manure wherein the biomass from aquatic weeds such as water hyacinth are used, its use enhances soil productivity and soil microorganisms, increases the organic content and nutrient status of the [67]. Aquatic weeds have been recognized to have significant effects on agricultural yield [68–70]. Manure and biofertilizers consisting water hyacinth in all combinations have been observed to have a positive impact on the growth of crop plants like rice, corn, sesame, brinjal and onion [71].

An increase in the yield of all these plants with the use of both compost and manure of water hyacinth in various combinations has been reported in various studies [72, 73]. The positive impacts of on the soil fertility conditions post application of manure or biofertilizers consisting of water hyacinth in different proportions have also been observed in several studies in various parts of the world [74–76].

Enhanced nitrogen availability which results due to the release of nitrogen from water hyacinth during the process of mineralization has been recorded as one of the prominent reasons for increase in soil fertility and plant growth [69, 77].

4.2 *Comparison of the Proposed Integrated Plan with Other Water Hyacinth and Waste Management Methods*

In a developing country like ours, where the quantum of work in waste management is large and human resources willing to work in it are steadily decreasing, the need to develop a waste management plan which requires minimum human resource and caters to process or treat maximum amount of waste is evident. In cases when any one of the four methods described above had been applied to cater the water hyacinth issue in Futala Lake, some of the other drawbacks have been observed. The proposed integrated management plan not only caters to the water hyacinth problem in the lake but also treats the municipal solid waste from the restaurants and hotels surrounding the lake along with the aquatic weeds to produce a better quality compost.

Table 5 shows the comparison of the proposed method for water hyacinth control with the methods described above.

Table 5 Comparison of the proposed integrated plan with other water hyacinth and waste management methods

Water hyacinth control methods					
Parameters	Mechanical	Chemical	Biological	Environment al	Integrated (proposed)
Operational feasibility	Special equipment and skilled labour to use them are required	Herbicides are available in the market	Detailed outline of the method and its impact analysis on other stakeholders to the lake is required	Very difficult to implement in the case of Futala Lake	Special automatically operated equipment is required
Requirement of human resources	High	Low	Low	Low	Moderate
Capital Investment	Moderate	Moderate	Low	Low	High
Financial feasibility (returns on investment)	The chopped or crushed weeds are then buried, dried or incinerated. Thus, ROI is nil	The plants are killed inside the lake Lower level infestations may occur. Hence, ROI is nil	It is naturally sustainable but does not yield any useful by-product	Catalyses the growth of invertebrates and small fishes. This can be considered as an added advantage of the method	Processes the weeds to compost along with the municipal solid wastes from the surrounding area, thereby generates a tangible by-product

(continued)

Table 5 (continued)

Water hyacinth control methods					
Parameters	Mechanical	Chemical	Biological	Environment al	Integrated (proposed)
Sustainability	As the weeds only removed mechanically, it is not a sustainable solution	Depends on the chemicals used and the method of application	Completely sustainable solution to the water hyacinth problem	Cannot be considered as sustainable as in the Futala Lake maximum weeds grow on the surface of water	It uses the biomass available from the water hyacinth and municipal solid wastes to process better quality compost. Hence, it is sustainable
Other tangible benefits	Nil	Nil	Nil	Nil	Compost can be sold in the market or could be used by the municipal corporation. Thereby generating or saving a considerable amount of money

4.3 Comparison of Proposed Waste-to-Compost Unit with Other Waste-to-Energy and Waste-to-Compost Methods

There are a number of decentralized waste-to-compost and waste-to-energy solutions which can be combined with the mechanical method to control water hyacinth. Table 6 shows the comparison of the decentralized waste-to-compost and waste-to-energy technologies on the basis of technology, operations and labour to be engaged in the technology, volume of waste reduced with respect to time, restrictions and reservations on waste processing, quality of compost, operation expense and space required.

4.4 Comparison of GHG Emissions After the Implementation of the Integrated Waste Management Plan with the Common Practices Like Incineration and Landfill

The calculation of greenhouse gas emissions as tonnes of CO_2 equivalent released from the treatment, dumping, disposal or incineration of the harvested aquatic weeds and the solid wastes from the restaurants have been compared under two scenarios. Fig. 3—Practice as Usual—represents the emissions due to dumping,

Table 6 Comparison of waste-to-compost technologies available in the market

Waste-to-compost technologies				
Parameters	Vermicomposting	Organic waste converter	Biogas plant	Thermophilic decomposition machine (proposed)
Technology	Use of earthworms in specially designed underground pits to convert wet waste to compost	Crushes/grinds the waste into small particles and then mixes it with old manure, sawdust and bioculture mix in trays for two weeks in a curing system	Anaerobic digestion of organic matter to produce a gas which principally contains methane and carbon dioxide	Use of thermophilic bacteria in a specially designed unit (which provides optimum conditions to the bacteria to function) to convert waste to compost within 24 h
Operation and Labour	Labour-intensive	Requires continuous supervision and hence it increases the operations cost	Requires skilled labour for operations, monitoring and supervision	Fully automatic operation. Minimum labour requirement
Replenishing microorganism s, sawdust, etc.	Not applicable	Each time waste is added, the supervisor must add the bioculture with respect to design specifications	Not applicable	Generally added on an annual basis.
Volume Reduction	40% (100 kg of fed waste would produce 60 kg of compost)	50–55%	60–65%	80–85%
Time for composting	60–90 days	15 days	Not applicable	24–36 h
Restriction on waste processing	Non-vegetarian food waste or municipal wastes having high oil contents cannot be processed	It takes longer period of time to compost if the waste composition varies regularly	Types of waste that can be processed depend on design specification s	All types of organic wastes (in varying proportions) can be easily processed
Quality of compost	Depends on the operations and management of the unit	Depends on the operations and management of the unit. Usually good	Slurry cannot be used directly as compost. Compost quality depends on a number of factors	High-quality compost. As all the parameters like microbial activity like oxygen, temperature and moisture content are automatically regulated
Foul smell/ odour	Odour problem persists	Odour problem occurs if site not maintained properly	Odour and flies problem occurs when slurry is dumped in the near vicinity	Odour problem occurs only in the case of machine failure

(continued)

Table 6 (continued)

Waste-to-compost technologies				
Parameters	Vermicomposting	Organic waste converter	Biogas plant	Thermophilic decomposition machine (proposed)
Operation expense	Rs. 2–5 per kg on maintenance, bioculture, dry leaves (excluding labour charges)	Around Rs. 3–5/kg as labour charges + sawdust and bioculture + maintenance	Rs. 5–7/kg (all inclusive)	Rs. 1.5–3.5/kg. (all inclusive)
Capital investment	Low	Moderate	Moderate	High (initially)
Space required	Large	Moderate	Large	Very less

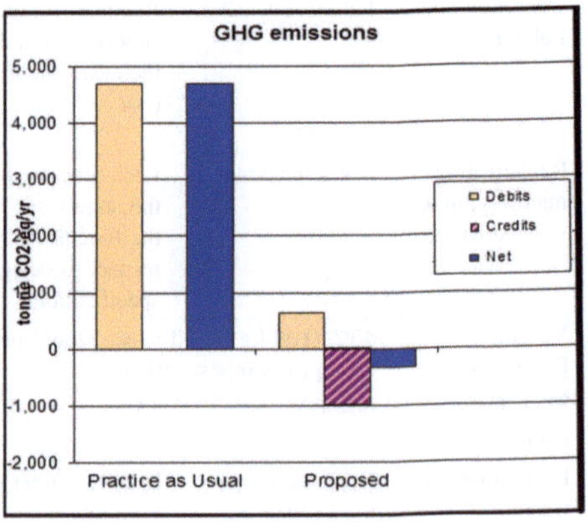

Fig. 3 Greenhouse gas emissions in the present and proposed scenario

disposal or incineration of the harvested weeds and collected or dumped solid wastes and—Integrated Plan—represents the emissions which would arise after the proposed plan is operational.

The CO_2-equivalent debits and credits have been calculated on the basis of the Intergovernmental Panel for Climate Change guidelines (1996), and a comprehensive and comparative excel freeware developed by Institute of Energy and Environmental Research (IFEU—Institut für Energie und Umweltforschung Heidelberg GmbH gemeinnütziges ökologisches Forschungsinstitut) has been used. From the comparison of the greenhouse gas emissions debits and credits, it is clear that the proposed integrated water hyacinth and waste management plan emits lesser greenhouse gases as compared to the present practices.

5 Conclusion

The results obtained after comparing the integrated water hyacinth and waste management plan with alternative methods of water hyacinth control and waste treatment technologies clearly state that the proposed plan encompasses benefits for all the stakeholders engaged with Futala Lake, Nagpur, in one or the other manner. The inferences from the results and discussions also suggest that such a plan could be implemented in any freshwater body having similar land use classification.

References

1. International Energy Agency (IEA): Biogas production and utilization. IEA Bio-Energy (2005). www.iea-biogas.net
2. Mitchell, D.S.: The growth and management of *Eichhornia crassipes* and *Salvinia* spp. in their native environment and in alien situations. Aquatic weeds Southeast Asia (1976)
3. Ludwig, S.: Biogas plants, a publication of the Deutsches Zentrum für Entwicklungstechnologien-gATE. In: Deutsche Gesellschaft für Technische Zusammenarbeit (1988)
4. Langeland, K.A., Burks, K.C.: Identification and Biology of Non-Native Plants in Florida's Natural Areas, p. 165. UF/IFAS (1998)
5. Holm, L.G., Plucknett, D.L., Pancho, J.V., Herberger, J.P.: The World's Worst Weeds: Distribution and Biology. University Press of Hawaii, Honolulu (1977)
6. Bartodziej, W., Weymouth, G.: Waterbird abundance and activity on water-hyacinth and Egeria in the St-Marks river, Florida. J. Aquat. Plant Manag. **33**, 19–22 (1995)
7. Brendonck, L., Maes, J., Rommens, W., Dekeza, N., Nhiwatiwa, T., Barson, M., Callebaut, V., Phiri, C., Moreau, K., Gratwicke, B., Stevens, M., Alyn, N., Holsters, E., Ollevier, F., Marshall, B.: The impact of water hyacinth (*Eichhornia crassipes*) in a eutrophic subtropical impoundment (lake Chivero, Zimbabwe). II. Species diversity. Arch. Fur Hydrobiol. **158**, 389–405 (2003)
8. Charudattan, R., Dinoor, A.: Biological control of weeds using plant pathogens: accomplishments and limitation. Crop Prot. (2000)
9. Wilson, J.R., Rees, M., Holst, N., Thomas, M.B., Hill, G.: Water hyacinth population dynamics. In: Biological and Integrated Control of Water Hyacinth, *Eichhornia crassipes*. ACIAR Proceedings, No. 102 (2001)
10. Mangas-Ramirez, E., Elias-Gutierrez, M.: Effect of mechanical removal of water hyacinth (*Eichhornia crassipes*) on the water quality and biological communities in a Mexican reservoir. J. Aquat. Health Manag. **7**, 161–168 (2004)
11. de Casabianca, M.-L., Laugier, T.: *Eichhornia crassipes* production on petroliferous wastewaters: effects of salinity. Bioresour. Technol. **54**, 39–43 (1995). Deutsche Gesellschaft für Technische Zusammenarbeit (GTZ)
12. Gopal, B.: Water Hyacinth. Elsevier, New York (1987)
13. Wolverton, B.C., McDonald, R.C.: Upgrading facultative wastewater lagoons with vascular aquatic plants. J. Water Pollut. Control (1979)
14. Flora (2003)
15. Shoeb, F., Singh, H.J.: Kinetic studies of biogas evolved from water hyacinth. In: 2nd International Symposium on New Technologies for Environmental Monitoring and Agro—Applications, p. 138 (2002)

16. Osumo, Wilfred M.: Effects of Water Hyacinth on Water Quality of Winam Gulf. The United Nations University, Lake Victoria (2001)
17. Kunatsa, T., Mufundirwa, A.: Biogas production from water hyacinth case of lake Chivero—Zimbabwe a review. Int. J. Recent Technol. Eng. IJRTE **2**(2) (2013). ISSN 2277-3878
18. Gargstad, Edward: Freshwater Vegetation Management. Thomas Publications, Fresno (1986)
19. Sajn, S.A., Bulc, T.G., Vrhovsek, D.: Comparison of nutrient cycling in a surface flow constructed wetland and in a facultative pond treating secondary effluent. Water Sci. Technol. **51**, 291–298 (2005)
20. Szczeck, M.M.: Suppressiveness of vermicompost against fusarium wilt of tomato. J. Phytopathol (Phytopathologische Zeitschrift) **47**, 155–161 (1999)
21. Meshandete et al. (2004)
22. Ali, N., Chaudhary, B.L., Khandelwal, S.K.: Better use of water hyacinth for fuel, manure and pollution free environment. Indian J. Environ. Prot. **24**, 297–303 (2004)
23. Malik, A.: Environmental challenge vis a vis opportunity: the case of water hyacinth. Environ. Int. **33**, 122–138 (2007)
24. Aboud, A.A.O., Kidunda, R.S., Osarya, J.: Potential of water hyacinth (*Eicchornia crassipes*) in ruminant nutrition in Tanzania. Livest. Res. Rural Dev. **5**, 17 (2005)
25. Goswami, T., Saikia, C.N.: Water hyacinth—A potential source of raw material for greaseproof paper. Bioresour. Technol. **50**, 235–238 (1994)
26. Oudhia, P.: Medicinal weeds in rice fields of Chhattisgarh (India). Int. Rice Res. Notes **24**, 40 (1999)
27. Oudhia, P.: Studies on allelopathy and medicinal weeds in chickpea fields. Int. Chickpea Pigeonpea Newsl. **6**, 29–33 (1999)
28. Charudattan, R., Labrada, R., Center, T.D., Kelly-Begazo, C. Report of a Panel of Experts Meeting 11–14 September, Fort Lauderdale, Florida, USA
29. Chilton II, E.: Freshwater Fishes of Taxes. Texas Parks and Wildlife (1998)
30. Tung, K.S.: The Effectiveness of Mechanical Control of Water Hyacinth (*Eichhornia crassipes*) (2014)
31. Madsen, J.D.: Methods for management of nonindigenous aquatic plants. In: Luken J.O., Thieret, J.W. (eds.) Assessment and Management of Plant Invasions, pp. 145–171. Springer, New York (1997)
32. Wittenberg, R., Cock, M.J.W. (eds.).: Invasive Alien Species: A Toolkit of Best Prevention and Management Practices. CAB International, Wallingford, Oxon, UK, xvii–228
33. Campbell, P.L., Naude, D.C., Cillers, C.J.: Development of management plans for control of water hyacinth in South Africa (2002)
34. Greenfield, B.K., Siemering, G.S., Andrews, J.C., Rajan, M., Andrews, S.P., Spencer, D.F.: Mechanical shredding of water hyacinth (*Eichhornia crassipes*): effects on water quality in the Sacramento-San Joaquin river Delta, California. Estuaries Coasts **30**, 627–640 (2007)
35. Hutto, D.T., Sabol, M.B.: Application of harvest, mechanical, bioaccumulation models, as an operational aquatic macrophyte management decision tool. In: North American Lake Management Society, Fifty Annual Conference and International Symposium on Applied Lake and Watershed Management, vol. 2, pp. 267–279. Lake Geneva, Wisconsin, 13–15 Nov 1985 (1986)
36. Wilson, A.L., Dehaan, R.L., Watts, R.J., Page, K.J., Bowmer, K.H., Curtis, A.: Proceedings of the 5th Australian Stream Management Conference. Australian rivers: making a difference. Charles Sturt University, Thurgoona, New South Wales (2007)
37. Wade, P.M.: Physical control of aquatic weeds. In: Pieterse, A.H., Murphy, K.J. (eds.). Aquatic Weeds: The Ecology and Management of Nuisance Aquatic Vegetation, pp. 94–137. Oxford University Press, New York (1990)
38. Petroeschevsky, A.: NSW Department of Primary Industries. Weed Management Guide—Water Hyacinth. License from Common Wealth of Australia under a Creative Commons Attribution 3.0 Australia license
39. Helfrich, L.A., Neves, R.J., Libey, G., Newcomb, T.: Extension Specialist, Fisheries and Wildlife Sciences, Virginia Tech. Control Methods for Aquatic Plants in Ponds and Lakes

40. Gargstad, Edward: Freshwater Vegetation Management. Thomas Publications, Fresno (1986)
41. Carlock, M.: California Department of Boating and Waterways. Pers. Interview 5 May 2003
42. Gutiérrez, L.E., Huerto, R.D., Saldaña, F.P., Arreguin, F.: Strategies for waterhyacinth (*Eichhornia crassipes*) control in Mexico. Hidrobiologia **340**, 118–185 (1996)
43. Environmental Planning and Climate Protection Department, eThekwini Municipality, 2004. Water Hyacinth Control. Guideline Document-Insight into Best Practice, Removal Methods, Training & Equipment
44. Westerdahl, H.E., Getsinger, K.D.: Aquatic Plant identification and herbicide use guide, Volume II: Aquatic plants and susceptibility to herbicides. Technical report A-88-9. Department of the Army, Waterways Experiment Station, Corps of Engineers, Vicksburg, MS (1988)
45. Misra, G., Das, N.: Response of Water Hyacinth to Two Hormone Herbicides, 2,4-D and 2,4,5-T. Department of Botany, Ravenshaw College (2012)
46. Hitchcock, A.E., Zimmerman, P.W., Kirkpatrick Jr., H., Earle, T.T.: Growth and reproduction of water hyacinth and alligator weed and their control by means of 2,4-D. Contrib. Boyce-Thompson Inst. **16**, 91–130 (1950)
47. Mitra, G.P.: Eradication of water hyacinth by methoxone on a field scale trial. Sci. Cult. **14**, 143–145 (1948)
48. Murphy, T.R., Shelton, J.L.: Aquatic Weed Management Herbicides. Southern Regional Aquaculture Center, Publication No. 361 (1996)
49. Findlay, J.B.R, Jones, D.: Integrated Control of Water Hyacinth, *Eichhornia Crassipes*, in Africa based on Roundup Herbicide Treatment. In: 9th International Symposium on Biological Control of Weeds, pp. 435–440 (1996)
50. Willoughby, N.G., Watson, I.G., Lauer, S., Grant, I.F.: The effects of water hyacinth on the biodiversity and abundance of fish and invertebrates in Lakes Victoria, Uganda. Final technical report to ODA/NRI, p. 48 (1993)
51. Akinyenmiju O.A. (ed.): Hebicidal control of water hyacinth at Abiala Creek, Delta State, Nigeria, Project report, Department of Plant Sciences, Obafemi Awolowo University, Ile-Ife, Nigeria (1994)
52. Madsen, J.D.: Advantages and Disadvantages of Aquatic Plant Control Techniques. Mississippi State University (2012)
53. Paradoxes about Liquid Versus Granular Fertilizers, 2001. Separating fact from fiction
54. Julien, M.H., Hill, M.P., Center, T.D., Jianqing, D. (ed.).: Biological and integrated control of water hyacinth, *Eichhornia crassipes*. In: Proceedings of the Second Meeting of the Global Warming Working Group for Biological Integrated Control of Water Hyacinth, Beijing, China. ACIAR Proceedings No. 102, 152 p. (2001) 9–12 Oct 2000
55. Pipalova, I.: A review of grass crap use for aquatic weed control and its impact on water bodies. J. Aquat. Plant Manag. **44**, 1–12 (2006)
56. Masser, M.P.: Using grass carp in aquaculture and private impoundments. Southern Regional Aquaculture Center, Publication No. 3600 (2002)
57. M.O. Beshir, F.D. Bennett, 1985. Biological control of water hyacinth on the White Nile, Suda. In: Delfosse, E.S. (ed.) Proceedings of the VI International Symposium on Biological Control of Weeds, 1984, Vancouver, Canada. Agric. Can., pp. 491–496
58. Wright, A.D.: Biological control of water hyacinthin Australia. In: Delfosse, E.S. (ed.) Proceedings of the Fifth International Symposium on the Biological Control of Weeds, 1981, Brisbane, Australia, CSIRO, Melbourne, pp 529–535
59. Saha, S.S., Rajemahadik, C.F.: Control of water hyacinth—a case study. The 12th World Lake Conference, pp 1051–1057 (2007)
60. Jones, R.W.: Integrated control of water hyacinth on Nseleni/Mposa rivers and lake Nsezi, Kwa Zulu-Natal, South Africa. ACIAR Proceeding 102, Resources Institute Publication No. 10066 and A0328 (2001)
61. Charudattan, R.: Integrated control of waterhyacinth (*Eichhornia crassipes*) with a pathogen, insects and herbicides, vol. 34. Supplement 1: Microbiological Control of Weeds Symposium, pp. 26–30 (1986)

62. Sonke, D.: Tithonia weed a potential green manure crop. Echo. Dev. Notes **57**, 5–6 (1997)
63. Gachengo, C.N.: Phosphorus release and availability on addition of organic materials to phosphorus fixing soils. M. Sc. Thesis. Moi University, Eldoret, Kenya (1996)
64. Jama, B., Palm, C.A., Buresh, R.J., Niang, A., Gachengo, C., Nziguheba, G., Amadalo, B.: Tithonia diversifolia. A green manure for soil fertility improvement in western Kenya: a review. Agrofor. Syst. **49**, 201–221 (2000)
65. Nziguheba, G., Merckx, R., Palm, C.A., Mutuo, P.: Combined uses of Tithonia diversifolia and inorganic fertilizers for improving maize production in a phosphorus deficient soil in Western Kenya. Agrofor. Syst. **55**, 165–174 (2002)
66. Chukwuka, K.S., Omotayo, O.E.: Effects of Tithonia green manure and water hyacinth compost application on nutrient depleted soil in South-Western Nigeria. Int. J. Soil Sci. **3**(2), 69–74 (2008)
67. Beckman, E.O.: Organic fertilization: vegetable farming luxury or necessity. Tech. Commun. ISHA. **29**, 247 (1973)
68. Oyetunji, O.I., Ekanakaye, I.J., Osonubi, O.: Influence of arbuscular mycorrhizae fungus, mulch and fertilizer application on yield of-yams in an agroforestry system in South Western Nigeria. Maurik Bull. **6**, 75–82 (2003)
69. Widjajanto, D.W., Matsushila, K., Honmura, T., Miyanch, N.: Studies on the release of N from water hyacinth incorporated into soil-crop systems using 15N—labeling techniques. Part J. Biol. Sci. **2001**(4), 1075–1077 (2001)
70. Widjajanto, D.W., Honmura, T., Miyauchi, N.: Nitrogen release from green manure of water hyacinth in rice cropping systems. Pak. J. Biol. Sci. **2002**(5), 740–743 (2002)
71. Majid, F.Z.: Aquatic weeds and algae, the neglected natural resources of Bangladesh (Booklet) (1983)
72. Majid, F.Z.: Aquatic Weeds, Utility and Development. In: S.S. Purohit (ed.). Agro Botanical Publishers, Dhaka Bangladesh, India (1992)
73. Majid, F.Z., Lutfun, N., Sajeda, B., Rahima, K., Akhatar, N.: Use of naturally occurring algae as biofertilizer or organic manure. National Workshop on Algal Systems Proceedings, Madras, India, vol. 26, pp. 11–14, 3–4 Oct 1980
74. Gunnarsson, C.C., Petersen, C.M.: Water hyacinth as a resource in agriculture and energy production: a literature review. Waste Manage. **27**, 117–129 (2006)
75. Kayum, M.A., Asaduzzaman, M., Haque, M.Z.: Effects of indigenous mulches on growth and yield of tomato. J. Agric. Rural. Dev. **6**(1–2), 1–6 (2008)
76. Amitava, R., Sarkar, N.C., Debashish, S.: Influence of organic manures on productivity of two varieties of rice. J. Cent. Eur. Agric. **9**(4), 629–634 (2008)
77. Contantinides, M., Fownes, J.H.: Nitrogen mineralization from leaves and litter of tropical plants: relationship to nitrogen, lignin and soluble polyphenol concentrations. Soil Biol. Biochem. **26**, 49–55 (1994)

Analyzing the Strength Behavior of Cement Composites with Waste Glass Fibers

Nikhil Kale, Mangesh Madurwar and Rahul Ralegaonkar

Abstract Cement composites fail due to tensile strains exceeding their limiting strain capacity. This results into development of cracks and gradual loss of strength. The controlling of cracks is vital for better fatigue performance of concrete construction. The paper presents a study of the suitability of waste glass fibers for utilization in cement mortar and concrete specimens in order to promote sustainable construction. These fibers were added to the composites by weight fraction, and relevant tests were performed on them. Test results indicate that cement mortar showed an increase in strength and a pseudo-ductile behavior due to addition of fibers. while concrete specimens showed loss in compressive and flexural strength.

Keywords Tensile strain · Fatigue · Waste glass · Sustainable Pseudo-ductile

1 Introduction

Construction sector has been tried and tested extensively use of fibers over a long period of time. The use of fibers to reinforce brittle materials can be traced back to ancient times when straws were used for brick manufacturing. Fibers have been widely used in cement composites prominently to assess their strength behavior, crack resistance, energy absorption, permeability, and elongation properties. Due to sudden and at times heavy application of loads, premature cracking occurs in cement composites. Cement being a brittle material, development of cracks leads to

N. Kale · M. Madurwar · R. Ralegaonkar (✉)
Department of Civil Engineering, Visvesvaraya National Institute of Technology, South Ambazari Road, Nagpur, India
e-mail: sanvan28@yahoo.com

N. Kale
e-mail: nikhilkale1989@gmail.com

M. Madurwar
e-mail: mangesh_bits@yahoo.com

A. S. Kalamdhad et al. (eds.), *Advances in Waste Management*,
https://doi.org/10.1007/978-981-13-0215-2_14

gradual loss of its strength. Fibers help in controlling of cracks by accommodating into the very fine gaps of the microstructure. They impart additional energy absorbing capability, thus helping in transforming a brittle material into a pseudo-ductile one. They also serve as a crack arrestor creating a stage of slow crack propagation and gradual failure.

There is a considerable amount of research done on the use of glass fibers. Cahn et al. [1] studied a durability of Portland cement mortar having epoxy-phenol coatings with fiberglass protection. Their tests of the duration of two years on stress corrosion and aging showed that normal glass fiber requires a complete protection of pin hold coating to endure the alkali attacks over longer duration and a coating to give better bond strength, because the coating will emigrate into the mortar to give better bond strength. Choi et al. [2] used the glass fiber (GF) and carbon fiber (CF) to enhance the properties of non-ductileflexural concrete components. They determined a volume ratio of (CF/GF) to be approximately (8.8/1) to produce pronounced hybrid effects and pseudo-ductility. Ho and Woodhams [3] examined the compressive strength and flexural strength of fiber-reinforced sulfur concrete for a polyethylene terephthalate (PET) fiber and a glass fiber (alkali-resistant). The glass fibers increased maximum flexural load by a factor of 3.5 and the compressive strength by 50%, whereas the polyester fibers increased maximum flexural load marginally by a factor of 1.3 and compressive strength by 33% at the optimum fiber length of 4–5 cm. Shah and Naaman [4] studied the mechanical properties of steel and glass fiber-reinforced mortar. Compressive, flexural, and tensile tests were conducted on mortar samples reinforced with various volumes and lengths of glass and steel fiber. The flexural or tensile strength of the reinforced samples was 2–3 times that of control mortar, while the resultant deflection or strain was nearly 10 times that of mortar. Failed flexural sample surfaces were observed to have extensive micro-cracking; significant contribution of the matrix was observed even after the first cracking. Messan et al. [5] studied the early shrinkage of cement mortar due to the influence of cellular ether, ethylene-vinyl acetate (EVA), and glass fibers. Optical instrumentation system found the free plastic shrinkage. Active restrained system performed the restrained shrinkage tests. The residual stress and the overall shrinkage were decreased because of the use of EVA. The shrinkage was reduced by the glass fibers and the heterogeneity of the surface shrinkage. Khan et al. [6] studied the consequence of aging on the physico-chemical properties of glass fiber under the different environmental circumstances. Indoor and outdoor atmosphere, 95% humidity, low temperature, water soaking treatment, and the chemical environment were the conditions selected. The results showed that glass fiber is good in resisting corrosive atmosphere, sunlight, shady atmosphere, and low temperature, but the detrimental outcome on the glass fiber was caused because of the influence of prolonged time exposure to water. Nunes and Reis [7] used digital image correlation method to estimate crack extension and the crack-tip-opening displacement of glass fiber-reinforced polymer mortars to study their fracture behavior. The percentage blending of fiber in mortar mix was 0, 2, 5, 10% by the weight of aggregate. Curing of all the samples was done for 7 days at room temp. and the post-curing for 3 h at 800 °C. Study results showed that the fractured resistance of polymer mortar was increased because of glass fibers. Lva et al. [8]

studied that the glass fiber-reinforced concrete (GFRC) in flexure has the fatigue performance. The fatigue lives of GFRC at various stress levels were obtained by testing the beam specimens (size of 100 × 100 × 400 mm) under four-point flexural fatigue loading. 0.6, 0.8, and 1% volume fraction of glass fiber was incorporated into the samples. They obtained results for the coefficients of the fatigue equation corresponding to different survival probabilities so as to predict the flexural fatigue strength of GFRC for the desired level of survival probability.

Fibers are broadly classified as natural and synthetic. Use of natural fibers like coir, jute, sisal depends upon their regional availability. Synthetic fibers like steel, polypropylene, polyester, nylon-66 which incur costs are being effectively used in cement composites. In recent times, growth in industrialization has resulted into generation of more and more waste by-products. These materials are non-degradable, difficult to dispose off. Thus, it becomes equally important to consider alternative ways for their utilization in order to promote sustainability.

The main objective of this research paper is to study the strength behavior of cement mortar and concrete specimens with the use of glass fibers that are obtained as an industrial waste.

2 Materials and Methods

2.1 *Materials*

Portland Pozzolana Cement (fly ash based), conforming to IS 1489-1991, part 1 was used. Clean river sand, conforming to IS 650-1991, was used. The properties were determined as per IS 2116-1980. Aggregates confirming to IS 383-1970 were used. The properties of the materials used were as mentioned in Table 1.

A melamine resin-based superplasticizer (anionic) having specific gravity 1.23 and pH 7–8 was used, confirming to IS-9103:1999. Recommended dosage percentage (v/w of cement) is 0.5–1%. The glass fiber (Fig. 1) sample was collected from an industry which comes as a waste after specific processing. It is available in white color with non-uniform thread length varying from 10 to 50 mm and diameter of around 0.02 mm (20 microns). Water absorption was observed as 0%.

2.1.1 Characterization of Glass Fiber

For identification and quantification of crystalline phases and understanding the surface morphology of the fibers, X-ray diffraction (XRD, Fig. 2) and scanning electron microscope (SEM, Fig. 3) tests were conducted, respectively. The frequent occurrence of peaks (Fig. 2) shows that the glass fiber has a definite crystalline cell structure. The compounds in the fiber are quartz (SiO_2), calcium and aluminum oxide ($Ca_3Al_2O_6$, i.e., $3CaO{\cdot}Al_2O_3$), along with minute amounts of titanium oxide

Table 1 Properties of materials

Property / Material	Specific gravity	Standard consistency (*P* %)	28-days compressive strength (Mpa)	Sieve analysis	Water absorption (%)	Free surface moisture (%)	Shape, nominal size (mm)
Cement	3.00	32	47.45	–	–	–	–
Sand	2.61	–	–	Zone 3 (as per IRC 44: 2008, Table 2)	1.01	0	–
Aggregates	2.91	–	–	–	2	0	Angular,10 and 20

Fig. 1 Waste glass fibers

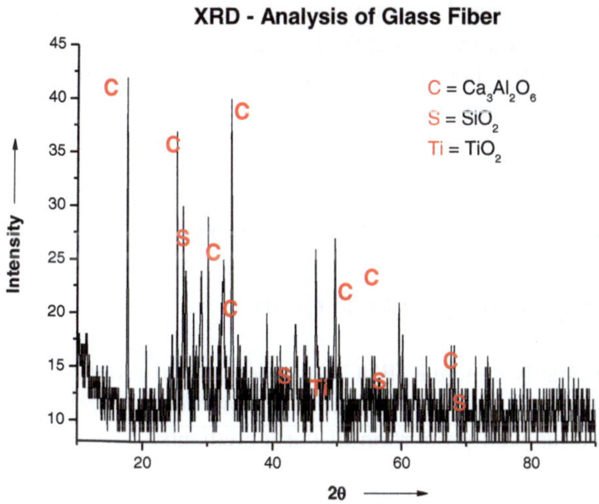

Fig. 2 XRD analysis of glass fiber

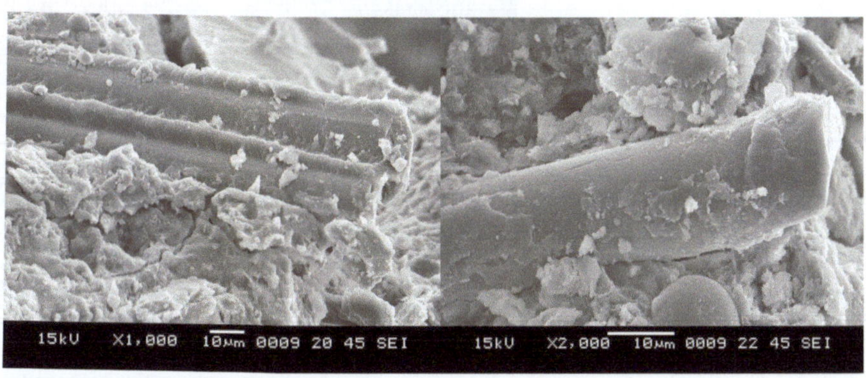

Fig. 3 SEM micrographs for glass fiber

(TiO_2). TiO_2 is mostly used as a white pigment and physical blocker to ultraviolet light. The surface texture (Fig. 3) of the fibers was found smooth.

3 Experimental Procedure

3.1 *Preparation of Cement Cortar Cubes*

Cement and sand were mixed in 1:3 proportions. The fibers were chopped into length of 10 mm and added to the mixture in quantities of 0.2, 0.4, 0.8, 1, and 1.5% by weight of cement. The fibers were randomly oriented during dry mixing. Water of amount (P/4 + 3) % of combined weight of cement and sand was then continuously added to the mix while mixing to provide the homogeneous mix without twine or lump as per standard procedure. A total of six sets of specimens of size 70.6 × 70.6 × 70.6 mm were prepared using standard molds (Fig. 4). Demoulding was done after 24 h followed by curing for 28 days at constant temperature of 270 °C and relative humidity of 90%.

3.2 *Preparation of Concrete Specimens*

3.2.1 Mix Proportioning for M-40 Grade

For pavement quality concrete (PQC), mix design and proportioning were done as per IRC 44-2008 [9]. Target design strength was set as 48.25 MPa. Table 2 indicates the weight as well proportion of the ingredients.

Fig. 4 Set of 18 cubes casted

Table 2 Mix proportions for M40 grade per m^3 of concrete

Component	Weight (kg)	Proportion (%)
Cement	391.6	1
Sand	629.21	1.61
Coarse aggregate	1386.83	3.54
Water	148.8	0.38 (water/cement ratio)

3.2.2 Concrete Cube Specimens

A total casting of 18 cubes of size 150 × 150 × 150 mm was done using standard molds with fibers added in proportions of 0, 0.2, 0.4, 0.6, 0.8, and 1% by weight of concrete (Fig. 5). After 24 h, demoulding was done and curing for 28 days was carried out under controlled environment for studying the change in ultimate strength of concrete as per IS 516-1959 [10].

3.2.3 Preparation of Concrete Beam Specimens

Casting of 18 beams of size (500 × 100 × 100) mm was done for M-40 grade using standard molds with fibers added in 0, 0.2, 0.4, 0.6, 0.8, and 1% by weight of concrete (Fig. 6). After 24 h, demoulding was done and curing for 28 days was carried out under controlled environment for studying the change in flexural strength of concrete.

Fig. 5 Cubes after 28 days curing

Fig. 6 Specimen after filling and compaction

4 Test Results and Discussions

4.1 Test Results of Cement Mortar Specimens

The tests were carried out with compression testing machine at a fixed loading rate of 0.228 MPa/s (Fig. 7). Due to the smooth surface texture of fiber, bond slippage occurs between cement paste and fiber. At 0.2% fiber content, lower fiber concentration causes ineffective dissipation of stresses, so strength decreases. With the increase of fiber content, its density in the matrix increases leading to more number of fibers contributing in dissipation of stresses. Hence, strength goes on increasing till it reaches a maximum value of 57.02 MPa which is 20% more than strength

Fig. 7 Gradual failure under compression

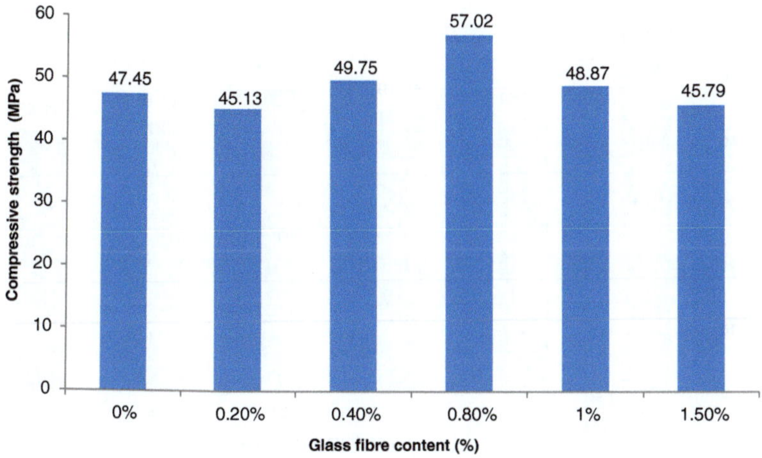

Fig. 8 Estimated compressive strength for mortar with varying glass fiber content

obtained at 0% fiber (Fig. 8). Beyond 0.8% fiber added, the mortar mix becomes stiff due to increased fiber content making it less workable. Hence, the strength achieved is less compared to value of 57.02 Mpa.

4.2 Test Results on Concrete Cube Specimens

Initially with addition of 0.2% fibers, strength decreases by 17.65% as compared to conventional specimen at 0% fiber (Fig. 9). This happens due to the "bond slippage" between cement paste and fiber. The fiber's smooth surface ineffectively transfers the stress at the interface that leads to considerable loss instrength. With further gradual increase in fiber content, more fibers contribute to stress dissipation that results in increase of strength till it reached a value of 48.63 MPa. Further, increase in fibers causes decrease in workability of concrete leading to lesser efficient compaction and hence strength decreases.

4.3 Test Results on Concrete Beam Specimens

For the tested concrete beams, it was observed that flexural strength decreases with the addition of fibers (Fig. 10). This shows that these fibers are fragile and have poor tensile strength leading to loss in flexure strength.

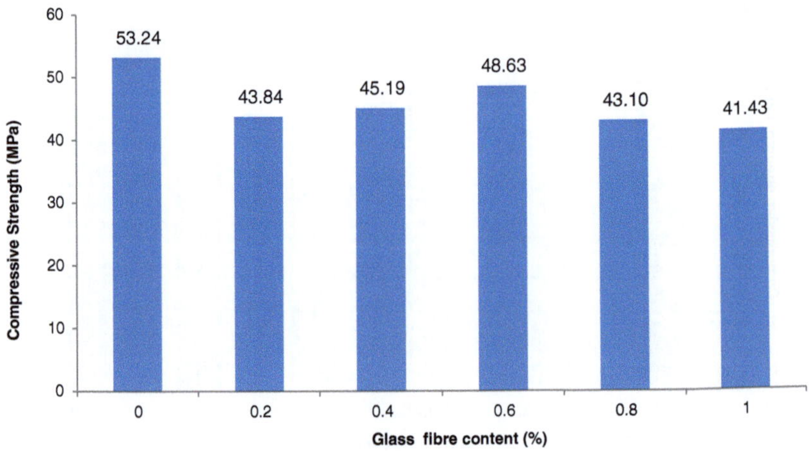

Fig. 9 Estimated compressive strength for concrete cube with varying glass fiber content

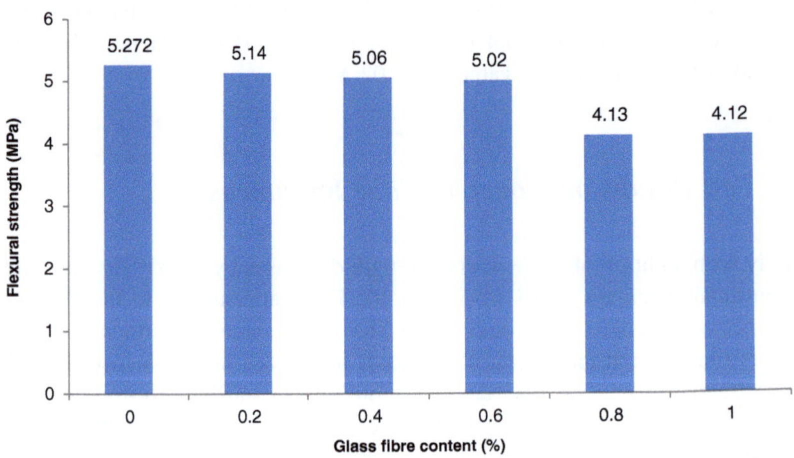

Fig. 10 Glass fiber content (%) versus flexure strength for concrete beam

5 Conclusion

Effective transfer of stresses from cementitious matrix to fibers is influenced by interfacial bonding and dissipation of stresses is influenced by fiber density matrix. Interfacial bonding is influenced by surface texture of fibers, whereas fiber density matrix depends upon the amount of fiber content available in the matrix.

In cement mortar, initially at 0.2% fiber content, ineffective stress transfer phenomenon is more pronounced and so strength decreases. At 0.8% fiber content, optimum fiber density helps in stress dissipation, and hence, higher strength is achieved. Also, addition of fibers causes gradual failure of specimens thus imparting them pseudo-ductility.

In concrete cubes, similar trend was observed. The strength achieved at 0.6% fiber content (i.e., 48.63 MPa) was slightly more than target design strength (i.e., 48.25 MPa) but less than 53.24 MPa at 0% fiber. This is because concrete is more heterogeneous material than cement mortar.

It is concluded that the waste fibers have been found fit for utilization in mortars.

References

1. Cahn, D.S., Phillips, C., Ishai, O., Aroni, S.: Durability of fiber glass Portland cement composites. ACI Mater. J. **70**(3), 187–189 (2013)
2. Choi, D.U., Kang, T.H.K., Sang-Su, H., Kill-Hee, K., Woosuk, K.: Flexural and bond behavior of concrete beams strengthened with hybrid carbon-glass fiber reinforced polymer sheets. ACI Struct. J. **108**(1), 90–98 (2011)
3. Ho, J.L.K., Woodhams, R.T.: Fracture toughness of fiber reinforced sulfur concrete. ACI J. Proc. **79**(4), 288–295 (1982)
4. Shah, S.P., Naaman, A.E.: Mechanical properties of glass and steel fiber reinforced mortar. ACI J. Proc. **73**(1), 50–53 (1976)
5. Messan, A., Ienny, P., Nectoux, D.: Free and restrained early age shrinkage of mortar: influence of glass fiber, cellulose ether and EVA. Cem. Concr. Compos. **33**, 402–410 (2011)
6. Khan, M., Akhtar, G., Saeed, A., Ullah, N., Ahmed, I.: Effect of aging on physico chemical properties of glass fiber under different environmental conditions. J. Chin. Chem. Soc. **55**, 307–310 (2008)
7. Nunes, L.C.S., Reis, J.M.L.: Estimation of crack tip opening displacement and crack extension of glass fiber reinforced polymer mortars using digital image correlation method. Mater. Des. **33**, 248–253 (2012)
8. Yan, L., Cheng, H., Ma, Z.: Fatigue performances of glass fiber reinforced concrete in flexure. Procedia Eng. **31**, 550–556 (2012)
9. IRC44-2008: Guidelines for cement concrete mix design for pavements. Indian Road Congress (2008)
10. IS 516-1959: Methods of tests for strength of concrete. Bureau of Indian Standard (1959)

Surface Water Quality Modeling by Regression Analysis and Artificial Neural Network

Kamal Uddin Ahamad, Pratik Raj, Najmul Haque Barbhuiya and Abhinay Deep

Abstract The major objective of the present study is to develop water quality prediction models after evaluation of water quality to predict water pollution of two lakes situated inside Tezpur University. The water quality parameters were analyzed using linear regression analysis and artificial neural network to predict the water quality. Correlation studies suggested a highly positive correlation between TS with turbidity and EC for both lakes. Modeling of TS and BOD by regression analysis suggests a good correlation between actual and predicted TS but a very poor correlation between actual and predicted BOD. Modeling of TS and BOD by ANN shows a very good correlation between the actual and predicted values for both TS and BOD for both the lakes studied. The error between the experimental and estimated ANN model is less than regression model, suggesting suitability of ANN model for prediction of studied parameters.

Keywords Artificial neural network · BOD · Regression analysis
Total solid · Water quality

K. U. Ahamad (✉) · P. Raj · N. H. Barbhuiya · A. Deep
Department of Civil Engineering, Tezpur University, Napaam 784028, Assam, India
e-mail: kahamad@tezu.ernet.in

P. Raj
e-mail: Pratik_cib13@tezu.ernet.in

N. H. Barbhuiya
e-mail: najmul_cib13@tezu.ernet.in

A. Deep
e-mail: abhinaya_cib13@tezu.ernet.in

A. S. Kalamdhad et al. (eds.), *Advances in Waste Management*,
https://doi.org/10.1007/978-981-13-0215-2_15

1 Introduction

The water quality plays a significant role in the natural ecosystem as all the living organisms depend on it. In the recent years, due to rapid urbanization, growth in population and due to various anthropogenic activities, the surface water is getting continuously polluted as it received large amount of wastewater from various sources. The quality of the water bodies is continuously degrading due to continuous disposal of municipal, industrial, and agricultural wastewater. The increase in the surface water pollution causes an imbalance in aquatic ecosystem due to the deteriorated water quality. The polluted water will also be a hindrance in economic development and social prosperity of the nearby society. Therefore, it requires attention and effective information regarding the quality of the water body to prevent and control the surface water pollution, which will also be helpful for effective management of the water body. Surface water quality is expressed by a broad range of parameters that can be included within physical, chemical, and biological parameters. Some of the parameters that control the water quality are temperature, pH, total solids, total suspended solids, total dissolved solids, turbidity, BOD, COD, pathogens, nutrients, organics, etc. The water quality also depends on the natural parameters like precipitation, climate along with anthropogenic activities like domestic wastewater, industrial wastewater, and agricultural runoff.

To assess and manage the surface water, water quality evaluation can be a valuable tool where by measuring the overall water quality of a location in a time-based manner on several water quality parameters. The water quality data can be utilized for water quality modeling, which will help in modifying the policies formulated by different environmental monitoring agencies. Water quality models are very important tools and developed by taking information from number of sources and combining them into a model. The developed model helps and increases the ability of understanding of the policy makers as well as the general users of the water about the issues and the quality of the highlighted water. The increase in level of pollution in various water bodies has affected the developed as well as the developing countries. One of the major causes for such environmental crisis is due to the exploration of virgin natural resources for the sake of economic growth [1]. In the recent past, many researchers have carried out studies in this direction to understand and model the process [2–9]. Adamu and Ado [10] have carried out a study to identify the pollution sources and their contribution toward water quality variation using principal component analysis and multiple linear regressions. Potential of the artificial neural network (ANN) on simulating interrelation between water quality parameters was examined by Bahaa et al. [11]. They studied the Nile Delta drainage system by evaluating monthly records of 33 parameters, monitored at 102 locations of the drainage system.

The major objective of the present study is to develop water quality prediction models after evaluating the water quality of two lakes situated inside Tezpur University, using mathematical simulation techniques. The water quality parameters were analyzed using statistical techniques—linear regression analysis and artificial neural network to predict the water quality. Linear regression examines the relationship between single-dependent variables and a set of independent variables to best represent relationship in water quality parameters. ANN modeling has the potential to reduce the computation time, effort and the possibility of errors in the calculation. Therefore, this study illustrates design of a neural network model for rapid, direct calculation of the water quality. Water sample for the experimental purpose was collected from the lake in front of honorable vice-chancellors bungalow and the lake near Academic Building II of Tezpur University and analyzed in the environmental engineering laboratory in civil engineering department of Tezpur University. The present study has been made to evaluate the current status of pollutants and an effort has been made to study the inter-relationship between different components of surface water using statistical techniques.

2 Materials and Methods

2.1 *Sampling Location*

Water sample for the present study was collected from two natural lakes situated inside Tezpur University, Assam, India, and designated as Lake 1 and Lake 2. Both the lakes have its importance from both scenic as well as fishing point of view. There is substantial raise in suspended solids along with water level during the rainy season and a low water level was observed during the winter season. Throughout the year, a high turbidity has been observed in both the lakes.

2.2 *Sampling Strategy*

Water samples for the experimental purpose were collected at an interval of 5 days from the sampling site in the month of January, February, March, and April 2015. 5 L plastic container was used for sample collection. The sampling containers were washed carefully prior to the sampling to remove any type solids or impurities which may be present earlier. All samples were collected well away from the edges of the water body. Temperature of the sample was recorded at the site itself.

2.3 *Analysis of Sample*

The water quality depends on various physical and chemical parameters and their concentration, which are estimated in the laboratory from the collected water sample.

All the experiments were carried out in environmental engineering laboratory of civil engineering department of Tezpur University, Assam, India. All the experiment carried out as per the method specified in standard methods [12]. The estimated water quality parameters are pH, alkalinity, turbidity, conductivity, total solids (TS), chloride, phosphorous, dissolved oxygen (DO), and biological oxygen demand (BOD).

2.4 Predictive Model Study

Regression Models: In order to present an actual water quality situation, regression models are used which requires two parameters—dependent and independent. The closeness between dependent and independent variables are suggested by correlation analysis. A correlation coefficient between +1 and −1, suggests the probability of linear relationship between variables *X* and *Y*. However, the correlation between +0.8 to 1.0 and −0.8 to −1 is characterized as strong, between +0.5 to 0.8 and −0.5 to −0.8 as moderate and between +0.0 to 0.5 and −0.0 to −0.5 as weak. Correlation analysis provides or suggests a mechanism for prediction or forecasting based on the relationship between the variables, i.e., *X* and *Y*. For the present study, the relationship between various estimated water quality parameters on each other was determined by calculating Karl Pearson's correlation coefficient, *R*, by using the formula given:

$$R = \frac{\Sigma(X - \bar{X})(Y - \bar{Y})}{\sqrt{\Sigma(X - \bar{X})^2 \Sigma(Y - \bar{Y})^2}} \quad (1)$$

where *X* (*X* = value of *X*-variable, $\bar{X}$ = average value of *X*) and *Y* (*Y* = value of *Y*-variable, $\bar{Y}$ = average value of *Y*).

The variation of a single-dependent variable as function of several independent variables can be described by multiple linear regressions. For the present study, a single water quality parameter is the dependent variable and its variation is dependent on two or more independent variables of physical, chemical, or biological water characteristics. The general equation is:

$$Y' = a + b_1X_1 + b_2X_2 + b_3X_3 + \cdots + b_kX_k \quad (2)$$

where Y' is the dependent variable, X's are the independent variables, k is the number of independent variables in the equation, and a is the regression constant. By choosing appropriate independent variables (water parameters), aim is to maximize the correlation between the predicted value of our water quality variable and the actual value of the variable.

Neural Network Model: Artificial neural network (ANN) is a simulation of the real nervous system; in other words, it is a mathematical model based in biological neural networks. A neural network consists of a set of interconnected individual neurons organized into several layers, the first layer being the input layer, which produces the network output. Numerical data used in the study moves from connection to each unit where it is processed. Processing takes place locally at each unit and between connections in a parallel fashion. ANN can be configured for specific applications, such as pattern recognition or data classification using learning process. This learning process as well as the biological system provides adjustments of developed models. Because of the need of robust nonlinear models, the importance of artificial neural networks (ANN) is increasing, as the ANN models can be used to improve the drinking water quality, and at the same time, reduces the operating costs with the advanced process control. The total data of 50 samples of 2 sampling locations were used for testing. MATLAB Neural Network tool has been used to run the model. For the present study, Levenberg Marquardt method has been used for the modeling, using standard 2-layer feedforward neural network training.

3 Results and Discussion

3.1 *Analysis of Sample*

Water quality in general is mostly characterized by its physical and chemical parameters, which changes widely depending upon factors like water source, pollution type (industrial, municipal, agricultural, etc.), seasonal fluctuation. Statistical analysis viz. descriptive statistics, correlation, regression analysis, and neural network of the physicochemical properties of water give a fairly good amount of information like their average values and possibly prediction of one variable (usually the one which is difficult to evaluate). Ten parameters were analyzed for two batches of aqueous samples from two lakes over a four months' time span (January 2015–April 2015). The observed water quality values for both the lakes are summarized and presented in Table 1 along with the basic statistics of the dataset. Maximum and minimum temperature varies from 20 °C (in January) to 28 °C (in April). Lake's surface temperature measurements are confined to the top portion of the water. During the daytime in high sunshine conditions may lead to the formation of a warm layer at the lake's surface. Turbidity mainly indicates the pollution level of the water bodies. The turbidity varied between 5 and 35 NTU for both lakes. The turbidity is higher in monsoons seasons, which resulted in increase in silt, clay, and other suspended particles, brought in the reservoir by surface runoff. The conductivity was found to be in the range between 190–500 μs/cm at Lake 1 and 150–200 μs/cm at Lake 2. The maximum conductivity was recorded during the April month while minimum during January month. The conductivity is mostly affected by the bedrock and soil in the watershed. It is also affected by human influences like agricultural runoff. The pH value varied between 6.2 and 7.3

Table 1 Summary of water sample analyses and basic statistics

Parameter	Maximum	Minimum	Average	Std. Dev.	Variance
Lake 1					
Turbidity	32.00	5.00	20.68	5.83	34.04
Temperature	28.00	20.50	24.27	2.45	5.98
Conductivity	497.30	190.20	384.73	77.07	5939.04
TS (mg/L)	356.00	125.00	258.35	58.34	3403.22
Alkalinity (mg/L)	2.40	1.80	2.01	0.12	0.01
Chloride (mg/L)	41.20	7.99	28.85	6.55	42.92
BOD	6.20	4.00	5.53	0.59	0.35
DO (mg/L)	9.00	3.00	5.00	1.90	3.61
pH	7.23	6.28	6.71	0.25	0.06
Phosphate (mg/L)	0.16	0.00	0.04	0.04	0.00
Lake 2					
Turbidity	35.00	7.00	18.91	8.33	69.42
Temperature	28.00	19.00	23.86	2.47	6.08
Conductivity	302.00	142.40	225.85	48.74	2375.63
TS (mg/L)	213.00	112.00	166.56	30.41	924.73
Alkalinity (mg/L)	4.00	1.80	2.19	0.44	0.20
Chloride (mg/L)	48.01	10.01	31.49	9.77	95.53
BOD	7.30	4.10	5.84	0.85	0.73
DO (mg/L)	9.00	2.00	4.74	2.09	4.35
pH	7.20	6.11	6.64	0.27	0.07
Phosphate (mg/L)	0.16	0.00	0.05	0.05	0.00

for both lakes. Alkalinity of water samples varied from 1.8 to 2 mg/L and 1.8 to 2.4 mg/L of Lake 1 and Lake 2, respectively. The mild alkalinity may be due to the seepage of effluent, domestic sewage, etc. Fluctuation in alkalinity is found out in the month of February for both lakes may be because of unseasonal rainfall. The total solid values were ranged between 125 to 360 mg/L at Lake 1 and 110 to 220 mg/L at Lake 2, respectively. The maximum values of total solid occurred during April while minimum during January month. One of the reasons for possible increase in TS is because of rainfall which may bring the nearby soil and debris into the lakes. For the present study, a positive correlation was observed between total dissolved solid and conductivity. Chloride levels in the lakes have ranged from 25 to 50 mg/L. A huge fluctuation has been observed in chloride concentration for 5–45 mg/L which has been observed in Lake 1. This is may be due to the fact that Lake 1 receives a small portion of hostel as well as residential wastewater which may contributes chloride in the water. The average concentration of BOD for both the lakes varied in the range of 4–6 mg/L. A low BOD concentration indicates high DO. The concentration of BOD in the both lakes is very low, which may be contributed by the organic decomposition of nearby vegetation which may fall in the lakes. During January, the average DO was found as 9 mg/L and suggests lesser pollution and better aquatic life for animals and plants. A stream must have a minimum of about 4 mg/L of dissolved oxygen to maintain higher life forms. DO of Lake 1 varies from 3 to 8 mg/L and of Lake 2 varies from 1 to 8 mg/L, respectively. Average concentration of phosphorous for both lakes was observed as 0.002–0.163 mg/L. The increase in concentration of available phosphorous allows plants to assimilate more nitrogen before the depletion of phosphorous. The phosphorous in the water is contributed by the inflow of nearby residential and hostel wastewater into the lakes.

3.2 Correlation of Physicochemical Parameters

In order to meet the objective of the present work, i.e., accurate water quality prediction, by appropriately reducing the number of parameters needed without the loss of significant information. Therefore, a correlation analysis has been carried out to find out the relationship between various estimated water quality parameters. Correlation analysis is used to establish a relationship between two or more variables among each other, i.e., how one variable is dependent on other and how one variable can predict the other. The correlation matrices for 10 variables were prepared for the study area and presented in Table 2 for both the lakes. A highly positive correlation is observed between TS with turbidity and EC for both lakes. Correlation coefficient (R) of EC and TS for Lake 1 and 2 is 0.953 and 0.957, respectively. It is observed from Table 2 that the pH has significant correlation ($0.5 < r < 0.7$) with DO but poor correlation with other parameters.

The poor correlation of electrical conductivity with pH as well as total dissolved solids indicates a low dissociation capacity of the dissolved solids. For the present study, a poor correlation among most of the estimated parameters was observed for both the lakes. However, some correlations were observed between various parameters like turbidity and DO, chloride and DO, chloride and turbidity, phosphorus and turbidity, and phosphorus and temperature for both lakes. The analysis suggests that all parameters having very poor correlations are weakly dependent on each other. Table 3 shows the R-value (linear correlation coefficient) and the corresponding regression equation for the pairs of parameters which have R-value greater than 0.5 for both the Lake 1 and 2, respectively.

3.3 Predictive Model Analysis

Biochemical oxygen demand (BOD) is considered as the most critical and important parameter to find the pollution strength or the usage conditions of surface water as it measures the organic biodegradable matters present in water sample. Therefore, BOD may be defined as the amount of oxygen required to decompose the organic matter present in water under aerobic condition. Excess concentration of

Table 2 Correlation coefficient among various water quality parameters

Parameters	Conductivity	TS (mg/L)	Turbidity (NTU)	pH	Alkalinity (mg/L)	Chloride (mg/L)	BOD (mg/L)	Temp.	Phosphorus (mg/L)	DO (mg/L)
Lake 1										
Conductivity	1.000									
TS (mg/L)	0.953	1.000								
Turbidity (NTU)	0.788	0.846	1.000							
pH	−0.137	−0.065	0.016	1.000						
Alkalinity (mg/L)	−0.417	−0.324	−0.279	0.084	1.000					
Chloride (mg/L)	−0.065	0.009	−0.097	0.326	0.404	1.000				
BOD (mg/L)	0.091	−0.047	−0.166	−0.182	0.289	−0.057	1.000			
Temp.	0.509	0.438	0.143	−0.541	−0.341	−0.161	0.285	1.000		
Phosp.	0.224	0.197	0.061	0.364	0.036	0.102	−0.041	−0.047	1.000	
DO (mg/L)	−0.307	−0.256	0.096	0.666	0.191	−0.059	−0.053	−0.759	−0.108	1.000
Lake 2										
Conductivity	1									
TS (mg/L)	0.957	1								
Turbidity (NTU)	0.857	0.875	1							
pH	0.255	0.180	0.265	1						
Alkalinity (mg/L)	−0.422	−0.428	−0.231	−0.191	1					
Chloride (mg/L)	−0.471	−0.416	−0.359	0.263	−0.150	1				
BOD (mg/L)	−0.443	−0.505	−0.421	−0.202	0.135	0.359	1			
Temp.	−0.355	−0.382	−0.270	−0.472	0.059	0.289	0.619	1		
Phosp.	0.306	0.220	0.166	−0.258	0.183	−0.633	−0.077	−0.182	1	
DO (mg/L)	0.201	0.242	0.222	0.599	−0.112	0.104	−0.553	−0.719	−0.129	1.00

Table 3 Linear correlation coefficient and regression equation

Pairs of parameters	*R*-value	Regression equation
Lake 1		
EC–TDS	0.953	EC = 0.7211 (TDS) − 19.089
Turbidity–TDS	0.846	*T* = 0.0846 (TDS) − 1.1747
Turbidity–EC	0.788	*T* = 0.0596 (EC) − 2.2668
DO–Temperature	−0.759	DO = −0.5896 (Temp.) + 19.316
DO–pH	0.666	DO = 0.0875 (pH) + 6.2699
Temperature–pH	−0.541	pH = −0.0551 (Temp.) + 8.0459
Temperature–EC	0.509	Temp. = 0.0162 (EC) + 18.058
Lake 2		
EC-TDS	0.957	EC = 1.5341 (TDS) − 29.662
Turbidity–TDS	0.879	*T* = 0.2396 (TDS) − 21
Turbidity–EC	0.862	EC = 5.0142 (*T*) + 130.81
DO–Temperature	−0.721	DO = −0.8501 (Temp.) + 27.886
DO–pH	0.643	DO = 4.674 (pH) − 26.275
Temperature–pH	−0.501	Temp. = −4.3521 (pH) + 52.735
Phosphorus–Chloride	−0.633	Chloride = −136.71 (*P*) + 37.947

BOD in the water means there is depletion of DO level as dissolved oxygen in water is consumed by the microorganisms to oxidize the organic matter present in water, and thereby, a critical and dangerous situation arises for aquatic lives present in the water body. BOD of water body depends upon DO level, water temperature, water flow (continuous or stagnant), aquatic plants, and nutrient levels (ammonia, nitrite, and nitrate). Therefore, in order to find out the pollution strength or suitability of the water, testing BOD concentration in water is very important and it requires 5 days with sample collection and evaluation. The difference in dissolved oxygen concentration of the initial sample and after 5 days of incubation gives the BOD concentration in the water sample. Inorganic and organic matter present in water sample gives the total solid (TS) present. Since TS is not easily measured and requires approximately 24 h in the laboratory under controlled conditions. Therefore in the present work, an attempt has been made to model TS and BOD along with its dependent water quality parameters using statistical method—regression analysis and artificial neural network. Further comparison will be made between the model outcomes with the actual experimental values to validate the results obtained by the models.

3.3.1 Regression Model

Regression analysis provides an indirect method to monitor the water quality w.r.t. a specific parameter with the calculation of coefficient of correlation between the estimated water quality parameters. A good or high value of coefficient of correlation means a higher degree of dependence between the dependent and the independent variables. A higher value of regression coefficient suggests a better fit and more useful regression variable. Correlation is the mutual relationship between two variables.

Total Solid Model: For development of this model, TS is taken as output, i.e., dependent variable and electrical conductivity and turbidity are taken as input, i.e., independent variable for both lakes because there is strong correlation between independent and dependent variable. From these data, multiple regression analysis is done using data analysis tool of MS Excel at 95% confidence level and the regression equation relating TS with electrical conductivity and turbidity has been obtained. Summary of regression analysis and their result is presented in Table 4 for Lake 1 and 2. A scatter is plotted between actual TS and predicted TS for both lakes shown in Fig. 1 for Lake 1 and 2. *R*-square value of scatter plot shows the probability of existence of error. If R-square is approximately 1, then probability of existence of error is negligible that means the predicted model is valid. Regression coefficient for Lake 1 and 2 is 0.9315 and 0.9274, respectively, which shows that the analysis carried out by regression analysis matches the experimental values very accurately.

Biological oxygen demand model: For development of this model, BOD is taken as output, i.e., dependent variable and reaming all nine parameters (parameter measured in laboratory) has taken as input, i.e., independent variable for both lakes.

Table 4 Summary of multiple regression analysis for TS

	Independent parameter	*P* value	*t*-stat	Adjusted R^2	Significant *F*	Equation
Lake 1	Turbidity	0.018	2.57	0.924	8.7×10^{-12}	TS = 2.5 × turbidity + 0.57 × EC
	EC	2.74×10^{-7}	7.73			
Lake 2	Turbidity	0.01	1.722	0.919	1.5×10^{-11}	TS = 0.76 × turbidity + 0.48 × EC+ 42.71
	EC	4.70×10^{-6}	6.73			

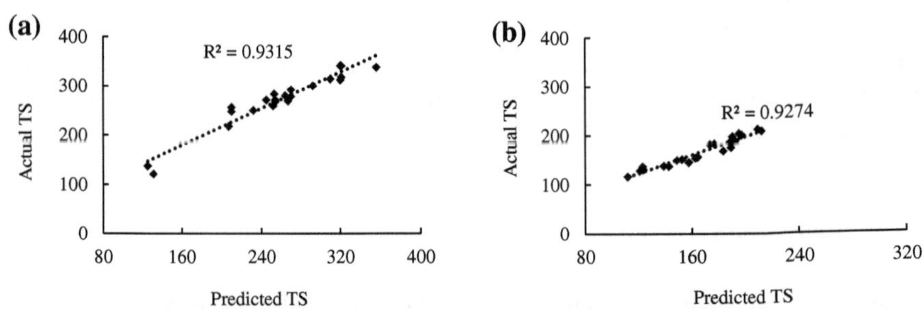

Fig. 1 Scatter plot between actual and predicted TS for **a** Lake 1 and **b** Lake 2

Table 5 Summary of multiple regression analysis for BOD

	Independent parameter	*P* value	*t*-stat	Adjusted R^2	Significant *F*	Equation
Lake 1	Turbidity	0.2	−1.37	0.41	0.06	BOD = 2.95 × Alkalinity + 0.15 × Temp + 0.25 × DO −0.01 × *T* + 0.013 × EC
	EC	0.03	2.45			
	DO	0.19	1.38			
	Temperature	0.1	1.81			
	Alkalinity	0.05	2.2			
Lake 2	DO	0.20	1.18	0.27	0.15	BOD = 0.25 × Temp + 0.15 × DO
	Temperature	0.15	1.78			

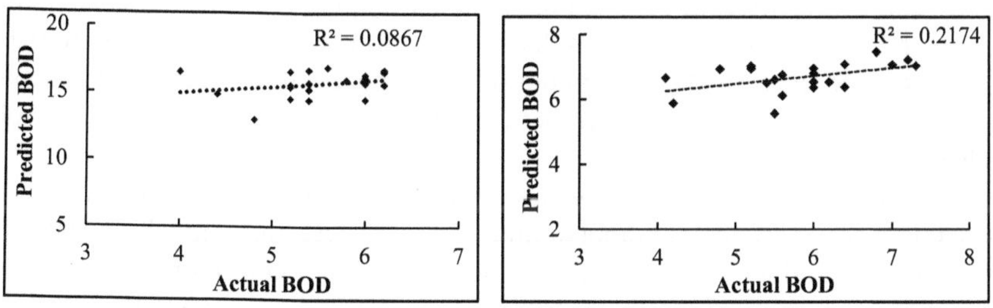

Fig. 2 Scatter plot between actual and predicted BOD **a** Lake 1 and **b** Lake 2

From these data, multiple regression analysis is done using data analysis tool of MS Excel at 80% confidence level and the regression equation relating BOD with its dependent water quality parameters has been obtained. Summary of regression analysis and their result is presented in Table 5 for Lake 1 and 2. A scatter is plotted between actual predicted BOD for both lakes shown in Fig. 2 for Lake 1 and 2. The regression model analysis resulted in very poor R-square value between the actual and predicted BOD values for both the lakes, which suggests regression model cannot be used for the present analysis.

3.3.2 Neural Network Model

Artificial neural network (ANN) models can work even with not-so-correlated predictors; all the values can be considered on the same foot while constructing single hidden layer ANN models. To test the data, samples were collected and analyzed in the laboratory. For the present study project, feedforward Levenberg algorithm was used by the MATLAB Neural Network tool. The data in neural networks are categorized into three sets: training or learning sets, validation, and test or over-fitting test set. A total of 22 data samples were divided into a training sets consisting of 16 samples (70% of the total), 3 samples (15% of the total) were used for validation, and the remaining 3 samples (15% of the total) were used for testing. Goodness of the fit on the basis of statistical indices suggests the performance of input layer. Goodness of the fit statistics changes with the increase of decrease in the hidden layer of the network algorithm of the model. For the present study, hidden layers are fixed to 10 during modeling process and goodness of fit statistics has been checked for training, validation, and testing of time series data. The input–output and error correlation, autocorrelation is simultaneously calculated by the model. Root mean square error (RMSE) was estimated which is the error between output and given target values by the model. Here, R is the measure of regression between output and target data.

TS Model: To simulate and predict the TS of Lake 1 and 2, ANN-based predictive model was developed. A very small amount of organic matter along with inorganic matters in water resulted in TS concentration. In this model, turbidity and conductivity is consider as input and total dissolved solid is as output. The regression output on ANN results of total solid for Lake 1 and 2 is shown in Figs. 3 and 4, respectively. The output of ANN model shows very good R-square value and shows very less error between the actual and predicted total solid values.

Biological oxygen demand model: For the BOD model using ANN, the remaining nine parameters (temperature, alkalinity, DO, turbidity, TDS, chloride, phosphorus, conductivity, pH) are considered as input and BOD is as output. The output fitting curves BOD using ANN for Lake 1 and 2 are shown in Figs. 5 and 6, respectively. ANN output shows a very close fit of model BOD value to actual the actual BOD value (R-square value ranges from 0.69 to 0.82) for both the lakes.

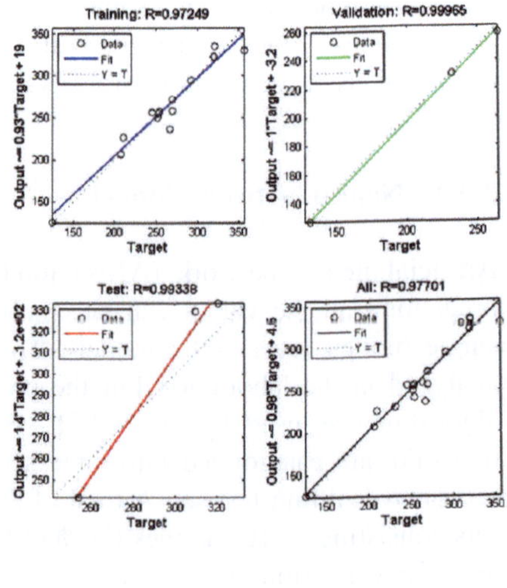

Fig. 3 Output on ANN results of TS model for Lake 1

Fig. 4 Output on ANN results of TS model for Lake 2

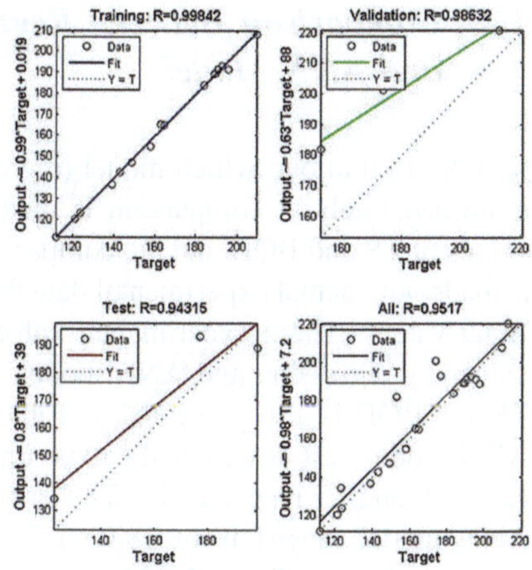

Fig. 5 Output on ANN results of BOD model for Lake 1

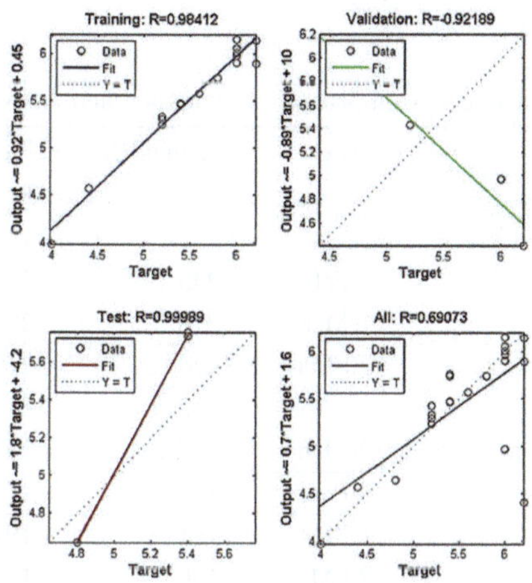

Fig. 6 Output on ANN results of BOD model for Lake 2

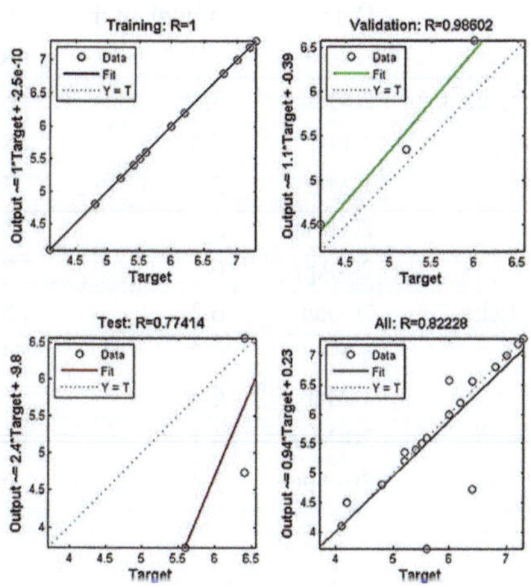

3.4 Comparison Between Regression Model and ANN Model

In order to find out, which model (regression or ANN) fits the experimental values more accurately, a comparison is made between the experimental and estimated values of TS and BOD and the comparison is presented in Table 6. The comparison is made with actual experimental data that was obtained in the laboratory for TS and BOD values. These experimental values were then compared with the calculated values by regression and ANN model. The calculated error for both the model for TS and BOD is also presented in Table 6. It has been observed, from the Table 6, ANN model best describes the experimental values for both TS and BOD for both Lakes 1 and 2, respectively. The error between the experimental and estimated ANN model ranges from 18.0 to 2.7 and 6.2 to 2.6 for Lake 1 and Lake 2, respectively, for TS and 22.9 to 3.7 and 12.2 to 1.0 for Lake 1 and Lake 2, respectively, for BOD. However, the calculated error percentage for regression

Table 6 Comparison of regression and ANN model with experimental values

	Date	Actual TS	Predicted TS		Error % TS	
			Regression	ANN	Regression	ANN
Lake 1	20-Jan	131	110.2	120.9	15.9	7.7
	4-Feb	245	279.0	271	13.9	10.6
	1-Mar	210	252.2	247.7	20	18.0
	20-Mar	254	279.0	266.8	9.8	5.0
	20-Apr	252	279.0	258.9	10.7	2.7
Lake 2	20-Jan	177.1	182.3	172.2	2.9	2.8
	4-Feb	123.5	130.4	127.6	5.6	3.3
	1-Mar	191	197.1	202.8	3.2	6.2
	20-Mar	190	174.7	194.9	8.1	2.6
	20-Apr	121	128.1	116.3	5.9	3.9
	Date	Actual BOD	Predicted BOD		Error % BOD	
			Regression	ANN	Regression	ANN
Lake 1	20-Jan	4.8	6.8	5.9	41.7	22.9
	4-Feb	5.2	8.5	5.5	63.5	5.8
	1-Mar	5.4	7.0	5.6	29.6	3.7
	20-Mar	5.4	8.2	5.6	51.9	3.7
	20-Apr	6	7.2	5.6	20.0	6.7
Lake 2	20-Jan	6.2	5.5	5.8	11.3	6.5
	4-Feb	6.2	4.6	6.5	25.8	4.8
	1-Mar	6	5.9	6.2	1.7	3.3
	20-Mar	4.1	5.1	4.6	24.4	12.2
	20-Apr	7	6.7	7.07	4.3	1.0

models lies in the range of 20–2.9 for TS and 63.5–1.7 for BOD for both the lakes. The results indicated that ANN can be best-described model for prediction of studied parameters.

4 Conclusions

1. Correlation studies suggested a highly positive correlation between TS with turbidity and EC for both lakes. Correlation coefficient (*R*) of EC and TS for Lake 1 and Lake 2 is 0.953 and 0.957, respectively.
2. A poor correlation was observed among the various estimated parameters of both lakes. However, some correlations were observed between various parameters like turbidity and dissolved oxygen, chloride and dissolved oxygen, chloride and turbidity, phosphorus and turbidity, and phosphorus and temperature for both lakes.
3. Modeling of TS and BOD by regression analysis suggests a good correlation between actual and predicted TS but a very poor correlation between actual and predicted BOD, suggesting regression model cannot be used to model BOD for the present case.
4. Modeling of TS and BOD by ANN shows a very good correlation between the actual and predicted values for both total solid and BOD (R-square value ranges from 0.69 to 0.82) for both the lakes studied.
5. ANN model best describes the experimental values for both TS and BOD for both Lakes 1 and 2, respectively. The error between the experimental and estimated ANN model ranges from 18.0 to 2.7 and 6.2 to 2.6 for Lake 1 and Lake 2, respectively, for TS and 22.9 to 3.7 and 12.2 to 1.0 for Lake 1 and Lake 2, respectively, for BOD, which is less than estimated error by regression model. Thus, it can be concluded that ANN can be best-described model for prediction of studied parameters.

References

1. Gibbs, J.: Water chemistry of Amazon river. Geochim. Cosmochim. Acta **36**, 1061–1066 (1972)
2. Aziz, R., Muhammad, O.C.: Regional interpretation of river Indus water quality data using regression model. Afr. J. Environ. Sci. Technol. **8**(1), 86–90 (2014)
3. Ustun, B., Ekercin, S., Oztopal. A.: Investigation of water quality parameters by using multiple regression and fuzzy logic in the Istanbul Strait, Turkey. In: Bochenek, Z. (Ed.), New Developments and Challenges in Remote Sensing, pp. 603–611. Millpress, Rotterdam (2007)
4. Emrah, D., Bulent, S., Rabia, K.: Modelling biological oxygen demand of the Melen river in Turkey using an artificial neural network technique. J. Environ. Manag. **90**(2), 1229–1235 (2009)

5. Kunwar, P.S., Ankita, B., Amrita, M., Gunja, J.: Artificial neural network modelling of the river water quality—a case study. Ecol. Model. **220**(6), 888–895 (2009)
6. Majid, H., Ehsan, O., Hamid, M., Ozgür, K.: Development of a neural network technique for prediction of water quality parameters in the Delaware river. Pa. Middle East J. Sci. Res. **13** (10), 1367–1376 (2013)
7. Qing, Z., Stephen, J.S.: Forecasting raw-water quality parameters for the North Saskatchewan river by neural network modelling. Water Res. **31**(9), 2340–2350 (1997)
8. Simeonov, V., Stratis, J.A., Samara, C., Zachariadis, G., Voutsa, D., Anthemidis, A., Sofoniou, M., Kouimtzis, T.: Assessment of the surface water quality in Northern Greece. Water Res. **37**(17), 4119–4124 (2003)
9. Wang, H., Hondzo, M., Xu, C., Poole, V., Spacie, A.: Dissolved oxygen dynamics of streams draining an urbanized and an agricultural catchment. Ecol. Model. **160**, 145–161 (2003)
10. Adamu, M., Ado, A.: Application of principal component analysis and multiple regression models in surface water quality assessment. J. Environ. Earth Sci. **2**(2), 16–23 (2012)
11. Bahaa, M.K., Ayman, G.A., Hussein, K., Ashraf, E.S.: Application of artificial neural networks for the prediction of water quality variables in the Nile Delta. J. Water Resour. Prot. **4**, 388–394 (2012)
12. APHA: Standard Methods for the Examination of Water and Wastewater, 20th edn. American Public Health Association/American Water Works Association/Water Environment Federation, Washington, DC, USA (1998)

Removal of Fluoride from Groundwater by Adsorption onto Brick Powder–Alum–Calcium-Infused Adsorbent

Kamal Uddin Ahamad, Abhigyan Mahanta and Sayeesta Ahmed

Abstract Removal of fluoride from groundwater onto BAC—an adsorbent made by infusing brick powder, alum, and calcium chloride has been investigated in the present study. Individually, the three materials can take up fluoride, but in very low concentration and therefore these have been selected and fused together to enhance the overall fluoride removal capacity. The three materials were taken in the ratio of 1:1:1, fused together with 20% conc. HCl, oven dried at 105 °C for 24 h. The material was then crushed and powder and was kept in airtight container. For initial fluoride concentration of 4 mg/L, an equilibrium time of 120 min. was observed at dose of 7.5 g/L of BAC at 280 rpm shaking speed. Pseudo-second-order kinetic model fits experimental data very well suggested the applicability of the same. The results of the isotherm studies showed that fluoride removal by BAC followed the Freundlich isotherm model.

Keywords Adsorption · Alum · Batch studies · Brick powder
Low cost

K. U. Ahamad (✉) · A. Mahanta · S. Ahmed
Department of Civil Engineering, Tezpur University, Napaam 784028, Assam, India
e-mail: kahamad@tezu.ernet.in

A. Mahanta
e-mail: abhigyanmahanta@gmail.com

S. Ahmed
e-mail: sayeesta@gmail.com

A. S. Kalamdhad et al. (eds.), *Advances in Waste Management*,
https://doi.org/10.1007/978-981-13-0215-2_16

1 Introduction

Fluoride in the drinking water may cause serious health issues, if found more than the limits set by regulatory agencies like WHO [1, 2]. Fluoride in drinking water has both beneficial and harmful effect on human health. The excessive fluoride found in the groundwater of major part of the India is a matter of greater concern. One of the major sources of fluoride in groundwater and even in surface water is the leaching of same from the earth crust or the rocks containing fluoride. It is reported that many states like Andhra Pradesh, Rajasthan, Punjab, Haryana, Maharastra, Tamilnadu, Karnataka, Madhya Pradesh, Gujrat, Uttar Pradesh, and few districts of West Bengal have fluoride concentration in the range of 1–35 mg/L in the groundwater [3]. However, the permissible limit in drinking water is 1–1.5 mg/L [1, 2]. Unlike iron in the groundwater which gives visible coloration to the water, fluoride does not give any coloration to the water. People are aware of the presence of iron, but unaware of presence of fluoride in the groundwater and therefore they do not put any effort or treat the water containing fluoride before using it for drinking and cooking purposes in the rural and semi-urban areas. Adsorption is most commonly used technique for fluoride removal, in which fluoride is adsorbed onto an adsorbate, or a fixed bed packed with resin or other mineral particles. Reverse osmosis, electrodialysis, Donnan dialysis, ion exchange, limestone reactor, and activated alumina column [4, 5] are among the few techniques that are been used for fluoride removal. Ion exchange, electrodialysis, and membrane processes are very effective and can remove the fluoride to the prescribed limits. However, these methods are very expensive and require frequent regeneration apart from cleaning of the scaling and fouling [5–8]. Many researchers in the recent years have tried to explore various low cost but effective adsorbents for fluoride removal. Many materials such as calcite [9], clay [10], charcoal [11], bleaching earth [10], red mud [12] have been tested for fluoride removal. There are numerous technologies and techniques available for removal of fluoride from groundwater but none or very less of the techniques and technologies has reached rural and semi-urban population of the country. People in these parts were not able to use these techniques and technologies mostly due to lack of knowledge, awareness and might be due to cost economics also. This paper concentrates on investigating the fluoride adsorption behavior of a laboratory prepared low-cost adsorbent and evaluates its effectiveness for removal of fluoride from aqueous solutions to the prescribed limit. The potential of laboratory prepared adsorbent would be assessed for removal of F^- through kinetics, and equilibrium studies in batch mode.

2 Experimental

2.1 *Adsorbent*

For the present study, three adsorbents, namely alum, brick powder, and calcium carbonate, have been infused together and tested for its fluoride removal capacity from drinking water. Calcium carbonate ($CaCO_3$) and alum used in this study were procured from Merck-Paris, France, and used as procured. Brick Powder (BP) has been obtained by crushing brickbats into finer size followed by sieving, washing, and then oven drying at a temperature of 105 °C for approximately 24 h. Particle passing through 300 µm sieve and retained on 150 µm sieve is used for the adsorption studies. 4 g each of BP, Alum, and $CaCO_3$ was taken in a ratio of 1:1:1 in a 500 mL beaker. 100 mL of distilled water was then added to the beaker and the components were mixed thoroughly. 5 mL of conc. HCl was then added to the mix and the resulting mixture was kept in the oven for 24 h at 105 °C. After 24 h the mixture was dried and is in the form of a cake, which was again grounded and kept in airtight container to be used further in subsequent adsorption studies and the prepared adsorbent was termed as BAC, i.e., brick powder, alum, and calcium-infused adsorbent.

2.2 *Methods*

Analytical grade sodium fluoride (NaF) was dissolved in distilled water to prepare stock fluoride (F^-) solution of 100 µg/mL concentration. Properly diluting the stock solution with distilled water has given the working F^- solution. The residual F^- concentration was measured as per the standard SPADNS Method [13] using a digital spectrophotometer (Model: 167, M/S Systronics India Ltd., India).

2.3 *Batch Experiments*

Batch mode kinetic and equilibrium adsorption studies were carried out to evaluate the potential of BAC for F^- removal. The effect of contact time on F^- adsorption onto BAC was investigated using ten specimen tube of 50 mL capacity. 0.2 g of BAC was added to each of the specimen tube having 40 mL of initial F^- solution of 4 mg/L concentration. The specimen tubes were then shaken in a mechanical shaker at a speed of 280 rpm for 3 h at room temperature (22 ± 1 °C). Specimen tubes containing F^- solution along with BAC were taken out one at a time at a predetermined time interval and residual F^- concentrations were estimated after separating the liquid and solid phases using a 0.40 μm filter paper. Dose test is carried out to determine the optimum dose of BAC to meet the drinking quality standard, i.e., F^- concentration of ≤ 1 mg/L [1, 3]. Four different doses of BAC—2.5, 5.0, 7.5 and 10 g/L were selected for the present study. The doses were tested separately with initial F^- concentration of 4 mg/L. The tests were conducted using 50 mL capacity specimen tube with 40 mL of F^- solution. The specimen tubes were then shaken in a mechanical shaker at a speed of 280 rpm for 2 h at room temperature (22 ± 1 °C). Specimen tubes containing F^- solution along with BAC were taken out at a time at a predetermined time interval and residual F^- concentrations were estimated after separating the liquid and solid phases using a 0.40 μm filter paper. The BAC dose that meets the objective of the present test was selected a dose for subsequent studies. Four different shaking speeds of 160, 200, 240, and 280 rpm were selected to check the effect of shaking speed on F^- uptake. BAC dose of 7.5 g/L was selected for initial F^- concentration of 4 mg/L, in a 50 mL capacity specimen tube with 40 mL of F^- solution. Each of specimen tube containing BAC and F^- solution was then shaken separately at different shaking speed. Samples were taken out at a time at a predetermined time interval and residual F^- concentrations were estimated after separating the liquid and solid phases using a 0.40 μm filter paper.

Adsorption kinetics was tested using the experimental values as obtained from the variation in contact time test. A fixed BAC dose of 7.5 g/L, at an initial F^- concentration of 4 mg/L, at a shaking speed of 280 rpm up to 2 h was selected for the test.

For adsorption equilibrium isotherm studies, BAC dose was varied in fixed fluoride solution of 40 mL having initial F^- concentration of 4 and 5 mg/L at a fixed shaking speed of 280 rpm for 3 h. All batch experiments were conducted in duplicate and average values were used for analysis of the results. The amount of F^- uptake and the F^- uptake percentage were calculated using:

$$q_t = \frac{(C_o - C_t)V}{W} \tag{1}$$

and

$$\text{uptake } (\%) = \frac{C_o - C_t}{C_o} \times 100 \tag{2}$$

where q_t, V, W, C_o and C_t are the amount of F^- adsorbed (mg/g) at time t, volume of the solution (L), weight of the BAC (g), F^- concentration (mg/L) at initial time and remaining at a time t, respectively.

3 Results and Discussion

3.1 Effect of Variation in Contact Time

The effect of variation in contact time on F^- uptake by BAC is presented in Fig. 1. The F^- uptake by BAC was observed to be rapid for first 50 min as indicated by the slope of the curve (Fig. 1) and most of the F^- uptake occurred within 100 min indicating achievement of equilibrium. Based on the results, an equilibrium time of 120 min was adopted for F^- uptake for all subsequent batch studies.

3.2 Effect of Variation in Adsorbent Dose

The effect of variation in adsorbent dose on the extent of F^- removal by BAC is shown in Fig. 2. Increasing BAC dose over the tested doses showed that dose of BAC has a definite role in adsorption of F^- from the water sample.

The results of the present study suggested that with increase in adsorbent doses, F^- removal is increasing (55, 96, 98, and 100% for 2.5, 5.0, 7.5, and 10 g/L of BAC doses), however the F^- uptake is decreasing with increase in BAC doses (0.88, 0.77, 0.52, and 0.40 mg/g for 2.5, 5.0, 7.5, and 10 g/L of BAC doses). The increase in F^- removal and decrease in F^- uptake with increase in BAC dosage clearly suggested that the dose of BAC has definite influence in the adsorption process. In the present study, BAC doses were varied from 2.5 to 10 g/L but initial F^- concentration was fixed at 4 mg/L. Increasing BAC doses with fixed F^- concentration will result in more surface area for adsorption. Therefore, for the present study amount of adsorbate was kept fixed at 4 mg/L while the dose of adsorbent has varied from 2.5 to 10 g/L. The result of the present study will be helpful in achieving the objective by selecting a dose of BAC to meet the guideline value set for the drinking water, i.e., reducing the F^- concentration to ≤ 1.0 mg/L [1, 2]. The dose to meet the F^- concentration of ≤ 1.0 mg/L may be selected either based on higher F^- removal or higher F^- uptake. Considering a dose of 10 g/L or higher—selected based on higher F^- removal is having F^- uptake of 0.4 mg/g or even lower than this. However, if a dose of 2.5 g/L or lower is selected based on

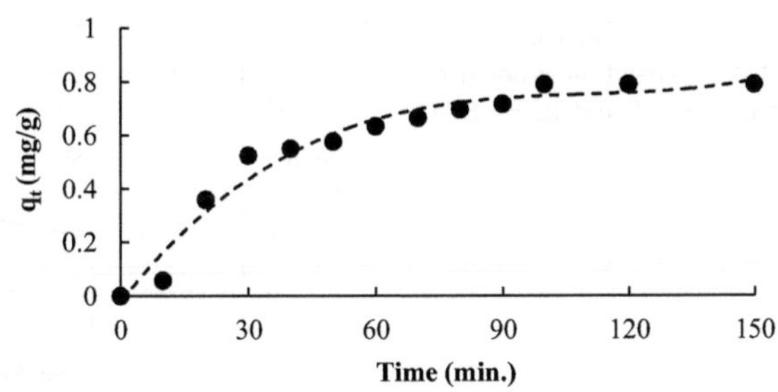

Fig. 1 Effect of variation in contact time on F^- uptake by BAC

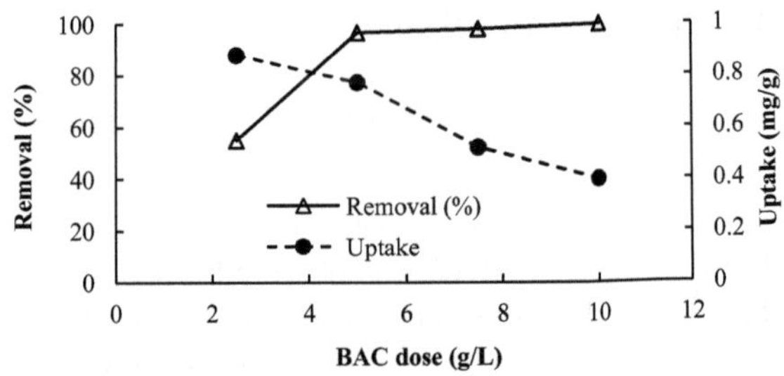

Fig. 2 Effect of variation in PWC dose on extent of F^- removal and F^-uptake

higher F^- uptake it failed to meet the drinking water standard set for F^- removal, i.e., ≤ 1 mg/L even though the uptake was 0.88 mg/g or more. BAC is prepared in laboratory and is a low-cost adsorbent, therefore selection of higher dose, to meet the drinking water standard set for F^- removal of ≤ 1 mg/L is justified even though the same will result in lower uptake. Hence, a dose of 7.5 mg/g was selected for further studies.

3.3 *Effect of Variation in Shaking Speed*

Adsorption tests were carried out by varying the shaking or agitation speed for the adsorption of F^- by BAC to optimize the same for maximum adsorption. The shaking speed was varied from 160 to 280 rpm and effect on adsorption capacity of BAC was studied at different shaking speeds and results are presented in Fig. 3. It is evident from the test results (Fig. 3) that the uptake of F^- by BAC at 160 rpm was approximately 0.613 mg/g. However, the uptake of F^- by BAC increased to approximately 0.8 mg/g at 280 rpm yielding residual F^- concentration of ≤ 1 mg/L at BAC dose of 0.3 g. The increase in uptake capacity of BAC at 280 rpm can be explained by the fact that the increased agitation or shaking speed probably reduces the boundary layer resistance and consequently bulk diffusion into the pores of BAC may be increased due to increase in the driving force of F^- ions. Therefore, it may be concluded that both the pore diffusion along with the film diffusion is

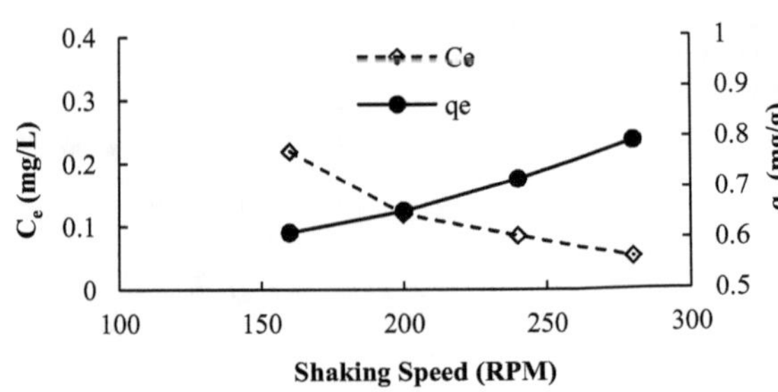

Fig. 3 Effect of variation in shaking speed on residual F^- conc. and F^-uptake by BAC

responsible for the uptake of F^- by BAC [14–16]. Therefore, a shaking speed of 280 rpm was selected for subsequent batch studies.

3.4 Adsorption Kinetic Studies

Adsorption kinetics was tested for pseudo-first-order, pseudo-second-order, and intraparticle diffusion models to explain the mechanism of adsorption of F^- onto BAC. Adsorption kinetics was tested using the experimental values as obtained from the variation in contact time test (as described in Sect. 3.1). The most widely used rate equation to explain the adsorption kinetics is pseudo-first-order rate equation [17, 18] which is expressed in linearized form as:

$$\log(q_e - q_t) = \log q_e - \frac{k_1}{2.303}t \tag{3}$$

where, q_e is the F^- adsorbed at equilibrium (mg/g), q_t is the F^- adsorbed at time t (mg/g) and k_1 is the rate of first-order adsorption (min^{-1}). The adsorption kinetic data were also tested using pseudo-second-order kinetic model [19] which is expressed in linearized form as:

$$\frac{t}{q_t} = \frac{1}{k_2 q_e^2} + \frac{t}{q_e} \tag{4}$$

where q_e is the F^- adsorbed at equilibrium (mg/g), q_t is the F^- adsorbed at time t (mg/g) and k_2 is the rate constant of the second-order adsorption (gm/mg min). The first-order and second-order constants were estimated and presented in Table 1 for the initial F^- concentration studied.

In order to find out the best fit kinetic model, i.e., pseudo-first-order or pseudo-second-order kinetic model with the experimental data, error analysis was carried. The error analysis was carried out using the chi-square (χ^2) and the root

Table 1 Summary of estimated kinetics model parameters and error analysis

First-order model		Second-order model	
$q_{e:\text{experimental}}$ (mg/g)	0.789	$q_{e:\text{experimental}}$ (mg/g)	0.789
k_1 (min^{-1})	0.0177	$q_{e:m}$ (mg/g)	0.0312
$q_{e:m}$ (mg/g)	1.3079	R^2	0.9768
R^2	0.9768	χ^2	0.9910
χ^2	0.8476	RMSE	0.1468
RMSE	0.2558		0.0565

$q_{e:\text{experimental}}$: q_e value obtained from experiment; $q_{e:m}$: q_e value obtained using kinetic model mean squared error (*RMSE*) tests. The mathematical equation of χ^2 test is represented as [20]:

$$\chi^2 = \sum \left(\frac{(q_t - q_{tm})^2}{q_{tm}} \right) \tag{5}$$

where q_{tm} and q_t are adsorption capacities (mg/g) obtained by using the model and the experimental data respectively. If the experimental data are similar to the data obtained from the kinetic models, the value of chi-square test will be smaller, if they are not similar the value will be larger. Root mean squared error (*RMSE*) test was carried out using the following expression [21]:

$$\text{RMSE} = \sqrt{\left(\frac{1}{N} \sum_{i=1}^{N} \left(q_{tm,i} - q_{t,i} \right)^2 \right)} \tag{6}$$

where N is the number of experimental data and q_{tm} and q_t represent the adsorption capacities (mg/g) obtained by using the model and the experimental data, respectively. For RMSE test also, if the experimental data are similar to the data obtained from the kinetic models, the RMSE value will be smaller, if not, it will be larger. The estimated values of error analysis are also presented in Table 1. For F^- adsorption with BAC for the initial F^- concentration studied, pseudo-second-order rate model shows relatively lower values of χ^2 and RMSE compared to the first-order kinetic model, which suggests the applicability of pseudo-second-order kinetic model which is in line with the observations made on the basis of Fig. 4.

To find out the adsorption mechanism of F^- ions, i.e., film or pore diffusion onto the BAC, it was tested against the following equation [22]:

$$q_t = K_p t^{0.5} \tag{7}$$

where K_p is the intraparticle diffusion rate constant (mg/g min$^{0.5}$). For the present study the mechanism of adsorption is shown in Fig. 5 for the system investigated.

Fig. 4 Variation in F^- uptake capacity of BAC obtained experimentally and by kinetic models

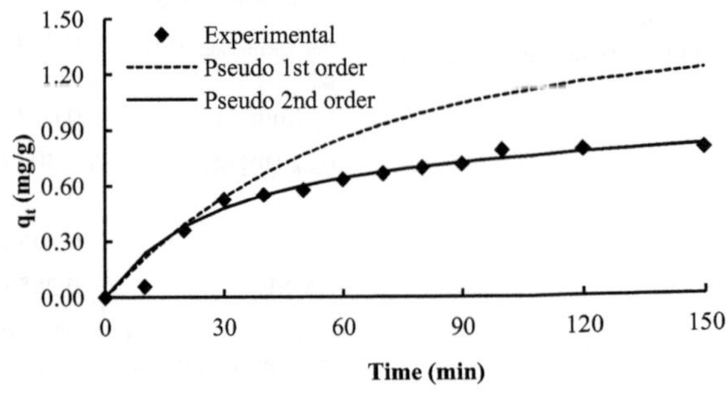

Fig. 5 Intraparticle diffusion model for adsorption of F^- on BAC

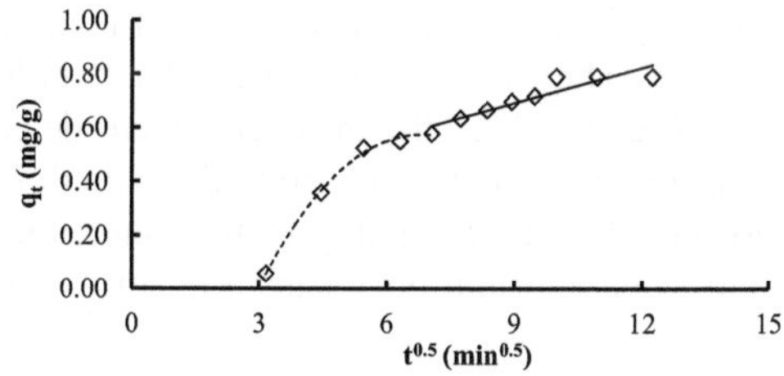

It has been observed from the plot of q_t versus $t^{0.5}$ for initial F^- concentration studied was having initial curved followed by a linear portion. The curve portion of the plot indicates the bulk diffusion to the external surface of the adsorbent or boundary layer diffusion, whereas the linear portion of the plot describes the gradual pore diffusion. Therefore, for the present study initial curve portion of the plot indicated bulk diffusion or boundary layer diffusion of F^- ions and the linear portion suggest pore diffusion [23, 24]. Therefore, for the present study it may be concluded that adsorption by BAC was due to both boundary layer diffusion and pore diffusion. The concentration gradient that exists between the concentration of F^- ions in adsorbent and adsorbate might be the probable reason for both boundary layer as well as pore diffusion.

3.5 Adsorption Equilibrium Studies

Adsorption equilibrium studies were carried out with BAC at an initial F^- concentration of 4 and 5 mg/L at room temperature of 22 ± 1 °C. The adsorption equilibrium data were fitted into both Langmuir and Freundlich isotherm model. The Langmuir model is derived on the assumption that it is valid for monolayer sorption with finite number of identical sites. The linearized form of Langmuir isotherm model [25] can be represented as:

$$\frac{C_e}{q_e} = \frac{1}{ab} + \frac{C_e}{a} \tag{8}$$

where b, a, and q_e are the Langmuir constant, maximum adsorption capacity (mg/g), and the amount of metal adsorbed at equilibrium (mg/g), respectively. The adsorption equilibrium data were also fitted into the Freundlich isotherm model, which is represented [26] in its linearized form as:

$$\log q_e = \log k_f + \frac{1}{n} \log C_e \tag{9}$$

where k_f is Freundlich constants w.r.t adsorption capacity and n is adsorption intensity. The various isotherm constants (a, b for Langmuir isotherm model; k_f and n for Freundlich isotherm model) were determined and the estimated values are presented in Table 2 for both the initial F^- concentration of 4 and 5 mg/L, respectively. The feasibility of the Langmuir isotherm was checked using dimensionless equilibrium parameter, R_L represented by [27]:

$$R_L = \frac{1}{1 + bC_o} \tag{10}$$

where b is the Langmuir isotherm constant and C_o is the initial F^- concentration (mg/L). The R_L value for the two initial F^- concentration of 4 and 5 mg/L were calculated and are presented in Table 2. The R_L values between 0 and 1 indicated favorable F^- adsorption on BAC.

The adsorption equilibrium data were fitted into both the Langumuir and Freundlich isotherm equations, variables of the models were estimated and are presented in Table 2. Both Langmuir and Freundlich models are empirical in nature and derived using certain assumptions. It has been observed that, the values of correlation coefficient (R^2) were higher for Langmuir model compared with the same for Freundlich model suggesting applicability of Langmuir model. To find out the suitable adsorption equilibrium model among the two, i.e., Langmuir and Freundlich models, error analysis was carried out for both experimental and model values. Chi-square (χ^2) and the root mean squared error (*RMSE*) tests (using Eqs. (2) and (4) but replacing q_t by q_e and q_{tm} by q_{em} where q_e and q_{em} are the equilibrium adsorption capacities (mg/g) obtained experimentally) were carried out to find out the suitable adsorption equilibrium model. The corresponding values of χ^2 and RMSE are also presented in Table 2. It has been observed that the χ^2 and RMSE values for Freundlich model were lower than the values of the same for Langmuir model, suggesting applicability of Freundlich isotherm model. The adsorption isotherm plots using experimental data and fitted isotherm models are shown in Fig. 6. It is evident from (Fig. 6a, b) that the Freundlich model represents the equilibrium data with BAC at initial F^- concentrations of 4 and 5 mg/L.

Table 2 Estimated Langmuir and Freundlich adsorption isotherm model parameters

Langmuir isotherm			Freundlich isotherm		
Parameters	4 mg/L	5 mg/L	Parameters	4 mg/L	5 mg/L
a (mg/g)	1.8145	0.4717	$1/n$	0.1056	0.1925
b (L/mg)	5.9707	1.8744	k_f	1.5360	0.3207
R_L	0.0402	0.1176	R^2	0.9870	0.9891
R^2	0.9989	0.9971	χ^2	4.8×10^{-6}	9.9×10^{-6}
χ^2	4.2×10^{-4}	1.3×10^{-4}	RMSE	6.0×10^{-3}	3.6×10^{-3}
RMSE	4.7×10^{-2}	1.0×10^{-2}			

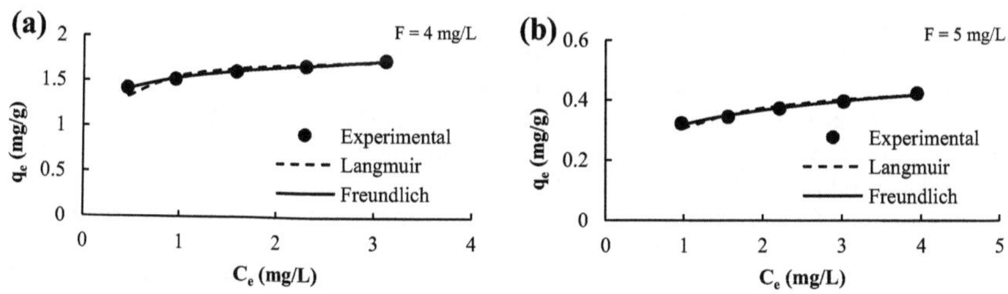

Fig. 6 Adsorption isotherm plots using experimental data and fitted isotherm models for BAC at initial F^{-} conc. of **a** 4 mg/L and **b** 5 mg/L

4 Conclusion

Following conclusion has been drawn from the present study

- Fluoride adsorption on BAC achieves equilibrium within 120 min.
- The kinetic studies suggest applicability of pseudo-second-order kinetic model that fits the experimental data.
- Fluoride uptake on BAC is due to surface adsorption as well as intraparticle diffusion.
- Fluoride adsorption on BAC is favorable and the equilibrium data fitted into Freundlich isotherm model.
- BAC, i.e., Brick powder—Alum—Calcium carbonated infused adsorbent can be used as a low-cost adsorbent for fluoride removal.

References

1. IS 10500: Specification for Drinking Water. Bureau of Indian Standards, New Delhi (1991)
2. WHO: Guidelines for Drinking Water Quality. Health Criteria and Supporting Information, Recommendations, 2nd edn., vol. I, p. 41. World Health Organization, Geneva (1993)
3. Meenakshi, Maheshwari R.C.: Fluoride in drinking water and its removal. J. Hazard. Mater. **137**(1), 456–463 (2006)
4. Hichour, M., Persin, F., Molenat, J., Sandeaux, J., Gavach, C.: Fluoride removal from diluted solutions by Donnan dialysis with anion-exchange membranes. Desalination **122**, 53–62 (1999)
5. Amor, Z., Malki, S., Taky, M., Bariou, B., Mameri, N., Elmidaoui, A.: Optimization of fluoride removal from brackish water by electrodialysis. Desalination **120**, 263–271 (1998)
6. Srimurali, M., Pragathi, A., Karthikeyan, J.: A study on removal of fluorides from drinking water by adsorption onto low-cost materials. Environ. Pollut. **99**, 285–289 (1998)
7. Cohen, D., Conrad, H.M.: 65000 GPD Fluoride removal membrane system in Lakeland California USA. Desalination **117**, 19 (1998)
8. Pervov, A.G., Dudkin, E.V., Sidorenko, O.A., Antipov, V.V., Khakhanov, S.A., Makarov, R.I.: RO and RF membrane systems for drinking water production and their maintenance techniques. Desalination **132**, 315–321 (2000)

9. Yang, M., Hashimoto, T., Hoshi, N., Myoga, H.: Fluoride removal in a fixed bed packed with granular calcite. Water Res. **33**(16), 3395–3402 (1999)
10. Mahramanlioglu, M., Kizilcikli, I., Bicer, I.O.: Adsorption of fluoride from aqueous solution by acid treated spent bleaching earth. J. Fluor. Chem. **115**, 41–47 (2002)
11. Bhargava, D.S., Killedar, D.J.: Fluoride adsorption on fishbone charcoal through a moving media adsorbent. Water Res. **26**(6), 781–788 (1992)
12. Cengeloglu, Y., Kir, E., Ersoz, M.: Removal of fluoride from aqueous solution by using red mud. Sep. Purif. Technol. **28**, 81–86 (2002)
13. APHA: Standard Methods for the Examination of Water and Wastewater, 20th edn. American Public Health Association, American Water Works Association, Water Environment Federation, Washington, DC, USA (1998)
14. Kundu, S., Gupta, A.K.: Adsorptive removal of As(III) from aqueous solution using iron oxide coated cement (IOCC): evaluation of kinetic, equilibrium and thermodynamic models. Sep. Purif. Technol. **51**, 165–172 (2006)
15. Weber, W.G.: Physicochemical Process for Water Quality Control. Wiley-Interscience, New York (1972)
16. Mckay, G., Otterburn, M.S., Sweeny, A.G.: Surface mass transfer processes during colour removal from effluent using silica. Water Res. **15**, 327–331 (1981)
17. Lagergren, S.: About the theory of so-called adsorption of soluble substances. Kungliga Svenska Vetenskapsakademiens. Handlingar **24**(4), 1–39 (1898)
18. Bhattacharjee, S., Chakravarty, S., Maity, S., Kar, S., Thakur, P., Bhattacharyya, G.: Removal of lead from contaminated water bodies using sea nodule as an adsorbent. Water Res. **37**, 3954–3966 (2003)
19. Ho, Y.S., Mckay, G.: The kinetics of sorption of divalent metal ions onto sphagnum moss peat. Water Res. **34**, 735–742 (2000)
20. Costodes, V.C.T., Fauduet, H., Porte, C., Ho, Y.S.: Removal of lead (II) ions from synthetic and real effluents using immobilized Pinus sylvestris sawdust: adsorption on a fixed-bed column. J. Hazard. Mater. **123**(1–3), 135–144 (2005)
21. Shihabudheen, M.M., Sanjay, S., Philip, L., Indumathim, M.N.: Enhanced fluoride removal from drinking water by magnesia-amended activated alumina granules. Chem. Eng. J. **140**, 183–192 (2008)
22. Weber, W.J., Morris, J.C.: Kinetics of adsorption on carbon from solution. J. Sanit. Eng. Div. (ASCE) **89**, 31–59 (1963)
23. Yadav, A.K., Kaushik, C.P., Haritash, A.K., Kansal, A., Rani, N.: Defluoridation of groundwater using brick powder as an adsorbent. J. Hazard. Mater. **B128**, 289–293 (2006)
24. Namasivayam, C., Yamuna, R.T.: Waste biogas residue slurry as an adsorbent for the removal of Pb(II) from aqueous solution and radiator manufacturing solution and radiator manufacturing industry wastewater. Environ. Pollut. **52**, 125–131 (1995)
25. Langmuir, I.: The adsorption of gases on plane surfaces of glass, mica and platinum. J. Am. Chem. Soc. **40**, 1362–1403 (1918)
26. Freundlich, H.: Colloid and Capillary Chemistry. Methuen and Co., Ltd., London (1926)
27. Hall, K.R., Eagleton, L.C., Avrivos, A., Vermeulen, T.: Pore and solid diffusion kinetics in fixed bed adsorption under constant pattern conditions. Ind. Eng. Chem. Fundam. **5**, 212–215 (1966)

Credibilistic TOPSIS Model for Evaluation and Selection of Municipal Solid Waste Disposal Methods

Jagannath Roy, Krishnendu Adhikary and Samarjit Kar

Abstract Municipal solid waste management (MSWM) is a challenging issue of urban development in developing countries. Each country having different socio-economic-environmental background might not accept a particular disposal method as the optimal choice. Selection of appropriate disposal method of MSW under vague and imprecise information usually considered and solved as a multi-criteria decision-making problem (MCDM). In the present paper, Technique for Order Preference by Similarity to Ideal Solution (TOPSIS) methodology is extended based on credibility theory for evaluating the performances of MSW disposal methods against some criteria fixed by experts. A numerical example is provided to demonstrate the usefulness of our proposed model. From the final outcomes, decision makers can recommend the preferable choice for MSW disposal methodologies for their municipal area. A sensitivity analysis is conducted to verify the robustness of our proposed model.

Keywords Waste disposal · MCDM · TOPSIS · Fuzzy numbers Credibility theory

---

J. Roy · K. Adhikary (✉) · S. Kar
Department of Mathematics, National Institute of Technology Durgapur, Durgapur 713209, India
e-mail: krish.math23@hotmail.com

J. Roy
e-mail: jaga.math23@gmail.com

S. Kar
e-mail: dr.samarjitkar@gmail.com

A. S. Kalamdhad et al. (eds.), *Advances in Waste Management*,
https://doi.org/10.1007/978-981-13-0215-2_17

1 Introduction

Municipality local body officials or a civic agency plays a vital role in municipal solid waste management (MSWM) since it has become a major issue of urban development. Solid waste disposal and its proper management problem concern about the social, economic, environmental, technical, political factors of a municipality. Due to inadequate consciousness among citizens, local body officials often struggle to make consensus with them in selecting the site for solid waste treatments and the methodologies that are to be implemented in the disposing of solid wastes [1]. These factors are conflicting in nature. For example, we need to maximize technical reliability, feasibility while net cost per ton, air pollution and emission levels are of minimizing criteria. Many factors (population growth, competitive economy, urbanization and the rise in community living standards) have caused exponential generation rate of municipal solid waste in developing countries [2]. Improper management of MSW can cause adversarial impacts on environment, civic healthiness and other socio-economic factors [3]. A long-term deficient disposal of municipal solid waste will cause poverty, poor governance and standards of living, major health, environmental and social issues [4]. Thus, selection of disposal methods for municipal solid waste is complex decision-making problem. These kind of decision-making problems are generally ill-structured and often behavioural decision approach. Researchers have shown that human intelligence is usually relatively unsuccessful to solve such problems [5]. As an important area of knowledge-based decision theory and decision support tools (DST), multicriteria decision-making (MCDM) techniques can serve strategic decisions in the framework of MSWM [6, 7]. Moreover, the DST makes it easier to communicate among the decision experts and shareholders to yield a systematic, transparent and documented process in decision-making.

In this context, MCDM has become a very crucial in management research and decision theory with many methods developed, extended and modified in solving problems in the present and past few decades. Among them, TOPSIS model developed by Hwang and Yoon continues to work satisfactorily across different application areas due to its simple computation and inherent characteristics. For determining the best alternative solution, TOPSIS uses two reference concepts which are positive ideal solution (PIS) and negative ideal solution (NIS). The best alternative must be the closest to PIS and farthest from the NIS. TOPSIS model outlines a performance index known as closeness coefficient index (CCI) to the ideal solution. An alternative with maximum CCI value is preferred to be chosen [8].

Earlier researchers dealt with precise and certain information-based MCDM methods. A decision support system depends on human acquaintance about the particular measure of a physical or intangible detail of a decision-making problem. The decision may be persuaded from practical data or knowledge-based opinion [9]. The data and information in decision-making are usually imprecise based on experts' subjective judgment where a group of experts is invited to implement the final performance evaluation, ranging from individual judgments to aggregate and rank the alternatives. Many theories, such as fuzzy set theory [10], Dempster–Shafer theory of evidence [11], Rough set theory [12], have been developed to deal with imprecise and incomplete data, human subjective judgment and real-life uncertainty. Suitable extensions of TOPSIS method, based on these theories, can be found in the literature. Few of them can be found in [13–15]. Ye and Li [16] extended the classical TOPSIS model introducing the Markowitz's portfolio mean–variance model in fuzzy domain and proposed the Possibilistic TOPSIS model. Although possibility degree serves well in describing a fuzzy event, sometimes a fuzzy event may not hold even if its possibility attains 1 and may hold even though its necessity is 0. Thus, there is a drawback of using possibility of a fuzzy event [17]. On the other hand, credibility theory removes this shortcoming and enables us to make more accurate decisions and avoid this drawback; credibility of a fuzzy event is more preferred than possibility of it. So, we utilize credibility of a fuzzy event in TOPSIS model. To the best of our knowledge, there is no work on TOPSIS which uses credibility theory in the literature. With these considerations, the paper proposes credibilistic TOPSIS, a novel MCDM model, to assist decision makers for evaluation and selection of the suitable disposal methods for municipal solid waste in urban areas in a country.

The rest of the paper is organized as follows. Section 2 introduces the basic concepts on the fuzzy numbers, and credibility of fuzzy variables. Section 3 presents the proposed credibilistic TOPSIS model based on fuzzy variables. The implementation of the proposed model for evaluating the municipal solid waste disposal methods is provided in Sect. 4. Result discussion is given in Sect. 5. Finally, Sect. 6 concludes the paper.

2 Preliminaries

2.1 *Fuzzy Sets and Fuzzy Numbers*

Fuzzy set theory was introduced by Zadeh [10] to resolve the impreciseness and ambiguity of human knowledge and judgments. Since its inception, fuzzy set theory has been used successfully in data processing by providing mathematical tools to cope with such uncertainties inherited from human thinking and reasoning. Fuzzy MCDM can build a strong comprehensiveness and reasonableness of the decision-making process [18]. Some applications could be found in Bellman and Zadeh [19].

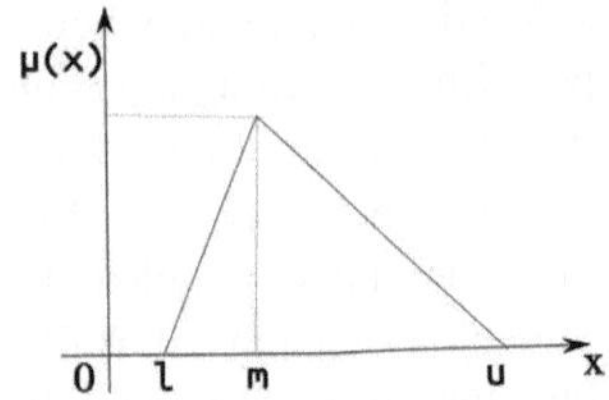

Fig. 1 Triangular fuzzy number

Definition 1 [18] Triangular Fuzzy Numbers (TFN)

A triangular fuzzy number, $\widetilde{A} = (l, m, u)$ is defined by the following membership function

$$\mu_{\widetilde{A}}(z) = \begin{cases} 0, z < l \\ \frac{z-l}{m-l}, & l \le z \le m \\ \frac{u-z}{u-m}, & m \le z \le u \\ 0, z > u \end{cases} \quad (1)$$

With $l \le m \le u$ and is shown in Fig. 1.

2.2 Arithmetic Operation of Fuzzy Numbers

Let $\widetilde{A}$ and $\widetilde{B}$ be two TFNs parameterized by triplets (a_1, a_2, a_3) and (b_1, b_2, b_3) respectively, then the fuzzy arithmetic operations of these two TFN can be defined as follows [18]:

1. Addition operation $(+)$ of two TFNs $\widetilde{A}$ and $\widetilde{B}$:

$$\widetilde{A}(+)\widetilde{B} = (a_1 + b_1, a_2 + b_2, a_3 + b_3) \quad (2)$$

2. Subtraction operation $(-)$ of two TFNs $\widetilde{A}$ and $\widetilde{B}$:

$$\widetilde{A}(-)\widetilde{B} = (a_1 - b_3, a_2 - b_2, a_3 - b_1) \quad (3)$$

3. Multiplication operation $(\times)$ of two TFNs $\widetilde{A}$ and $\widetilde{B}$:

$$\widetilde{A}(\times)\widetilde{B} = (a_1 b_1, a_2 b_2, a_3 b_3) \quad (4)$$

4. Division operation $(\div)$ of two TFNs $\widetilde{A}$ and $\widetilde{B}$:

$$\widetilde{A}(\div)\widetilde{B} = \left(\frac{a_1}{b_3}, \frac{a_2}{b_2}, \frac{a_3}{b_1}\right) \tag{5}$$

5. Scalar multiplication in TFNs:

$$kA = (ka_1, ka_2, ka_3) \tag{6}$$

For any real constant $k > 0$.

2.3 Aggregation of Fuzzy Numbers

Consider a group of K decision experts and the fuzzy rating of each decision expert, DM_k $(k = 1, 2, \ldots, K)$ can be represented by a positive TFNs $\widetilde{R}_k = (a_k, b_k, c_k)$ for $k = 1, 2, \ldots, K$.

Then, the aggregated fuzzy rating can be defined as:

$R = (a, b, c)$, where, $a = \min_{1 \le k \le K}\{a_k\}$, $b = (b_1 \times b_2 \times \cdots \times b_k)^{\frac{1}{K}}$ and $c = \max_{1 \le k \le K}\{c_k\}$.

2.4 Expected Value (Credibilistic Mean) of Fuzzy Variables

Definition 2 [17] Let ξ be a fuzzy variable. Then the credibilistic mean value of is defined as follows

$$E(\xi) = \int_0^{\infty} Cr\{\xi \ge r\}\mathrm{d}r - \int_{-\infty}^{0} Cr\{\xi \le r\}\mathrm{d}r \tag{7}$$

where at least one of the two integrals exists finitely.

Let ξ be a continuous nonnegative fuzzy variable with membership function μ. If μ is decreasing on $[0, \infty)$, then

$$Cr\{\xi \ge r\} = \frac{\mu(x)}{2} \quad \text{for any } x > 0 \text{ and } E(\xi) = \frac{1}{2}\int_0^{\infty} \mu(x)\mathrm{d}x$$

Theorem 1 [17] *Let ξ be a continuous fuzzy variable with membership function μ. If its expected value exists, and there is a point x_0 such that x is increasing on $(-\infty, x_0)$ and decreasing on (x_0, ∞), then*

$$E(\xi) = x_0 + \int_{x_0}^{\infty} \mu(x)\mathrm{d}x - \int_{-\infty}^{x_0} \mu(x)\mathrm{d}x \tag{8}$$

Example 1 Let ξ be a triangular fuzzy number, i.e. $\xi = (a, b, c)$, then it has an credibilistic mean value

$$E(\xi) = b + \int_{x_0}^{\infty} \frac{c-x}{c-b}\mathrm{d}x - \int_{-\infty}^{x_0} \frac{x-a}{c-b}\mathrm{d}x = b + \frac{c-b}{4} + \frac{a-b}{4} = \frac{a+2b+c}{4} \tag{9}$$

2.5 Credibilistic Variance of Fuzzy Variables

The variance of a fuzzy variable conveys the amount of the range of the distribution about its mean value. A lower amount of variance denotes that the fuzzy variable is strongly concentrated about its mean value, and a larger amount of variance denotes that the fuzzy variable is widely distributed about its mean value.

Definition 3 [17] Let ξ be a fuzzy variable with finite mean value e. Then the variance of ξ is defined by $V[\xi] = E[(\xi - e)^2]$.

Definition 4 [17] Let ξ be a triangular fuzzy variable and k be a positive real number. Then the mean value $E[\xi^k]$ is called the kth moment.

Theorem 2 [17] *Let ξ be a triangular fuzzy number, i.e. $\xi = (a, b, c)$ then its variance*

$$\mathrm{Var}[\xi] = \begin{cases} \frac{33\alpha^2 + 11\alpha\beta^2 + 21\alpha^2\beta - \beta^3}{384\beta}, & \text{if } \alpha > \beta \\ \frac{\alpha^2}{6}, & \text{if } \alpha = \beta \\ \frac{33\beta^2 + 11\alpha^2\beta + 21\alpha\beta^2 - \alpha^3}{384\beta}, & \text{if } \alpha < \beta \end{cases} \tag{10}$$

where $\alpha = b - a$ and $\beta = c - b$.

Fuzzy standard deviation,

$$\sigma = \sqrt{\mathrm{Var}[\xi]} \tag{11}$$

3 The Proposed Credibilistic TOPSIS Model

The basic steps of TOPSIS are:

(i) Determine the weights of criteria
(ii) Construction of initial decision matrix
(iii) Computation of normalized decision matrix and weighted normalized decision matrix
(iv) Finding ideal and anti-ideal solutions for each criterion
(v) Calculate the distance of each alternative from the ideal and anti-ideal values
(vi) Finally, rank the alternatives based on the closeness coefficient of each alternative to the ideal solution

Here, we extend TOPSIS model based on credibility theory.

Step 1: *Determining the weights of assessment criteria.*

Decision makers (DMs) are asked to decide significant assessment criteria for selecting the appropriate waste disposal method(s). Experts used suitable linguistic rating and their corresponding fuzzy variables (shown Table 1) for expressing their opinions about the relative performances of the MSW disposal methodologies against those criteria. Here, we systematically describe an extended TOPSIS model using credibilistic mean and variance as follows.

Step 2: *Construct the fuzzy performance/decision matrix*

Choose the appropriate linguistic variables for the alternatives with respect to criteria and form fuzzy decision matrix, $\widetilde{X} = (\tilde{x}_{ij})_{p\times q}$ where $\tilde{x}_{ij} = (l_{ij}, m_{ij}, u_{ij})$ is the value of the ith alternative according to the jth criterion $i = 1, 2, \ldots, p; j = 1, 2, \ldots q$.

$$\widetilde{X} = \begin{bmatrix} \tilde{x}_{11} & \ldots & \tilde{x}_{1q} \\ \ldots & \ldots & \ldots \\ \tilde{x}_{p1} & \ldots & \tilde{x}_{pq} \end{bmatrix}$$

Table 1 Fuzzy evaluation scores for the alternatives

Linguistic terms	Fuzzy score
Very poor (VP)	(1, 1, 1)
Poor (P)	(1, 1, 3)
Medium poor (MP)	(1, 3, 5)
Fair (F)	(3, 5, 7)
Medium good (MG)	(5, 7, 9)
Good (G)	(7, 9, 10)
Very good (VG)	(9, 10, 10)

$\tilde{x}_{ij}^{k} = (l_{ij}^{k}, m_{ij}^{k}, u_{ij}^{k})$ is the rating of the alternative A_i w.r.t. criterion c_j evaluated by expert k and $l_{ij} = \min_{1 \leq k \leq K}\left\{l_{ij}^{k}\right\}, m_{ij} = \left(m_{ij}^{k} * m_{ij}^{k} * \cdots * m_{ij}^{k}\right)^{1/k}$ and $u_{ij} = \max_{1 \leq k \leq K}\left\{u_{ij}^{k}\right\}$ where p denotes the number of the alternatives and q stands for the total number of criteria.

Step 3: *Normalize the elements from the initial fuzzy decision matrix* $(\widetilde{X})$.

$$\widetilde{N} = \begin{bmatrix} \tilde{n}_{11} & \ldots & \tilde{n}_{1q} \\ \ldots & \ldots & \ldots \\ \tilde{n}_{p1} & \ldots & \tilde{n}_{pq} \end{bmatrix}$$

The elements of the fuzzy normalized matrix $(\widetilde{N})$ are computed using the Eqs. (12) and (13)

- For benefit-type criteria (maximizing type)

$$\tilde{n}_{ij} = \left(l'_{ij}, m'_{ij}, u'_{ij}\right) = \left(\frac{l_{ij}}{u_j^*}, \frac{m_{ij}}{u_j^*}, \frac{u_{ij}}{u_j^*}\right) \tag{12}$$

- For cost-type criteria (minimizing type)

$$\tilde{n}_{ij} = \left(l'_{ij}, m'_{ij}, u'_{ij}\right) = \left(\frac{l_j^-}{u_{ij}}, \frac{l_j^-}{m_{ij}}, \frac{l_j^-}{l_{ij}}\right) \tag{13}$$

Let x_{ij}, x_j^* and x_j^- are the elements from the initial fuzzy decision matrix $\widetilde{X}$ for which x_j^* and x_j^- are defined

$$u_j^* = \max_{1 \leq i \leq p}(u_{ij}) \quad \text{and} \quad l_j^- = \min_{1 \leq i \leq p}(l_{ij}), \;\; j = 1, 2, \ldots, n$$

The normalization approach mentioned above is to make the ranges of normalized triangular fuzzy numbers belong to [0, 1].

Step 4: *Construct the credibilistic mean value matrix*

According to Eq. (9), we can get the expected value (credibilistic mean) of triangular fuzzy number

$$E[\widetilde{N}] = \begin{bmatrix} E(\tilde{n}_{11}) & \ldots & E(\tilde{n}_{1q}) \\ \ldots & \ldots & \ldots \\ E(\tilde{n}_{p1}) & \ldots & E(\tilde{n}_{pq}) \end{bmatrix}$$

Step 5: *Construct the credibilistic standard deviation matrix*

Calculate the credibilistic standard deviation matrix $\sigma[\widetilde{N}]$ using Eqs. (10)–(11) about the fuzzy normalized matrix $\widetilde{N}$ which is given by

$$\sigma[\widetilde{N}] = \begin{bmatrix} \sigma(\tilde{n}_{11}) & \ldots & \sigma(\tilde{n}_{1q}) \\ \ldots & \ldots & \ldots \\ \sigma(\tilde{n}_{p1}) & \ldots & \sigma(\tilde{n}_{pq}) \end{bmatrix}$$

Step 6: *Calculating the positive ideal solution (PIS) and negative ideal solution (NIS) about credibilistic mean value matrix*

Here, we determine the PIS $E(\widetilde{N})^+$ and NIS $E(\widetilde{N})^-$ about the *credibilistic mean value matrix* $E(\widetilde{N})$ for the decision maker as:

$$E(\widetilde{N})^+ = (E(\tilde{n}_1)^+, E(\tilde{n}_2)^+, \ldots, E(\tilde{n}_q)^+) \tag{14a}$$

$$E(\widetilde{N})^- = (E(\tilde{n}_1)^-, E(\tilde{n}_2)^-, \ldots, E(\tilde{n}_q)^-) \tag{14b}$$

$$E(\tilde{n}_j)^+ = \max_{1 \le i \le p}\{E(\tilde{n}_{ij})\} \tag{14c}$$

$$E(\tilde{n}_j)^- = \min_{1 \le i \le p}\{E(\tilde{n}_{ij})\} \; j = 1, 2, \ldots, q \tag{14d}$$

Step 7: *Determine the positive ideal solution (PIS) and negative ideal solution (NIS) about credibilistic standard deviation matrix*

Furthermore, we identify PIS $\sigma(\widetilde{N})^+$ and NIS $\sigma(\widetilde{N})^-$ around the expected value matrix $E(\widetilde{N})$ for the decision maker as:

$$\sigma(\widetilde{N})^+ = (\sigma(\tilde{n}_1)^+, \sigma(\tilde{n}_2)^+, \ldots, \sigma(\tilde{n}_q)^+) \tag{15a}$$

$$\sigma(\widetilde{N})^- = (\sigma(\tilde{n}_1)^-, \sigma(\tilde{n}_2)^-, \ldots, \sigma(\tilde{n}_q)^-) \tag{15b}$$

$$\sigma(\tilde{n}_j)^+ = \max\{\sigma(\tilde{n}_{ij})\} \tag{15c}$$

$$\sigma(\tilde{n}_j)^- = \min\{\sigma(\tilde{n}_{ij})\} \quad j = 1, 2, \ldots, q \tag{15d}$$

Step 8: *Finding the distance of every alternative's credibilistic expected value from PIS and NIS*

Using the distance between two points in Euclidean q-space, the distance measures of each alternative's expected value from the PIS $E(\widetilde{N})^+$ and expected value from the PIS $\sigma(\widetilde{N})^+$ are given as:

$$d_i(E(\widetilde{N})^+) = \left\{ \sum_{j=1}^{q} \left[(E(n_j)^+ - E(\tilde{n}_{ij}))w_j\right]^2 \right\}^{\frac{1}{2}} \tag{16}$$

$$d_i(E(\widetilde{N})^-) = \left\{ \sum_{j=1}^{q} \left[(E(n_j)^- - E(\tilde{n}_{ij}))w_j\right]^2 \right\}^{\frac{1}{2}} \tag{17}$$

Step 9: *Calculate the distance of every alternative's credibilistic standard deviation from PIS and NIS*
Using the distance between two points in Euclidean q-space, the distance measures of each alternative's expected value from the NIS $E(\widetilde{N})^-$ and expected value from the NIS $\sigma(\widetilde{N})^-$ are given as:

$$d_i(\sigma(\widetilde{N})^+) = \left\{ \sum_{j=1}^{q} [(\sigma(n_j)^+ - \sigma(\tilde{n}_{ij}))w_j]^2 \right\}^{\frac{1}{2}} \tag{18}$$

$$d_i(\sigma(\widetilde{N})^-) = \left\{ \sum_{j=1}^{q} [(\sigma(n_j)^- - \sigma(\tilde{n}_{ij}))w_j]^2 \right\}^{\frac{1}{2}} \tag{19}$$

Step 10 *Calculate the closeness coefficient (CC) of alternative around its credibilistic expected value and credibilistic standard deviation*
A closeness coefficient is defined to determine the ranking order of all alternatives once $d_i(E(\widetilde{N})^+)$, $d_i(E(\widetilde{N})^-)$ $d_i(\sigma(\widetilde{N})^+)$ and $d_i(\sigma(\widetilde{N})^-)$ of each alternative A_i are found. Now, an alternative A_i is best if it is closest to the PIS and farthest from the NIS. Thus for making optimal choice, we need to find relative closeness coefficient of each alternatives. Here, in this case, relative closeness coefficient about its expected value and standard deviation are defined as

$$\mathrm{CC}_i(E(\widetilde{N})) = \frac{d_i(E(\widetilde{N})^-)}{d_i(E(\widetilde{N})^+) + d_i(E(\widetilde{N})^-)} \tag{20}$$

and

$$\mathrm{CC}_i(\sigma(\widetilde{N})) = \frac{d_i(\sigma(\widetilde{N})^-)}{d_i(\sigma(\widetilde{N})^+) + d_i(\sigma(\widetilde{N})^-)} \tag{21}$$

Step 11: *Ranking of the alternatives according to the final relative closeness coefficient*
Thus, the final relative CC of alternative A_i is computed with Eq. (22) as follows

$$\mathrm{CC}(A_i) = \sqrt{\mathrm{cc}_i(E(\widetilde{N})) \times \mathrm{cc}_i(\sigma(\widetilde{N}))} \tag{22}$$

The alternatives are ranked in decreasing value of integrated relative closeness coefficients.

4 Implementation of the Proposed Method for Ranking Solid Waste Disposal Methods

In this section, we present a numerical example adopted from Ekmekçioğlu et al. [20] to demonstrate the applicability of our model proposed in this paper. The problem aims to find the most appropriate solid waste disposal methodology in a particular region/country. Criteria and alternative disposal methods are briefly described in Tables 2 and 3, respectively. Readers are referred to Ekmekçioğlu et al. [20] for detailed description of criteria and alternatives.

To supplement and support Ekmekçioğlu et al. [20] for identification of potential methodologies of MSW disposal and significant criteria, a literature review, although not completely systematic, was completed. The accountable standards preferred for this survey are as follows:

(1) Chosen articles should take account of key aspects from environment, ecology, society and technology in the MSWM, and evaluation of suitable solid waste disposal methodologies. The articles are collected using the significant keywords including "municipal solid waste management", "solid waste disposal methods and evaluation", "environmental and economic and social and technical and solid waste disposal" while implementing a keyword search on Google Scholar.

(2) Selected literature included peer-reviewed articles, conference proceedings and books. For each search, the first three pages of results were identified, and based on an initial investigation on abstract and main text of those published works, the usefulness of the article was determined in this work.

(3) We have enlisted the assessment criteria as net cost per ton (C_1), technical reliability (C_2), feasibility (C_3), air pollution control (C_4), emission levels (C_5), separation of waste materials (C_6), surface water dispersed releases (C_7), number of employees (C_8), waste recovery (C_9) and energy recovery (C_{10}). In this study, solid waste disposal methodologies are reflected; then the most appropriate disposal method is determined using credibilistic TOPSIS model.

Table 2 Criteria and their brief description (Ekmekçioğlu et al. [20])

Criteria	Criteria name	Type
C_1	Technical reliability	Maximizing
C_2	Feasibility	Maximizing
C_3	Separation of waste materials	Maximizing
C_4	Waste recovery	Maximizing
C_5	Energy recovery	Maximizing
C_6	Net cost per ton	Minimizing
C_7	Air Pollution Control	Minimizing
C_8	Emission levels	Minimizing
C_9	Surface water dispersed releases	Minimizing
C_{10}	Number of employees	Minimizing

Table 3 Alternatives and their brief description (Ekmekçioğlu et al. [20])

Alternative	Disposal methods	Brief description
A_1	Landfilling	Landfilling is the most commonly employed method in MSWM. It integrates an engineering-based method for disposing the solid wastes by means of land-use keeping in mind that environmental hazards to be minimized while scattering the solid wastes as tinny layers
A_2	Composting	The four tasks are crucial for designing a modern MSW composting system: collection, contaminant separation, sizing and mixing, and biological decomposition
A_3	RDF combustion	Refuse-derived fuel (RDF) combustion is technologically advanced avoiding the step in which solid wastes were immediately burned. As a substitute of immediate burning, in this disposal technique, MSW is crooked into a fuel which can be stored and transported for further processing later. Despite the fact that conventional incineration focuses a little on arranging or treating of the MSW, RDF combustion uses a number of intermediate stages for waste disposal
A_4	Conventional incineration	Conventional incineration which is extremely effective way of managing solid wastes and significantly practiced since the 1890s to reduce volume of wastes. In this process, MSW disposal is comparatively easy, without sorting wastes being served into a blast furnace and, by scorching, condensed to 10% of its actual size

Alternative disposal technologies are comprised by examining popular MSW disposal methods including landfilling (A_1), composting (A_2), refuse-derived fuel (RDF) combustion (A_3) and conventional incineration (A_4).

We cross-check and validate the selected criteria and MSW disposal alternatives with the experts/decision makers (an environmental engineer (DM_1), an urban-regional planner (DM_2) and an academician (DM_3) working MSWM) in our study by asking them whether they feel the criteria set will characterize the various dimensions/objectives covering economic, technical, environmental and social issues. The decision experts were accepting of this set and did not recommend any others for consideration.

Although the proposed model is applied to the data from Ekmekçioğlu et al. [21] who discussed a case of MSW disposal technologies in Turkey but also it can be utilized for evaluating similar problems according to the socio-economic-technical scenario of the case country.

Step 1: Weight priorities of the criteria given by the decision makers are shown in Table 4.

Step 2: Decision maker's opinion for evaluation of alternatives is presented in Table 5. Next, the aggregated decision matrix (Table 6) is calculated using the aggregation rule of fuzzy numbers (refer to Sect. 2.3).

Table 4 Weight priorities of criteria (experts' opinion)

Criteria	C_1	C_2	C_3	C_4	C_5	C_6	C_7	C_8	C_9	C_{10}
Criteria weights	0.020	0.014	0.136	0.200	0.240	0.033	0.146	0.121	0.003	0.087

Table 5 Decision experts' opinion for evaluation of alternatives

		C_1	C_2	C_3	C_4	C_5	C_6	C_7	C_8	C_9	C_{10}
A_1	DM_1	P	MG	F	MP	MG	VG	F	G	P	MP
	DM_2	MP	G	F	P	F	VG	F	MG	VP	P
	DM_3	MP	MG	F	F	F	VG	MG	F	P	MP
A_2	DM_1	F	VP	F	F	F	VG	MG	G	MP	P
	DM_2	MP	MP	F	P	MG	VG	F	MG	VP	MP
	DM_3	F	VP	F	F	G	VG	F	F	MP	MP
A_3	DM_1	G	VP	G	G	G	VP	F	F	MG	F
	DM_2	MG	P	MG	MG	G	VP	MG	MG	G	MG
	DM_3	G	VP	G	G	G	VP	F	MG	G	F
A_4	DM_1	F	P	F	P	P	VP	P	P	F	F
	DM_2	F	P	F	P	P	G	P	MP	F	F
	DM_3	F	P	F	P	P	G	P	MP	F	F

Table 6 Aggregated initial fuzzy decision matrix ($\widetilde{X}$)

	A_1	A_2	A_3	A_4
C_1	(1.000, 2.080, 5.000)	(1.000, 4.217, 7.000)	(5.000, 8.277, 10.00)	(3.000, 5.000, 7.000)
C_2	(5.000, 7.612, 10.00)	(1.000, 1.442, 5.000)	(1.000, 1.000, 3.000)	(1.000, 1.000, 3.000)
C_3	(3.000, 5.000, 7.000)	(3.000, 5.000, 7.000)	(5.000, 8.277, 10.00)	(3.000, 5.000, 7.000)
C_4	(1.000, 2.466, 7.000)	(1.000, 2.924, 7.000)	(5.000, 8.277, 10.00)	(1.000, 1.000, 3.000)
C_5	(3.000, 5.593, 9.000)	(3.000, 6.804, 10.00)	(7.000, 9.000, 10.00)	(1.000, 1.000, 3.000)
C_6	(9.000, 10.00, 10.00)	(9.000, 10.00, 10.00)	(1.000, 1.000, 1.000)	(1.000, 4.327, 10.00)
C_7	(3.000, 5.593, 9.000)	(3.000, 5.593, 9.000)	(3.000, 5.593, 9.000)	(1.000, 1.000, 3.000)
C_8	(3.000, 6.804, 10.00)	(3.000, 6.804, 10.00)	(3.000, 6.257, 9.000)	(1.000, 2.080, 5.000)
C_9	(1.000, 1.000, 3.000)	(1.000, 2.080, 5.000)	(5.000, 8.277, 10.00)	(3.000, 5.000, 7.000)
C_{10}	(1.000, 2.080, 5.000)	(1.000, 2.080, 5.000)	(3.000, 5.593, 9.000)	(3.000, 5.000, 7.000)

Step 3: For normalization, we have used Eqs. (12) and (13) for maximization or minimization criteria, respectively, as shown in Table 7.

Step 4: The credibilistic mean value matrix about the normalized fuzzy decision matrix is calculated using Eq. (9) as shown in Table 8.

Step 5: The credibilistic standard deviation matrix about the normalized fuzzy decision matrix is formed using Eqs. (10) and (11) (Table 9).

Table 7 Normalized fuzzy decision matrix $(\widetilde{N})$

	A_1	A_2	A_3	A_4
C_1	(0.100, 0.208, 0.500)	(0.100, 0.422, 0.700)	(0.500, 0.828, 1.000)	(0.300, 0.500, 0.700)
C_2	(0.500, 0.761, 1.000)	(0.100, 0.144, 0.500)	(0.100, 0.100, 0.300)	(0.100, 0.100, 0.300)
C_3	(0.300, 0.500, 0.700)	(0.300, 0.500, 0.700)	(0.500, 0.828, 1.000)	(0.300, 0.500, 0.700)
C_4	(0.100, 0.247, 0.700)	(0.100, 0.292, 0.700)	(0.500, 0.828, 1.000)	(0.100, 0.100, 0.300)
C_5	(0.300, 0.559, 0.900)	(0.300, 0.680, 1.000)	(0.700, 0.900, 1.000)	(0.100, 0.100, 0.300)
C_6	(0.100, 0.100, 0.111)	(0.100, 0.100, 0.111)	(1.000, 1.000, 1.000)	(0.100, 0.231, 1.000)
C_7	(0.111, 0.179, 0.333)	(0.111, 0.179, 0.333)	(0.111, 0.179, 0.333)	(0.333, 1.000, 1.000)
C_8	(0.100, 0.147, 0.333)	(0.100, 0.147, 0.333)	(0.111, 0.160, 0.333)	(0.200, 0.481, 1.000)
C_9	(0.333, 1.000, 1.000)	(0.200, 0.481, 1.000)	(0.100, 0.121, 0.200)	(0.143, 0.200, 0.333)
C_{10}	(0.200, 0.481, 1.000)	(0.200, 0.481, 1.000)	(0.111, 0.179, 0.333)	(0.143, 0.200, 0.333)

Table 8 Credibilistic mean value matrix $E(\widetilde{N})$

	C_1	C_2	C_3	C_4	C_5	C_6	C_7	$C8$	C_9	C_{10}
A1	0.254	0.756	0.500	0.323	0.580	0.103	0.201	0.182	0.833	0.540
A2	0.411	0.222	0.500	0.346	0.665	0.103	0.201	0.182	0.540	0.540
A3	0.789	0.150	0.789	0.789	0.875	1.000	0.201	0.191	0.135	0.201
A4	0.500	0.150	0.500	0.150	0.150	0.391	0.833	0.540	0.219	0.219

Table 9 Credibilistic standard deviation matrix $\sigma(\widetilde{N})$

	$C1$	$C2$	$C3$	$C4$	$C5$	$C6$	$C7$	$C8$	$C9$	$C10$
A_1	0.097	0.102	0.082	0.148	0.129	0.003	0.052	0.059	0.195	0.183
A_2	0.123	0.109	0.082	0.140	0.144	0.003	0.052	0.059	0.183	0.183
A_3	0.108	0.059	0.108	0.108	0.065	0.000	0.052	0.056	0.025	0.052
A_4	0.082	0.059	0.082	0.059	0.059	0.238	0.195	0.183	0.045	0.045

Table 10 Positive ideal solution and negative ideal solution around the credibilistic mean value matrix

	C_1	C_2	C_3	C_4	C_5	C_6	C_7	C_8	C_9	C_{10}
$E(N)^+$	0.789	0.756	0.789	0.789	0.875	1.000	0.833	0.540	0.833	0.540
$E(N)^-$	0.254	0.150	0.500	0.150	0.150	0.103	0.201	0.182	0.135	0.201

Table 11 PIS and NIS about the standard deviation matrix

	C_1	C_2	C_3	C_4	C_5	C_6	C_7	C_8	C_9	C_{10}
$\sigma(N)^+$	0.082	0.059	0.082	0.059	0.059	0.000	0.052	0.056	0.025	0.045
$\sigma(N)^-$	0.123	0.109	0.108	0.148	0.144	0.238	0.195	0.183	0.195	0.183

Step 6: Then calculate the positive ideal solution (PIS) and negative ideal solution (NIS) about credibilistic expected value matrix by applying the Eqs. (14a), (14b), (14c), (14d), given in Table 10.

Step 7: Next, PIS and NIS about credibilistic standard deviation matrix is calculated according to Eqs. (15a), (15b), (15c), (15d), shown in Table 11.

Step 8: Find the distance of every alternative's credibilistic mean value from PIS and NIS by using Eqs. (16) and (17) presented in Table 12.

Step 9: Table 13 is formed after computing the distance measure of each alternative's credibilistic standard deviation from PIS and NIS using Eqs. (18) and (19).

Step 10: Closeness coefficient of every alternative about its expected value and standard deviation are find by Eqs. (20) and (21).

Step 11: The alternatives are ranked in decreasing value of final relative closeness coefficients by using Eq. (22) shown in the Table 14.

Table 12 Distance of each alternative with respect to credibilistic mean value from PIS and NIS

	A_1	A_2	A_3	A_4
$d_i(E(N)^+)$	0.163	0.153	0.106	0.223
$d_i(E(N)^-)$	0.113	0.133	0.222	0.103

Table 13 Distance of each alternative with respect to standard deviation matrix from PIS and NIS

	A_1	A_2	A_3	A_4
$d_i(\sigma(N)^+)$	0.027	0.029	0.011	0.027
$d_i(\sigma(N)^-)$	0.027	0.027	0.036	0.030

Table 14 Closeness coefficient and ranking of each alternative

Alternatives	$CC_i(E(\widetilde{N}))$	$CC_i(\sigma(\widetilde{N}))$	$CC(A_i)$	Rank
A_1	0.409	0.500	0.452	3
A_2	0.465	0.485	0.475	2
A_3	0.676	0.771	0.722	1
A_4	0.316	0.525	0.407	4

5 Result Discussion

5.1 Comparison with Other Models

In order to verify the validity of our proposed method, we perform a comparison of our proposed method with two other previous methods including fuzzy TOPSIS Chen et al. [21] and Possibilistic TOPSIS Ye and Li [16] which also deal with fuzzy numbers. The results are shown in Table 15.

It is clear from Table 15 that the three methods have the similar results. Note that RDF combustion (A_3) is best alternative according to all three models under fixed preference of criteria weights. These shows the method we proposed in this paper is reasonable.

5.2 Sensitivity Analysis

An MCDM method is dependent on the weight priorities of the criteria, that is on the relative importance involved to particular criteria. Occasionally, the final evaluation and ranking change whenever there is a very small variation in the relative preference of the criteria. Hence, a good MCDM method should act as a sensitive to the change in criteria weights.

In this research work, a sensitivity analysis is conducted to investigate whether the priorities/ranking of MSW disposal alternatives will change if the criteria weights are changed. Different weight priorities are given to the criteria (see Table 16) corresponding to different scenarios (1)–(6).

Applying the proposed credibilistic TOPSIS model, we obtain ranking priorities for all the six scenarios (see Table 17). The results show that assignment of different criteria weights leads to different rankings, i.e. the proposed model is sensitive to these weights, which is necessary for any MCDM models. On comparison of all scenarios, we find A_3 is best choice except in *scenario 5*. However, *scenario 1* and *scenario 4* provide same ranking order though weight priorities have large change. From scenarios (2) and (3), it is clear that a slight change in criteria weights will lead to alter A_2 and A_3 as second best choice. Similar arguments could be done

Table 15 Comparison with other models

	Fuzzy TOPSIS [21]		Possibilistic TOPSIS [16]		Proposed TOPSIS model	
	CC_i	Rank	CC_i	Rank	CC_i	Rank
A_1	0.5570	3	0.2180	3	0.452	3
A_2	0.5701	2	0.1871	4	0.475	2
A_3	0.7429	1	0.2746	1	0.722	1
A_4	0.3178	4	0.2384	2	0.407	4

Table 16 Weight priorities of criteria for different scenarios

	Scenario 1	Scenario 2	Scenario 3	Scenario 4	Scenario 5	Scenario 6
C_1	0.1117	0.0613	0.1458	0.1692	0.1024	0.2367
C_2	0.0061	0.1510	0.0495	0.0512	0.2161	0.1726
C_3	0.1446	0.1456	0.0982	0.1639	0.0432	0.0974
C_4	0.1590	0.0361	0.1356	0.0490	0.0626	0.1424
C_5	0.1156	0.0264	0.1728	0.1870	0.0345	0.1115
C_6	0.1290	0.1107	0.1861	0.0704	0.0323	0.0211
C_7	0.1265	0.2133	0.1062	0.0396	0.2063	0.0666
C_8	0.0668	0.0756	0.0269	0.0505	0.1376	0.0342
C_9	0.1116	0.1301	0.0290	0.1240	0.1305	0.0510
C_{10}	0.0291	0.0497	0.0500	0.0952	0.0344	0.0666

Table 17 Closeness coefficient and ranking of each alternative in different scenarios

	Scenario 1	Scenario 2	Scenario 3	Scenario 4	Scenario 5	Scenario 6
	Ranking	Ranking	Ranking	Ranking	Ranking	Ranking
A_1	2	2	3	2	*1*	3
A_2	3	4	2	3	4	4
A_3	*1*	*1*	*1*	*1*	2	*1*
A_4	4	3	4	4	3	2

for other pairwise comparisons among the scenarios. These comparisons confirm that using different weights to the criteria under consideration may help to choose the best design alternative in different context if needed (see Fig. 2).

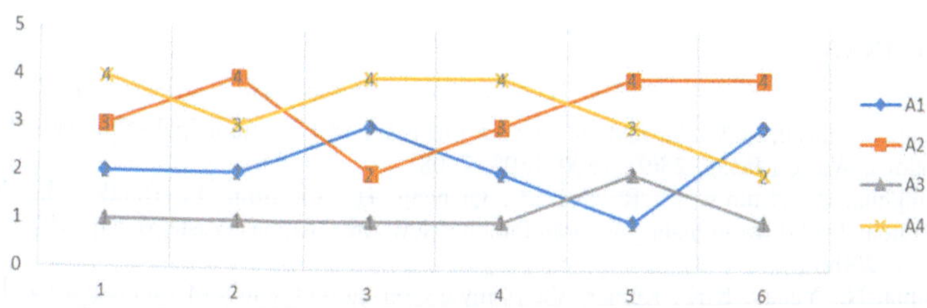

Fig. 2 Rankings of the alternatives in various scenarios

6 Conclusion

This study proposes the credibilistic TOPSIS model based on expected value (credibilistic mean value) operator for a fuzzy variable to facilitate a more precise analysis of the alternatives, considering several criteria in imprecise environment. The proposed method is applied successfully to select the most preferable disposal method(s) for municipal solid waste for imprecise data. Also, this model provides the expected optimal choice for disposal method(s) after effectively avoiding vague and ambiguous judgments.

In many practical MCDM problems, different relative weights of criteria must be taken into consideration since these are influenced by the socio-economic-environmental-technical condition of a country (for municipal solid waste disposal). So, a particular disposal method might not be accepted as optimal solution to all countries. It is evident that the sensitivity analysis performed (Tables 16 and 17) through the proposed model shows that RDF combustion A_3 is not the optimal choice in all cases (scenarios 4 and 5). In future, credibilistic TOPSIS would produce interesting hybrid MCDM methods with the combination of other MCDM techniques like, ANP, DEMATEL and Shannon Entropy.

Acknowledgements We wish to express our deepest appreciation to the Editors and the anonymous reviewers. The first author extends his sincere thanks to the Department of Science and Technology, India, for their financial support under INSPIRE Fellowship for Ph.D. program with the research grant no. DST/INSPIRE Fellowship/2013/544.

Conflict of interest The authors declare that there is no conflict of interest regarding the publication of this paper.

References

1. Khan, S., Faisal, M.N.: An analytic network process model for municipal solid waste disposal options. Waste Manag. **28**(9), 1500–1508 (2008)
2. Minghua, Z., Xiumin, F., Rovetta, A., Qichang, H., Vicentini, F., Bingkai, L., Yi, L.: Municipal solid waste management in Pudong New Area, China. Waste Manag. **29**(3), 1227–1233 (2009)
3. Gupta, N., Yadav, K.K., Kumar, V.: A review on current status of municipal solid waste management in India. J. Environ. Sci. **37**, 206–217 (2015)
4. Ogu, V.I.: Private sector participation and municipal waste management in Benin City, Nigeria. Environ. Urban. **12**(2), 103–117 (2000)
5. Promentilla, M.A.B., Furuichi, T., Ishii, K., Tanikawa, N.: Evaluation of remedial countermeasures using the analytic network process. Waste Manag. **26**(12), 1410–1421 (2006)
6. Charnpratheep, K., Zhou, Q., Garner, B.: Preliminary landfill site screening using fuzzy geographical information systems. Waste Manage. Res. **15**(2), 197–215 (1997)

7. Hung, M.L., Ma, H.W., Yang, W.F.: A novel sustainable decision making model for municipal solid waste management. Waste Manag. **27**(2), 209–219 (2007)
8. Hwang, C.L., Yoon, K.: Multiple attribute decision making: methods and applications. In: Lecture Notes in Economic and Mathematical Systems. Springer, Berlin (1981)
9. Khoo, L., Tor, S., Zhai, L.: A rough-set based approach for classification and rule induction. Int. J. Adv. Manuf. Technol. **15**(6), 438–444 (1999)
10. Zadeh, L.: Fuzzy sets. Inf. Control **8**(3), 338–353 (1965)
11. Shafer, G.: A Mathematical Theory of Evidence (Vol. 1, pp. xiii + 297). Princeton University Press, Princeton (1976)
12. Pawlak, Z.: Rough sets. Int. J. Comput. Inf. Sci. **11**(5), 341–356 (1982)
13. Jahanshahloo, G.R., Lotfi, F.H., Izadikhah, M.: Extension of the TOPSIS method for decision-making problems with fuzzy data. Appl. Math. Comput. **181**(2), 1544–1551 (2006)
14. Chen, T.Y., Tsao, C.Y.: The interval-valued fuzzy TOPSIS method and experimental analysis. Fuzzy Sets Syst. **159**(11), 1410–1428 (2008)
15. Chatterjee, K., Kar, S.: Multi-criteria analysis of supply chain risk management using interval valued fuzzy TOPSIS. OPSEARCH, 1–26 (2016)
16. Ye, F., Li, Y.: An extended TOPSIS model based on the possibility theory under fuzzy environment. Knowl.-Based Syst. **67**, 263–269 (2014)
17. Liu, B.: Uncertain Theory: An Introduction to Its Axiomatic Foundation. Springer, Berlin (2004)
18. Chen, C.T.: Extensions of the TOPSIS for group decision-making under fuzzy environment. Fuzzy Sets Syst. **114**(1), 1–9 (2000)
19. Bellman, R.E., Zadeh, L.A.: Decision-making in a fuzzy environment. Manag. Sci. **17**(4), B-141 (1970)
20. Ekmekçioğlu, M., Kaya, T., Kahraman, C.: Fuzzy multicriteria disposal method and site selection for municipal solid waste. Waste Manag. **30**(8), 1729–1736 (2010)
21. Chen, C.T., Lin, C.T., Huang, S.F.: A fuzzy approach for supplier evaluation and selection in supply chain management. Int. J. Prod. Econ. **102**(2), 289–301 (2006)

Influence of Recycling on the Thermal Properties of Municipal Solid Waste

Roshni Mary Sebastian, Dinesh Kumar and Babu Alappat

Abstract A multitude of conventional and non-conventional techniques are adopted all over the world to handle the huge quantities of municipal solid waste (MSW) generated. Waste reduction, reuse and recycling are instrumental in curbing the issues related to MSW management and are being implemented in many countries including India. Most cities in India have active waste recycling programs. Not only do the recycling programs help in material recovery, but also aid reduction in greenhouse gas emissions by replacing virgin items with recycled items. In the present study, the variation in the calorific values of the MSW generated in the capital city, Delhi, is investigated with possible recycling options for various components. Biodegradables, paper, plastic, metals, etc., from MSW are assumed to be recycled at different rates. Recycling of different combination of MSW components is also considered to analyse the variation in heat content per kilogram. Resultant changes in the combustibility of the MSW and its impact on the efficiency of thermal waste-to-energy facilities are examined. The Tanner diagram and conventional thumb rules are used to interpret the results. While removal of biodegradables, ferrous metals and inert fraction increased the calorific value of the MSW that was left behind, by 18.4, 0.37 and 39.6%, respectively, recycling of paper and plastic fraction of high inherent calorific value reduced the net heat

R. M. Sebastian (✉) · B. Alappat
Department of Civil Engineering, IIT Delhi, Hauz Khas, New Delhi, India
e-mail: roshnimary@gmail.com

B. Alappat
e-mail: bjalappat@yahoo.com

D. Kumar
North Delhi Municipal Corporation, New Delhi, India
e-mail: dineshkathpalia@gmail.com

A. S. Kalamdhad et al. (eds.), *Advances in Waste Management*,
https://doi.org/10.1007/978-981-13-0215-2_18

content of the remaining MSW by 5.2 and 18.7%, respectively. Consequently, the position in the Tanner diagram moved towards the areas of high combustibility in the former case and away from the area of high combustibility in the latter case.

Keywords Calorific value · Incineration · MSW · Recycling · Waste to energy

1 Introduction

India is a steadily developing economy, with high GDP growth, as well as population growth. World Bank reports an annual population growth rate of 1.2% and GDP growth rate of 7.3%, in the period of 2014–15 in India. Consequently, the municipal solid waste (MSW) generation also increased and is expected to reach 0.7 kg/capita/day by 2025, from the value of 0.34 kg/capita/day in 2012, in urban areas alone [1]. With the urbanization rate of the country expected to reach 26% by 2026, combined with economic growth, the MSW generation is anticipated to become alarmingly high. An increase in urban population and economic prosperity results in a subsequent increase in MSW generated [1]. The uninhibited urban migration had added to the quantity and complexity of the generated MSW. MSW generation in Delhi alone has increased from about 7000 tons/day in 2007–08 to about 9000 tons/day, presently [2, 3]. It is expected to reach up to 17,000–25,000 tons/day by 2021, according to New Delhi Municipal Council (NDMC) studies [3]. Figure 1 depicts the increase in MSW generation for major metros from 1999 to 2011 [4]. Huge increase was observed in all cases, and even higher values anticipated in the future, due to the aforementioned reasons. Despite being a crucial issue, the MSW management strategies adopted continue to be abysmal and the resultant problems are innumerable.

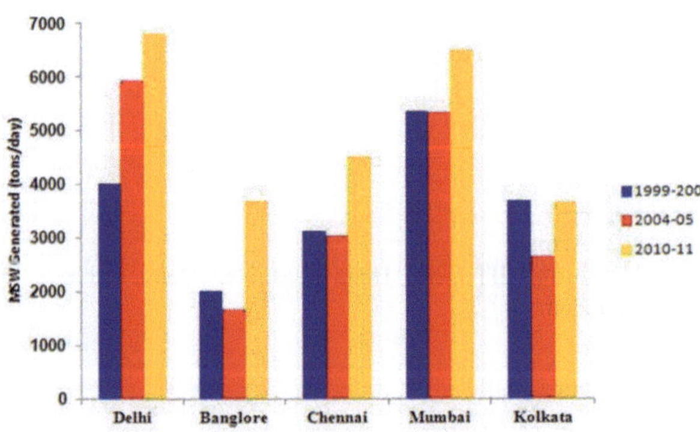

Fig. 1 Variation in the MSW generated per day in major metros

The waste collection efficiency is at an average of 72% in major metros and at a dismal 50% or below in smaller cities [5, 6]. Besides, conventional management techniques like composting, vermi-composting and biomethanation cannot handle the huge quantities of MSW generated per day and disposal techniques like land-filling fail due to shortage of land. Most of the wastes get dumped unscientifically in land-dumps, later causing groundwater pollution and release of greenhouse gases (GHG). It is in this scenario where the need for a prompt and well-devised technique of MSW management, like thermal waste-to-energy techniques, arises. Thermal waste-to-energy units were not successful in the past due to the high moisture and inert content and the subsequent low calorific value of the MSW. The Timarpur waste-to-energy plant in Delhi, set up in 1989, was a state-of-the-art incineration and power generation unit, which the Government of India (GoI) set up in collaboration with the Government of Denmark. However, the plant which was designed to take in 300 tons/day of waste and generate 3.7 MW power shut down exactly after 21 days of operation. This was attributed to the poor calorific value of the MSW, making it unfit for incineration [3].

Ideal solution is integrated solid waste management (ISWM), which according to Tchobanoglous et al. is the selection and application of appropriate technologies and management programs to achieve specific waste management objectives [7]. Reduction, reuse and recycling of wastes are instrumental in ISWM, in addition to energy recovery and waste disposal. Recognizing the role of recycling in sustainable solid waste management, the USA has been performing active recycling and composting of wastes, and the recycling rate has increased from less than 10% in 1980 to about 34% in 2012. The disposal of wastes in the landfills during the same period had decreased from 89 to about 54%. An added benefit of recycling is the reduction in GHG emissions. Annual recycling and composting of 87 million tons of waste helped in the reduction of emissions up to 168 million metric tons of CO_2 equivalents [8]. Unlike the developed countries, recycling in developing countries like India remains an informal activity, where the waste collected by recyclers, through a hierarchical process, ends up in recycling units. The number of workers involved in the recycling sector has increased from 84,000 to 1,60,000 in the city of Delhi alone. The informal sector collects about 15–20% of the total waste generated in Delhi and recycles about 2000 tons of waste each day [9].

The Municipal Solid Waste (Management and Handling) Rules, 2000 (presently the Solid Waste Management Rules, 2016), recommend the biodegradable fraction of MSW to be processed by composting/vermi-composting or biomethanation. The mixed wastes with recoverable resources on the other hand are recommended to follow the recycling route, followed by other techniques. The Plastics (Manufacture, Usage and Waste Management) Rules, 2009, require the recyclable plastic fraction to be segregated and channelized to registered plastic recycler and recycled conforming to Indian Standards. As each component of MSW is recycled, the properties of the remaining MSW vary accordingly. When components like paper and plastics, which have comparatively higher heat content, are recycled, overall heat content varies significantly. This further affects the efficiency of the incineration units.

This study is aimed at understanding the effect of recycling on the heat content of the leftover MSW and how it affects the efficiency of thermal waste-to-energy facilities. Paper, plastics, biodegradables, ferrous metals and inert fractions are assumed to be recycled at different recycling rates. In addition to this, a few combinations of physical components of MSW are also assumed to be recycled and the resultant calorific value per kilogram of the wastes is evaluated. The subsequent effect on the efficacy of the thermal waste-to-energy units is also examined.

2 MSW in Delhi

The National Capital Territory of Delhi, located between the latitude of 28° 24′ 17″ to 28° 53′ 00″ North and longitude of 76° 50′ 24″ to 77° 20′ 37’″ East, spans over an area of 1483 km^2. While Municipal Corporation of Delhi (MCD) has 94% of the area and 96% of the population under its jurisdiction, New Delhi Municipal Council (NDMC) and Delhi Cantonment Board (DCB) look after civic amenities of the rest of the area and population. The average annual exponential growth rate of population was recorded as 1.92%, for the period 2001–11, as per Census 2011. The Gross State Domestic Product (GSDP) of Delhi recorded an impressive 15.35% growth rate over 2014–15 period, according to MCD [10].

Figure 2 indicates the MSW generated in the areas under the three MCDs (North, East and South), NDMC and DCB for the year 2001 and projected to 2021 [10]. A very steady increase in MSW generation can be observed from the same. Table 1 presents the average physical composition of MSW generated in the MCD area of Delhi in 2004. All the values are reported on wet weight basis. Table 2 presents the average chemical properties of the aforesaid MSW, reported on dry weight basis, by MCD authorities [11]. In the following sections, the variation in the calorific value of MSW, when various components are recycled, is presented.

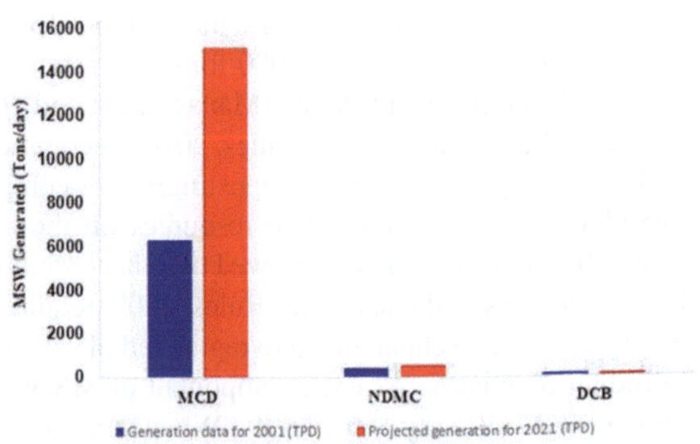

Fig. 2 MSW generation data for Delhi

Table 1 Physical composition of MSW generated in Delhi

Composition (%)	HIG areas	MIG area	LIG and slum area	Market area	Commercial area	Industrial area	Mixed waste	Landfill site
Biodegradable	44.45	39.96	36.39	58.09	38.54	16.10	38.60	30.62
Paper	7.25	5.01	4.48	5.13	7.92	8.90	5.57	3.78
Plastic	6.76	6.17	5.28	4.86	7.81	7.13	6.03	4.45
Metal	0.34	0.30	0.35	0.11	1.46	0.20	0.23	0.10
Glass and crockery	1.05	1.12	1.84	0.4	3.02	0.66	0.99	1.45
Bioresistant[a]	14.84	13.41	13.32	8.25	16.45	25.10	13.87	19.73
Inert[b]	25.31	34.03	38.34	23.16	24.79	42.77	34.71	32.70

[a]Includes leather, rubber, bones, and other synthetic material
[b]Includes stones, bricks, ashes.

Table 2 Chemical characteristics of MSW generated in Delhi

Property (%)	HIG areas	MIG area	LIG and slum area	Market area	Commercial area	Industrial area	Mixed waste	Landfill site
Moisture	50.70	47.78	51.00	53.77	50.61	32.00	43.65	25.58
Organic carbon	28.15	23.74	20.32	34.92	22.19	22.13	20.47	14.82
Nitrogen as N	0.96	0.92	0.78	1.17	0.87	0.90	0.85	0.65
Phosphorous as P_2O_5	0.42	0.35	0.30	0.44	0.26	0.47	0.34	0.27
Potassium as K_2O	0.80	0.65	0.62	0.95	0.58	0.80	0.69	0.91
C/N ratio	29.32	25.80	26.05	29.84	25.58	24.50	24.08	22.80

2.1 Recycling of MSW Generated in Delhi: A Theoretical Approach

The viability of energy recovery from MSW, as a part of integrated solid waste management system, can be evaluated by performing the proximate and ultimate analyses and the determination of calorific value henceforth. Proximate analysis is used to determine the weight percentage of ash, moisture, volatiles, while ultimate analysis is used to determine weight percentage of chemical elements. Kaiser (1966) has given the proximate and ultimate analyses of 20 MSW components along with their calorific values [12]. This method was opted to perform the proximate and ultimate analyses of the MSW generated in Delhi with aforesaid properties. The results of ultimate and proximate analyses performed on Delhi MSW, using the average values of the physical properties, were given in Table 3.

Using the approximate heat contents and moisture contents of various components, the calorific value of the MSW was found mathematically. The high heat contents of various components were aggregated and divided by discarded weight. The net calorific of the MSW on discarded basis was found as 2010 kcal/kg.

With the original heat content of MSW at 2010 kcal/kg, recycling of different components was assumed to be done at different rates, and corresponding heat contents were evaluated. Figure 3 illustrates the different components removed/recycled and corresponding energy recovered from them.

2.1.1 Composting of Biodegradables

If the Municipal Solid Waste (Management and Handling) Rules, 2000 (presently the Solid Waste Management Rules, 2016), were to be implemented completely, then entire biodegradable fraction is to be stabilized by composting, vermi-composting or anaerobic digestion. The biodegradable fraction has to be segregated at the source and then sent to compost plants, where they will be effectively composted. Assuming that the rules were followed to the word, and varying the rates at which the biodegradables were composted between 0 and 100%,

Table 3 Proximate and ultimate analyses of Delhi MSW

Proximate analysis	
% Moisture	37.6
% Volatiles	23.3
% Inert	38.9
Ultimate analysis (including inert fraction)	
% Carbon	40.36
% Hydrogen	5.09
% Oxygen	14.00
% Nitrogen	2.77
% Sulphur	0.24

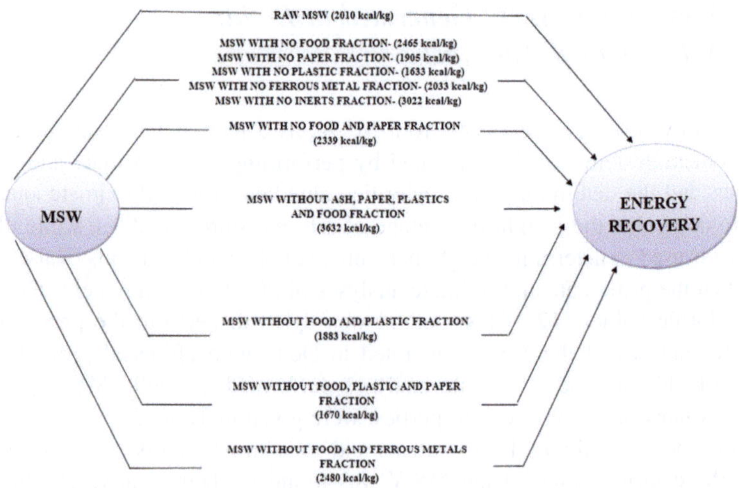

Fig. 3 A flow chart showing the various routes of energy recovery from MSW, by removal/recycling of different components

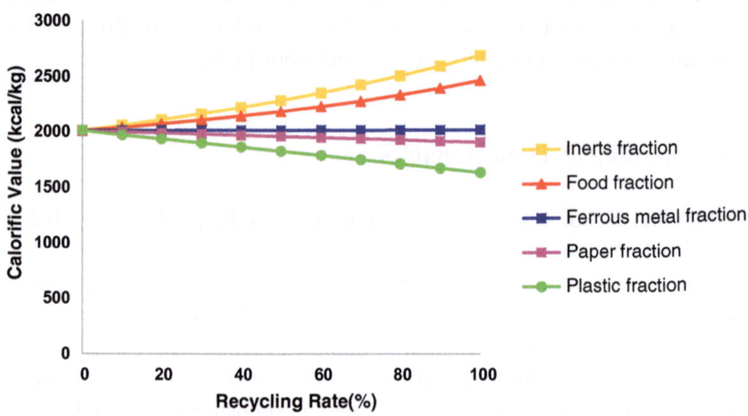

Fig. 4 Variation in the calorific value of MSW generated in Delhi when individual components are recycled

the calorific values were computed. Figure 4 demonstrates the increasing trend in the calorific values of the MSW left, when biodegradables were composted, with the maximum when 100% of the biodegradables were composted. This can be explained by the reduction in the net moisture content of the MSW, when biodegradables with high moisture content were removed.

Currently, Delhi has two compost plants, one at Okhla constructed, operated and maintained by M/S IELIS through South Delhi Municipal Corporation and the second one at Narela-Bawana constructed, operated and maintained by M/S

DMSWSL through North Delhi Municipal Corporation. The plants have capacities of 300 and 500 tons/day waste processing capacity, respectively.

2.1.2 Recycling of Plastics

The municipal authorities are given the responsibility to segregate, transport and process the plastic wastes. The recycling, recovery and disposal of plastic wastes are carried out adhering to IS 14534: 1998. Delhi, as per Delhi Government statistics, presently generates about 650 tons/day of plastic waste, and this has been creating nuisance over the years. With increasing awareness, campaigns like "Say No To Plastic Bags", and the government directives imposing interim ban on the production, sale, storage, usage and transport of plastic carry bags, the use of plastics has been reduced over time and is now being replaced by biodegradable plastics and other alternatives. As majority of retail outlets are opting for paper carry bags and jute bags instead of plastic bags, it can be conveniently assumed that plastic components are recycled at rates between 0 and 100%. Moreover, chlorinated plastics are not to be incinerated, as per the Municipal Solid Waste (Management and Handling) Rules, 2000 (presently, the Solid Waste Management Rules, 2016).

The variation in the calorific value, when plastic content was removed from MSW and recycled at different rates, is as depicted in Fig. 4. At 100% recycling rate, when the entire plastic fraction was removed, the calorific value reduced by 18.7%. The reduced value can be explained by the removal of plastic, which in itself has high calorific value.

2.1.3 Recycling of Paper

Paper content of Indian MSW has been increasing over the years. Despite vast digitization, India's paper consumption is anticipated to rise in the future, owing to the increasing literacy rate. Paper is usually picked up by informal waste collectors/recyclers and ends up in recycling units. Recycled paper results in reduced GHG emissions to atmosphere. However, when paper fraction was recycled at rates from 0 to 100%, a mild reducing trend was observed in the calorific value, as evident from Fig. 4. This was due to the removal of paper fraction, which has high heat content.

2.1.4 Recycling of Ferrous Metals

Ferrous metal fraction, which can be easily recovered from heterogeneous MSW using a magnetic separator, may be recovered, reused and recycled. A negligible increase in calorific value was observed (Fig. 4) when entire ferrous metal fraction was removed from the MSW, which can be explained by the removal of the comparatively heavier fraction of MSW.

2.1.5 Removal of Inert Fraction

Inert fraction, which includes ash, dirt, construction and demolition waste, glass, metals, stones, can be removed by manual separation or by air classifiers. As per Municipal Solid Waste (Management and Handling) Rules, 2000 (presently the Solid Waste Management Rules, 2016), inert and non-biodegradable fraction should be disposed off in landfills. When the inert fraction, which can be recovered, reused or recycled, was removed at various rates, the calorific value was found to increase as apparent in Fig. 4 and at 100% removal, rose from 2010 kcal/kg, by 40%. Inert fraction, which has low innate calorific value, forms a major part of the generated MSW and when removed boosts up the net calorific value of the MSW.

The North Delhi Municipal Corporation and the East Delhi Municipal Corporations now operates two Construction and Demolition (C&D) Waste Recycling Plants at Burari and Shastri Park, respectively. The two plants have a capacity of processing 2000 and 500 TPD of C&D waste. They are operating to their full capacity, and the rejects from the two plants are less than 10% of the input.

2.1.6 MSW Without Biodegradables and Variable Removal of Paper

When the entire fraction of biodegradables was removed, the heat content of the remaining MSW was found to increase by 18.4%, due to lowered net moisture content of the MSW. The corresponding heat content was found to be 2465 kcal/kg. When the paper fraction of the MSW that was left behind was recycled at different rates from 0 to 100%, a declining trend was observed (Fig. 5) in the heat content

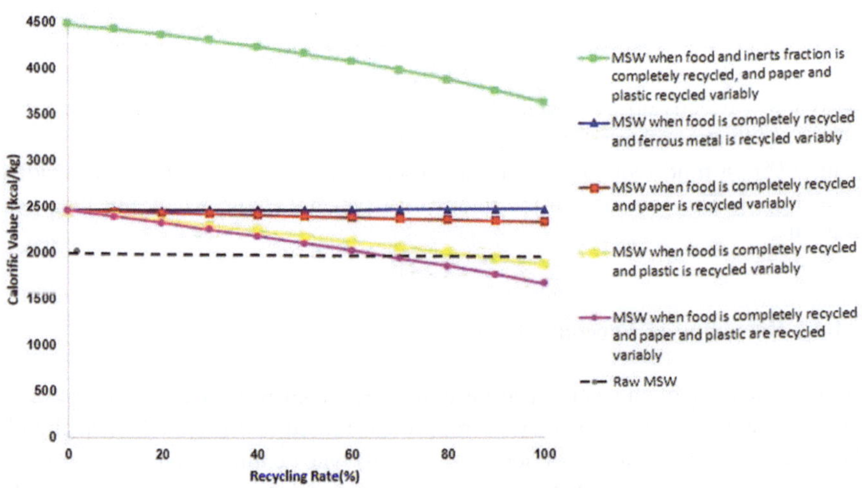

Fig. 5 Variation in calorific value of MSW generated in Delhi when various components are recycled together

and the value reduced by 5%, at 100% removal of the paper fraction, the reason same as stated above.

2.1.7 MSW Without Biodegradables and Variable Removal of Plastic Fraction

When plastic fraction, which has high inherent calorific value, was removed from the MSW with all of its biodegradables removed (for composting), the heat content reduced by about 24%. A steep negatively sloped graph was obtained as shown in Fig. 5, when calorific value was plotted against recycling rate.

2.1.8 MSW Without Biodegradables and Variable Removal of Ferrous Metals

The removal of entire ferrous metals from the MSW left after removing the biodegradables resulted in a 0.6% increase in its calorific value. The slightly increasing trend, due to the removal of the comparatively heavier fraction of MSW, is evident from Fig. 5.

2.1.9 MSW Without Biodegradables and Simultaneous Variable Removal of Paper and Plastic Fraction

The calorific value of the MSW reduced by about 32%, when entire paper and plastic fraction was removed from the MSW devoid of biodegradables, for recovery, reuse and recycling. The negatively sloped graph (Fig. 5) was largely due to the removal of high heat content fractions of the MSW.

2.1.10 MSW with no Biodegradables and Inert Fraction and Variable Removal of Paper and Plastic Fraction

The calorific value of the MSW left over when the entire biodegradable and inert fraction were removed was estimated as 4483 kcal/kg. The calorific value of this MSW increased by about 55% of the raw MSW, when the net moisture content got reduced, and the bulk of the ash content with low heat content got eliminated. When paper and plastic fraction was removed simultaneously from this MSW, at rates varying from 0 to 100%, for recycling and recovery, the heat content was found to follow a decreasing trend (Fig. 5) and at 100% removal, the heat content was found to have lowered to 3632 kcal/kg. This was due to the removal of components like paper and plastic, which have high inherent heat content.

2.2 *Reduction in GHG Emissions*

Recycling of the wastes is a better solution, as opposed to landfilling, and waste-to-energy technologies, as far as GHG emissions and energy conservation are considered. More than 65% of the MSW in Delhi is land-dumped, resulting in GHG emissions. However, when virgin materials are replaced by recycled materials, reduction in the emissions of GHG can be achieved [2, 13, 14].

According to a comprehensive model Waste Reduction Model (WARM), developed by US Environmental Protection Agency (USEPA), reduction in GHG emission when the virgin materials were replaced by recycled materials was estimated and plotted as shown in Fig. 6 [13]. The highest reduction was observed when novel paper was replaced by recycled paper, with a reduction of 18.6 tons CO_2 equivalent/tons of recycled waste, at 100% recycling of paper present in the Delhi MSW. This is followed by biodegradables, plastics and ferrous metals.

2.3 *Effect of MSW Recycling on Thermal Waste-to-Energy Facilities*

When the various components of MSW were removed for recycling, reuse or recovery, the heat contents of the MSW which was left were found to vary considerably.

A popularly followed thumb rule states that if the heat content of the MSW is below 1500 kcal/kg, the MSW is not fit to be incinerated for energy recovery. If the value lies between 1500 and 2400 kcal/kg approximately, the MSW can be

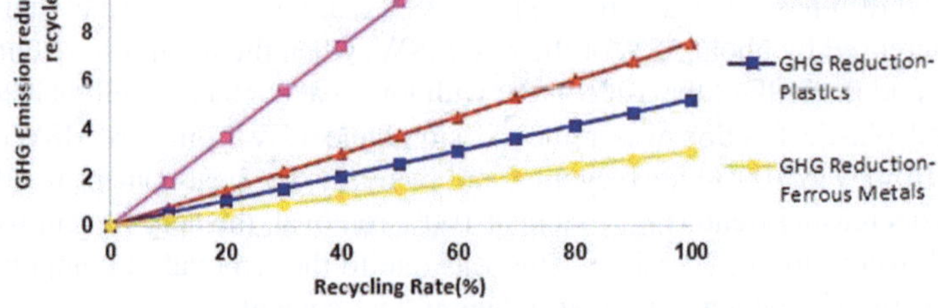

Fig. 6 Reduction in the GHG emission with recycling rate, when virgin materials are replaced with recycled materials

incinerated with auxiliary fuel, for energy recovery. If the heat content was above 2400 kcal/kg, then it can be subjected to autogenous incineration, for energy recovery. Raw MSW remained unfit for incineration with energy recovery over the years; however, the theoretically computed heat content showed that the MSW generated in MCD area of Delhi could be incinerated with an auxiliary fuel. Assessing the combustibility of the remnant MSW in different cases, from Fig. 7, it is evident that the MSW becomes acceptable for autogenous incineration, when all of the biodegradables are removed, or the entire inert fractions are removed. Similar is the case when there is a combined removal of ferrous metal and biodegradables or biodegradables, inert, paper and plastic fraction.

On the basis of this study, it can be concluded that while the segregation and removal/recycling of biodegradables, inert and ferrous metals enhance the combustibility of MSW, removal/recycling of paper and plastic content downgrades the combustibility of the leftover MSW. Consequently, the former boosts the efficiency of thermal waste-to-energy facility, while the latter may necessitate the addition of an auxiliary fuel.

Figure 8 illustrates the Tanner diagram, which is used to represent the combustibility of a fuel, diagrammatically based on its proximate analysis. Within the darkened zone, the MSW has higher combustibility which goes on increasing towards the right apex of the triangle, as the percentage of moisture and ash reduces and that of volatiles increases. Raw MSW is theoretically present inside the combustible zone, towards the boundary, as depicted in the figure. As fractions like ferrous metals, paper and plastics are removed individually, the MSW remained in the combustible zone. When the biodegradable and inert fractions are removed from

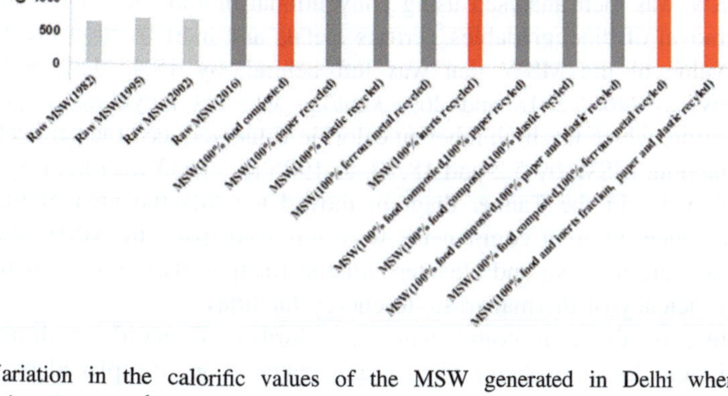

Fig. 7 Variation in the calorific values of the MSW generated in Delhi when different components are removed

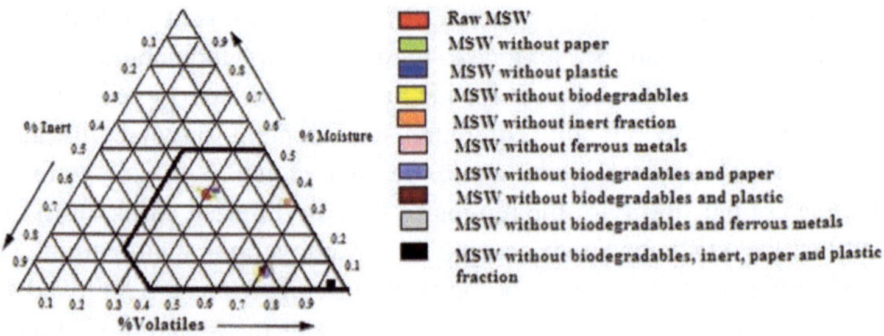

Fig. 8 Tanner diagram representing the various cases

the MSW, the combustibility increases considerably, and the position shifted closer to the right apex of the triangle. This can be explained by reduced ash and moisture content and increased volatiles fraction. A corresponding increase in the calorific value was computed earlier. When the entire fractions of paper and plastic are also recycled from the remaining MSW from the previous case, the position in the Tanner diagram shifted to the right apex and marked the highest comparative calorific value of all the cases.

3 Conclusion

Reduce, reuse and recycle are instrumental in curbing MSW management issues. The Municipal Solid Waste (Management and Handling) Rules, 2000 (presently the Solid Waste Management Rules, 2016), recommend composting of entire biodegradable fraction, whereas the Plastics (Manufacture, Usage and Waste Management) Rules, 2009, advocate the recycling of plastic wastes. Assuming the various components like biodegradables, paper, plastic, ferrous metals and inert fraction are removed/recycled at different rates, the heat content of the leftover waste was computed using the method provided by Kaiser [12]. The combustibility of the MSW was then analysed using conventional thumb rules and Tanner diagram. Removal of biodegradables, ferrous metals and inert fraction increased the calorific value of the MSW that was left behind, by 18.4, 0.37 and 39.6%, respectively, to 2466, 2018 and 2691 kcal/kg, whereas recycling of paper and plastic fraction, which has high inherent calorific value, reduced the net heat content of the remaining MSW by 5.2 and 18.7% to 1906 and 1633 kcal/kg, respectively. The position within the Tanner diagram moved towards the area of high combustibility, when different components were removed from the MSW, due to the reduced moisture and ash and elevated volatile fraction. This further helped augment the efficiency of thermal waste-to-energy facilities.

Recycling of different components was further observed to diminish the GHG emissions to atmosphere, when virgin materials were replaced by recycled materials. Thus, recycling activities offer wide socio-economic and environmental benefits and should be encouraged.

Simultaneous adherence to the various laws and promotion of recycling could not only improve the efficiency of thermal waste-to-energy facilities, but also prove to be advantageous as far as economic, environmental and employment opportunities are considered.

References

1. Hoornweg, D., Bhada-Tata, P.: What a Waste: A Global Review of Solid Waste Management. Urban Development and Local Government Unit, World Bank, Washington DC (2012)
2. Jain, P., Handa, K., Paul, A.: Studies on waste-to-energy technologies in India & a detailed study of waste-to-energy plants in Delhi. Int. J. Adv. Res. **2**(1), 109–116 (2014)
3. Talyan, V., Dahiya, R.P., Sreekrishnan, T.R.: State of municipal solid waste management in Delhi, the Capital of India. Waste Manag. **28**, 1276–1287 (2008)
4. CPCB: Status Report on Municipal Solid Waste Management. Ministry of Environment & Forests, Parivesh Bhawan, East Arjun Nagar, New Delhi (2011)
5. Kaushal, R.K., Varghese, G.K., Chabukdhara, M.: Municipal solid waste management in India—current state and future challenges: a review. Int. J. Eng. Sci. Technol. **4**(4), 1473–1489 (2012)
6. Parvathamma, G., An analytical study on problems and policies of solid waste management in India—special reference to Bangalore City. IOSR J. Env. Sci. Toxicol. Food Technol. **8**(10), 6–15 (2014)
7. Tchobanoglous, G., Theisen, H., Vigil, S.A.: Integrated solid waste management: engineering principles and management issues. McGraw-Hill, New York (1993)
8. EPA: Municipal Solid Waste Generation, Recycling, and Disposal in the United States: Facts and Figures for 2012, pp. 1–13 (2014)
9. Agarwal, A., Singhmar, A., Kulshrestha, M., Mittal, A.K.: Municipal solid waste recycling and associated markets in Delhi, India. Resour. Conserv. Recycl. **44**(1), 73–90 (2005)
10. JNNURM: City Development Plan. Delhi, Department of Urban Development, Government of Delhi, New Delhi (2006)
11. Municipal Corporation of Delhi: Feasibility Study and Master Plan for Optimal Waste Treatment and Disposal for the Entire State of Delhi based on Public Private Partnership Solutions, vol. 2(1), pp. 1–274 (2004)
12. Kaiser, E.R.: Chemical Analyses of Refuse Components. Department of Chemical I Engineering, New York University, New York (1966)
13. King, M.F., Gutberlet, J.: Contribution of cooperative sector recycling to greenhouse gas emissions reduction: a case study of Ribeiro Pires, Brazil. Waste Manag. **33**(12), 2771–2780 (2013)
14. Halim, D.: Impact of Municipal Solid Waste Recycling Programs on Climate Change. Solid Waste Reuse/Reclamation, City College of New York, New York (2014)

Trihalomethanes Occurrence in Chlorinated Treated Effluents at Sewage Treatment Plants of North-Indian Region

Surbhi Tak and Arun Kumar

Abstract Chlorination of treated wastewater effluent generally leads to the formation of disinfection byproducts (DBPs) which are a matter of great concern because of their carcinogenic nature. Trihalomethanes and Haloacetic acids are two major classes of DBPs. Chloroform being the predominant one. THM has also been classified as "possible human carcinogen" (group B2) by US Environmental Protection Agency. Total trihalomethanes (TTHM) concentration were determined in seven treated effluents from sewage treatment plants (STPs) in North-Indian region (Haryana and Uttrakhand) over a span of 10 months (July 2014–April 2015). STPs were selected based on different treatment technology, viz. up-flow anaerobic sludge blanket (UASB), sequential batch reactor (SBR), activated sludge process (ASP), and oxidation pond (OP). Low TTHM concentrations were obtained in all the samples analyzed (as per international standards for drinking water) though higher concentrations were seen in UASB plants (lower than the standard limits). The samples were evaluated only for winter season; effect of warm season on TTHM formation was not evaluated.

Keywords Disinfection · Disinfection byproducts · Chlorination
UV disinfection · Trihalomethanes · Secondary effluent · Water reuse

S. Tak (✉) · A. Kumar
Indian Institute of Technology Roorkee, Roorkee 247667, Uttrakhand, India
e-mail: surbh.dce2015@iitr.ac.in

A. Kumar
e-mail: aheciitr.ak@gmail.com

A. S. Kalamdhad et al. (eds.), *Advances in Waste Management*,
https://doi.org/10.1007/978-981-13-0215-2_19

1 Introduction

Chlorination is the most commonly used method for disinfection of treated wastewater effluent before discharging them into rivers, lakes, or any other receiving streams. Chlorination is still the most prevailing form of disinfection because of its well-established practices, broad-spectrum germicidal efficiency, and low cost of chlorine [1]. Chlorine is highly reactive in nature and due to its high reactivity it readily reacts with a variety of organic matter present in water and oxidizes them into disinfection byproducts (DBPs) [2]. Natural organic matters (NOMs) present in water mainly acts as a precursor for the formation of these DBPs. These NOMs mainly include humic acids and fulvic acids which readily react with chlorine leading to the formation of DBPs [3]. These DBPs can be further classified as regulated DBPs and emerging DBPs. Regulated DBPs mainly include trihalomethanes (THMs) and haloacetic acids (HAAs), these two are a major group of DBPs which constitute the major proportion of all [4]. THMs mainly include chloroform ($CHCl_3$), bromodichloromethane ($CHCl_2Br$), dibromochloromethane ($CHClBr_2$), bromoform ($CHBr_3$).THMs were first discovered in drinking water [5, 6], later they were regarded as a pollutant by National Pollutant Discharge Elimination System (NPDES) in treated wastewater effluents also [7]. Generally, water having high anthropogenic interference have higher DBP forming potential due to increased organic load [8, 9]. The formation of DBPs is also affected by the type of treatment technology being used (activated sludge process (ASP), up-flow anaerobic sludge blanket (UASB), sequential batch reactor (SBR), oxidation pond (OP), UASB being the major contributor to DBP formation.

THMs have adverse health impacts on humans if consumed directly or indirectly, many studies have detected their carcinogenic nature [10–12]. The mutagenicity, carcinogenicity, and teratogenicity of DBPs in drinking water have been tested in various studies [13–15]. THMs were also classified as possible human carcinogens (Group B2) [2]. Apart from having direct routes of exposure (ingestion), indirect routes (absorption through skin or lung inhalation) can also pose a adverse impact on human health, therefore use of chlorinated treated effluents for indirect uses like agricultural or irrigation for production of crops which are subsequently consumed by humans, can also act as a indirect source of DBP exposure [16]. There are various guidelines or standards available for control of these THMs in drinking water (World health Organization (WHO)—200 μg/l for chloroform, European Union (EU)—100 μg/l, US Environmental Protection Agency USEPA—80 μg/l, Bureau of Indian Standards (BIS)—100ug/l for chloroform [17]) but there are no such standards available for control of THMs in treated wastewater effluents, thus these drinking water standards can be used as surrogate standard for control of THMs in treated wastewater effluents.

Chlorination is by far the most cost-effective way of disinfection for treated wastewater effluents but little is known about its other reaction with NOMs and subsequently THM formation. Limited data are available on occurrence of THM in treated wastewater effluents. Apart from major issue of THM formation, chlorination imparts other onsite health and safety issues to working operators; another setback to chlorination [1, 2]. Various alternatives to chlorination have been suggested, best being ultraviolet (UV) disinfection, however, still there is very limited information available on the applicability of UV for control of THMs. There is still an ambiguity regarding the formation of DBPs during UV disinfection [18].

In India, THMs are considered as an important public health parameter mainly from drinking water perspective and continuous monitoring is being done at various regions [17, 19]. Central Pollution Control Board (CPCB, New Delhi) has also been monitoring THM levels in various water treatment plants of Delhi [20]. UV as a method for disinfection of treated water is used widely across the world for both drinking water and treated wastewater but in India, its use is much prevalent in drinking water as compared to treated wastewater.

The main objective of this study was to determine the levels of THMs formed at seven sewage treatment plants (STPs) of North-Indian region. Most of these STPs utilize the treated wastewater for agricultural or irrigational purposes. The effect of treatment technology being used in STP on THM formation was also evaluated. THM concentrations in chlorinated treated effluents were monitored for the period of 10 months, mainly in the winter season. The effect of the summer season on THM formation was not evaluated.

2 Materials and Methods

2.1 Chemicals

Calibration mixture (502/524 volatile organics calibration mix) used for the standardization purpose was obtained from Supelco. All the standards of different concentrations were prepared using ultrapure water (Millipore). Analytical grade sodium hypochlorite was used for spiking of non-chlorinated samples.

2.2 Sewage Treatment Plants

The STPs studied were located in Haryana and Uttrakhand region of North India as shown in Fig. 1. The description of STPs studied for the occurrence of THMs is shown in Table 1. STPs were selected based on different treatment technology, viz. UASB, SBR, ASP, and OP. Chlorination was employed as a method of disinfection

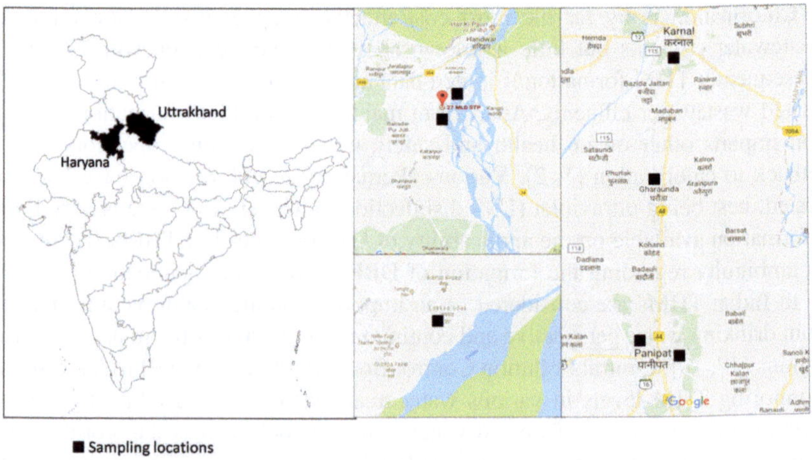

Fig. 1 Map showing STP location

at all the STPs in Uttrakhand, whereas for STPs of Haryana sample was spiked for estimating THM concentration.

2.3 Sampling Strategy

Water samples at each STP were collected in acid-washed amber-colored glass bottles, stored in ice coolers with temperature maintained at 4 °C till analysis (within 24 h), all the analysis work has been carried out at the National reference trace organic laboratory (NRTOL), CPCB, New Delhi. Seven sewage treatment plants in the state of Uttrakhand and Haryana have been covered. Sampling has been done in the month of January 2015 for STPs in Uttrakhand and in the month of March 2015 for STPs of Haryana.

2.4 THM Measurements

The samples were analyzed using gas chromatography-mass spectrometry (GC-MS-6890) supplied by Agilent technology with column SLB-5MS obtained from Supelco (60 m * 0.25 mm, 0.25 mm thickness) with purge and trap system for sample injection and USEPA protocol 524.2 was followed for the analysis of THMs in treated wastewater [21]. Wastewater samples for analysis of THM were collected in 50 ml amber colored bottles. Bottles were filled to the top (headspace free) and were stored at 4 °C until the analysis at the laboratory. All the samples were filtered using membrane filter before purging into the apparatus. 5 ml of a

Table 1 Description of STPs studied for THM analysis

STP location	Uttrakhand state			Haryana state			
	Haridwar	Haridwar	Rishikesh	Panipat	Karnal	Panipat	Gharaunda
Technology used	ASP	SBR	OP	UASB	UASB	UASB	OP
Treatment capacity (MLD)	18	27	6	10	40	35	3
Per day effluent (MLD) in peak hours	40	33–35	2–3	17	38–45	35–40	1–2
TOC (ppm)	17.01	10.03	8.49	41.05	31.15	30.78	18.80
Chlorination dose (ppm)	1.5 (gaseous form)	12 (gaseous form)	12 (gaseous form)	Sample spiked[a]	Sample spiked[a]	Sample spiked[a]	Sample spiked[a]
Treated water released to	For irrigation	River Ganga	River Ganga	For agriculture	For agriculture	For agriculture	Local lake
Carcinogenic activity reported	Yes	No	No	Yes	NA	NA	NA
Cost of chlorination	Approx. 2–3000 per month	Approx. 4–5000 per month	NA	NA	NA	NA	NA

[a]Chlorination performed at laboratory using bleaching powder/hypochlorite
NA not available

sample of each STP was injected using the syringe and after 11 min of purging, computer operated run of 60 min for each sample was performed. The sample is dry purged, cryo-cooled, and VOCs are desorbed at 250 °C and injected to GC. Oven temperature was programmed to increase by 1 °C/min from 35 to 42 °C (10 min) and then by 5–220 °C. The coolant and carrier gas used was liquid nitrogen and helium (15 ml/min), respectively. All the compounds present in the sample were detected and quantified on the basis of molecular weight and retention time.

3 Results and Discussion

3.1 THM Concentration

TTHM concentration was obtained in the range of 0–36.54 ppb and per MLD TTHM concentration was in the range of 0.004–3.654 ppb. The maximum concentration was seen in UASB plants and minimum in SBR plants. The concentration of THMs obtained at 7 STPs were shown in Table 2. Though the maximum concentration of THM, i.e., 36.54 ppb was obtained at UASB plant at Panipat this concentration is still lower than the drinking water standards of THMs i.e. 80 ppb for TTHMs (USEPA). However, at SBR plant Rishikesh THM concentration obtained was below detection limit. Despite having high total organic carbon (TOC) concentration, the value of THM obtained is quite low; this may be due to complexity in nature of influent wastewater constituents (organic matter) that are non-reactive toward chlorine in terms of THM formation, second reason can be lowered rate of reaction due to winter season. The absence or negligible concentration of brominated THMs may be due to the mere absence of bromide ion in influent water. Comparative analysis of different THM concentration obtained is shown in Fig. 2.

3.2 Seasonal Variability

The samples were studied only during the winter season (December 2014 and March 2015), the effect of the summer season on THM formation was not evaluated the therefore effect of seasonal variability and the water temperature was not included in the present findings. Generally, it was considered that THM formation increases with increasing water temperature [22].

Table 2 THM concentration obtained at 7 STP

STP	Treatment capacity (MLD)	Technology used	THM concentration (ppb)				TTHM (ppb)
			Chloroform	Bromodichloromethane	Dibromochloromethane	Bromoform	
Rishikesh	6	OP	0.38	0.13	0.26	0.25	1.02
Haridwar	27	SBR	BDL	BDL	BDL	BDL	–
Haridwar	18	ASP	0.08	BDL	BDL	0.56	0.64
Panipat	10	UASB	36.26	0.28	BDL	0.24	36.78
Panipat	35	UASB	26.4	1.25	BDL	0.91	28.56
Karnal	40	UASB	BDL	2.50	0.62	0.68	3.8
Gharaunda	3	OP	BDL	1.65	0.29	0.81	2.75

[a]*BDL* below detection limit

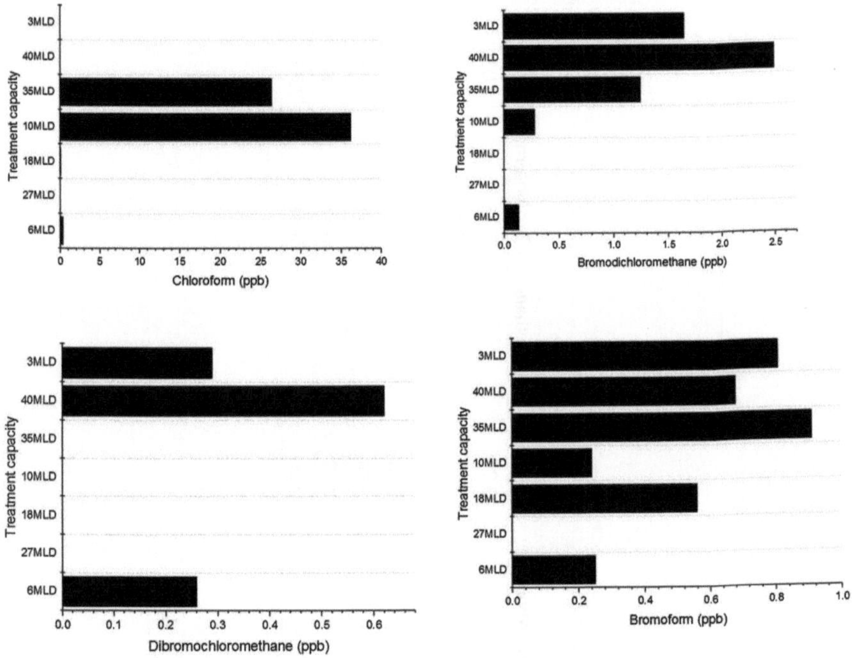

Fig. 2 THM concentrations at 7 STPs

4 Conclusions

Based on the sampling, testing, and analysis at seven STPs with different treatment technologies following conclusions are drawn:

1. In India number of STPs having disinfection as a part of wastewater treatment process is not very significant and out of all the STPs with disinfection, majority of them are chlorination based.
2. Chlorination has various risks associated with it including handling risk, storage risk, dosing issues, and other safety issues. Apart from these risks, there is a major problem of DBP (THM) formation.
3. THMs can be a potent carcinogen, various studies showed their carcinogenic nature in animals as well as humans.
4. In India data on THM formation during chlorination of treated wastewater is very limited, though few studies in various regions of India showed the occurrence of these THMs above standard limits.
5. THMs obtained in this study were below standard limits (USEPA standards for drinking water: 80 ppb) though comparatively higher concentrations were obtained in UASB plants than other ASP, SBR, and OP plants.

6. Based on different studies on UV disinfection for drinking water as well as wastewater treatment, it can be suggested that UV disinfection can act as a better alternative to chlorination in terms of performance as well as cost.

Acknowledgements Authors are grateful to Dr. C.S Sharma, Scientist "E", NRTOL, CPCB, New Delhi, for his permission in carrying out this work and Dev Prakash, NRTOL, CPCB, New Delhi, for his assistance and cooperation in conducting the analysis work. Authors are also thankful to the staff of Biomass System Lab, AHEC, IIT, Roorkee, for rendering their support during the collection of water samples.

References

1. Yang, X., Shang, C., Huang, J.C.: DBP formation in breakpoint chlorination of wastewater. Water Res. **39**, 4755–4767 (2005)
2. Matamoros, V., Mujeriego, R., Bayona, J.M.: Trihalomethane occurrence in chlorinated reclaimed water at full-scale wastewater treatment plants in NE Spain. Water Res. **41**, 3337–3344 (2007)
3. Nikolaou, A.D., Lekkas, T.D.: The role of natural organic matter during formation of chlorination by-products: a review. Acta hydrochemical et hydrobiologica **29**, 63–77 (2001)
4. Krasner, S.W.: The formation and control of emerging disinfection by-products of health concern. Philos. Trans. R. Soc. Lond. A Math. Phys. Eng. Sci. **367**, 4077–4095 (2009)
5. Rook, J.J.: Formation of haloforms during chlorination of natural waters. Water Treat. Exam. **23**, 234–243 (1974)
6. Bellar, T.A., Lichtenberg, J.J., Kroner, R.C.: The occurrence of organohalides in chlorinated drinking waters. American Water Works Association **66**, 703–706 (1974)
7. Krasner, S.W., Westerhoff, P., Chen, B., Rittmann, B.E., Amy, G.: Occurrence of disinfection byproducts in united states wastewater treatment plant effluents. Environ. Sci. Technol. **43**, 8320–8325 (2009)
8. Sirivedhin, T., Gray, K.A.: Comparison of the disinfection by-product formation potentials between a wastewater effluent and surface waters. Water Res. **39**, 1025–1036 (2005)
9. Singer, P.C., Chang, S.D.: Correlations between trihalomethanes and total organic halides formed during water treatment. Am. Water Works Assoc. **81**, 61–65 (1989)
10. Mishra, B.K., Gupta, S.K., Sinha, A.: Human health risk analysis from disinfection by-products (DBPs) in drinking and bathing water of some Indian cities. J. Env. Health Sci. Eng. **12**, 73 (2014)
11. Villanueva, C.M., Cantor, K.P., Cordier, S., Jaakkola, J.J., King, W.D., Lynch, C.F., Kogevinas, M.: Disinfection byproducts and bladder cancer: a pooled analysis. Epidemiology **15**, 357–367 (2004)
12. National Cancer Institute: Carcinogenesis bioassay of chloroform. Technical Report, National Technical Information Service, National Cancer Institute, Springfield, VA, USA (1976)
13. Singer, P.C., American Water Works Association: Formation and Control of Disinfection by-Products in Drinking Water. AWWA (1999)
14. Black, B.D., Harrington, G.W., Singer, P.C.: Reducing Cancer Risks by improving organic carbon removal. Am. Water Works Assoc. **88**, 40 (1996)
15. Kanitz, S., Franco, Y., Patrone, V., Caltabellotta, M., Raffo, E., Riggi, C., Ravera, G.: Association between drinking water disinfection and somatic parameters at birth. Environ. Health Perspect. **104**, 516 (1996)
16. Levine, A.D., Asano, T.: Recovering sustainable water from wastewater. Environ. Sci. Technol. **38**, 201A–208A (2004)

17. Upadhyay, N., Jain, R.K., Shrivastava, P.K., Puraiya, M.: Study on formation of trihalomethanes (THMs) in potable treated water of Gwalior City, Madhya Pradesh, India. J. Eng. Comput. Appl. Sci. **2**, 4–4 (2013)
18. Reckhow, D.A., Linden, K.G., Kim, J., Shemer, H., Makdissy, G.: Effect of UV treatment on DBP formation. Am. Water Works Assoc. **102**, 100 (2010)
19. Mishra, N.D., Dixit, S.C., Srivastava, H.C.: Evaluation of trihalomethane formation potential due to anthropogenic sources in the ground water of Kanpur. J. Chem. **9**, 693–699 (2012)
20. Central Pollution Control Board: Brief Report of Trihalomethanes (THMs) in Potable Treated Water of Delhi Jal Board Water Works. Technical Report
21. Eichelberger, J.W., Munch, J.W., Bellar, T.A.: Method 524.2 Measurement of Purgeable Organic Compounds in Water by Capillary Column Gas Chromatography/Mass Spectrometry. Environmental monitoring systems laboratory office of research and development, USEPA (1989)
22. Uyak, V., Toroz, I., Meric, S.: Monitoring and modeling of trihalomethanes (THMs) for a water treatment plant in Istanbul. Desalination **176**, 91–101 (2005)

Enzymatic Application in Anaerobic Digestion (AD) of Organic Fraction of the Municipal Solid Waste (OFMSW)

B. Chatterjee, S. Goswami and D. Mazumder

Abstract In the twenty-first century, anaerobic digestion (AD) is considered to be the most efficient method among various other classical municipal solid waste management (MSWM) practices, which stabilize the generated organic fraction of municipal solid waste (OFMSW). Hydrolysis in AD is where the insoluble polymers in the procured OFMSW are converted into simpler soluble monomers by the fermentative bacteria, which generate certain extracellular enzymes known as hydrolases. These enzymes catalyze all the reactions. The hydrolysis being the rate-limiting step thus plays one of the crucial roles in the anaerobic decomposition of OFMSW by converting it into stable by-products. Chemically, hydrolysis can be defined as the breakdown of long-chain biomolecules into corresponding short chains in the presence of water. Biologically, hydrolysis acts through the influence of enzymes. The present research reviewed the potential, functional routes, and behavior of hydrolytic enzymes and studied the suitability of the enzyme cellulase in improving the performance of AD of OFMSW using a laboratory-scale setup.

Keywords Organic fractions · Municipal solid waste · Anaerobic digestion
Enzymatic hydrolysis · Application perspective

B. Chatterjee (✉) · S. Goswami · D. Mazumder
Civil Engineering Department, Indian Institute of Engineering Science and Technology, Shibpur, Botanic Garden, Howrah 711103, India
e-mail: cbiswabandhu@yahoo.com

S. Goswami
e-mail: supriyogoswami.ju@gmail.com

D. Mazumder
e-mail: debabrata@civil.iiests.ac.in

A. S. Kalamdhad et al. (eds.), *Advances in Waste Management*,
https://doi.org/10.1007/978-981-13-0215-2_20

1 Introduction

The triggering of potential environmental problems associated with the classical municipal solid waste management (MSWM) practices such as sanitary landfilling, composting, and incineration (pyrolysis/gasification) have resulted in the need for biological treatment of the generated municipal solid waste (MSW). As such, the practice of AD has come up as the most innovative and attractive technology, not only in terms of stabilizing the generated quanta of OFMSW but also in terms of generating valuable by-products [1–4]. These by-products include biogas, which is an able substitute for fossil-based fuels, and solid sludge, which can be conditioned further for use as a good-quality soil conditioner. AD can be defined as a biochemical process, which consists of three major phases such as hydrolysis, where fermentative group of bacteria firstly degrade the complex organic matters or polymers to simply soluble organic matters or monomers, followed by acidogenesis, where the conversion of the monomers into long-chain fatty acids (LCFAs), short chain fatty acids (SCFAs), and alcohols occur, and finally methanogenesis, where the methanogens convert the acetate, alcohols, and $H_2 + CO_2$ into methane and CO_2 [1, 3]. However, there is an additional acetogenesis phase, where the LCFAs and alcohols are converted into CH_3COOH, $CH_3CH_2OH + H_2$, and $H_2 + CO_2$ [2].

The OFMSW comprises of biodegradable polymers, which include lignocelluloses, proteins, lipids, and starch. A group of specialized microbial population of the fermentative bacteria are responsible for depolymerization of these organic polymers into simple monomers. There are three main types of fermentative bacteria that are responsible for the liquefaction of OFMSW, and these are: (1) proteolytic bacteria—which produce an enzyme protease, capable of breaking down proteins and peptides into ammonia and amino acids; (2) lipolytic bacteria—which generate lipase enzyme that is responsible for the breakdown of saponifiable lipids into fatty acids and glycerol; and (3) cellulolytic bacteria—which hydrolyze polysaccharides into monosaccharides using cellulases. Since the action of these enzymes is a complex process, hydrolysis is deemed to be the slowest and rate-limiting step in the overall process of AD. So much so that the composition of methane in the biogas generated/ultimate methane yield is directly dependent on the efficiency of the hydrolysis phase [5]. Presently, enzymatic application as a pre-treatment step prior to AD for the stabilization of OFMSW has been of growing interest [6]. Zheng et al. [7] studied the scope of biological pretreatment like enzymatic hydrolysis as a cost-effective technology that enhances solid organic waste stabilization process as well as the post-biogas production. Lignocellulose refers to the three major components of the plant tissue: cellulose, hemicellulose, and lignin. While the phenolic groups in the lignin are resistant to enzymes, the celluloses and hemicelluloses are biodegradable and constitute more than 90% of the biochemical methane potential of the OFMSW [8]. These are degraded via hydrolysis to yield soluble disaccharides and cellobiose, which on further hydrolysis result in the formation of D-glucose.

Enzymatic hydrolysis of lignocellulosic matter present in the OFMSW involves several important steps, such as (i) transfer of enzyme from the bulk aqueous surface to the surface of the lignocellulosic particles, (ii) adsorption of the enzyme and subsequent formation of enzyme–substrate (ES) complexes, (iii) hydrolysis of lignocellulose, (iv) transfer of lignocellulosic material (glucose, cellodextrins, and cellobiose) from the surface of the substrate into the bulk aqueous phase, and (v) hydrolysis of the transferred lignocellulosic materials in the aqueous phase [9]. Among these steps, the process of enzyme adsorption and the subsequent formation of ES complexes are considered to be the most crucial in the enzymatic hydrolysis of lignocellulosic biomasses [10, 11]. It is noteworthy to mention that during the enzymatic hydrolysis of lignocelluloses, the physical, chemical, and morphological characteristics of the lignocelluloses play crucial roles, beside the effectiveness of the enzymes. The factors affecting the enzymatic hydrolysis of lignocellulosic materials can be subdivided into two factors: enzyme-related factors, which mainly focus on improving the enzyme-related activity, and substrate-related factors, which concentrate on improving the accessibility of enzymes to lignocelluloses [12]. The accessibility of enzymes to lignocelluloses depends on the crystallinity of the cellulose present, the contents and distributions of lignin and hemicelluloses in the OFMSW, and the surface area available [13].

The production of biofuels from cellulosic biomasses, via several biochemical steps, results in the yield of vital products at low costs. One of the most complex steps in the above process is the conversion of lignocellulosic biomasses into fermentable sugars via enzymatic hydrolysis. In terms of breakdown of complex lignocellulosic biomasses, enzymatic hydrolysis has the potentiality to offer higher yields, lower energy costs, higher selectivity, and milder operating conditions over various physicochemical processes. However, the mechanism of enzymatic hydrolysis, and the relation between the substrate structure and the function of different glycosyl hydrolase components are yet to be fully understood. This is one of the main hindering factors in the maximization of sugar yields from lignocellulosic biomass at low costs. The enzymatic deconstruction of lignocellulosic biomass becomes very difficult due to its many structural features. The cellulosic fibers, which remain bound together in a composite structure by lignin, reduce the accessibility of enzymes to cellulose. The composition and the specific structural feature(s) of lignin, which are responsible for binding these cellulosic fibers together, are not well explored yet. In order to deconstruct these heterogeneous structures, which are present in the plant cell wall, a complete understanding of the interactions between the lignin structure and the synergistic reactions of cellulases, hemicellulases, accessory enzymes, and lignin-modifying enzymes is required. The synergistic reactions of the aforesaid enzymes promote enzymatic hydrolysis of lignocellulosic biomass, thereby converting them into low-molecular weight products, which are capable of being utilized by the archaea [14–16].

Hydrolysis of cellulose results in the sequential breakdown of the linear glucose chains that are present in cellulose [17]. The enzymatic hydrolysis of cellulose comprises of adsorption of cellulase on to the surface of cellulose, biodegradation of cellulose to fermentable sugars, and desorption of the cellulase enzyme.

The synergistic action of three major classes, which are endo-glucanases, exo-glucanases, and beta-glucosidases, are responsible for the enzymatic breakdown of cellulose into glucose. Collectively, these enzymes are known as cellulolytic enzymes. Endo-glucanases are responsible for creating free-chain ends by attacking the low crystalline region of cellulose fibers. Removal of cellobiose units from the free-chain ends is carried out by exo-glucanases, who further degrade the sugar chains. Next, the cellobiose units are cleaved to glucose by beta-glucosidase. Beta-glucosidase is therefore responsible in bringing out complete depolymerization of cellulose to glucose. The breakdown of hemicelluloses, containing different sugar units, is carried out by the hemicellulolytic enzymes, which are even more complex than cellulolytic enzymes [18].

The present study concerns the suitability of different types of enzymes such as cellulase, amylase, xylanase, and pectinase in improving the performance of hydrolysis in anaerobic digestion which is very much essential for ultimate waste stabilization, i.e., biochemical oxygen demand (BOD) removal through methane formation in an oxygen-depleted environment. Fruit and vegetable waste (FVW) were considered as substrate in the present study due to their high biodegradability.

2 Materials and Methods

2.1 *Seed Acclimation for Hydrolysis Study of OFMSW*

The various aforementioned enzymes were cultured with special emphasis on cellulase. For this purpose, sludge from a laboratory-scale hybrid bioreactor treating composite tannery wastewater, sludge from another laboratory-scale hybrid bioreactor treating domestic wastewater, leachate from waste dumping grounds of two local markets, and leachate content from two different points of a sanitary landfilling site were collected. Apart from those, the leachate content from a bucket full of bamboo and vegetable pieces mixed with 1 g of carboxymethyl cellulose was also procured. Thereafter, 62.5 ml of soil solution each from the seven aforesaid sources were incubated with 125 ml carboxymethyl cellulose and 62.5 ml phosphate buffer in seven different 250 ml conical flasks. The conical flasks were kept closed using cotton plugging and placed in a bacteriological incubator at 40 °C for seven days. It was observed that there was an appreciable growth in all the seven conical flasks and subsequently these were centrifuged at 10,000 rpm for 10 min to procure the clear supernatant content. This was done in order to obtain the enzyme content from each of the aforesaid seven different sources.

2.2 Determination of Enzyme Concentration

To determine the concentration of different types of enzymes, DI NITRO SALICYLIC ACID or DNSA method was adopted. For each of the seven conical flasks, having a particular type of enzyme content, three test tubes were taken. The first test tube was used as a control tube, and the remaining two were used for studying the enzymatic concentration. Each test tube was firstly filled with 500 μL of phosphate buffer solution using micropipette followed by the addition of 1 ml substrate solution. For cellulase enzyme, carboxymethyl cellulose was used as a substrate. For amylase enzyme, starch was used. For xylanase enzyme, xylan was used, and for pectinase enzyme, pectin was used. For carrying out the enzyme assay method, 0.2 g of the respective substrate was added in 10 ml of commercially available distilled water followed by the addition of 3 ml of DNSA. Thereafter, the whole assembly of test tubes was placed in a boiling water bath for 10 min. This was followed by measurement of the absorbance of the test tube content (for the purpose of determining the enzyme concentration) via single-beam spectrophotometer at a wavelength of 540 nm.

2.3 Experimental Setup

The laboratory-scale setup was fabricated by arranging six buckets, each having 10 L capacities. The first containment had 8 L surface water (treated) and zero volume of sludge, which was basically the enzyme obtained from the laboratory-scale hybrid bioreactor treating domestic wastewater. The reason for selecting the sludge from the laboratory-scale hybrid bioreactor treating domestic wastewater source was the fact that maximum concentration of cellulase was obtained from the said source compared to the other six sources. The second containment had 300 mL sludge and 7.7 L of surface water (treated). The third one had 600 ml of sludge and 7.4 L of surface water (treated). The fourth one had 900 ml of sludge and 7.1 L of surface water (treated). The fifth one had 1200 ml sludge and 6.8 L surface water (treated). The sixth one had 1500 mL sludge and 6.5 L surface water (treated). Composite vegetable waste of 1 kg was filled in a nylon bag and placed inside each of the seven buckets to undergo hydrolysis. Figure 1 shows the schematic diagram of the laboratory-scale hydrolytic batch reactor.

2.4 Characterization of OFMSW Under Composite Basis

For characterization purpose, different types of fruit and vegetable waste (FVW) were collected and the following parameters were evaluated: specific

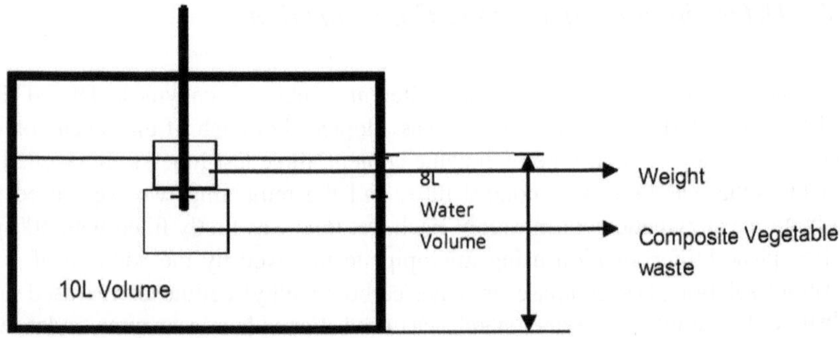

Fig. 1 Schematic diagram of hydrolytic batch reactor

gravity, loss of moisture, moisture content, fixed solids, and volatile solids. Different biological parameters of the hydrolyzed fraction, such as chemical oxygen demand (COD), mixed liquor suspended solids (MLSSs), and pH, were measured. The COD measured was basically the soluble COD (sCOD) obtained after filtering the raw sample. In addition, percentage reduction in volume (hydrolysis potential) and concentration of the enzyme cellulase were determined by testing the leachate collected from each of the eight nylon bags kept submerged in the eight different buckets of water having varying enzyme (sludge) volume and 1 kg of composite vegetable waste. Each of the aforementioned parameters was studied at 24 h interval.

2.5 Analytical Study

For the purpose of monitoring the different parameters such as COD, MLSS, and pH, the procedures described in the standard methods [19] were followed.

3 Results and Discussion

3.1 Characterization of Organic Fraction of Municipal Solid Waste (OFMSW)

The characterization was carried out by using FVW. This was attributed to the fact that a major portion of the generated OFMSW (over 60%) comprises of FVW. The FVW procured was mixed up thoroughly, and a composite mass was obtained. As mentioned, for the purpose of characterization, specific gravity, unit weight, loss of moisture, moisture content, fixed solids, and volatile solids were studied. The bulk density and specific gravity were measured by water displacement method.

Table 1 Characterization results of municipal solid waste

Day	Specific gravity	Bulk density (kg/m^3)	Loss of moisture (%)	Moisture content (%)	Fixed solids (%)	Volatile solids (%)
1	0.925–0.928	925–928	29.60–33.98	83.57–86.48	1.99–3.33	96.67–98.01
2	0.789–0.837	789–837	30.50–34.00	86.00–87.50	1.00–1.50	98.50–99.00
3	0.936–0.941	936–941	33.00–35.00	85.50–86.50	1.50–2.00	98.00–98.50
4	0.992–0.998	992–998	30.50–39.00	88.50	1.00–1.50	98.50–99.00
5	0.867–0.974	867–974	27.12–32.50	88.50	1.50	98.50

The loss of moisture, moisture content, fixed solids, and volatile solids were measured in accordance with the method adopted by Tchobanoglous et al. [20]. Table 1 shows the results of the characterization performed on five different days.

It was observed that the moisture content of FVW was in excess of 85% in almost all the above measured instances. It is evident that for wastes with high moisture content, adoption of stabilization techniques such as incineration and composting, which involve external drying as a part of preprocessing the waste, ultimately results in the overall waste stabilization process to be uneconomical. The present study on characterization of FVW corroborates past results and justifies the scope of selecting AD as the most appropriate technique to manage this sort of waste. This is because AD can be regarded as the only process where the presence of high moisture content in the waste can prove to be beneficial, since the rate of hydrolysis improves with the increase in moisture content of the waste. In addition, the presence of high percentages of volatile solid content in the waste results in the occurrence of high-rate AD, which is considered as a major precursor to elevated biogas production. However, in order for the AD process to be efficient, in terms of biogas production, it has to be ensured that the rate-limiting hydrolysis step is optimized to the maximum extent, mainly in terms of solubilization of the complex substrates that are present in the FVW.

3.2 *Determination of the Enzyme Concentration*

For finding the concentration of the different types of enzymes collected from the seven different sources, the aforementioned enzyme assay procedure was adopted. The concentrations of cellulases, amylases, xylanases, and pectinases were determined with an increased emphasis on measuring the concentration of cellulases, since it was easier to obtain in an increased concentration compared to the other three enzymes. In addition, past studies on enzymatic hydrolysis have shown that lignocellulose is primarily composed cellulose (38–50%), whose breakdown is mainly caused by the enzyme cellulase. Table 2 shows the name of the enzyme (cellulase), the collection point, the obtained concentration, and the volume of samples used in the analysis.

Table 2 Various sources of cellulase enzyme

Source of enzyme	Name of enzyme	Concentration of enzyme (micromole/min/ml)	Volume of sample (microliter)
Leachate of Dhapa dumping ground, Kolkata	Cellulase	Almost nil	500
Hybrid laboratory-scale aerobic bioreactor	Cellulase	(i) 3.148 (without centrifuge) (ii) 1.486 (with centrifuge)	500
Solution of soil collected from CA Market, Salt Lake, Kolkata	Cellulase	1.5 (with centrifuge)	500
Solution of soil Collected from EC Market, Salt Lake, Kolkata	Cellulase	Almost nil	500
Leachate from wasted bamboo pieces	Cellulases	Almost nil	500
Leachate from vegetable pieces	Cellulases	(i) 0.25 (ii) 1.07	500
Carboxymethyl cellulose solution	Cellulases	(i) 0.15 (ii) 1.22	500

As discussed earlier, a major percentage of the commingled OFMSW is composed of lignocellulosic materials (leaves, grasses, papers, wheat straw, nutshells, hardwood/softwood stems, sorted refuse, primary wastewater sludge, solid cattle dung, etc.) whose primary component is mostly cellulose. Cellulase enzymes extracted from the above-mentioned sources are responsible for catalyzing the hydrolysis of the cellulose. However, the accessibility of the cellulase enzymes to cellulose is affected by the relative composition of lignin and hemicellulose present in the waste. It has been observed that the efficiency of hydrolysis via application of cellulase decreases with the increasing presence of hemicellulose and lignin in the waste. This demands the need for some sort of pretreatment technique (e.g., mechanical comminution) that ensures the removal of lignin and hemicellulose besides reducing cellulose crystallinity [21]. In the present study, the collected FVW was chopped into small pieces of size 20 mm by the use of kitchen knife. This proved to be not only cost-effective but also avoided the loss of essential biodegradable carbohydrates and removal of the non-biodegradable fractions that are present in FVW.

3.3 *Results of Hydrolytic Batch Study*

Following hydrolysis, the following biological parameters of the hydrolyzed products were studied at 24 h interval: pH, COD, and MLSS. In addition, the percentage reduction in volume (hydrolysis potential) of the composite FVW

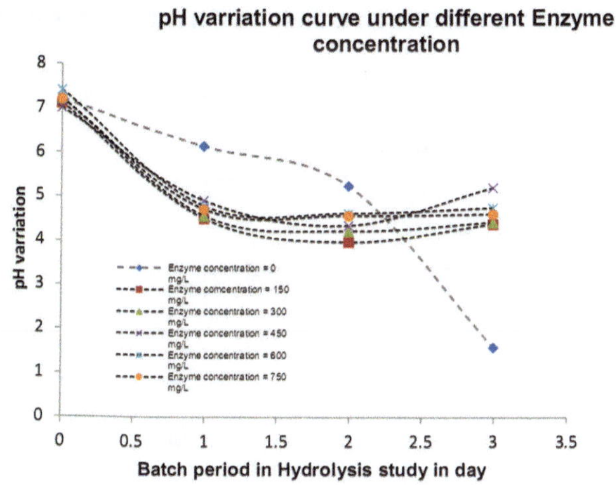

Fig. 2 pH variation curve under different enzyme concentrations

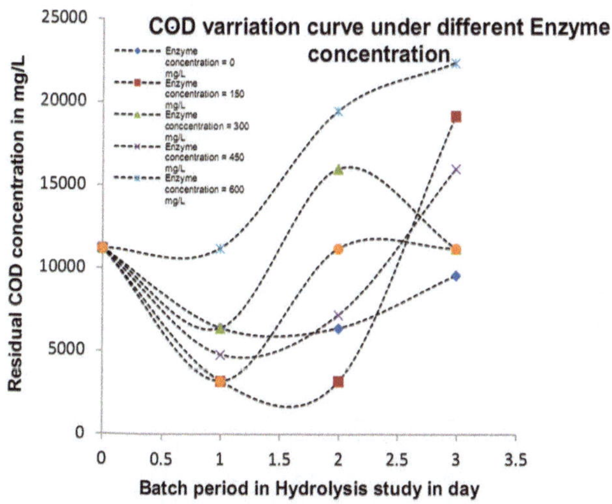

Fig. 3 COD variation curve under different enzyme concentrations

fractions those underwent hydrolysis were also measured. Figures 2, 3, 4, and 5 show the variation of pH, COD, MLSS, and hydrolysis potential under different enzyme concentrations, following hydrolysis of the composite FVW fraction.

The reduction in pH was mainly due to the formation of LCFAs and SCFAs. The LCFAs are responsible for lowering the pH in the range of 4.0–4.5, whereas the SCFAs start to produce from pH 5.0 onwards. It can be observed from Fig. 2 that the initial (day 0) pH readings of all the sets of FVW fractions, which underwent hydrolysis under different cellulase concentrations, were in between 7.0 and 7.5,

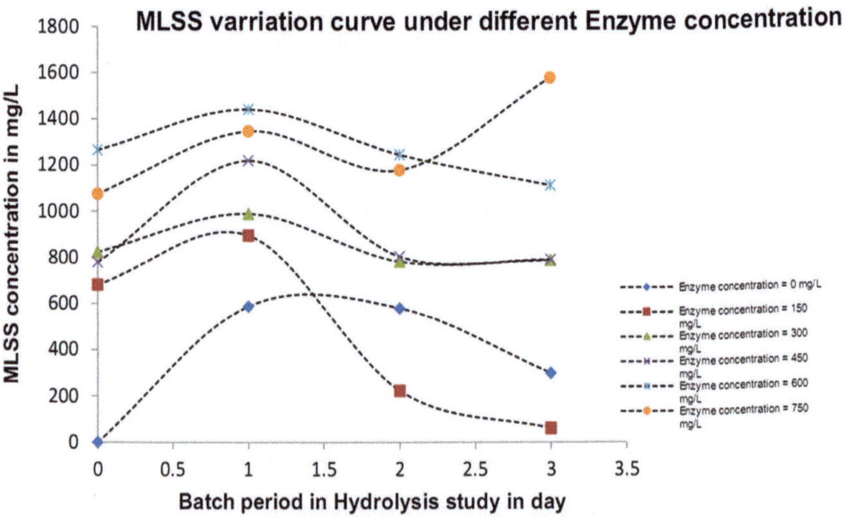

Fig. 4 MLSS variation curve under different enzyme concentrations

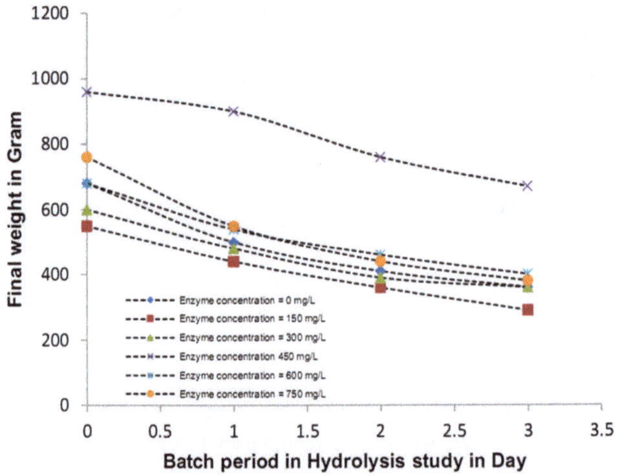

Fig. 5 Variation of weight loss due to hydrolysis curve under different enzyme concentrations

suggesting slow degradation of the FVW fraction and the absence of LCFAs and the SCFAs in the system. Again, it can be seen from Fig. 2 that the drop in pH values below 6.0, for all the six sets of FVW, had occurred after 12 h from the onset of hydrolysis. Hydrolysis of FVW, whose dominant fraction is carbohydrates among other, occurs in between pH 4.0 and 5.0 [22]. As such, a similar trend can be observed from the pH profile, shown in Fig. 2, after 36 h from the point of enzyme addition/initiation of hydrolysis. The pH optimum to the occurrence of acidogenesis lies in between 5.0 and 6.5 [23, 24]. Similarly, the steady rise in the pH values above 4.7, after 48 h from the onset of hydrolysis, suggests the activity of the acidogens, which had started degrading the LCFAs produced during the hydrolysis phase to SCFAs via acidogenesis.

Past studies indicate that inoculation of microorganisms plays a crucial role in the initiation of the substrate breakdown [25]. In this regard, the enzymes secreted by the inoculated microbes degrade the complex lignocellulosic substrates that are present in the FVW. It can be observed from the COD profile, shown in Fig. 3, that initially with the onset of hydrolysis the sCOD increased sharply, suggesting the breakdown of the easily biodegradable fractions present in the FVW. Thereafter, the sCOD values slowly decreased, suggesting the utilization of the produced sCOD by the microbial species present in the buckets. At the end, however, there was again an increase in the sCOD values, suggesting the possible degradation of the complex fraction present in the FVW. The enzymatic action of the added cellulase may be defined as the reason behind the sudden increase in the sCOD concentration after the 30 h duration. It can be observed from the sCOD profile that the greater the enzymatic concentration, the greater the extent of sCOD generation.

The "peeled"-off solids from the surface along with the particulate organic matter from the FVW were responsible for contributing to the MLSS concentrations. As such, it can be observed from Fig. 4 that the greatest MLSS concentration was in the bucket with the highest enzyme concentration. With the progress of time, the measured MLSS concentration for all the six buckets had increased. However, the presence of inconsistent suspended solids in the sample during collection can be the reason behind variable MLSS concentration, as observed.

The variation in weight loss due to hydrolysis, as shown in Fig. 5, suggests that with the increase in concentrations of the cellulase enzyme the percentage reduction in weight improved. The presence of lignocellulosic fraction in the FVW particularly procured from the guesthouse kitchen of IIESTS led to the noticeable degradation of the complex substrates in the later stages of the hydrolysis study, as observed from Fig. 5. It is well known from past studies [26–28] that a certain fraction of the usually cellulose-poor FVW is usually comprised of complex organic C-chains, the breakdown of which necessitates either a strong enzymatic action or other types of advanced pretreatment (thermal, chemical, mechanical) strategies. The present study corroborated this principle by enhancing the degradation of the lignocellulosic fraction with the addition of cellulase.

4 Conclusion

It is well established from various previously performed case studies by different researchers that the hydrolysis step in an AD process (of OFMSW) is significantly enhanced by addition of suitable enzymes such as cellulase, amylase, pectinase, xylanase. In this regard, various cellulose-degrading microbes were earmarked from seven different sources, and following culture of these microbes, the cellulose degrading cellulase was obtained and hydrolysis was carried out in varying cellulase concentration. It was observed that the addition of the enzyme cellulase resulted in significant increase in COD concentration of the treated waste and drop in pH following hydrolysis, which is the rate-limiting step in AD. Thus, it can be concluded that the isolation of the enzyme cellulase from the respective bacteria and thereafter applying the cellulase onto the complex substrate (FVW) for their rapid solubilization, prior to hydrolysis, can be suitable strategy to speed up the AD of OFMSW. Similar to cellulase, studies involving amylase, pectinase, and xylanase can be undertaken in the future.

Acknowledgements The first author is grateful to Mr. Somsekhar Bhattacharya, M.E. scholar, Department of Civil Engineering, IIEST Shibpur, for giving valuable effort in the collection of the different samples and performing the hydrolysis case studies.

References

1. Chatterjee, B., Mazumder, D.: Anaerobic digestion for the stabilization of the organic fraction of municipal solid waste: a review. Environ. Rev. **24**(4), 426–459 (2016)
2. Metcalf & Eddy, Inc. Revised by Tchobanoglous, G., Burton, F.L., Stensel, H.D.: Wastewater Engineering: Treatment and Reuse, pp. 630–631. Tata McGraw Hill Education Pvt. Ltd. (2003)
3. Khalid, A., Arshad, M., Anjum, M., Mahmood, T., Dawson, L.: The anaerobic digestion of solid organic waste. Waste Manag. **31**(8), 1737–1744 (2011)
4. Troschinetz, A.M., Mihelcic, J.R.: Sustainable recycling of municipal solid waste in developing countries. Waste Manag. **29**(2), 915–923 (2009)
5. Palmisano, A.C., Barlaz, M.A.: Microbiology of solid waste. CRC Press, Inc. (1996)
6. Fdéz.-Güelfo, L.A., Álvarez-Gallego, C., Márquez, D.S., García, L.I.R.: Biological pre-treatment applied to industrial organic fraction of municipal solid. Chem. Eng. J. **172**, 321–325 (2011)
7. Zheng, Y., Zhao, J., Xu, F., Li, Y.: Pretreatment of lignocellulosic biomass for enhanced biogas production. Prog. Energy Combust. Sci. **42**, 35–53 (2014)
8. Quiñones, T.S., Plöchl, M., Budde, J., Heiermann, M.: Results of batch anaerobic digestion test – effect of enzyme addition. Agric. Eng. Int: CIGR J. **14**(1), 38–50 (2012)
9. Walker, L.P., Wilson, D.B.: Enzymatic hydrolysis of cellulose: an overview. Biores. Technol. **36**(1), 3–14 (1991)
10. Mandels, M., Kostick, J., Parizek, R.: The use of adsorbed cellulase in the continuous conversion of cellulose to glucose. J. Polym. Sci. Part C: Polym. Symposia **36**(1), 445–459
11. Beldman, G., Voragen, A.G.J., Rombouts, F.M., Pilnik, W.: Synergism in cellulose hydrolysis by endoglucanases and exoglucanases purified from Trichoderma viride. Biotechnol. Bioeng. **31**(2), 173–178 (1988)
12. Sun, S., Sun, S., Cao, X., Sun, R.: The role of pretreatment in improving the enzymatic hydrolysis of lignocellulosic materials. Biores. Technol. **199**, 49–58 (2016)
13. Alvira, P., Tomás-Pejó, E., Ballesteros, M., Negro, M.J.: Pretreatment technologies for an efficient bioethanol production process based on enzymatic hydrolysis: a review. Biores. Technol. **101**(13), 4851–4861 (2010)
14. Sun, Y., Cheng, J.: Hydrolysis of ligcellulosic materials for ethanol production—a review. Biores. Technol. **83**(1), 1–11 (2002)
15. Taherazadeh, M.J., Karimi, K.: Enzymatic-based hydrolysis processes for ethanol from lignocellulosic materials: a review. Bio Resour. **2**(4), 707–738 (2007)

16. Crine, D.G., Agbor, V.B., Björnsson, L.: Enhanced solubilisation of the residual of municipal solid waste. in: 11th IWA World Congress on Anaerobic Digestion, 23–27 Sept (2007)
17. Rao, S.S.: Enzymatic hydrolysis of cellulosic fiber. Georgia Institute of Technology (2009)
18. Romano, R.T., Zhang, R., Teter, S., McGarvey, J.A.: The effect of enzyme addition on anaerobic digestion of Jose Tall Wheat Grass. Biores. Technol. **100**(20), 4564–4571 (2009)
19. Eaton, A.D., Clesceri, L.S., Greenberg, A.E.: Standard methods for the examination of water and wastewater (1995)
20. Tchobanoglous, G., Theisen, H., Vigil, S.: Integrated Solid Waste Management: Engineering Principles and Management Issues, pp. 69–72. McGraw-Hill Inc., New York (1993)
21. McMillan, J.D.: Pretreatment of lignocellulosic biomass. In: Himmel, M.E., Baker, J.O., Overend, R.P. (eds.) Enzymatic Conversion of Biomass for Fuels Production, pp. 411–437. American Chemical Society, Washington, DC (1994)
22. Gerardi, M.H.: The Microbiology of Anaerobic Digesters. Wiley, New York (2003)
23. Yu, H.Q., Fang, H.H.P.: Acidogenesis of dairy wastewater at various pH levels. Water Sci. Technol. **45**(10), 201–206 (2002)
24. Lv, W., Schanbacher, F.L., Yu, Z.: Putting microbes to work in sequence: recent advances in temperature-phased anaerobic digestion processes. Biores. Technol. **101**(24), 9409–9414 (2010)
25. Komemoto, K., Lim, Y.G., Nagao, N., Onoue, Y., Niwa, C., Toda, T.: Effect of temperature on VFA's and biogas production in anaerobic solubilization of food waste. Waste Manag. **29** (12), 2950–2955 (2009)
26. Hofenk, G., Lips, S.J.J., Rijkens, B.A. and Voetberg, J.W.: Two-phase anaerobic digestion of solid organic wastes yielding biogas and compost. Final EC-Report No. ESE-ER-040-NL, IBVL, Wageningen (1984)
27. Cecchi, F., Traverso, P.G., Cescon, P.: Anaerobic digestion of organic fraction of municipal solid wastes—digester performance. Sci. Total Environ. **56**, 183–197 (1986)
28. Mata-Alvarez, J., Llabres, P., Cecchi, F., Pavan, P.: Anaerobic digestion of the Barcelona central food market organic wastes: experimental study. Biores. Technol. **39**(1), 39–48 (1992)

Significance of the Presence of Asbestos in Construction and Demolition Wastes in India

Richa Singh, J. M. Vivek, Bakul Rao and Shyam R. Asolekar

Abstract Construction and demolition (C&D) wastes, typically referred to as "debris", is continuously generated and is being seen as an urban nuisance in the last three decades in the developing world, particularly in India. The unrestricted use of asbestos has led to the generation of high levels of friable asbestos, which is directly linked to mesothelioma, a form of lung cancer due to asbestosis. In the Indian context, exposure to asbestos can be attributed to asbestos mining, asbestos cement (AC) industries, asbestos product manufacturing and processing units and renovation and demolition of old asbestos-cemented building roofs and other insulation materials. A major portion of C&D waste comes from the demolition of old buildings. The present study deals with the analysis of current patterns of asbestos production, import and usage, thus examining, quantifying and forecasting asbestos-containing C&D wastes. Weathered 30-year-old AC roof sheets and pipes were analysed using analytical techniques such as scanning electron microscope (SEM) fitted with EDX prime energy-dispersive analysis system. The study points out that the asbestos fibres are encapsulated in cement matrix but the hazardous characteristic of the chrysotile is not altered, and this can be a matter of great concern as these fibres can be released into the atmosphere due to demolition activity or ageing.

R. Singh · J. M. Vivek · S. R. Asolekar (✉)
Centre for Environmental Science and Engineering, Indian Institute of Technology Bombay, Powai, Mumbai 400076, India
e-mail: asolekar@iitb.ac.in

R. Singh
e-mail: richasingh@iitb.ac.in

J. M. Vivek
e-mail: 144180005@iitb.ac.in

B. Rao
Centre for Technology Alternatives for Rural Areas, Indian Institute of Technology Bombay, Powai, Mumbai 400 076, India
e-mail: bakulrao@iitb.ac.in

A. S. Kalamdhad et al. (eds.), *Advances in Waste Management*,
https://doi.org/10.1007/978-981-13-0215-2_21

Keywords Asbestos · AW · ACM · ACW · Carcinogenic · Chrysotile
C&D wastes · Fibres · SEM

1 Introduction

The rapid industrialization and infrastructural development in Asia during the last few decades have resulted in the difficult task of finding means to manage the tremendous amount of C&D wastes generated. The developmental expansion is often derailed when a disaster strikes, causing causalities, loss of property and livelihood, leading to economic and social pressures [1]. As the environmental impacts from C&D wastes are gradually becoming a major bottleneck in the urban solid waste management, it has become a vital need for the environmental regulators to understand the C&D waste generation and handling. This will facilitate in developing accurate statistics and thus creating sustainable techniques to manage these construction wastes.

In this context, hazardous entities in C&D wastes such as asbestos have to be addressed and proper management solution should be put into practice. Asbestos is considered as an "industrial resource", which has historically been used in manufacture of around 3000 products due to its durability, good tensile strength and associated flexibility, low conductivity of heat and electricity, good mechanical thermal stability and high absorbance capacity [2, 3]. In European nations, it has been banned for most of its uses because of its unambiguous links to deleterious health impacts in the form of "mesothelioma" and "lung fibrosis", but it is still widely used in Asian countries and other nations because it is effective, yet relatively inexpensive. As per Indian bureau of mines, asbestos had an apparent demand that was estimated to be 393,000 tonnes in 2011–12 and is expected to touch 605,000 tonnes in 2016–17 (9% growth rate as per the report of the Working Group for 12th Plan) [4]. The unique mineralogical characteristic of asbestos minerals is their morphology or their crystallization property to form polyfilamentous bundles of fibre. The single fibrils developing into bundles of fibres tend to have a property of forming elongated structures having small diameter. The expansion of the long fibres is along an axis parallel to the entire bundle length. The fibres get moulded in rare conditions when the crystallization takes place for certain particular silicate minerals in numerous bundles of compact, flexible fibrils which resemble vegetable fibres [5]. Chrysotile (hydrated magnesium silicate, $3MgO_2$ SiO_2 $2H_2O$) accounts for nearly 95% of all the asbestos that is manufactured and widely used in diverse industrial applications, owing to its properties such as easily woven, easy to blend in and add to other inorganic substances and polymeric complexes to develop composite substances or products [6, 7].

Besides its useful properties, it is an established carcinogen [8]. The major havoc arises when the asbestos-containing materials (ACMs) undergo the dry cutting with abrasive equipment when an old building is being demolished. As per the definition of World Health Organization (WHO), the existing regulations focus on long asbestos fibre (LAF) (Length: $L \geq 5$ μm, Diameter: $D < 3$ μm and L/D ratio > 3). Nevertheless, in a recent study on air quality of buildings having asbestos-containing materials (ACMs), the air samples had a concentration of ≥ 10 fibres L^{-1}, mainly, short asbestos fibres (SAFs) [9]. Exposure to these microscopic asbestos fibres occurs through inhalation principally from contaminated air in the working environment, as well as from ambient air in the vicinity of point sources, or air from indoor locations (houses and buildings) [10]. Lack of adequate identification technology has led to the massive dumping of asbestos-containing C&D wastes along with the typical construction wastes or wastes generated from households. This may lead to long-term public health hazards. The need of the hour is in addressing the huge volume of untreated or unprocessed asbestos wastes (AWs) in the developing countries. The major obstruction is the inadequate waste treatment systems in such countries. Also, there is no stringent regulation for recycling and final disposal of C&D wastes in developing countries. Asbestos demand is massive and is currently approximated to be 2,000,000 tons per year globally, which was an alarming development over the last 120 years [3]. In India, activities such as asbestos mining and processing, asbestos cement (AC) manufacturing, restoration and demolition of the age-old asbestos contained cement roofs or any other constructions as well as repair and dismantling of various asbestos-containing equipment, both mechanical as well as electrical, open up pathways of exposure to asbestos. Eventually, the workers involved in these above-mentioned activities as well as others getting exposed to these friable asbestos often encounter severe damage to the lung and respiratory tract [2]. There are a number of studies conducted by various researchers which has advocated the hazardous impacts of asbestos on living beings. Studies have identified stabilization processes which can minimize the threats of asbestos-containing waste (ACW) by entrapping the friable material in a composite matrix of cement or resin. Few studies have reported treatment techniques which can alter the fibrous nature of asbestos. This will help in transforming the fibre matrix into a substance of inert nature [11].

2 Potential Health Hazards

Asbestos waste is generally categorized into two types: one tailings originating from the mining of asbestos as well as the process of manufacturing asbestos-based products such as roofing sheets and the other generated from the demolition of the buildings/equipment constructed out of materials containing asbestos [12]. Furthermore, the health hazard by ACWs generated at the end of life of buildings/ equipment, or after natural calamities like earthquakes often not given due consideration. Simultaneously, heaps of AW generated from mining of asbestos as well as construction activities are still discarded unscientifically due to lack of proper management and supervision. There are various reasons for these actions which may include the absence of ample management from the local regulatory bodies and inadequate treatment technologies [13]. As per the reports of WHO (2014), on an average, 107,000 people die every year from diseases occurring due to exposure to asbestos which includes mesothelioma, lung cancer and asbestosis. Asbestos is a well-known dust carcinogen and should be taken very seriously by building deconstructing individuals or agencies. Many studies have claimed for the carcinogenicity of asbestos fibres [14–16]. Asbestosis is a condition of diffuse interstitial fibrosis caused to the lungs triggered by inhalation of dust particles of asbestos. It is assumed to be related to the pleural fibrosis, but the pleural disease cannot be believed to constitute the asbestosis condition alone [17]. The amphiboles constitute major types of asbestos which include chrysotile (white), crocidolite (blue) and amosite (brown). All these types of asbestos minerals have the capacity to induce asbestos-related diseases, but the amphiboles are considered to be more carcinogenic than the chrysotile form of asbestos mineral.

3 Sources of Asbestos Wastes Generation

India is rising as the primary consumer of asbestos after China, whereas the developed world has imposed ban on most of its uses. There are around 25 asbestos mines in India, which are in vigorous operation. Approximately 3000 tonnes/month is the production rate of these asbestos mines in India. In addition, the quantity of asbestos imported from Canada to India is nearly more than 70% [18]. As per US geological survey report (2003) in a span of 100 years (1900–2003), around 182.2 million metric tonnes of asbestos was mined out. In present epoch, the major producers and exporters of asbestos are Russia, China, Brazil and Kazakhstan. The main targeted export countries are the low- and middle-income countries, where the health, safety and environment regulations are weak or do not exist [19]. Asbestos imports to India are in excess of 300,000 metric tons/annum as per US geological survey report [20]. The asbestos cement industry is the major consumer of asbestos in India. From 2010 to 2011, the consumption of asbestos dropped by 36% in India. 407,000 MT was the asbestos consumption in 2010 which declined to 303,000 MT in 2011. This can be owed to the growing awareness among public about the hazardous nature of asbestos. This can be assumed as a progressive step towards asbestos management, yet this is not enough to minimize the potential health hazards due to asbestos fibres [21]. This paper focuses mainly on C&D wastes, as infrastructure sector is often undergoing rapid transit to consume more and more resources and produce magnanimous quantities of wastes. The average lifetime of a building is 40–50 years, and the old-age building particularly in developing countries like India had employed low-cost lightweight durable asbestos and asbestos-containing construction materials during the Victorian era. As stated earlier, since the life has expired for such buildings, it is expected that the dismantling and disposal of such buildings will give rise to asbestos pollution. Moreover in the latter half of the twenty-first century, it has become a common practice to venture upon the new innovative lightweight sustainable materials for construction thus

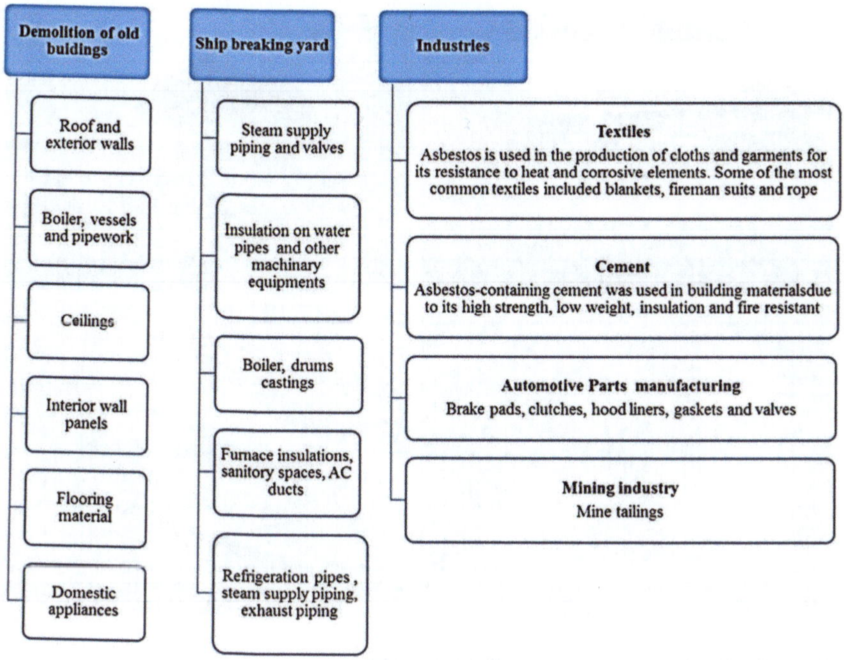

Fig. 1 Sources of asbestos wastes from different sectors

using asbestos for AC cement products. Thus, new avenue of pollution has been opened upon by this product. Its disposal at a later stage is going to be an issue of serious concern. Huge volumes of AWs are produced while old buildings are demolished, creating hazardous waste debris. In the construction industry, asbestos is generally found in installed products such as cement corrugated sheets, pipe insulation, floor tiles, cement pipe and sheet, roof and exterior walls and shingles, boiler vessels, interior wall panels. These days, asbestos is widely used as a major component in corrugated roofing sheets. Workers get exposed to carcinogenic fibres during the removal of ACM and the renovation and maintenance of buildings and structures containing asbestos. Various sources of asbestos waste generation in India can be broadly classified into three categories (Fig. 1):

- Construction and demolition of old buildings;
- From industrial operations; and
- From ship dismantling.

All asbestos slates are manufactured with a composition of about 90% cement and 10% chrysotile (white asbestos) [22]. That is why a huge amount of ACW is produced during demolition of asbestos roofs (Fig. 2). The huge quantum of asbestos consumption in various industries such as textile, cement and automotive parts manufacturing plants also produce a considerable amount of waste residues which is termed as asbestos-containing wastes. Wastes majorly include asbestos

Fig. 2 Asbestos in construction and demolition wastes

residues, discarded asbestos, used asbestos break liner, asbestos cement, corrugated sheets. In a study conducted by Dave and Becket [21], it was reported that there are approximately 673 small-scale asbestos mining and milling units and 33 large-scale asbestos manufacturing plants (17 AC product manufacturing plants and 16 other than AC product plants) in India. A considerable fraction of asbestos-containing wastes is generated by ship recycling and dismantling sector also [23].

Production and processing of asbestos industry are organized as follows: the initial step of the process is mining of the asbestos ore, followed by milling to ensure a homogeneous product which is then distributed to other primary units/industries. In the primary units/industries, the raw asbestos fibre is processed further and modified to produce an intermediate or finished good/product. A final processing of the product may be required to obtain the finished product. These finished products are then brought to the market and put for sale to consumer industries which may then apply, mount/install or utilize the obtained product without any further alteration. In all of the above-mentioned activities, there are chances of release of enormous amount of asbestos fibres in the working environment. The unique properties of asbestos as discussed earlier make it a preferable commercial product in regions where the production is in huge quantities. Asbestos has been used for making building materials which are fireproof which includes products like roof sheets, cement pipe or ceiling tile and other long-lasting cement products and insulation materials [21]. An illustrative representation of source-classified AW and ACM has been presented in Fig. 3.

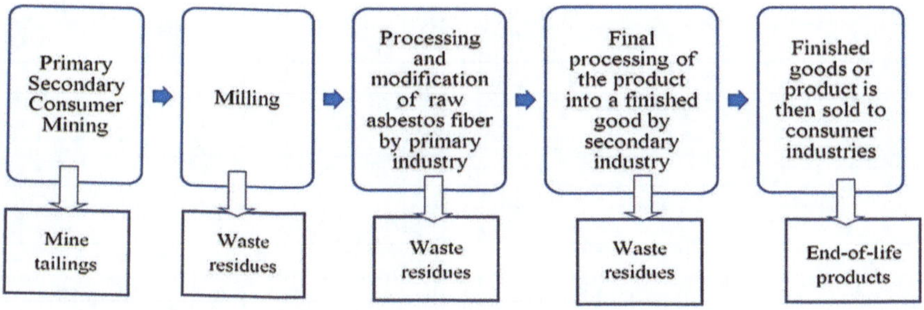

Fig. 3 Use of asbestos products in primary, secondary and consumer industries

4 Analysis of Construction Materials and Results Obtained

Asbestos cement roof sheets and pipes were analysed by using scanning electron microscope, and quantification of chrysotile present was estimated for calculating the quantum of asbestos-containing construction and demolition wastes. The sample was collected from demolition sites. The collected sample was crushed uniformly using a hammer. A very small amount of crushed sample was subjected to ultrasonic treatment for 1 min allowing a uniform dispersion of the fibres. A single drop was then pipetted onto a carbon film and allowed to dry. The sample was coated with titanium for around 200 s and observed under Field Emission Gun-Scanning Electron Microscope (FEG-SEM) fitted with an Energy-dispersive X-ray (EDX) analysis system.

In an EDX system, the electrons are made to focus into a small area of the asbestos fibre. The generated electrons in the beam interact with the targeted material inside the probe which leads to the emissions of X-ray. Some of these emissions are caused because of the interruption of the electrons in the atomic shells of an atom. This leads to the release of a quantum of energy in the form of X-ray which is distinctive feature of a particular element. Thus, the transitions of electrons happen within the atom. A graph of X-ray energy (in KeV) versus frequency of occurrence is plotted on the basis of EDAX technique of measuring energy of X-rays produced. The EDAX spectrum is characteristic of the number of typical X-ray peaks which is associated with the compositional elements of the matrix under analysis.

The fibres in all the bulk samples of construction material (asbestos roof sheets) were found to have the distinctive morphology of “chrysotile asbestos” (Fig. 4). When viewed at higher magnification, the extremely fine fibrils exhibited the peculiar “tubular structure” which is usually associated with the chrysotile fibres. There was no indication of alteration to the surface. The single fibres when subjected to analysis by EDX system showed the chemistry which was comparable to the reference standard of fibres of chrysotile studied in previous literature (Fig. 5). It can be inferred that the fibres are encapsulated in cement matrix but the hazardous

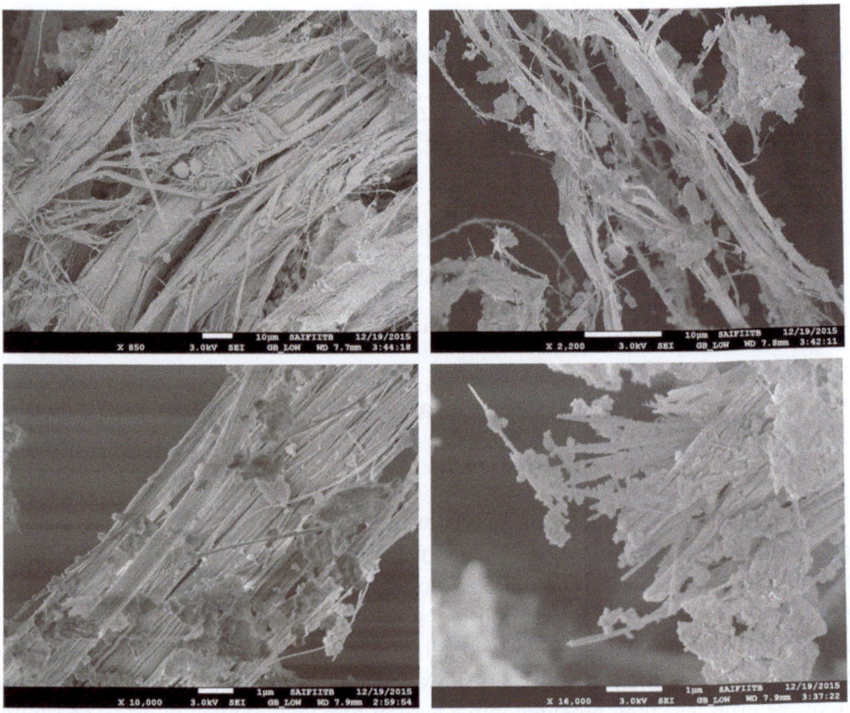

Fig. 4 Chrysotile fibre bundle projecting from a broken segment of the asbestos cement sample (scanning electron microscopy image at magnifications 850, 2200, 10,000 and 16,000X). Fine cement particles and larger pieces of matrix material can also be seen

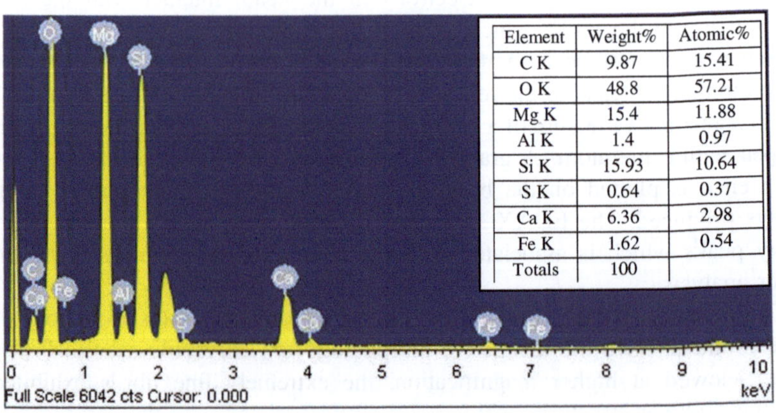

Element	Weight%	Atomic%
C K	9.87	15.41
O K	48.8	57.21
Mg K	15.4	11.88
Al K	1.4	0.97
Si K	15.93	10.64
S K	0.64	0.37
Ca K	6.36	2.98
Fe K	1.62	0.54
Totals	100	

Fig. 5 EDS peaks and quantitative elemental analysis of construction material sample

characteristics of the chrysotile are not altered, and this can be a matter of great concern if these fibres are into atmosphere due to demolition activity or due to ageing.

The analysis undertaken showed that the asbestos cement contained fibres of chrysotile and can potentially release chrysotile asbestos fibres in the environment when adequately disturbed. However, it is also reported that the Portland cement-based products like concrete and mortars can undergo deterioration due to attack from chemical agents, stresses by mechanical effects such as freezing–thawing, transportation, stockpiling, excessive loading and unloading cycles, exposure to natural forces like wind, moisture, heat or other environmental disturbances. Similarly, the fibre-cement corrugated made of asbestos fibres can also undergo deterioration which may lead to the release of these fibres [24]. The leaching of water can also lead to the weakening of the cementitious matrix of fibre-cement corrugated sheets. Few literature studies also suggest the leaching effect of Portland cement-based materials. It is a well-recognized fact that this mechanism of corrosion of the system can lead to a weakened composite structure.

It is evident from the analysis that a considerable amount of chrysotile is being used in the manufacture of cement corrugated sheets. These sheets (or other construction material made up of asbestos) when demolished after completion of their lifespan or due to climatic impacts will release hazardous chrysotile fibres into the environment and generate a massive quantum of wastes containing asbestos which comes under construction and demolition (C&D) wastes; considered as a major urban nuisance in the present era. Therefore, it has become very imperative to device various methods through which the exposure to the workforce to hazardous asbestos fibres could be minimized as well as for various technologies which can be used for conversion of asbestos-containing wastes into some useful products (construction products, as the fibres have high tensile strength and durability). Quantitative analysis of the sample showed the presence of major components such as magnesium, silicon, aluminium, calcium and iron.

5 Inventorization of Asbestos-Containing C&D Wastes in India

Method for predicting and quantifying the asbestos wastes generated from C&D sector involved review of the literature presenting patterns of asbestos-apparent consumption as well as review on documents from UN agencies, national agencies and NGOs on asbestos production and usage. The inventorization was mainly done with regard to wastes generated by demolition activity. For ACW generated from demolition of old buildings, the quantum of ACW was calculated by taking into account the apparent consumption of chrysotile asbestos in India. The consumption data were collected from US geological survey. Assuming that all the asbestos were utilized to make construction materials (considering the asbestos content as 20% in the bulk material), the asbestos wastes were generated at the end of the service life (considering 15 years adopted from the literature). The quanta of AW produced in the region can be calculated by Eq. (1) given by Li et al. [12].

$$Q_n = \frac{C_{n-15}}{k} \tag{1}$$

where Q_n quantity of asbestos waste in year n
$C_{n\text{-}15}$ consumption of asbestos in year n-15
k an approximate rate of asbestos-containing materials

The quantity of asbestos-containing hazardous wastes can be projected in the coming years using the above-mentioned assumptions (presented in Table 1). It is found that there is regular increase in the consumption of asbestos from the year

Table 1 Asbestos consumption in India and projection of asbestos wastes generated

Year	Production (in MT)	Import	Export	Apparent consumption	Rate of ACM	Year of waste generation	Quantity of asbestos waste generated (in MT)
1950	221	10,957	8	11,160	20	1965	558,000
1960	1,711	21,967	26	652	20	1975	32,600
1970	10,056	39,766	30	49,792	20	1985	2,489,600
1975	20,312	41,514	–	61,826	20	1990	3,091,300
1980	33,716	63,176	–	96,892	20	1995	4,844,600
1985	29,450	78,075	–	107,525	20	2000	5,376,250
1990	26,053	93,165	254	118,964	20	2005	5,948,200
1995	23,844	91,909	14	115,739	20	2010	5,786,950
1996	27,180	84,378	275	111,283	20	2011	5,564,150
1997	25,537	83,356	282	108,611	20	2012	5,430,550
1998	20,000	109,036	348	128,688	20	2013	6,434,400
1999	21,000	115,220	172	136,048	20	2014	6,802,400
2000	21,000	124,433	403	145,030	20	2015	7,251,500
2001	21,000	130,291	1129	150,161	20	2016	7,508,050
2002	18,000	150,461	169	168,292	20	2017	8,414,600
2003	19,000	175,581	2548	192,033	20	2018	9,601,650
2004	18,000	–	–	190,000	20	2019	9,500,000
2005	19,000	–	–	255,000	20	2020	12,750,000
2006	20,000	–	–	240,000	20	2021	12,000,000
2007	21,000	–	–	302,000	20	2022	15,100,000
2008	20,000	–	–	310,000	20	2023	15,500,000
2009	–	–	–	322,000	20	2024	16,100,000
2010	–	–	–	407,000	20	2025	20,350,000
2011	19,000	302,915	−112	321,803	20	2026	15,150,000
2012	–	–	–	473,000	20	2027	23,650,000

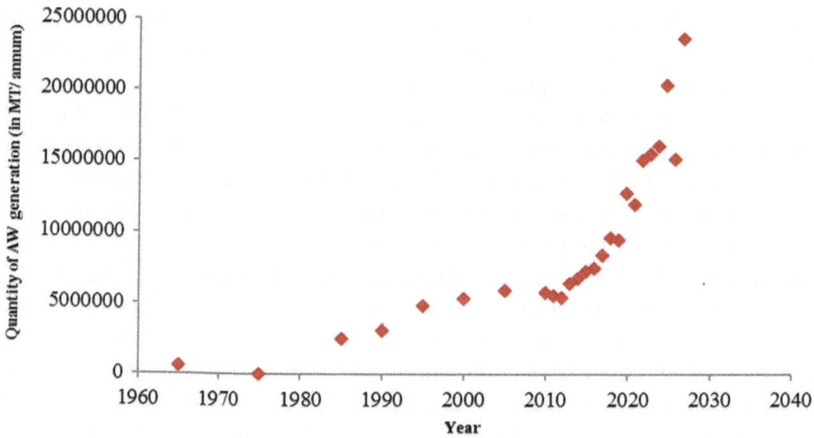

Fig. 7 Quantity of asbestos waste projected in coming years

1950 to 2012 [20, 25–45], thereby a considerable increase in generation of asbestos wastes too.

On the basis of asbestos consumed per annum in India (data obtained from US geological survey), the quantity of asbestos wastes has been projected as illustrated in Fig. 7. There is a continuous increase in the asbestos wastes produced with increasing consumption. In the year 2015, it has been estimated that the quanta of ACW generated is 7,251,500 MT.

6 Management of Asbestos-Containing C&D Wastes

Recycling of wastes and converting them into secondary raw materials in concrete industry is one of the basic principles of the circular economy and green concrete concept. This has become a major area of research due to the adverse environmental impacts associated with concrete industries and growing global interests in sustainable approach of resource consumption. Recycling asbestos-containing C&D wastes as a secondary raw material (SRM) as a replacement for cement in concrete is the need of the hour, due to its physical properties such as high tensile strength and durability [46]. This has shown viable recycling prospects for this range of hazardous wastes. In thermal treatment studies, simultaneous destruction of the fibres of asbestos and subsequent release of reactive MgO were reported. This method is found to be highly profitable in terms of energy usage as well as conserving natural resources. In few studies, ACW was used in the production of magnesium phosphate cements. The samples were mixed uniformly with magnesium carbonate and subjected to thermal conditions of 1100–1300 °C. The treatment of ACW consisted of the complete and permanent conversion of the hazardous fibres of asbestos and the possibility of recycling and reusing of these materials as SRM in various industrial applications [47]. Microwave thermal treatment studies on asbestos have exhibited the conversion of the fibres into non-hazardous silicate phases. As reported in some literature, the microwave-enabled inertization of asbestos-containing waste (ACW) and its use in ceramic bricks, porcelain stoneware tiles and porous single-fired wall tiles has also been conducted [48]. The energy-intensive milling process of asbestos cement wastes results in powdered product which can effectively be reused for making various range of building products [49]. Destruction or immobilization of hazardous characteristic of AW through chemical–physical transformations, such as mechano-chemical treatment, vitrification process, recrystallization process, hydrothermal technique, and the recycle/reuse of the altered material as SRM ensures a lesser environmental footprint and minimizing the use of naturally occurring raw materials. Asbestos-containing material (ACM) being transformed into SRM will offer sustainable recycling methods. There are plenty of literature available wherein inert ACM-derived SRM used for multitude purposes.

- The product formed as a result of thermal transformation of AC may be utilized as a secondary raw material of great importance, which has Mg-rich clinker-like chemical properties [50].
- The pozzolanic property of treated asbestos-containing wastes (roof sheets) was studied, and it was found suitable to be used as construction material [49].
- Study done on asbestos tailings claimed that they lack heavy metals and hence less toxic pollution. Therefore, these tailings can be utilized as aggregate material for building purposes [51].
- A variety of building materials which include glass, glass-ceramics, ceramic pigments, clay bricks, plastic materials has been successfully made out of products of transformation of cement-asbestos.
- Similarly, asbestos tailings can be recycled and can be applied as strengthening filler material which can enhance the mechanical properties of polypropylene [52].
- Inactivating AC wastes by thermo-chemical processes and utilizing the mineral residues in a wide range of products made out of cement were conducted by few researchers. It was found that strength parameters like mechanical strength and tensile stress of products incorporated with the SRM up to 10 %wt.

Analogous works on conversion of different types of hazardous wastes into advanced construction material are reported by many researchers. One such classic work involves the use of jarosite wastes generated by zinc metallurgy industries which was blended with other waste products such as coal combustion residue and marble processing residues and finally converted into construction bricks [53].

7 Conclusion

As reported in this study, there is a continuous increase in the asbestos wastes production, with an increasing consumption rate over the years. In the year 2015, it has been estimated that the quanta of ACW generated is 7,251,500 MT in India. This study revealed that C&D wastes account for a major portion of the asbestos waste in India. Another large fraction of the asbestos waste comes from the ship dismantling industry. The source of a major portion of the asbestos waste is as yet unidentified. A proper methodological framework is needed to provide a practical assistance to the construction and demolition waste managers and handlers to minimize the potential risk of exposure to asbestos waste. When buildings are torn down, workers, neighbours and the environment may be exposed to hazardous fibres. In European countries, there are regulations and norms for collection, disposal and recycling of C&D wastes but in India and other developing nations there is lack of solid waste regulations particularly for dealing with hazardous entities of C&D wastes, due to which a major portion of asbestos wastes remains unidentified and hence unattended. Therefore, it is advisable to label the buildings and/or its components prior to the activity of demolition. Sampling in a large pile of heterogeneous building debris is much more complicated as compared to a standing structure. In addition, sampling before demolition activity will help in identifying and removing specific constituents (such as siding with lead paint, asbestos) that may be designated as hazardous wastes.

A major portion of asbestos wastes in India remains unidentified, and rest of the wastes fraction, which is generated from ship dismantling sector and asbestos product manufacturing industries, is disposed through landfilling as per the CPCB norms and Hazardous Waste (Management, Handling and Transboundary) Rules-2008. Keeping in mind the integrated approach of wastes management, it is required to work on various recycling technologies for asbestos wastes in order to process and convert them into some constructive material and minimize the health risk to the population.

Acknowledgements Authors acknowledge the co-funding from Department of Science and Technology, New Delhi, and Indian Institute of Technology Bombay for this work.

References

1. Bisarya, R.K., Puri, S.: The Bhopal gas tragedy—a perspective. J. Loss Prev. Process Ind. **18** (4), 209–212 (2005)
2. Subramanian, V., Madhavan, N.: Asbestos problem in India. Lung Cancer **49**, S9–S12 (2005)
3. Ross, M., Langer, A.M., Nord, G.L., Nolan, R.P., Lee, R.J., Van Orden, D., Addison, J.: The mineral nature of asbestos. Regul. Toxicol. Pharmacol. **52**(1), S26–S30 (2008)
4. Indian Bureau of Mines: Indian Mineral year book-2014, Asbestos, pp. 2–10 (2015)
5. Habashi, F.: Chemisorption of organic dyes on chryostile asbestos and the toxicity issue. Eur. J. Minineral Process. Environ. Prot. **1**, 1–9 (2001)

6. Thompson, S.K., Mason, E.: Asbestos: mineral and fibers. Chem. Health Saf. **9**, 21–23 (2002)
7. Foresti, E., Fornero, E., Lesci, I.G., Rinaudo, C., Zuccheri, T., Roveri, N.: Asbestos health hazard: a spectroscopic study of synthetic geoinspired Fe-doped chrysotile. J. Hazard. Mater. **167**(1), 1070–1079 (2009)
8. WHO, Neira, M.: Chrysotile asbestos. World Health Organization (2014)
9. Boulanger, G., Andujar, P., Pairon, J.C., Billon-Galland, M.A., Dion, C., Dumortier, P., Brochard, P., Sobaszek, A., Bartsch, P., Paris, C., Jaurand, M.C.: Quantification of short and long asbestos fibers to assess asbestos exposure: a review of fiber size toxicity. Environ. Health **13**(59), 10–1186 (2014)
10. Paglietti, F., Malinconico, S., Della Staffa, B. C., Bellagamba, S., De Simone, P.: Classification and management of asbestos-containing waste: European legislation and the Italian experience. Waste Manage. (2016)
11. Plescia, P., Gizzi, D., Benedetti, S., Camilucci, L., Fanizza, C., De Simone, P., Paglietti, F.: Mechanochemical treatment to recycling asbestos-containing waste. Waste Manage **23**(3), 209–218 (2003)
12. Li, J., Dong, Q., Yu, K., Liu, L.: Asbestos and asbestos waste management in the Asian-Pacific region: trends, challenges and solutions. J. Clean. Prod. **81**, 218–226 (2014)
13. Sigon, F., Servida, A., Grassi, S.: Process for the hydrothermal treatment of asbestos and/or asbestos containing materials in supercritical water and relative production plant. US Patent, 2006/0149118 A1 (2006)
14. Rahman, Q., Viswanathan, P.N., Zaidi, S.H.: Some new perspectives on the biological effects of asbestos. Environ. Res. **14**(3), 487–498 (1977)
15. Gidarakos, E., Anastasiadou, K., Koumantakis, E., Nikolaos, S.: Investigative studies for the use of an inactive asbestos mine as a disposal site for asbestos wastes. J. Hazard. Mater. **153** (3), 955–965 (2008)
16. Scarselli, A., Corfiati, M., Di Marzio, D.: Occupational exposure in the removal and disposal of asbestos-containing materials in Italy. Int. Arch. Occup. Environ. Health, 1–9
17. Rudd, R.: Asbestos and the lung. Medicine **36**(5), 261–264 (2008)
18. Ramanathan, A.L., Subramanian, V.: Present status of asbestos mining and related health problems in India. a survey. Ind. Health **39**(4), 309–315 (2001)
19. Frank, A.L., Joshi, T.K.: The global spread of asbestos. Ann. Global Health **80**(4), 257–262 (2014)
20. Virta, R.L.: Asbestos in Metals and Minerals: U.S. Geological Survey Minerals Yearbook 2001, vol. 1, pp. 8.1–8.7 (2003)
21. Dave, S.K., Beckett, W.S.: Occupational asbestos exposure and predictable asbestos-related diseases in India. Am. J. Ind. Med. **48**(2), 137–143 (2005)
22. Kim, S.Y., Kim, Y.C., Kim, Y., Hong, W.H.: Predicting the mortality from asbestos-related diseases based on the amount of asbestos used and the effects of slate buildings in Korea. Sci. Total Environ. **542**, 1–11 (2016)
23. Hiremath, A.M., Tilwankar, A.K., Asolekar, S.R.: Significant steps in ship recycling vis-a-vis wastes generated in a cluster of yards in Alang: a case study. J. Clean. Prod. **87**, 520–532 (2015)
24. Dias, C.M.R., Cincotto, M.A., Savastano, H., John, V.M.: Long-term aging of fiber-cement corrugated sheets—the effect of carbonation, leaching and acid rain. Cement Concr. Compos. **30**(4), 255–265 (2008)
25. Virta, R.L.: Asbestos, U.S. Geological Survey Minerals Yearbook 1994, vol. 1, pp. 1-68.1-8.5 (1995)
26. Virta, R.L.: Asbestos, U.S. Geological Survey Minerals Yearbook 1995, vol. 1, pp. 1–7 (1996)
27. Virta, R.L.: Asbestos, U.S. Geological Survey Minerals Yearbook 1996, vol. 1, pp. 1–5 (1997)
28. Virta, R.L.: Asbestos, U.S. Geological Survey Minerals Yearbook 1997, vol. 1, pp. 8.1–8.5 (1998)

29. Virta, R.L.: Asbestos, U.S. Geological Survey Minerals Yearbook 1998, vol. 1, pp. 8.1–8.5 (1999)
30. Virta, R. L.: Asbestos, U.S. Geological Survey Minerals Yearbook 1999, vol. 1, pp. 8.1–8.6 (2000)
31. Virta, R.L.: Asbestos, U.S. Geological Survey Minerals Yearbook 2001, vol. 1, pp. 9.1–9.6 (2002)
32. Virta, R.L.: Asbestos, U.S. Geological Survey Minerals Yearbook 2002, vol. 1, pp. 8.1–8.6 (2003)
33. Virta, R.L.: Asbestos, U.S. Geological Survey Minerals Yearbook 2003, vol. 1, pp. 8.1–8.12 (2004)
34. Virta, R.L.: Asbestos, U.S. Geological Survey Minerals Yearbook 2006, vol. 1, pp. 8.1–8.6 (2005)
35. Virta, R.L.: Asbestos, U.S. Geological Survey Minerals Yearbook 2005, vol. 1, pp. 8.1–8.6 (2006)
36. Virta, R.L.: Worldwide Asbestos Supply and Consumption Trends from 1900 through 2003. U.S. Geological Survey, pp. 1–22 (2006)
37. Virta, R.L.: Asbestos, U.S. Geological Survey Minerals Yearbook 2006, vol. 1, pp. 8.1–8.7 (2007)
38. Virta, R.L.: Asbestos, U.S. Geological Survey Minerals Yearbook 2007, vol. 1, pp. 8.1–8.5 (2008)
39. Virta, R.L.: Asbestos, U.S. Geological Survey Minerals Yearbook 2009, vol. 1, pp. 8.1–8.6 (2009)
40. Virta, R.L.: Asbestos, U.S. Geological Survey Minerals Yearbook 2009, vol. 1, pp. 8.1–8.7 (2010)
41. Virta, R.L.: Asbestos, U.S. Geological Survey Minerals Yearbook 2010, vol. 1, pp. 8.1–8.5 (2011)
42. Virta, R.L.: Asbestos, U.S. Geological Survey Minerals Yearbook 2011, vol. 1, pp. 8.1–8.6 (2012)
43. Virta, R.L.: Asbestos, U.S. Geological Survey Minerals Yearbook 2012, vol. 1, pp. 8.1–8.7 (2013)
44. Virta, R.L.: Asbestos, U.S. Geological Survey Minerals Yearbook 2013, vol. 1, pp. 8.1–8.7 (2014)
45. Virta, R.L., Flanagan, D.M.: Asbestos, U.S. Geological Survey Minerals Yearbook 2014, vol. 1, pp. 8.1–8.8 (2015)
46. Viani, A., Gualtieri, A.F.: Preparation of magnesium phosphate cement by recycling the product of thermal transformation of asbestos containing wastes. Cem. Concr. Res. **58**, 56–66 (2014)
47. Dellisanti, F., Rossi, P.L., Valdrè, G.: Remediation of asbestos containing materials by Joule heating vitrification performed in a pre-pilot apparatus. Int. J. Miner. Process. **91**(3), 61–67 (2009)
48. Leonelli, C., Veronesi, P., Boccaccini, D.N., Rivasi, M.R., Barbieri, L., Andreola, F., Lancellotti, I., Rabitti, D., Pellacani, G.C.: Microwave thermal inertisation of asbestos containing waste and its recycling in traditional ceramics. J. Hazard. Mater. **135**(1), 149–155 (2006)
49. Colangelo, F., Cioffi, R., Lavorgna, M., Verdolotti, L., De Stefano, L.: Treatment and recycling of asbestos-cement containing waste. J. Hazard. Mater. **195**, 391–397 (2011)
50. Gualtieri, A.F., Veratti, L., Tucci, A., Esposito, L.: Recycling of the product of thermal inertization of cement-asbestos in geopolymers. Constr. Build. Mater. **31**, 47–51 (2012)
51. Xu, Linrong, Asbestos tailings as aggregates for asphalt mixture. J. Wuhan Univ. Technol. Mater. Sci. Ed. **26**(2), 335–338 (2011)
52. Zhai, W., Wang, Y., Deng, Y., Gao, H., Lin, Z., Li, M.: Recycling of asbestos tailings used as reinforcing fillers in polypropylene based composites. J. Hazard. Mater. **270**, 137–143 (2014)
53. Pappu, A., Saxena, M., Asolekar, S.R.: Solid wastes generation in India and their recycling potential in building materials. Build. Environ. **42**(6), 2311–2320 (2007)

Development of GIS-Based Optimization Method for Selection of Transportation Routes in Municipal Solid Waste Management

Sameer Singh and Sailesh N. Behera

Abstract The municipal bodies in Indian megacities are facing several problems to maintain a good management practice with the massive solid wastes generated at all corners of large cities. This unique study used inbuilt optimization techniques and self-modified algorithms of the network analyst tool in the platform of Geographical Information System (GIS) to propose the optimized paths for municipal solid waste (MSW) collections at Kanpur, India. The input parameters used in modeling exercise were geographical location of waste points, existing road network, proposed transport routes, and locations of the dumping yards. The zone with eighteen wards having thirty-three locations for collection and two transfer stations for disposal of MSW was used as a test bed for the simulation exercise. All the locations were modeled separately to find respective optimized collection routes for solid waste management. This study found the reductions in haul distance as $27.78 \pm 10.2\%$ for the selected network.

Keywords Municipal solid wastes · Route optimization · ArcGIS
Hauling distance · Waste collection points · Disposal sites

1 Introduction

In recent years, different optimization techniques such as network analysis tool, optimization software, and self-modified algorithms are frequently used by the researchers to solve the complex route network problems arising in collection and

S. Singh · S. N. Behera
Department of Civil Engineering, Shiv Nadar University, Gautam Buddha Nagar, Greater Noida, Uttar Pradesh 201314, India

S. N. Behera (✉)
Centre for Environmental Science and Engineering, Shiv Nadar University, Gautam Buddha Nagar, Greater Noida, Uttar Pradesh 201314, India
e-mail: sailesh.behera@snu.edu.in

A. S. Kalamdhad et al. (eds.), *Advances in Waste Management*,
https://doi.org/10.1007/978-981-13-0215-2_22

disposal of municipal solid waste (MSW) in urban environments [1–3]. The regulatory bodies, municipal authorities, and practicing engineers have been putting enormous efforts to improve the current scenario of municipal solid waste management system in which the cost of maintenance would be reduced [4]. From the recent report on MSW, it is clear that US $410 billion is spent every year across the world [5] for management of wastes of four billion tons MSW in both urban and rural sectors [6]. The cost of transportation management of MSW constitutes 50% of the total expenditure in the cities of developed countries and 85–90% in the cities of developing countries [7, 8]. During this decade, it becomes a challenge for the administrative authorities to optimize the resources meant for management of wastes in a city. Hence, necessary improvements in the existing practice are required to provide a system that can optimize the routine work of the municipal solid waste management to save the cost and time.

In the past, several studies have modeled the transportation paths of municipal solid waste management to propose the routes of collection from user ends to disposal sites that can reduce the cost of management [1, 2, 7, 9–13]. Specifically, Apaydin and Gonullu [10] used the GIS platform that utilized the Route View Pro™ software as an optimization tool in Trabzon city, Turkey, with area of 40 km^2. By applying this method, Apaydin and Gonullu [10] reported a reduction in distance meant for collection of wastes in the range from 4 to 59%. Karadimas et al. [11] modeled the transportation path using ant colony system algorithm at Attica's prefecture (a suburb of Athens) and reported a reduction in collection tour length by 25.6%. Singh et al. [2] proposed Dijkstra's algorithm-based optimization model at Kanpur city through testing for eight waste points connected in a mesh network, and the results obtained had shown a reduction in traveled distances in the range from 12 to 15%. Although the above-mentioned studies have proposed the informatics system based models for optimization of collection routes, finding the optimal path for the directed route network on user-friendly platform is lacking in literature. To fill the knowledge gap in the literature and propose a modeling technique that addresses the optimization method for collection routes in a larger city, this unique study has aimed at identifying the optimal routes for transportation of MSW in the platform of ArcGIS.

Similar to other cities in North India, Kanpur (26°26′52″ N, 80°20′46″ E) is facing several challenges for a proper solid management practice. To conduct this research, Kanpur was considered as the study domain, and out of six zones, the zone with eighteen wards having twelve locations for collection of MSW was treated as the test bed for the modeling exercise. The specific objectives of this study were: (1) map digitization and production of digitized information on GIS platform through various steps including base map generation and production of thematic layers; (2) identification of zones and collection points in the digitized map and later these digitized information to be used in modeling exercise; (3) setting up GIS-based modeling technique for route optimization from collection points to disposal sites meant for solid waste management; (4) model validation through relevant approach; and (5) model application with the same purpose from model setup expanding in other routes or collection paths.

2 Characteristics of the Study Area

2.1 *Existing Solid Waste Management Practice*

Kanpur is the 75th largest city in the world and also is the center of commercial and industrial activities, sprawling over an area of 605 km^2 in the state of Uttar Pradesh, India. The municipal corporation of Kanpur covers an area of 302 km^2 [14] and has the responsibility to manage 1600 tonnes of generated waste per day [15]. According to census 2011, the population of entire city was 5.03 million (605 km^2) which is increasing rapidly. The average rate of solid waste generation in municipality of Kanpur is about 0.623 kg per capita per day [16]. Kanpur city is administratively divided into 6 zones and 110 wards with a ward population ranging from 19,000 to 30,000. There are 203 active collection points in several wards of six zones [14]. A fleet of 132 vehicles and 3000 municipal workers is employed in collection and transport of the solid wastes generated in the city. These vehicles are only allowed to transfer the generated solid waste to an authorized site, which is situated in the periphery of the city [14]. The total expenditure for improving solid waste management was 43.25 crore during 2006–2011, and it is expected that this budget will be increased during subsequent years due to increase in quantity of generated solid wastes [17]. Table 1 provides the details of collection and transportation facilities in six zones. A total of 70 municipal trucks and 535 waste collecting rickshaws are collecting the daily generated waste in different wards of six zones in urban Kanpur.

2.2 *Population Growth in Kanpur*

The annual growth rate during 1951–1971 at city-level population was 1.6%. Similarly, the annual growth rate was 1.2% during 1971–1991. However, during 1991–2001 the growth rate was substantially higher compared to the previous

Table 1 Details of collection points and transportation facility in Kanpur

Zone name	Number of wards	Number of collection points	Municipal trucks	Rickshaw
Zone 1	18	41	10	55
Zone 2	18	42	14	90
Zone 3	18	24	10	110
Zone 4	18	20	10	70
Zone 5	19	33	14	130
Zone 6	19	43	12	80
Total	110	203	70	535

Source [14]

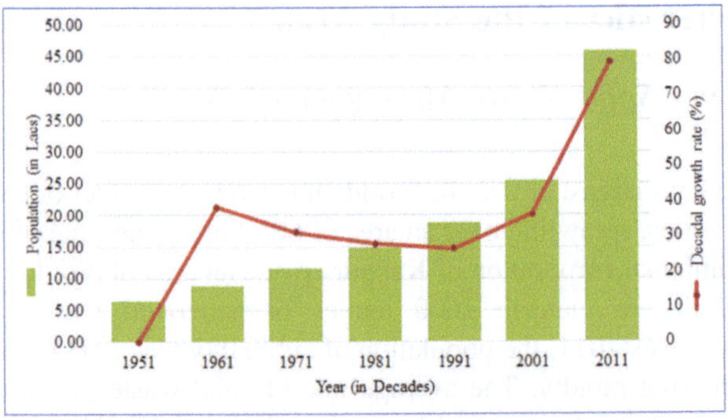

Fig. 1 Population growth versus decadal growth rate in Kanpur

decades. Hence, population data during 1991–2011 were considered to estimate the annual growth rate for this study. Figure 1 shows the trends of population data based on decadal scenario.

2.3 *Composition of Generated Solid Wastes*

The composition and characteristics of MSW are presented in Fig. 2 based on reported data from [15, 17, 18] during 1988, 1999, and 2014, respectively. The trend shows that quantity contributions of paper, textiles, plastics, and inert wastes increased substantially during 2014 compared to 1988 (see Fig. 2). The composition of waste was usually defined by the weight of solid waste and can be influenced by economic development, geographical location, energy sources, and climate.

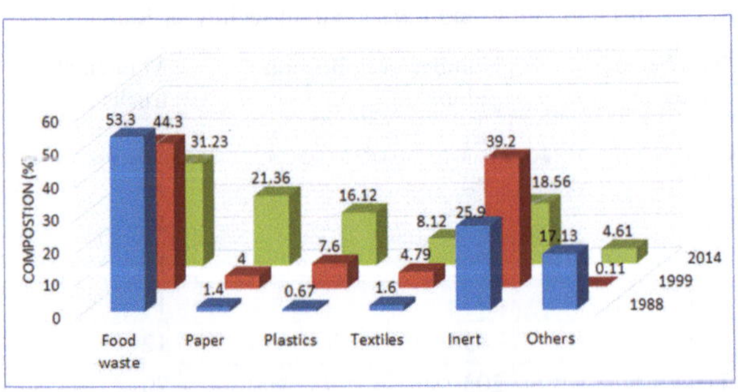

Fig. 2 Change in the solid waste characteristics of Kanpur

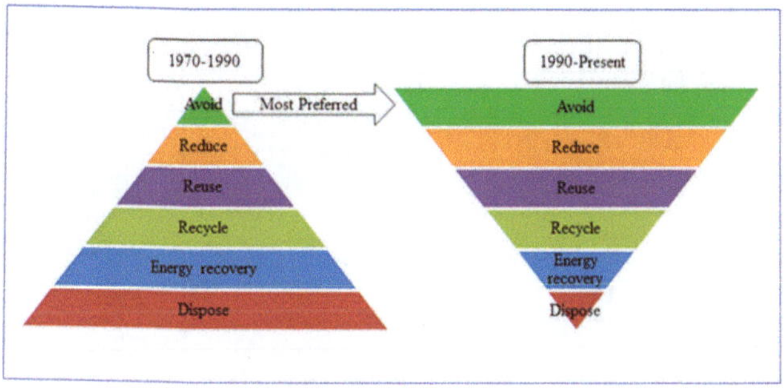

Fig. 3 Change in organization structures of the solid waste management pyramid

2.4 *Change in Solid Waste Management Trends*

During the late 1970s, the tendency of people of Kanpur toward the disposal of solid waste was to dump it on disposal sites with utmost priority. There were no efforts initiated to reduce, reuse, recycle, and treatment of wastes due to lack of awareness. However, during recent years, the hierarchy of waste management has been changed in a reverse order, as summarized in Fig. 3. Because of fast growth in population and land scarcity, the researchers and public bodies are searching for alternative methods on solid waste management that can reduce the volume of wastes from the generation points at street levels.

3 Methodology

In this study, the GIS-based methodology followed three steps for achieving the optimized outcomes. In the first step, the thematic layers for the location of collection points, transfer stations, administrative boundary, zonal boundaries, and road network consisting both major and minor roads were created in GIS interface, representing the current scenario and trends in the city (see Fig. 5). For this step, the location of waste collection points is taken as the centroid of ward by allocating the coordinates of ward area and symbolized by circular feature and in red color. The road network consists of both major and minor roads of study area which are symbolized by line features, in brown and red color simultaneously. The waste transfer station is marked as black color and symbolized by triangular features, assumed to be on the corner of zonal boundaries.

In the second step, the computations were carried out by using GIS-based network analyst tool, to identify the optimal route lengths in the study network. This tool contains inbuilt optimization techniques named as Dijkstra's algorithm [19], which is a greedy algorithm and solves the single-source shortest path problem when all edges have nonnegative weights. The evaluation of generated results

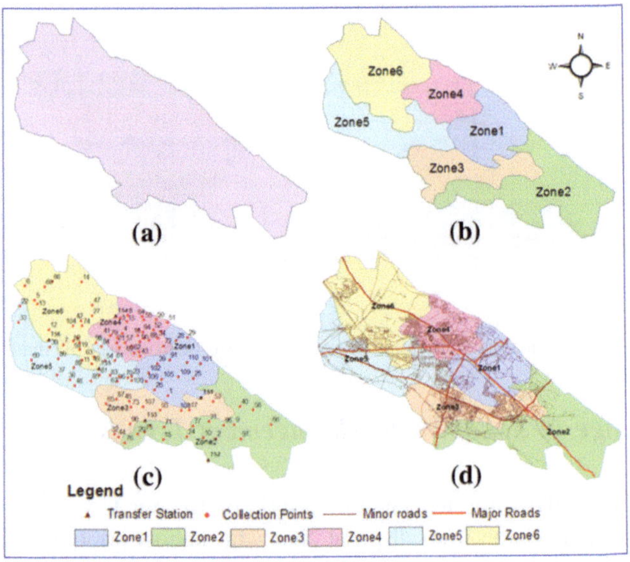

Fig. 4 **a** Kanpur administrative boundary. **b** Zonal boundaries. **c** Location of collection points and transfer stations. **d** Major and minor roads

obtained from inbuilt ArcGIS network analyst tool based on Dijkstra's algorithm approach has been performed (Fig. 4).

For optimization purpose, this tool was applied in this study, considering the starting node as '*s*' and the destination node as '*d*'. After selecting the 'solve feature' in the 'network analyst tool' of ArcGIS, the shortest route between the nodes was calculated, and subsequently, this information was displayed on the interface. In third step, the validation of generated results was verified manually by following the steps of Dijkstra's algorithm. The detailed description on all these steps is mentioned in the flowchart presented in Fig. 5.

3.1 Uniqueness of GIS for This Study

The GIS is a platform which allows its user to capture, store, manage, analyze, and manage a large volume of spatially or geographical reference data collected from various sources [7, 21]. The GIS enables the reader to visualize and interpret the data for better understanding of relationships and patterns in a user-friendly manner [12, 13]. The optimization technique within the ArcGIS is based on Dijkstra's algorithm [19] which estimates the shortest route from a single source to rest of the sources in a network. This algorithm is often used in routing and as a subroutine in other graph algorithm in a complex routing network; the algorithm identifies the optimized route between the sources vertex to every other vertex [20].

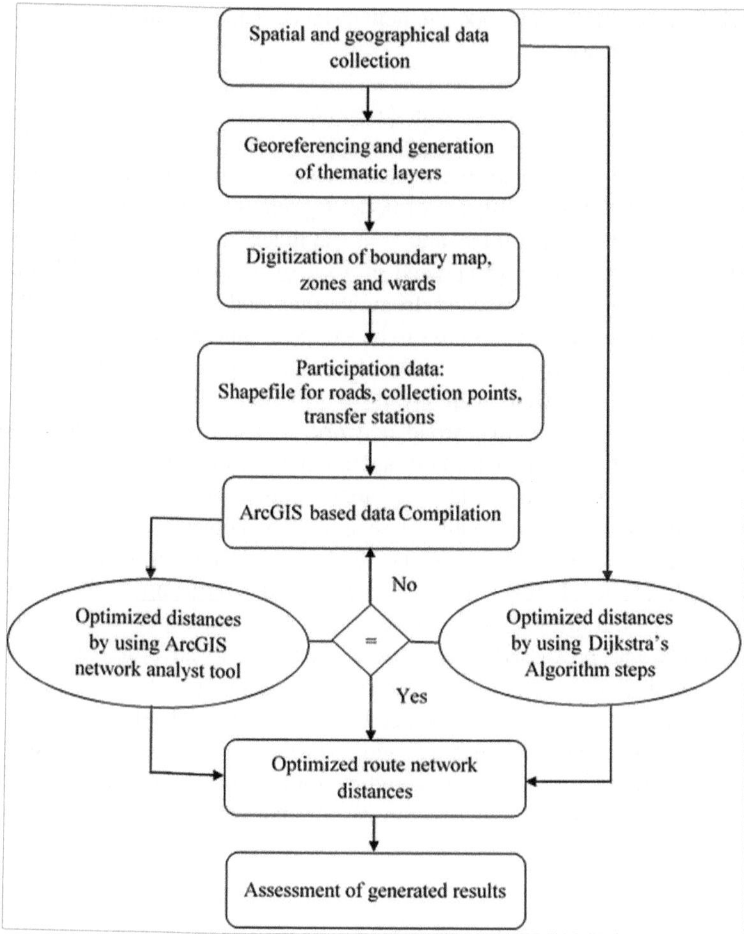

Fig. 5 Summary of sequential steps used as the methodology in this study to provide optimized haul distance of the selected network

3.2 *Inbuilt Optimization Techniques in Network Analyst Tool*

To find the optimized distance in a network, the algorithm follows various steps as follows: It first assigns the zero distance value to starting node '*s*', labels it as permanent node, and assigns a distance value of infinite to other nodes '*d*', marked them as 'temporary nodes.' After that, all other distance values are updated in set of temporary labeled nodes '*J*' that can be reached from the current node [22–24].

For each link among the network, the new distance value of nodes is updated using the following expression:

$$N_d = \text{Min}\left\{D_j, (D_i + C_{ij})\right\} \quad (1)$$

where N_d is the new updated distance, D_j is the distance of nearest node before updating, D_i is the current distance of node, C_{ij} represents the distance of existing link, 'i' is the index current node, and 'j' is the index of temporary node. By using Eq. (1), the new distances are updated in a set of temporary nodes. This process continues until all nodes reached from starting node, marked as permanently labeled, and the updating process has to be stopped. Otherwise, it further directs the updating process following Eq. (1) for the remaining nodes in a network. The primary data included were map layers of different utilities of the study area. The layers were boundary map, road network, location of collection points, and transfer station. The data contain both spatial and attribute information.

The following network having six nodes connected by directed lines with assigned distances gives the details of several steps between each iteration of the Dijkstra's algorithm (Fig. 6). In the specified network, the shortest route from the starting node (1) to all other nodes can be found by tracing back predecessors represented by bold arrows, while the cost of path is noted above the nodes.

Each node was processed exactly once according to an order that was being specified below. Node 1 (i.e., origin node) was processed first. A record of the nodes that were processed was kept, named it as Queue (Table 2). So, initially Queue = {1}. When node k was processed, the following task was performed: If the path's cost from the origin node to j could be improved including the vertex (k, j) in the path, then an update followed both of new cost and Predecessors [j] with k,

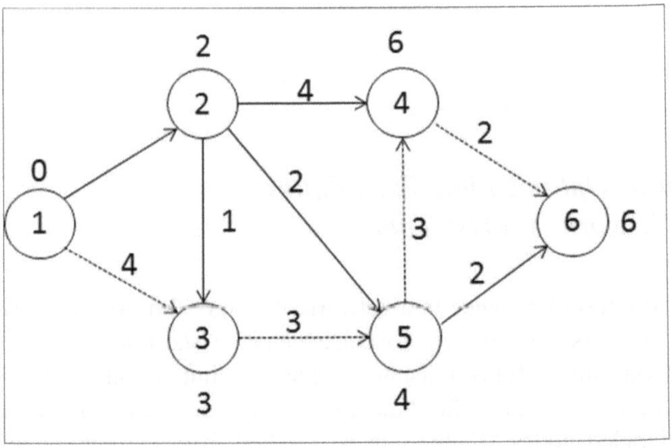

Fig. 6 An example that gives the details of several steps between each iteration of Dijkstra's algorithm [20]

Table 2 A record, named as queue, with all processed nodes

		Distance						Predecessors				
Queue	Next node	1	2	3	4	5	6	2	3	4	5	6
1	2	–	2	4	∞	∞	∞	–	–	–	–	–
1, 2	3	–	–	3	6	4	∞	–	2	2	2	–
1, 2, 3	5	–	–	–	6	4	∞	–	–	–	–	5
1, 2, 3, 5	4	–	–	–	6	–	6	–	–	–	–	–
1, 2, 3, 5, 4	6	–	–	–	–	–	6	–	–	–	–	–
1, 2, 3, 5, 4, 6	–	–	–	–	–	–	–	–	–	–	–	–

where j is any of the unprocessed nodes and Distance [] is the path's cost from the origin node to j. The next node to be processed was the one with the minimum Distance []; in other words, it was the nearest to the origin node among all the nodes that were yet to be processed. The shortest route was found by tracing back predecessors.

4 Results and Discussion

To achieve the objectives of this study, a selected part of KMA was considered with the area of 80.9 km^2, involving 39 routes connecting the centroid of thirty-three wards and two transfer locations. The GIS provided the detailed information on road network connectivity among different study points. After applying ArcGIS 'network analyst tool,' and giving the starting nodes, i.e., various wards one by one, and the destination node, i.e., the transfer station 6 (TS6) and transfer station 5 (TS5), the algorithm identified the shortest path and provided the same on a GIS map, along with route distance, for each of the desired links. The optimized routes generated from this approach were compared with the current routes that measured distance between individual ward to disposal site at TS6 and TS5. For each ward, the current route length was estimated in the ArcGIS-optimized route length. The detailed information of each of the links generated from the current approach is presented in Table 2.

The ArcGIS network analyst tool was applied by making the ward number 7 as a starting point and transfer station 6 (TS6) as a destination as shown in Fig. 7 and for the ward number 83 to transfer station (TS5) as shown in Fig. 8. The shortest route between TS6 and centroid of ward number 7 was found as 6.4 km (optimized distance), compared to 8.3 km (existing route distance) that is practiced by municipal vehicles for transportation purpose. Simultaneously, this process was applied on the waste collection point of thirty-three wards and two transfer stations (TS5 and TS6) to find the shortest route distances among the selected network of study area as mentioned in Table 3. The ArcGIS-optimized route showed the

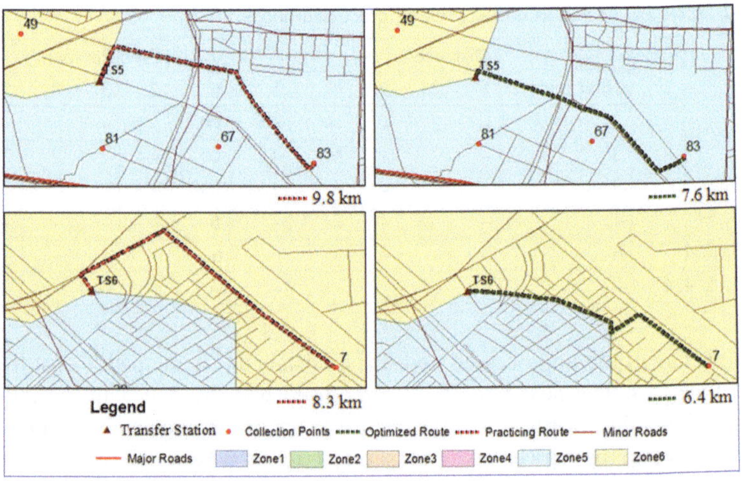

Fig. 7 ArcGIS-optimized route from the ward number 83 to TS5 and ward number 7 to TS6

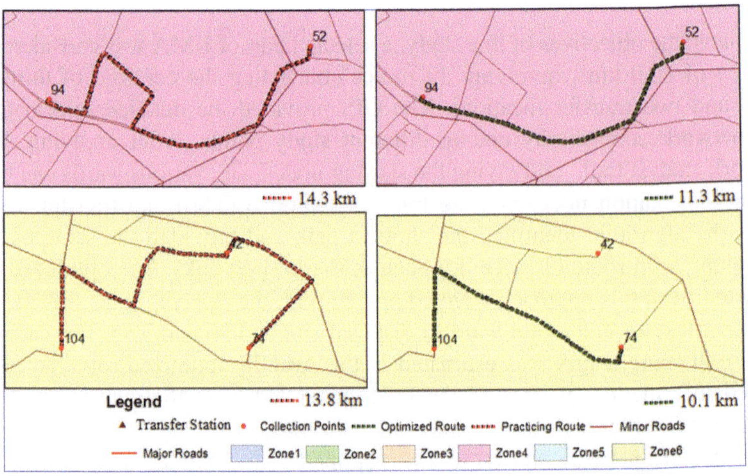

Fig. 8 ArcGIS-optimized route from the ward number 94–52 and 104–74

weighted average saving of 27.78% of the total haul distance (km) in a study network, connecting the centroid of thirty-three wards (as a starting location) and two transfer stations as well as among the centroid of wards (as a destination points) in a mesh network.

Table 3 Results from the approach by comparing current distance (km) versus optimal distance (km)

Source (ward number)	Destination Points	Practicing distance (km)	Optimized distance (km)	Saving in distance (km)	Saving in km (%)
5	TS6	10.6	7.4	3.2	30.19
6	TS6	11.2	9.3	1.9	16.96
7	TS6	8.3	6.4	1.9	22.89
11	TS6	13.2	9.5	3.7	28.03
12	TS6	15.1	11.4	3.7	24.50
13	TS6	15.3	11.6	3.7	24.18
14	TS6	19.6	13.4	6.2	31.63
83	TS5	9.8	7.6	2.2	22.45
85	TS5	16.4	13.2	3.2	19.51
93	50	12.5	8.2	4.3	34.40
94	52	14.3	11.3	3	20.98
95	54	10.2	6.5	3.7	36.27
96	55	8.9	8.9	0	0.00
97	64	7.8	5.4	2.4	30.77
98	67	8.4	6.1	2.3	27.38
99	68	9.6	5.4	4.2	43.75
100	69	11.4	7.5	3.9	34.21
101	71	9.3	6.2	3.1	33.33
102	72	12.2	7.4	4.8	39.34
103	73	9.5	6.1	3.4	35.79
104	74	13.8	10.1	3.7	26.81
Total		247.4	178.9	68.5	27.78[a]

[a]Weight average savings in km (*Source* ArcGIS network analyst computation)

5 Conclusions

This study presents an image processing technique that utilized the information of MSW management to propose the optimized transport paths for collection of wastes from end users to disposal sites in a large city. The network analyst tool in the GIS was utilized in modeling the optimal routes for MSW collection at Kanpur. Out of 110 existing wards (smallest political boundary of the city) and six zones with different land-use patterns, the zone with eighteen wards having thirty-three locations for collection of MSW was used as a test bed for the simulation exercise. All thirty-three locations were modeled separately to find respective optimized collection routes for solid waste management. This study finds the reductions in haul distance as 27.78 ± 10.2% for thirty-three locations. In this study, the distance between the nodes, location of collection points, location of intermediate stations, and location of transfer sites were provided as input parameters in the attribute table of modeling system. The results were obtained in terms of optimized routes considering the one-way and two-way communication, and even if there is no direct communication between the nodes. Future scope of this study may include savings of hauling time taken for both the current and shortest paths by considering traffic conditions. The approach in this study can be used as a tool for MSW management meant for the entire city in the study area, and the researchers can investigate the management scenario of MSW in other similar cities in the developing countries.

References

1. Baetz, Brian W.: Optimization/simulation modeling for waste management capacity planning. J. Urban Planning Develop. **88**, 59–79 (1990)
2. Singh, G., Singh, B., Rathi, S., Haris, S.: Solid waste management using shortest path algorithm. Int. J. Eng. Sci. Invention Res. Develop. **1**(2), 60–64 (2014)
3. Rodionov, M., Toshihiko, N.: Design of an optimal waste utilization system: a case study in St. Petersburg, Russia. Sustainability **3**, 1486–1509 (2011)
4. Sumathi, V.R., Natesan, U., Sarkar, C.: GIS-based approach for optimized siting of municipal solid waste landfill. Waste Manag. **28**(11), 2146–2160 (2008)
5. UNEP: Taking the pulse of the planet: connecting science with policy", United Nations Environment Program (UNEP) Global Environment Alert service (GEAS) (2013)
6. Chalmin, P., Gaillochet, C.: Waste to Resource, an Abstract of World Waste Survey, Cyclope, Veolia. Environmental Services, Edition Economica, France (2009)
7. Ghose, M.K., Dikshit, A.K., Sharma, S.K.: A GIS based transportation model for solid waste disposal—a case study on Asansol municipality. Waste Manag. **26**, 1287–1293 (2006)
8. Lu, T.H., Hsiao, T.Y., Yu, Y.H., Ma, H.W.: MSW management for waste minimization in Taiwan: the last two decades. J. Waste Manag. **26**, 661–667 (2006)
9. Aremu, A.S.: In-town tour optimization of conventional mode for municipal solid waste collection. Nigerian J. Technol. **32**(3), 443–449 (2013)
10. Apaydin, O., Gonullu, M.T.: Route optimization for solid waste collection: Trabzon (Turkey) case study. Global Nest J. **9**, 6–11 (2007)
11. Karadimas, N.V., Papatzelou, K., Loumos, V.G.: Optimal solid waste collection routes identified by the ant colony system algorithm. Waste Manag. Res. **25**, 139–147 (2007)
12. Liu, J.: A GIS-based tool for modelling large-scale crop–water relations. Environ. Model. Softw. **24**, 411–422 (2009)
13. Vijay, R., Gautam, A., Kalamdhad, A., Gupta, A., Devotta, S.: GIS-based locational analysis of collection bins in municipal solid waste management systems. J. Environ. Eng. Sci. **7**, 39–43 (2008)
14. KNN: Municipal report. "Kanpur Nagar Nigam", http://www.kanpurnagar.nic.in/ (2012). Accessed on April 2016
15. MOUD: Management of solid waste in Indian cities", Ministry of Urban Development, Government of India, New Delhi. http://moud.gov.in/ (2014). Accessed on April 2016
16. CPCB: Status report on municipal solid waste management", Central Pollution Control Board, http://cpcb.nic.in/ (2012). Accessed on April 2016
17. KCDP: A report on Kanpur city development plan under Jawaharlal Nehru National Urban Renewal Mission (JNNURM) scheme. http://www.kanpurnagar.nic.in/ (2006). Accessed on April 2016
18. NEERI: Strategy paper on solid waste management in India. Nagpur, India: National Environmental Engineering Research Institute. http://www.neeri.res.in/ (1996). Accessed on April 2016

19. Dijkstra, E.W.: A note on two problems in connection with graphs. Numer. Math. **1**, 269–271 (1959)
20. Orlin, J.: Dijkstra's algorithm animation, MIT open course ware, network optimization (2003)
21. Behera, S.N., Sharma, M., Dikshit, O., Shukla, S.P.: GIS-based emission inventory, dispersion modeling, and assessment for source contributions of particulate matter in an urban environment. Water Air Soil Pollut. **218**, 423–436 (2011)
22. Chang, N., Lu, H., Wei, Y.: GIS technology for vehicle routing and scheduling in solid waste collection systems. J. Environ. Eng. **123**(9), 901–910 (1997)
23. Huang, G.H., Baetz, B.W., Patry, G.G.: Grey dynamic programming for solid waste management programming under uncertainty. J. Urban Planning Develop. **120**, 132–156 (1994)
24. Sanjeevi, V., Shahabudeen, P.: Optimal routing for efficient municipal solid waste transportation by using ArcGIS application in Chennai, India. Waste Manag. Res. 1–11 (2015)

Layout Optimization of Sewer Network Using Minimum Cumulative Flow in the Sewer Network

Praveen K. Navin, Yogesh P. Mathur and Dilip Kumar

Abstract The optimal design of sewer network includes two subproblems: (1) finding the feasible sewer layouts and (2) optimal design of sewer layout components. The topic of sewer network optimization has been broadly studied since the concept was first proposed in the late-1960s. Most of the design practices have focused on sewer network component sizing with a fixed plan layout. The latter includes pipe diameters, slopes, and excavation depths, all of which is considerably influenced by the layout configuration. This works goal at introducing a method to solve the layout subproblem. A new approach to the selection of feasible sewer layouts has been developed, in which generation of all spanning tree algorithm is introduced to generate all possible sewer layouts from the base sewer network. After that, these sewer layouts are sorted in ascending order of total cumulative flow and a cost function is applied to determine the optimal sewer layout. The proposed method has been applied to solve two test examples at different scales, and the results have been discussed. The results clearly reveal the efficiency of the proposed method which can effectively solve the problem of optimal layout determination of a sewer network.

Keywords Sewer networks · Spanning tree · Layout optimization

P. K. Navin (✉) · Y. P. Mathur
Department of Civil Engineering, Malaviya National Institute of Technology,
Jaipur 302017, Rajasthan, India
e-mail: navin.nitj@gmail.com

Y. P. Mathur
e-mail: ypmathur.ce@mnit.ac.in

D. Kumar
Department of Civil Engineering, G. B. Pant Engineering College, Pauri, Uttarakhand, India
e-mail: Jhadilip27@gmail.com

A. S. Kalamdhad et al. (eds.), *Advances in Waste Management*,
https://doi.org/10.1007/978-981-13-0215-2_23

1 Introduction

Sewer network is an essential part of any urban infrastructure. The sewer network design consists of creating a suitable sewer layout that conforms to connect all buildings, street layouts, and local developing area. The alignment of sewer layouts is highly dependent on the location of sewage treatment plant or outlet, network size, and topography of the area. Finding the minimum length layouts among too many alternatives is the first step in designing a new urban sewer network. The capital cost of sewer network, installing, replacing, or modifying is very high. Reduction in the length of sewer line leads to a substantial saving in the capital cost. For this reason, significant researchers have attempted to the development of appropriate optimization techniques for sewer network design in recent years [1–6].

The sewer network design optimization problem is divided into two subproblems which are feasible layout selection and optimal component size determination. Due to the complex nature of the problem, most of the researchers have focused on the easier problem of component size optimization [7–26]. Only a few researchers have addressed the problem of feasible layout generation of networks by using different techniques of optimization [2–6, 11, 27–30].

This work describes a method for the selection of an optimal layout of a base sewer network. An algorithm generation of all spanning tree has applied to the generation of all minimum length layouts of a base sewer network. After that, all layouts are then sequenced in ascending order of total cumulative flow (Q) while the cost function of a layout based on that proposed by Walters and Smith [30] is applied to find the optimal sewer layouts. The proposed method is applied to two examples, and results are presented.

2 Feasible Sewer Layout Selection Method

The sewer network is a graph with specific properties. Therefore, it is necessary to review some basic definitions and principles of the graphs [31–33].

An undirected graph $G = (V, E)$ consists of a set of objects V ($V = v_1, v_2, \ldots, v_n$) called vertices and another set E ($E = e_1, e_2, \ldots, e_n$) called edges, such that each edge e_{ij} is identified with an unordered pair (v_i, v_j) of vertices. If in a graph G there is one path between every pair of vertices, G is a tree.

A weighted graph is a graph G in which every edge e has been allotted a real number w (e) called the weight of e. The weight of an edge in the sewerage system has been taken as its length.

A spanning tree of a graph G is a tree containing all vertices of G. A minimum spanning tree of an undirected weighted graph G is a spanning tree of which the sum of the edge weights is minimal.

There are several algorithms for finding a minimum spanning tree (MST) of a graph. Kruskal's algorithm is one of the optimized ways to generate a minimum length spanning tree for every connected undirected graph *G*. But layout optimization problem needs to generate all the spanning trees (layouts) of a base network (graph). Therefore, an algorithm generation of all spanning tree is introduced to find all sewer layouts (spanning trees) of a base network. The algorithm is based on the assumption that a base graph or sewer network, including all possible edges of the network, is available. In the base sewer network, there are *n* vertices (manholes or nodes) and *m* edges (sewers or links). A simplified representation of the algorithm is given below.

2.1 Algorithm—Generation of All Spanning Trees

Input—*n*, *m*, *E*, weights and nodal flow contribution;

n is the number of vertices, *m* is the number of edges, and *E* is the set of edges of graph *G*. Weight is the length of edges (sewers);

e_old is an edge of the current minimum spanning tree (MST), and e_new is a new edge from the remaining graph (base network);

Current_MST_weight is the total weight of current MST, and new_MST_weight is the total weight of newly generated MST;

```
Output – Output_ file (all spanning trees of G);
Step 1: Calculate MST from a base network by using Kruskal's algorithm;
Step 2: Print MST edges and weight;
Step 3: For each edge e_old in current_MST
Step 4:         Remove selected edge e_old
Step 5:         For each edge e_new in base graph (network) except current_MST edges
Step 6:                 New_MST_weight = current_MST_weight − e_old edge weight + e_new edge
                        weight;
Step 7:         If new_MST_weight = current_MST_weight
Step 8:                 Store the edges along with total weight in the final output for MST
Step 9:                 Else if new_MST_weight > current_MST_weight;
Step 10:        Store temporary new edges, and corresponding weight;
Step 11: Repeat step 5;
Step 12:        Find minimum from all temporarily stored edges and their corresponding weight to get a next
                Spanning tree;
Step 13: Print Spanning trees (new edges and weight);
Step 14: Repeat step 3 for next Spanning tree;
Step 15: Calculate discharges (flows) in each spanning tree;
Step 16: End;
```

2.2 Optimal Sewer Layout Selection

As described above, by using the generation of all spanning tree algorithm determined all sewer layouts of a base network. As a result, a large number of alternative layouts are available, and it is very difficult to identify directly the true optimal layout. Therefore, a strategy to sequence these alternatives needs to be introduced.

Discharge q_{ij} in ith link of the jth layout is calculated. After that, the sum of the cumulative discharges Q_j in all links of the jth layout is calculated as shown in Eq. (1).

$$Q_j = \sum_{i=1}^{i=n} q_{ij} \tag{1}$$

where n is the total number of links of the jth layout.

These layouts are then sequenced in ascending order by the sum of the cumulative discharges Q_j. The layouts are then investigated for optimality in this sequence. Walters and Smith [30] proposed a cost function for the cost of network layout in terms of the length and concave function of flow rate of each link as given in Eq. (2):

$$C_j = \sum_{i=1}^{n} L_{ij}\sqrt{q_{ij}} \tag{2}$$

where C_j = layout cost of the jth alternative, L_{ij} = link length, and q_{ij} = sewer discharge in the ith link of the jth layout.

The cost of each layout is calculated by using Eq. (2), and the optimal layout of a base network is selected where the cost of the layout is minimal.

3 Application

For the sewer network problem, manhole represents vertices and sewer pipes represent edges. The length of the sewer is taken as the weight of an edge.

The applicability of the procedures described in the previous section is tested in this section against two examples. The first example (network 1) that has been considered is a simple network, which is shown in Fig. 1. The network 1 consists of 6 nodes (vertices) and 10 links (edges), and the outlet is located at the node number 3. The wastewater contribution at each node of the network 1 is shown in Table 1.

Fig. 1 Base layout of network 1

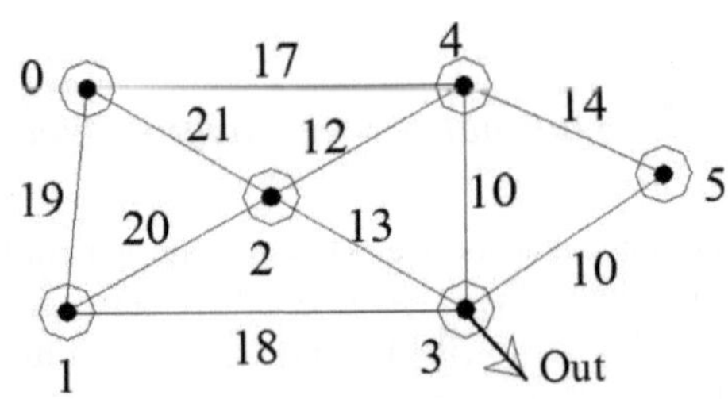

Table 1 Nodal wastewater contribution for network 1

Node No.	0	1	2	3	4	5
Flow contribution (l/s)	20	15	18	0	17	14

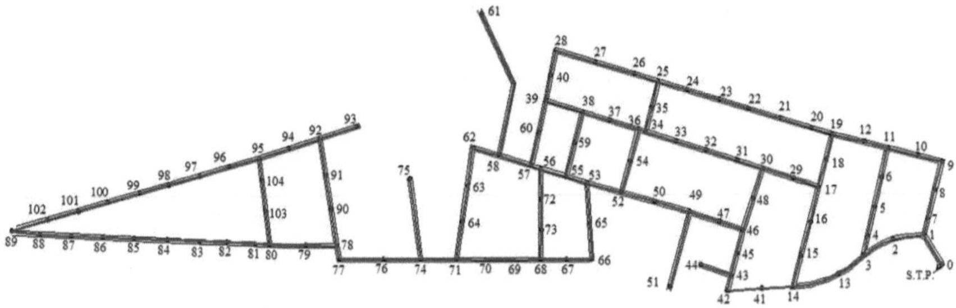

Fig. 2 Sudarshanpura base sewer network (network 2)

The second test example (sewer network of Sudarshanpura, Jaipur) is shown in Fig. 2. The Sudarshanpura base sewer network (network 2) consists of 105 nodes (vertices), 116 links (edges), and the outlet located at node number 0. Base graph data (link number, set of links, and its length) of network 2 are shown in Table 2. The wastewater contribution at each node of the network 2 is shown in Table 3.

The total number of nodes and links, set of links or edges, link length (weight), and nodal wastewater contributions are entered as an input detail for both the network.

Both problems network 1 and network 2 are solved by using above proposed procedure. The generation of all spanning tree algorithm is applied to find all layouts (spanning tree) of a base network; then, these layouts to be sequenced in ascending order of total cumulative flow Q (Eq. 1). After that, the cost function (Eq. 2) is applied to determine layout cost of the sequenced layouts. The top 4 optimal layouts of a network 1 are shown in Fig. 3.

The results of network 2 are illustrated in Table 4. It is clearly seen that the layout cost (C) generally increases with the total cumulative flow (Q). The minimum cost of 13111.46 was obtained corresponding to the layout with cumulative flow value (Q) of 3642.78 l/s (Fig. 4). The layout with minimum cumulative flow value of 3639.13 l/s has the cost of 13170.85 (Fig. 5) which is 0.45% above the minimum cost layout. The proposed method for selection of global optimal layout is very convenient to implement on the sewer network problem. It may be concluded that minimum cumulative flow layout is near global optimal layout.

Table 2 Base graph data for network 2

Link No.	Nodes		Length	Link No.	Nodes		Length	Link No.	Nodes		Length	Link No.	Nodes		Length
1	0	1	30	30	27	28	30	59	54	36	24	88	77	78	13
2	1	2	23	31	17	29	30	60	52	53	30	89	78	79	31
3	2	3	23	32	29	30	22	61	53	55	20	90	79	80	31
4	3	4	10	33	30	31	30	62	55	59	30	91	80	81	10
5	4	5	30	34	31	32	30	63	59	38	30	92	81	82	30
6	5	6	30	35	32	33	30	64	55	56	25	93	82	83	30
7	6	11	30	36	33	34	18	65	56	57	8	94	83	84	30
8	1	7	9	37	34	35	30	66	57	60	32	95	84	85	30
9	7	8	30	38	35	25	12	67	60	39	32	96	85	86	30
10	8	9	30	39	34	36	7	68	57	58	33	97	86	87	30
11	9	10	20	40	36	37	16	69	58	61	143	98	87	88	30
12	10	11	30	41	37	38	30	70	58	62	24	99	88	89	30
13	11	12	20	42	38	39	30	71	62	63	33	100	78	90	33
14	12	19	30	43	39	40	14	72	63	64	33	101	90	91	33
15	3	13	30	44	40	28	30	73	64	71	33	102	91	92	33
16	13	14	30	45	14	41	30	74	53	65	30	103	92	93	36
17	14	15	30	46	41	42	30	75	65	66	30	104	92	94	30
18	15	16	30	47	42	43	11	76	66	67	22	105	94	95	26
19	16	17	30	48	43	44	30	77	67	68	22	106	95	96	30
20	17	18	30	49	43	45	20	78	56	72	21	107	96	97	30
21	18	19	12	50	45	46	20	79	72	73	30	108	97	98	30

(continued)

Table 2 (continued)

Link No.	Nodes		Length	Link No.	Nodes		Length	Link No.	Nodes		Length	Link No.	Nodes		Length
22	19	20	18	51	46	48	30	80	73	68	30	109	98	99	30
23	20	21	30	52	48	30	24	81	68	69	26	110	99	100	30
24	21	22	30	53	46	47	26	82	69	70	26	111	100	101	30
25	22	23	30	54	47	49	26	83	70	71	26	112	101	102	30
26	23	24	30	55	49	51	72	84	71	74	34	113	102	89	30
27	24	25	30	56	49	50	30	85	74	75	76	114	80	103	27
28	25	26	27	57	50	52	30	86	74	76	38	115	103	104	27
29	26	27	30	58	52	54	30	87	76	77	38	116	104	95	27

Table 3 Nodal wastewater contribution for network 2

Node No.	Flow (l/s)	Node No.	Flow (l/s)	Node No.	Flow (l/s)	Node No.	Flow (l/s)	Node No.	Flow (l/s)
0	0.000	21	0.380	42	0.380	63	0.418	84	0.380
1	0.380	22	0.380	43	0.139	64	0.672	85	0.380
2	0.292	23	0.380	44	0.380	65	0.380	86	0.380
3	0.292	24	0.380	45	0.254	66	0.380	87	0.380
4	0.127	25	0.380	46	0.444	67	0.279	88	0.380
5	0.380	26	0.342	47	0.330	68	0.279	89	50.380
6	0.380	27	0.380	48	0.507	69	0.355	90	0.418
7	0.114	28	0.380	49	0.330	70	0.330	91	0.837
8	0.380	29	0.380	50	0.380	71	0.507	92	0.380
9	0.254	30	0.279	51	0.913	72	0.570	93	0.456
10	0.444	31	0.380	52	0.570	73	0.570	94	0.330
11	0.570	32	0.380	53	0.380	74	0.431	95	0.342
12	0.254	33	0.380	54	0.507	75	0.963	96	0.380
13	0.380	34	0.418	55	0.444	76	0.482	97	0.380
14	0.380	35	0.380	56	0.317	77	0.482	98	0.380
15	0.380	36	0.089	57	0.317	78	0.165	99	0.380
16	0.380	37	0.203	58	0.418	79	0.393	100	0.380
17	0.570	38	0.570	59	0.570	80	0.393	101	0.380
18	0.342	39	0.177	60	0.608	81	0.127	102	0.761
19	0.380	40	0.380	61	51.80	82	0.380	103	0.342
20	0.228	41	0.380	62	0.304	83	0.380	104	0.342

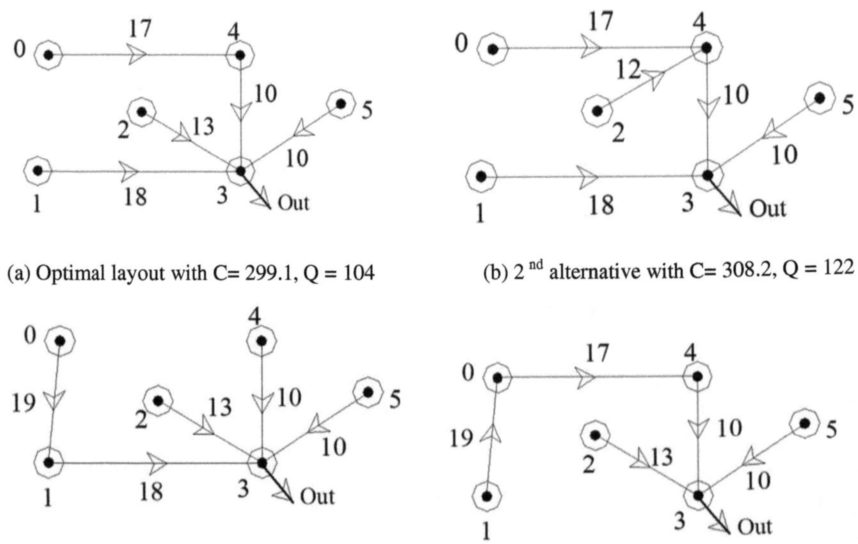

(a) Optimal layout with C= 299.1, Q = 104

(b) 2 nd alternative with C= 308.2, Q = 122

(c) 3rd alternative with C=325.2, Q = 104

(d) 4 th alternative with C=338.8, Q =134

Fig. 3 Top for optimal layout of a network 1

Table 4 Total cumulative flow versus layout cost comparison

Total cumulative flow (l/s)	Layout cost	Total cumulative flow (l/s)	Layout cost
3639.13	13170.85	4375.98	14411.83
3641.96	13114.59	4426.58	14543.76
3642.78	13111.46	4521.92	14993.24
3718.5	13183.71	4616.77	14816.62
3896.6	13915.56	4676.69	14619.86
4027.95	14022.27	4720.37	15102.32
4077.09	14135.68	4772.37	14946.37
4178.5	14562.22	4820.16	15163.73
4227.3	14476.68	4870.45	15113.36
4275.64	14739.47	4920.55	15321.61

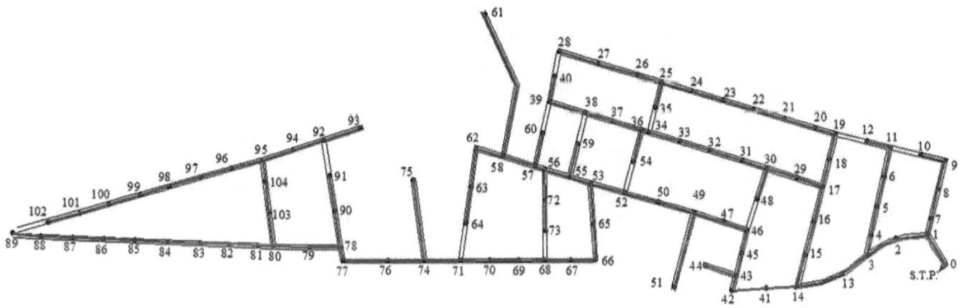

Fig. 4 Optimal sewer layout of network 2

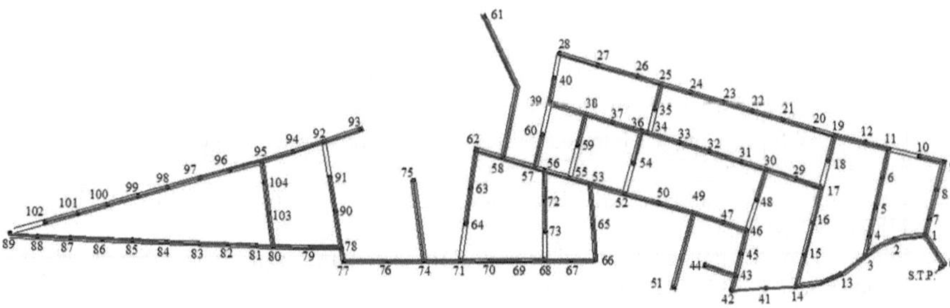

Fig. 5 Alternative sewer layout of network 2

4 Conclusions

A method for the selection of an optimal layout of a base network is introduced, in which an algorithm is applied to generate all spanning trees of a network (graph). As large numbers of possible spanning trees are available, these spanning trees are sorted in ascending order of total cumulative flow (Q). Further, the cost function is applied to determine layout cost of the sorted layouts. The optimal layout was selected where the total cost of layout found to be minimal. It may be concluded that the total cost of alternative generally increases with the total cumulative flow. The applicability and efficiency of the proposed method for layout optimization of sewer networks were tested against two examples. The result revealed that proposed method gives an optimal solution of the sewerage network.

References

1. Davidson, J.W.: Evolution program for layout geometry of rectilinear looped networks. J. Comput. Civ. Eng. **13**, 246–253 (1999)
2. Diogo, A.F., Graveto, V.M.: Optimal layout of Sewer systems: a deterministic versus a stochastic model. J. Hydraul. Eng. **132**, 927–943 (2006)
3. Afshar, M.H., Mariño, M.A.: Application of an ant algorithm for layout optimization of tree networks. Eng. Optim. **38**, 353–369 (2006)
4. Haghighi, A.: Loop-by-loop cutting algorithm to generate layouts for urban drainage systems. J. Water Resour. Plan. Manag. **139**, 693–703 (2013)
5. Haghighi, A., Bakhshipour, A.E.: Reliability-based layout design of sewage collection systems in flat areas. Urban Water J. 1–13 (2015)
6. Steele, J.C., Mahoney, K., Karlovic, O., Mays, L.W.: Heuristic optimization model for the optimal layout and pipe design of Sewer systems. Water Resour. Manag. **30**, 1605–1620 (2016)
7. Walsh, S., Brown, L.C.: Least cost method for Sewer design. J. Environ. Eng. Div. **99**, 333–345 (1973)
8. Dajani, J.S., Haiti, Y.: Capital cost minimization of drainage networks. J. Environ. Eng. Div. **100**, 325–337 (1974)
9. Price, R.K.: Design of storm water sewers for minimum construction cost. In: 1st International Conference on Urban Strom Drainage. pp. 636–647., Southampton, United Kingdom (1978)
10. Walters, G.A., Templeman, A.B.: Non-optimal dynamic programming algorithms in the design of minimum cost drainage systems. Eng. Optim. **4**, 139–148 (1979)
11. Mays, L.W., Wenzel, H.G.: Optimal design of multilevel branching sewer systems. Water Resour. Res. **12**, 913–917 (1976)
12. Kulkarni, V.S., Khanna, P.: Pumped wastewater collection systems optimization. J. Environ. Eng. **111**, 589–601 (1985)
13. Charalambous, C., Elimam, A.A.: Heuristic design of Sewer networks. J. Environ. Eng. **116**, 1181–1199 (1990)
14. Elimam, A.A., Charalambous, C., Ghobrial, F.H.: Optimum design of large Sewer networks. J. Environ. Eng. **115**, 1171–1190 (1989)
15. Li, G., Matthew, R.G.S.: New approach for optimization of urban drainage systems. J. Environ. Eng. **116**, 927–944 (1990)
16. Swamee, P.K.: Design of Sewer line. J. Environ. Eng. **127**, 776–781 (2001)
17. Liang, L.Y., Thompson, R.G., Young, D.M.: Optimising the design of sewer networks using genetic algorithms and tabu search. Eng. Constr. Archit. Manag. **11**, 101–112 (2004)
18. Afshar, M.H.: Application of a genetic algorithm to storm Sewer network optimization. Sci. Iran. **13**, 234–244 (2006)
19. Afshar, M.H.: Partially constrained ant colony optimization algorithm for the solution of constrained optimization problems: application to storm water network design. Adv. Water Resour. **30**, 954–965 (2007)

20. Izquierdo, J., Montalvo, I., Pérez, R., Fuertes, V.S.: Design optimization of wastewater collection networks by PSO. Comput. Math. Appl. **56**, 777–784 (2008)
21. Guo, Y., Walters, G.A., Khu, S.T., Keedwell, E.: A novel cellular automata based approach to storm Sewer design. Eng. Optim. **39**, 345–364 (2007)
22. Guo, Y.F., Walters, G.A., Khu, S.T., Keedwell, E.C.: Efficient multiobjective storm Sewer design using cellular automata and genetic algorithm hybrid. J. Water Resour. Plan. Manag. **134**, 511–515 (2008)
23. Pan, T.-C., Kao, J.-J.: GA-QP model to optimize Sewer system design. J. Environ. Eng. **135**, 17–24 (2009)
24. Afshar, M.H.: A parameter-free continuous Ant colony optimization algorithm for the optimal design of storm sewer networks: constrained and unconstrained approach. Adv. Eng. Softw. **41**, 188–195 (2010)
25. Haghighi, A., Bakhshipour, A.E.: Optimization of Sewer networks using an adaptive genetic algorithm. Water Resour. Manag. **26**, 3441–3456 (2012)
26. Karlovic, O., Mays, L.W.: Sewer system design using simulated annealing in excel. Water Resour. Manag. **28**, 4551–4565 (2014)
27. Argaman, Y., Shamir, U., Spivak, E.: Design of optimal sewerage systems. J. Environ. Eng. Div. **99**, 703–716 (1973)
28. Walters, G.A., Lohbeck, T.: Optimal layout of tree networks using genetic algorithms. Eng. Optim. **22**, 27–48 (1993)
29. Davidson, J.W., Coulter, I.C.: Evolution program for design of rectilinear branched networks. J. Comput. Civ. Eng. **9**, 112–121 (1995)
30. Walters, G.A., Smith, D.K.: Evolutionary design algorithm for optimal layout of tree networks. Eng. Optim. **24**, 261–281 (1995)
31. Deo, N.: Graph Theory with Applications to Engineering and Computer Science. Prentice Hall of India Pvt. Ltd., New Delhi (2005)
32. Sörensen, K., Janssens, G.K.: An algorithm to generate all spanning trees of a graph in order of increasing cost. Pesqui. Operational. **25**, 219–229 (2005)
33. Biswas, B., Basil, K., Naskar, S., Chakraborti, S., Sarma, S.S.: A combinatorial algorithm to generate all spanning trees of a weighted graph in order of increasing cost. In: arXiv preprint arXiv:1209.4206 (2012)

20. [illegible] detection networks by TS. Comput. Math. Appl. 26, [illegible] (1993)
21. Guo, Y., Walters, G.A., Khu, S.T., Keedwell, E.: A novel cellular automata based approach to storm sewer design. Eng. Optim. 39, 345–364 (2007)
22. Guo, Y., Walters, G.A., Khu, S.T., Keedwell, E.C.: Efficient multiobjective storm sewer design using cellular automata and genetic algorithm hybrid. J. Water Resour. Plan. Manag. 134, 511–515 (2008)
23. Pan, T.C., Kao, J.J.: GA-QP model to optimize sewer system design. J. Environ. Eng. 135, 17–24 (2009)
24. Afshar, M.H.: A parameter free continuous ant colony optimization algorithm for the optimal design of storm sewer networks: constrained and unconstrained approach. Adv. Eng. Softw. 41, 188–195 (2010)
25. Haghighi, A., Bakhshipour, A.E.: Optimization of sewer networks using an adaptive genetic algorithm. Water Resour. Manag. 26, 3441–3456 (2012)
26. Karovic, O., Mays, L.W.: Sewer system design using simulated annealing in Excel. Water Resour. Manag. 28, 4551–4565 (2014)
27. Argaman, Y., Shamir, U., Spivak, E.: Design of optimal sewerage systems. J. Environ. Eng. Div. 99, 703–716 (1973)
28. Walters, G.A., Lohbeck, T.: Optimal layout of tree networks using genetic algorithms. Eng. Optim. 22, 27–48 (1993)
29. Davidson, J.W., Goulter, I.C.: Evolution program for design of rectilinear branched networks. J. Comput. Civ. Eng. 9, 112–121 (1995)
30. Walters, G.A., Smith, D.K.: Evolutionary design algorithm for optimal layout of tree networks. Eng. Optim. 24, 261–281 (1995)
31. Deo, N.: Graph Theory with Applications to Engineering and Computer Science. Prentice Hall of India Pvt. Ltd., New Delhi (2003)
32. Sörensen, K., Janssens, G.K.: An algorithm to generate all spanning trees of a graph in order of increasing cost. Pesqui. Operacional 25, 219–229 (2005)
33. [illegible], Kumar, S., [illegible]: A [illegible] algorithm to generate all spanning trees of a weighted graph [illegible]

Investigating the Effect of Anaerobic Co-digestion of Poultry Manure with Lawn Grass Cuttings

M. J. Sukhesh and P. Venkateswara Rao

Abstract The effect of anaerobic co-digestion of poultry manure with lawn grass cuttings was studied to assess the biogas potential at different mixture proportions. The lawn grass cuttings at 18.76, 38.12, 58.11, 78.76, and 100% fractions were mixed with poultry manure based on volatile solids (VS) content. The anaerobic digestion experiments were conducted in 120-mL-reactor bottles at ambient conditions for 60 days. The biogas volume was measured by water displacement method. The reactor bottles were shaken daily before recording the biogas production volume. Biogas production potentials for 60 days were found to be 104.8, 49.8, 52.1, 63.7, 30.6 mL/g VS_{added} for the lawn grass cuttings proportions 18.76, 38.12, 58.11, 78.76, and 100%, respectively. Co-digestion of poultry manure with lawn grass cuttings of proportion of 18.76% has shown highest biogas production of 104.8 mL/g VS which is 83 and 242% higher than that of mono-digestion of poultry manure and lawn grass cuttings, respectively. This study observed the presence of synergistic effect between lawn grass cuttings and poultry manure during anaerobic digestion.

Keywords Anaerobic co-digestion · Lawn grass cuttings · Poultry manure Biogas potential

1 Introduction

Anaerobic digestion (AD) is a biological waste management method carried under the absence of oxygen and is associated with supplemental benefits of energy production in the form of methane and nutrient-rich digestate during stabilization of organic waste. Animal manures are generally being used as feed substrate for AD. The traditional mono-digestion of animal manures generally limited by lack of

M. J. Sukhesh (✉) · P. Venkateswara Rao
National Institute of Technology, Warangal, Warangal 506004, India
e-mail: m.sukhesh@gmail.com

A. S. Kalamdhad et al. (eds.), *Advances in Waste Management*,
https://doi.org/10.1007/978-981-13-0215-2_24

optimal nutrients and excess ammonia accumulation may lead to inhibition of the AD process [1]. The one way of tackling the limitation is the simultaneous digestion of two or more substrates with complementary characteristics which is generally termed as co-digestion [2]. However, proper selection of co-substrates and its composition is important for the improvement of process stability during anaerobic co-digestion [3].

The stability of the AD of animal manures having low carbon to nitrogen (C/N) ratio can be substantially improved by co-digesting with substrates having high C/N ratios such as vegetable waste, food waste, and agricultural residues. [4]. However, lignocellulose biomass is one such kind of available organic wastes that have high carbon content in the form of cellulose, hemicellulose, and lignin contents [5]. The co-digestion of animal manures with lignocellulose materials had improved efficiency of AD process with the mitigation of ammonia inhibition [6, 7]. Grass silage which is lignocellulose biomass has a high digestible organic matter and volatile solid (VS) content and was acted as an excellent feedstock for AD [8].

Lawn grass cuttings are lignocellulose biomass which is being generated widely in public lawn spaces. The AD of lawn grass cuttings is having hydrolysis as the rate-limiting step because of its complex structure [9]. The pretreatment methods for AD of lawn grass had overcome the hydrolysis but did not resulted in conversion of volatile fatty acids (VFAs) to biogas production due to formation of refractory compounds during pretreatment that inhibited methanogenic population [9]. However, the AD of grass silage co-digested with other wastes which offers an external alkalinity source was found to increase buffering capacity and resulted in good biogas yield [8]. The investigation of Lehtomaki et al. [10] noticed that co-digestion of different waste substrates which are very different in their characteristics, viz. cow manure with grass, straw and sugar beet tops under batch, semi-batch and continuous stirred tank reactors had shown significant synergistic effect and given substantiate amount of biogas when compared to their mono-digestion.

The rapid growth of poultry industry in India [11] results in the generation of poultry manure in large quantities and is traditionally been applied to agricultural field as a soil conditioner after stacking it for 6–8 weeks. The poultry manure can also be digested under anaerobic conditions that can yield biogas along with digestate acting as soil conditioner. The anaerobic co-digestion of poultry manure with lawn grass cuttings may improve the AD process efficiency due to the ability of supplementing the buffering capacity of the nitrogen-rich poultry manure [12]. Very few works have been reported studying the co-digestion of poultry manures with lawn grass cuttings. The present research work is confined to address the feasibility studies for co-digestion of lawn grass cuttings with poultry manure at different proportions and to assess its biogas potential.

2 Materials and Methods

2.1 *Substrates and Inoculum*

Lawn grass cuttings and poultry manure were used as co-substrates in the present study. Lawn grass cuttings were collected in lawn spaces at National Institute of Technology, Warangal campus. The collected lawn grass cuttings were sun-dried for one week and then ground to small particles in pulverizer. Thereafter, it was sieved through a mesh having a size of 1.18 mm. Poultry manure was collected from nearby poultry farm located in Kazipet Town, India. The poultry manure was cleaned from inorganic materials such as eggshells, stored at ambient conditions. Inoculum used in this study was anaerobic sludge and was collected from biogas plant that was being fed with food waste. Table 1 shows the characteristics of lawn grass cuttings, poultry manure, and inoculum used during the study.

2.2 *Batch Anaerobic Digestion Experiments*

Batch anaerobic digestion experiments were conducted in 120-mL-glass reactor bottles. Each reactor bottle was fed with lawn grass cuttings proportions of 0, 18.76, 38.12, 58.11, 78.76, and 100% with poultry manure based on volatile solids. Consistent weight of 35 g of inoculum and 15 mL of tap water was added in all the reactor bottles and then bottles were sealed with aluminum crimps. The initial characteristics of the reactor bottles are shown in Table 2. The experiments were performed in duplicates to confirm the repeatability and reproducibility of biogas volume measurement results. The experiments were conducted at ambient temperature (January and February months of 2017). Biogas volume was measured daily for first 30 days and thereafter once in two days for next 30 days due to low production of biogas volume. The biogas production volume was corrected by subtracting the biogas volume produced from control reactor containing inoculum.

Table 1 Characteristics of raw substrates and inoculum

Item	TS (%)	VS (% of TS)	pH
Inoculum	6.85	98.48	5.49
Lawn grass cuttings	87.94	75.09	ND
Poultry manure	73.86	41.92	ND

Table 2 Initial characteristics of anaerobic reactors

Reactor bottle	TS (%)	VS (% of TS)	pH	VFA (mg/L)	Alkalinity (mg/L as $CaCO_3$)	VFA/ alkalinity ratio
R_1	9.85	50.38	5.60	2560	6500	0.39
R_2	7.48	60.40	5.50	1300	6400	0.20
R_3	9.24	62.23	5.50	1840	7750	0.24
R_4	9.58	65.73	5.27	1760	5250	0.34
R_5	9.22	73.89	5.15	2000	5888	0.34
R_6	8.39	88.89	4.50	1260	3200	0.39

2.3 *Analytical Methods*

Total solids (TS), volatile solids (VS), pH were measured as per APHA standard methods [13]. Volatile fatty acids such as acetic acid, propionic acid, and butyric acid were determined with a gas chromatograph (YL Instruments Model 6500) equipped with flame ionization detector and capillary column of length 30 m 0.53 mm internal diameter ×0.1 μm film thickness. The samples were centrifuged at 12,000 rpm for 10 min followed by filtration through 0.22-μm membrane. The temperature of injection inlet and detector was 210 °C and 240 °C, respectively. Initial oven temperature is 100 °C for 3 min followed by an increase of 20 °C/min to the final temperature of 230 °C. Nitrogen gas was used as carrier gas. Biogas volume produced was measured by in-house water displacement method. Composition of biogas was analyzed with gas chromatograph (YL Instruments Model 6500) equipped with thermal conductivity detector and 4-m-stainless column packed with Porapak Q(80–100 mesh). The operational temperatures of the injection port, column oven, and detector were 120, 150, and 120 °C, respectively. Nitrogen gas was used as carrier gas (Figs. 1, 2, and 3).

Fig. 1 Poultry manure

Fig. 2 Lawn grass cuttings

Fig. 3 Biogas measurement by water displacement method

3 Results and Discussion

3.1 Characteristics of Substrates and Inoculum

The characteristics of substrates and inoculum were analyzed as shown in Table 2. The pH of inoculum was slightly acidic. The VS content of lawn grass cuttings (75.09%) was higher than that of poultry manure (41.92%) indicating more biologically degradable matter on lawn grass cuttings. The characteristics of each reactor containing different propositions of organic substrates were analyzed before the digestion and are presented in Table 1. The pH of all reactors is found to be between 4.5 and 5.6, indicating slight acidic conditions before the AD process. The AD system may be attained optimal neutral conditions few days after setup had initiated. Volatile fatty acids (VFAs) to alkalinity ratio of all the reactors were in the range of 0.2–0.4 indicating non-inhibiting conditions prior to the AD as considered by Borja et al. [14] (Fig. 4).

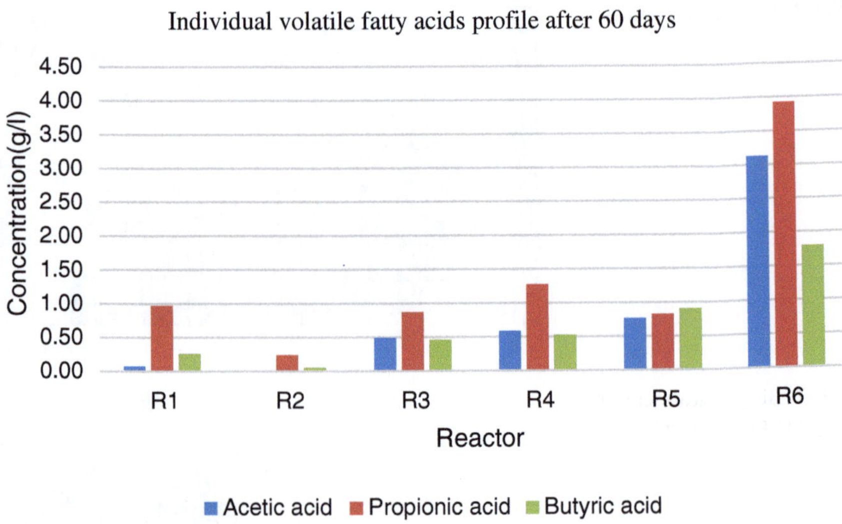

Fig. 4 Individual volatile fatty acids profile after 60 days

3.2 *Individual and Total Volatile Fatty Acids (VFAs) Profile After Digestion*

The accumulation of total VFAs demonstrates the methanogenic activity [15]. The concentration of organic acids is important in AD for two reasons that VFAs are the immediate precursors in the metabolic chain leading to biogas production and if present in high concentration, VFAs causes stress in the microbial population and leads to complete process failure [16]. As shown in Fig. 5, total VFAs accumulation in R_2 was found to be low when compared to other reactors. The possible inference is that there might have a symbiotic relationship between the acidogens and methanogens, and all the VFAs produced were being consumed by methanogens leading to stability without their accumulation, whereas VFAs in R_6 which was digested with only lawn grass cuttings were found to be 8.86 g/L indicating accumulation of VFAs. The accumulation of VFAs may be attributed to mono-digestion of lawn grass cuttings in R_6. The similar phenomenon of high accumulation of VFAs of 15.9 g/L and 12.48 g/kg was observed during mono-digestion of grass [8] and tomato residues [17], respectively. Thus, the low production of biogas in mono-digestion of lawn grass cuttings may be attributed to inhibition to methanogens due to VFAs accumulation.

All VFAs are transformed to acetic acid before being converted to methane. Conversion rate of propionic acid is slower than butyric acid. Propionic acid to acetic acid ratio can be used as an indicator of digester imbalance [16]. These ratios for R_1, R_2, R_3, R_4, R_5, and R_6 are 0.05, 0, 0.38, 0.43, 0.60, and 1.75, respectively. Hill et al. [16] reported that ratio of propionic acid to acetic acid greater than 1.4

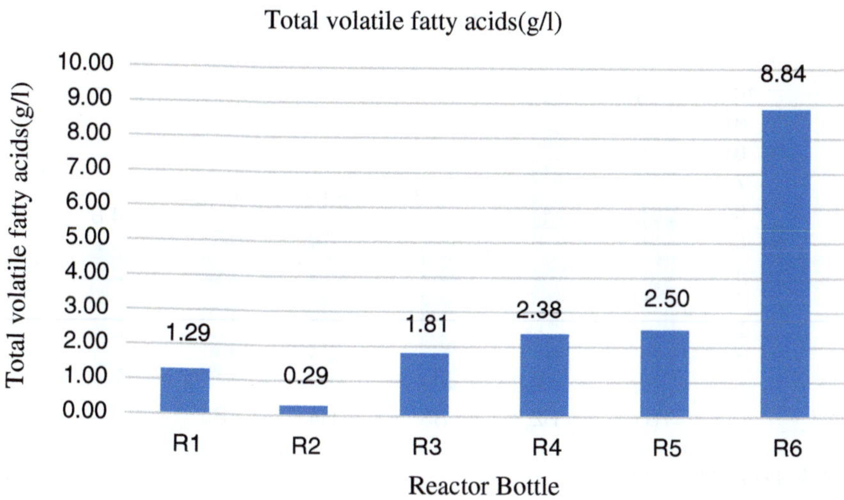

Fig. 5 Total volatile fatty acids (g/L)

indicates impending digester failure. Except R_6, all the reactor bottles had the ratio in the allowable range after stable digestion. The reactor bottle number R_6 consisting only lawn grass cuttings is having ratio of 1.75. The higher propionic acid to acetic acid ratio might be the reason for failure and low yield of biogas.

3.3 *pH Profile After Digestion*

Slight changes in pH away from optimum conditions adversely affect the AD process. Methanogens are more susceptible to pH change due to which it plays key role in methane production. Optimum range of pH for acidogens is 5.5–6.5 and for methanogens is 7.8–8.2. For combined system of AD, optimum pH range is 6.8–7.4. As shown in Fig. 6, pH in R_1 and R_2 reactors is near to optimal range of AD, whereas R_3, R_4, R_5, and R_6 are far away from the optimal pH range. The reason might be due to excess accumulation of volatile fatty acids caused pH drop leading to inhibition of methanogenic activity.

3.4 *Cumulative Volume of Biogas Production*

Biogas potentials for 60 days were found to be 57.3, 104.8, 49.8, 52.1, 63.7, 30.6 mL/g VS_{added} for the lawn grass cuttings proportions of 0, 18.76, 38.12, 58.11, 78.76, and 100%, respectively (Fig. 7). The methane content of the biogas in all the batch experiments was found to be in the range of 48–65%. Co-digestion of poultry

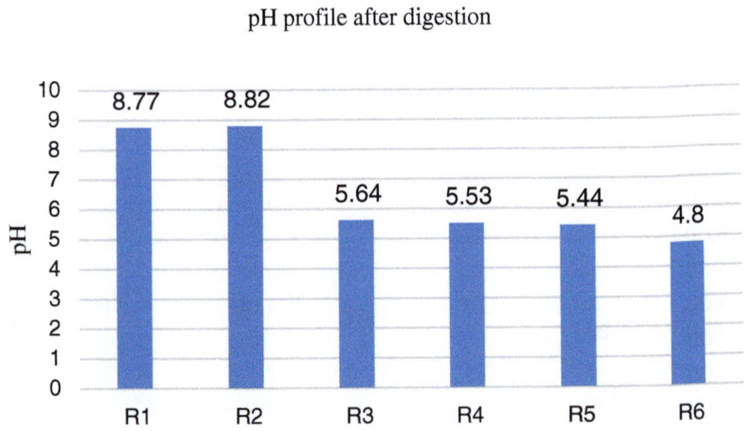

Fig. 6 pH profile after 60 days

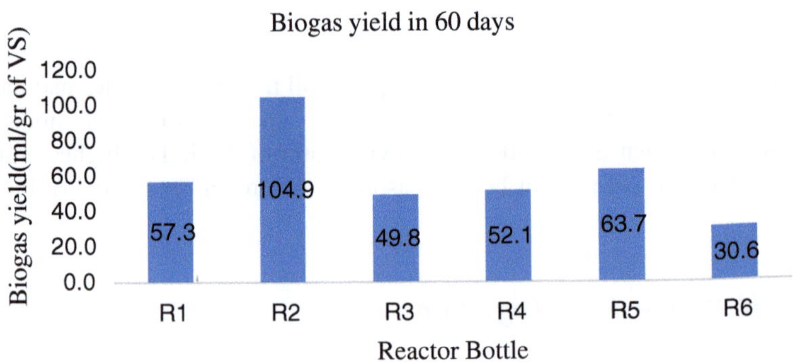

Fig. 7 Cumulative biogas production (mL/gr of VS)

manure with lawn grass cuttings of proportion of 18.76% has shown highest biogas production of 104.8 mL/g VS, and it is 83 and 242% higher than that of mono-digestion of poultry manure and lawn grass cuttings, respectively. The reason can be attributed to the balance of nutrients at this optimal substrate mixture proportion. Presence of less concentration of VFAs (0.29 g/L) in R_2 also supports that the synergism between methanogens and acidogens might have resulted higher biogas production. Among all reactors, R_6 yielded less biogas production and the reason can be attributed to insufficient nutrients and VFAs accumulation during mono-digestion of lawn grass cuttings. VFAs profile shows excess accumulation of VFAs happened in digestion of pure grass cuttings. This is in agreement with the earlier study carried out by Xie et al. [8] in which digestion of pure grass silage was unstable due to VFAs accumulation and stated that external source of alkalinity addition was recommended for enhancement of biogas production.

4 Conclusions

Biogas production of poultry manure was improved by co-digestion with lawn grass cuttings. Lawn grass cuttings' proportion of 18.76% (R_2) had shown highest biogas production of 104.8 mL/g VS_{added} which is 83% and 242% higher than that of mono-digestion of poultry manure and lawn grass cuttings, respectively. It is due to synergism of the co-digestion mixture at this proportion. However, pure lawn grass cuttings (R_6) yielded just 30.6 mL/g VS_{added} and high VFAs accumulation profile indicated that VFAs accumulation could have inhibited anaerobic digestion.

Acknowledgements The authors wish to thank Department of Biotechnology, Government of India for funding this work (Sanction order No: BT/PR6328/GBD/27/387/2012) and Department of Civil Engineering, National Institute of Technology, Warangal, for providing support to carry out present work.

References

1. Procházka, J., Dolejš, P., Máca, J., Dohányos, M.: Stability and inhibition of anaerobic processes caused by insufficiency or excess of ammonia nitrogen. Bioenergy Biofuels **93**(1), 439–447 (2012)
2. Shah, F.A., Mahmood, Q., Rashid, N., Pervez, A., Raja, I.A., Shah, M.M.: Co-digestion, pretreatment and digester design for enhanced methanogenesis. Renew. Sustain. Energy Rev. **42**, 627–642 (2015)
3. Rao, P.V., Baral, S.S.: Experimental design of mixture for the anaerobic co-digestion of sewage sludge. Chem. Eng. J. **172**(2–3), 977–986 (2011)
4. Callaghan, F.J., Wase, D.A.J., Thayanithy, K., Forster, C.F.: Continuous co-digestion of cattle slurry with fruit and vegetable wastes and chicken manure. Biomass Bioenerg. **22**(1), 71–77 (2002)
5. Zheng, Y., Zhao, J., Xu, F., Li, Y.: Pretreatment of lignocellulosic biomass for enhanced biogas production. Prog. Energy Combust. Sci. **42**(1), 35–53 (2014)
6. Hashimoto, G.: Conversion of straw-manure mixtures to methane at mesophilic and thermophilic temperatures. Biotechnol. Bioeng. **25**(1), 185–200 (1983)
7. Hills, D.J., Roberts, D.W.: Anaerobic digestion of dairy manure and field crop residues. Agric. Wastes **3**(3), 179–189 (1981)
8. Xie, S., Lawlor, P.G., Frost, J.P., Hu, Z., Zhan, X.: Effect of pig manure to grass silage ratio on methane production in batch anaerobic co-digestion of concentrated pig manure and grass silage. Bioresour. Technol. **102**(10), 5728–5733 (2011)
9. Yu, L., Bule, M., Ma, J., Zhao, Q., Frear, C., Chen, S.: Enhancing volatile fatty acid (VFA) and bio-methane production from lawn grass with pretreatment. Bioresour. Technol. **162**, 243–249 (2014)
10. Lehtomaki, A., Huttunen, S., Rintala, J.A.: Laboratory investigations on co-digestion of energy crops and crop residues with cow manure for methane production: effect of crop to manure ratio. Resour. Conserv. Recycl. **51**(3), 591–609 (2007)
11. Mohanty, S., Rajendran, K.: 2020 vision for Indian poultry industry. Int. J. Poult. Sci. **2**(2), 139–143 (2003)
12. Rahman, M.A., Møller, H.B., Saha, C.K., Alam, M.M., Wahid, R., Feng, L.: Optimal ratio for anaerobic co-digestion of poultry droppings and lignocellulosic-rich substrates for enhanced biogas production. Energy. Sustain. Dev. **39**, 59–66 (2017)

13. APHA/AWWA/WEF: Standard Methods for the examination of water and wastewater (2012)
14. Borja, R., Rincón, B., Raposo, F., Domınguez, J.R., Millan, F., Martın, A.: Mesophilic anaerobic digestion in a fluidised-bed reactor of wastewater from the production of protein isolates from chickpea flour. Process Biochem. **39**(12), 1913–1921 (2004)
15. Siles, J.A., Martín, M.A., Chica, A., Borja, R.: Kinetic modelling of the anaerobic digestion of wastewater derived from the pressing of orange rind produced in orange juice manufacturing. Chem. Eng. J. **140**(1–3), 145–156 (2008)
16. Hill, D.T., Cobb, S.A., Bolte, J.P.: Using volatile fatty acid relationships to predict anaerobic digester failure. Trans. Am. Soc. Agric. Eng. **30**(2), 496–501 (1987)
17. Li, Y., Li, Y., Zhang, D., Li, G., Lu, J., Li, S.: Solid state anaerobic co-digestion of tomato residues with dairy manure and corn stover for biogas production. Bioresour. Technol. **217**, 50–55 (2016)

Significance of Addressing Persistence of Pathogens and Micropollutants to Enhance Reuse of Treated Sewages Using Constructed Wetlands

Rahul S. Sutar, Vidyanand M. Motghare, Sharanappa C. Kollur, Yogen Parikh and Shyam R. Asolekar

Abstract This study addresses the occurrence, monitoring and treatment of micropollutants as well as pathogens. The persistent release of micropollutants with wastewater effluent may cause destructive effects on human health and environment. However, the investigation of undetected micropollutants requires more advanced analytical instrumentation and procedures due to their occurrence at trace concentration. If one intends to subject the treated effluents to reuse, there is a need for effective monitoring as well as elimination of pathogens and micropollutants from secondary and especially tertiary effluents. The mechanized treatment technologies including sequencing batch reactor, activated sludge process and extended aeration are less effective for the removal of the pathogens. On the other hand, significant removal of pathogens and micropollutants is exhibited by the natural treatment systems. Hence, constructed wetland is of distinctive importance among the natural treatment systems for the treatment as well as reuse of sewages and sullages in rural as well as peri-urban communities.

Keywords Micropollutants · Pathogens · Sewages · Reuse · Constructed wetland

R. S. Sutar · Y. Parikh · S. R. Asolekar (✉)
Centre for Environmental Science and Engineering, Indian Institute of Technology Bombay, Powai, Mumbai 400076, India
e-mail: asolekar@iitb.ac.in

R. S. Sutar
e-mail: rahulsutar@iitb.ac.in

Y. Parikh
e-mail: yogen.parikh@envadvisor.com

V. M. Motghare · S. C. Kollur
Maharashtra Pollution Control Board, Government of Maharashtra, Mumbai 400022, India
e-mail: jdair@mpcb.gov.in

S. C. Kollur
e-mail: airpams@mpcb.gov.in

A. S. Kalamdhad et al. (eds.), *Advances in Waste Management*,
https://doi.org/10.1007/978-981-13-0215-2_25

1 Introduction

Presence of pathogens and micropollutants are inevitable part of the sewage generation train and are observed to be persistently exhibited in the reuse of treated sewages. Thus, existence of pathogens and micropollutants in the wastewater is the real challenge. The micropollutants comprise of pharmaceuticals, surfactants, personal care products, pesticides, industrial chemicals and steroid hormones, etc., which are found in the varying concentration range in several kinds of wastewaters [1].

The major discharge points of micropollutants comprises of hospital effluents, industrial wastewater from cleaning discharges, agricultural run-off, domestic wastewater from laundry, dishwashing, bathing, inadequate cleaning, run-off from roadways, gardens and lawns [2]. In the water environment, even at trace levels, the micropollutants befitted impending risks to the health of human beings as well as aquatic life. The adverse effects of micropollutants on aquatic communities consist of impairment of liver and renal gill in fish, development of pathogen resistance, the feminization of male fish and decrease in plankton diversity. When exposed to micropollutants such as Diclofenac, the inflammatory and degenerative responses have been detected in the human liver when the toxic impact on human health has been studied. Furthermore, the proliferation inhibition effects are also detected for the development of embryonic cells of human [3].

The persistent release of micropollutants with wastewater effluent may cause destructive effects including long-term toxicity, antibiotic resistance of microorganisms, carcinogenic or teratogenic and endocrine disrupting effects even at trace concentrations [4–6]. These effects may be individualistic, synergistic or antagonistic. Owing to multifaceted structures and bio-accumulating nature, the micropollutants possess an ecological challenge for their degradation leading to negative effects on human health and the environment [7]. However, for the removal micropollutants, none of the technologies, i.e. existing biological wastewater treatment plants or mechanized treatment technologies, are specifically designed. The conventional wastewater treatment methods comprise trickling filters, activated sludge and slow sand filtration, etc. Conventional treatment methods are usually not equipped to remove the micropollutants as they are aimed with the primary objective of removal of moderately or easily biodegradable contaminants [3].

In recent years, to remove the micropollutants from wastewater, numerous advanced treatment technologies comprising advanced oxidation processes (Fenton, photo-Fenton, photolysis, ozonation, sonolysis, etc.), adsorption using activated carbon, membrane bioreactor and membrane separation have been studied to evaluate their effectiveness [3]. It has been recognized worldwide that the issues related to removal of micropollutants and pathogens from wastewaters would have to be addressed in concurrence with each other if one wants to enhance the reuse of treated sewages [8]. The most widely used method for pathogen disinfection is chlorination—which leads to production of carcinogenic trihalomethanes in the presence of natural organic matter. Also, these advanced treatment methods are exclusive which makes them cost prohibitive for application at larger scale.

Therefore, the treatment technologies having lower cost such as constructed wetlands (CWs) are the potential alternative for elimination of micropollutants and pathogens [3, 9]. Hence, the primary objective of this paper is to discuss "significance of addressing persistence of pathogens and micropollutants to enhance reuse of treated sewages using constructed wetlands".

The present study reviews the removal of micropollutants and pathogens addressing the current research accomplishments on the implementation of constructed wetlands. To investigate the performance of various types of CWs, the removal efficiencies of micropollutants and pathogens in CWs are furthermore summarized in the present work. Additionally, the issues and challenges faced for the detection of micropollutants are especially addressed in this paper in the context of their removal. The efforts are made to clearly understand the existence, monitoring and treatment of micropollutants as well as pathogens in the CWs. This study intends to provide an insight on removal of micropollutants and pathogens using CW for the future research.

2 Constructed Wetlands

For the removal of micropollutants and pathogens, the natural treatment systems have been found to be highly efficient. The removal of both micropollutants and pathogens can be enhanced using the low-cost treatment technologies as reported by several authors in the literature. For the removal as well as inactivation of pathogens from wastewaters, CWs have demonstrated to be an effective treatment substitute [10, 11]. One of the noteworthy features of CW eco-technology is its cost-effectiveness comprising the capital as well as operation and maintenance costs in comparison with conventional secondary or tertiary treatment methods which attributes to the glaring strength of this technology for attractive wastewater treatment alternative [12, 13].

The highly efficient treatment technologies comprising constructed wetland can be executed for the removal of micropollutants and pathogens as their partial removal takes place in conventional wastewater treatment. Also, for different types of wastewater, the CW has shown consistent removal of pathogens. Due to low cost, the CW eco-technology now became very popular for wastewater treatment among the rural and peri-urban communities which are unable to bear the cost of conventional treatment systems [14, 15].

2.1 *Micropollutant Removal Mechanisms in Constructed Wetlands*

2.1.1 Sorption

Sorption of micropollutants comprises of different mechanisms including hydrophobic partitioning, ion exchange, van der Waals interaction, surface complexation, etc. The exclusively rich organic matter including agricultural wastes, soil and compost exhibits adsorption of non-polar organic pollutants via hydrophobic process. On the other hand, the substrate materials (e.g. some kinds of clay) show adsorption of ionic or polar contaminants by ionic exchange or electrostatic interactions. In CWs, gravel is the most commonly used substrate material. Owing to lower cost, easy disposal by incineration as well as economic value of reuse, several biosorbents including pine bark, rice husk, granulated cork, etc., have shown better substitutions to common substrate constituents for CW in the recent times [3].

2.1.2 Phyto-degradation and Plant Uptake

For phytoremediation technology, the important mechanisms are accumulation, translocation and direct uptake of micropollutants by the plants. The cellular detoxification methods of plants have been termed as "green liver" as to counter the phytotoxicity of extensive range of xenobiotics, and they have versatile detoxification arrangements [16].

The organic pollutants may be transformed or metabolized to less toxic compounds or they may go through partial or complete degradation and may get rooted in plant tissues in unattainable forms following plant uptake. To accomplish the complete transformation of organic contaminants to end products which are non-toxic, the higher aquatic plants have metabolic proficiencies through mechanism of phyto-degradation. The plant enzymes acts on organic compounds leading to complete mineralization into inorganic compounds including CO_2 and H_2O or partly into stable intermediates which gets stored into the plants during the course of transformation or metabolic degradation. The organic chemicals commonly endure three transformation stages once taken up and translocated such as chemical modification including oxidations, hydrolysis and reductions; conjugation with glutathione, sugars and amino acids and compartmentation or sequestration [16].

2.1.3 Microbial Degradation

The micropollutants may undergo transformation or mineralization to more hydrophobic or hydrophilic compounds during biodegradation. The physico-chemical nature of the contaminants significantly affects the properties as well as magnitude of degradation of organic contaminants inside the CWs by the microbes.

On the other hand, the rates of biodegradation are varied as it involves structure specific enzymatic reactions to the chemicals. In aquatic plant-based systems, the improved insight about the microbial degradation mechanisms is significant as intermediates formed during biodegradation may have similar ecotoxicological effects and may be persistent [16].

2.1.4 Photolysis

In aquatic systems, the photolytic degradation of micropollutants is influenced by numerous factors including the availability of sunlight, light diminution by depth of water and light intensity [16]. In surface flow wetlands, photolysis plays a major role in the removal of micropollutants.

In summary, the foremost pollutant removal mechanisms in the CW include biological (absorption of nutrients and microbial degradation), physical (sedimentation and filtration) and chemical (adsorption) [16–19]. CW acts as biofilter for the effective removal of pathogenic bacteria and viruses from wastewater through amalgamation of physical and chemical as well as biological processes [20]. Several treatment technologies including biofiltration, advanced oxidation processes, ozonation, enzymatic treatment, etc, have been employed for the elimination of micropollutants and pathogens from the sewages.

To remain competitive, a considerable effort has been invested in addressing the multifarious benefits of macrophytes growth in CW for the removal of pathogens and micropollutants including absorption of nutrients, reduction in the flow velocity leading to enhanced adsorption and sedimentation, presence of large surface area for growth of microbes, secretion of toxic substances for removal of pathogenic micro-organisms and alteration of physical and chemical environment surrounding to the rhizosphere [16, 17, 19, 21]. In this study, for enabling recycling of treated sewages, considering the several benefits of CW among the natural treatment systems, such as polishing ponds, duckweed ponds, waste stabilization ponds, oxidation ponds, its importance has been significantly addressed for the removal of micropollutants and pathogens.

3 Removal of Micropollutants

In the recent years, the removal capability of numerous micropollutants using CW has been assessed by several authors. The several categories, different types and concentrations of some micropollutants are depicted in Table 1. However, the micropollutants belonging to the categories of pharmaceuticals and the personal care products are of major concern which may be discharged as a parent compound, conjugated compounds or their metabolites [22].

Table 1 Occurrence of micropollutants in the wastewater

S. No.	Categories	Compounds	Concentration (μg/L)	Reference
1	Pharmaceuticals	Norfloxacin	390–420	[34]
		Lomefloxacin	150–300	
		Enoxacin	150–300	
		Ofloxacin	150–160	
		Ranitidine	90–160	
		Caffeine	20.7 ± 6.8	[35]
		Salicylic acid	15.2 ± 7.3	
		Ibuprofen	9.8 ± 1.9	
2	Pesticides	Atrazine	0.02–28	[1]
		Diuron	0.03–1.96	
3	Surfactants	Nonylphenol	<0.03–101.6	[1]
		Octylphenol	<0.2–8.7	

3.1 *The Issues Faced While Analysing Micropollutants*

Regarding the detection of micropollutants, there are many problems associated due to their existence at trace concentrations hampering their detection and analysis which leads to generation of challenges for treatment processes. There are no regulatory limits present for the micropollutants till date in India.

In order to determine the micropollutants in aquatic environment, the most typically used techniques include solid phase extraction as well as chromatography technology combined with mass spectrometry—which are LC–MS, LC–MS/MS and GC–MS. The major issues associated with analysis of emerging micropollutants consists of sample preparation (loss of polar compounds and desorption of non-polar compounds), co-extraction of undesired intrusions, matrix effect and method validation as no standard validation protocols and no standardized methods exist [23].

3.2 *Removal Efficiencies of Micropollutants in Constructed Wetlands*

In CWs, the efficient reduction of most micropollutants under aerobic conditions has been already reported by several authors than anaerobic pathways. Under the aerobic conditions, some pharmaceuticals including ibuprofen are best reduced. On the other hand, the anaerobic conditions are favoured for the removal of other micropollutants such as Diclofenac, Clofibric acid. However, the halogenated pollutants show higher removal rates under anoxic conditions. Hence, the occurrence of aerobic, anaerobic and anoxic environments in the immediate vicinity of the plant rhizomes are the key advantages of horizontal and vertical subsurface flow wetland systems leading to significant reduction in the concentrations of different micropollutants [24].

In horizontal subsurface flow CW, the removal of Pentachlorobenzene, Pentachlorophenol, Endosulfan and Lindane is >94% [25]. Avila et al. also reported >90% removal of Naproxen, Diclofenac and Tonalide in the treatment scheme comprised of an anaerobic reactor followed by two 0.65 m^2 parallel wetlands and coupled to a 1.65 m^2 in series wetland. In a pilot-scale CW, a continuous injection experiment was executed to determine the behaviour of micropollutants [26].

Vertical subsurface flow CW is also implemented for the elimination of micropollutants [27, 28]. Matamoros et al. evaluated the removal efficiencies and elimination kinetics of several pharmaceuticals as well as personal care products (PPCPs) in a pilot-scale vertical subsurface flow constructed wetland (VFCW) planted with *Phragmites australis* and compared those with the sand filter. As compared to CWs of other configurations such as horizontal subsurface flow and sand filter (non-planted), the VFCW was found to be more efficient for most of the PPCPs. The contaminants including Ibuprofen, Salicylic acid, Naproxen and Caffeine have shown higher removal efficiencies (>90%) [3, 29].

Additionally, in surface flow systems, the photodegradation processes are mainly responsible for the elimination of several pharmaceuticals and personal care products such as Diclofenac and Ketoprofen from aquatic environments. In the removal of micropollutants, the biodegradation and photodegradation reactions are stimulated by high hydraulic retention times [24]. Breitholtz et al. evaluated the micropollutant removal in free water surface wetland which is dominated by emergent plants such as *Typha* spp., *P. australis, Carex* spp. and submerged plants including *Myriophyllum* spp., *Ceratophyllum demersum, Elodea* spp. The micropollutants such as Alfuzosin, Amitriptyline, Citalopram, Ibuprofen and Ranitidine have shown 96, 84, 84, 80 and 92% removals, respectively [30]. However, the complete removal has been reported in the surface flow wetland for Norfloxacin, Ciprofloxacin, Doxycycline, Azithromycin in the 18 similarly shaped, small (water surface 29 m^2) surface flow wetlands which were dominated by vegetation of *P. australis* and *Typha latifolia* [31].

The performance of numerous types of CWs for the removal of micropollutant has been overviewed and represented in Table 2.

4 Removal of Pathogens

A pathogen is a biological agent which is responsible for disease or illness in its host. Waterborne pathogens are classified into five classes consisting viruses, protozoa, bacteria, helminths and fungi. The foremost indicator micro-organisms of faecal contamination are the *Escherichia coli,* faecal coliforms, enterococci and *Clostridium perfringens*. Hence, these indicators are typically employed to investigate the microbiological quality of treated effluents as the presence as well as

Table 2 Removal of micropollutants in constructed wetland

S. No.	Type of constructed wetland	Compound	Removal (%)	References
1	Free water surface wetland	Alfuzosin	96	[30]
		Amitriptyline	84	
		Citalopram	84	
		Ibuprofen	80	
		Ranitidine	92	
2	Surface flow CW	Norfloxacin	100	[31]
		Ciprofloxacin	100	
		Doxycycline	100	
		Azithromycin	100	
		Clarithromycin	59	
3	Horizontal subsurface flow CW	Alachlor	80	[25]
		Chlorpyriphos	83	
		Pentachlorobenzene	>99	
		Pentachlorophenol	94	
		Endosulfan	>99	
		Lindane	>99	
4	Horizontal subsurface flow CW	Naproxen	52–93	[36]
		Ibuprofen	52–72	
		Caffeine	82–90	
		Salicylic acid	86–90	
		Ketoprofen	94–96	
5	Horizontal subsurface flow CW	Naproxen	93	[26]
		Diclofenac	93	
		Tonalide	94	
		Bisphenol A	83	
6	Horizontal subsurface flow CW	Doxycycline	71–79	[3]
		Sulfadimethoxine	99	
		Sulfamethoxazole	73–87	
		Trimethoprim	95–99	
7	Subsurface flow CW	Methyl dihydrojasmonate	71–96	[35]
		Galaxolide	67–82	
		Tonalide	55–74	
8	Microcosm subsurface flow CW	Ibuprofen	96	[37]
		Carbamazepine	97	
		Clofibric acid	75	
9	Hybrid CW system	Ibuprofen	42–89	[3]
		Diclofenac	78–87	
		Ketoprofen	77–81	
		Naproxen	73–82	
		Caffeine	95–96	

(continued)

Table 2 (continued)

S. No.	Type of constructed wetland	Compound	Removal (%)	References
10	Hybrid CW system	Sulfamethoxazole	91	[3]
		Sulfapyridine	76	
		Atenolol	93	
		Metoprolol	92	
		Nadolol	77	
		Gemfibrozil	61	
		Caffeine	>99	
		Cotinine	99	
11	Vertical subsurface flow CW	Ibuprofen	99	[3, 29]
		Diclofenac	73	
		Salicylic acid	98	
		Naproxen	89	
		Caffeine	99	
12	Vertical subsurface flow CW	Ciprofloxacin HCl	82–85	[3, 28]
		Oxytetracycline HCl	91–95	
		Sulfamethazine	68–73	
13	Vertical subsurface flow CW	Enrofloxacin	94	[3, 27]
		Tetracycline	94	

behaviour of the foremost human pathogens in wastewaters is evidenced by this determination [13]. Epidemiological research has shown that the higher concentration of indicator bacteria in the polluted water ensures positive connection with the gastrointestinal and respiratory diseases when it comes in contact [32].

Conventional methods of pathogen removal include trickling filter, slow sand filtration and activated sludge which are most frequently reported. However, the tertiary treatment (or polishing) methods for the elimination of pathogen comprise of UV disinfection, ozone and chlorination which are also most commonly cited in the literature. In the recent years, a growing concern over chlorination has been developed, although it is a popular method of disinfection, as chlorine residuals may produce harmful effects on several micro-organisms and fish in the receiving water bodies. Also, another disadvantage of chlorination is the formation of trihalomethane or other carcinogenic organochlorine compounds when free chlorine gets react with a substantial amount of natural organic matter [13].

The possible formation of harmful by-products has led to the development of UV and ozonation methods for the removal of pathogens. However, higher energy usage, cost and maintenance have limited their use for the pathogen treatment. Hence, in comparison with the conventional or tertiary treatment systems, the wetland technology has now become fascinating cost-effective treatment option owing to low capital and operating costs [13]. This low energy, non-chemical method of pathogen treatment now becomes popular in remote areas as well as developing countries like India.

The macrophytes (aquatic plants) in the wetlands possess several advantages such as enormous surface areas for the growth and attachment of microbes, stabilization of the surface of CWs and provision of physical filtration [13, 18]. The associated biofilm with the root-substrate complexes has the capacity for filtration as well as adsorption of pathogens in planted CW systems. CWs have already been implemented for the elimination of pathogens such as faecal coliforms, *Giardia,* and *Cryptosporidium* and parasites including *Ascaris suum, Toxocara vitulorum* and *Hymenolepis diminuta* [13].

4.1 Pathogen Removal Mechanisms in Constructed Wetlands

The key mechanisms responsible for the removal of pathogens in the CW include natural die off, sedimentation, inactivation or death related to temperature, predation, oxidation, adhesion to biofilm, unfavourable water chemistry, mechanical filtration, etc. [13]. In the wetland, pathogens either get removed or may inactivate within the sediments of wetland.

The removal and transformation of dissolved and particulate contaminants in the wetland are governed by the activity of microbial communities, predominantly those linked with biofilms. The biofilm presents in the immediate vicinity of bed media and in the rhizosphere region influences the elimination of pathogens. Oxidation is also major mechanism responsible for the elimination of pathogens. The presence of oxygen develops unfavourable situations for enteric bacteria (either obligate anaerobes or facultative). Enhanced removal of *E.* coli has been reported at higher dissolved oxygen levels [13].

4.2 Removal Efficiencies of Pathogens in Constructed Wetlands

The horizontal subsurface flow CW has shown higher removal of pathogens. Makvana and Sharma have reported 96, 99, 94, 87 and 94% removal of total coliform, faecal coliform, *Salmonella, Shigella* and *Vibrio,* respectively [14]. Chang et al. have reported the 98.93 and 99.92% removal of faecal coliform and *E. coli* in subsurface up-flow wetland [33]. Combination of surface and subsurface flow CW has shown almost complete removal of *C. perfringens, Giardia* cysts and *Cryptosporidium oocysts, E. coli* and total coliforms [32].

The performance of several types of CWs for the removal of pathogens has been overviewed and represented in Table 3.

Table 3 Removal of pathogens in constructed wetland

S. No.	Type of constructed wetland	Pathogen	Removal (%)	**References**
1	Surface flow wetland	Faecal coliform	96.8–99.7	[20]
2	Surface flow wetland	Faecal coliform	98.7	[38]
	Subsurface flow wetland	Faecal coliform	82	
3	Subsurface flow CW	*Escherichia coli*	95.5	[39]
		Total coliforms	74.4	
4	Horizontal subsurface flow CW	Faecal coliform	92.3	[40]
5	Horizontal subsurface flow CW	Total coliform	96	[14]
		Faecal coliform	99	
		Salmonella	94	
		Shigella	87	
		Vibrio	94	
6	Subsurface up-flow wetland	Faecal coliform	98.93	[33]
		Escherichia coli	99.92	
7	Combination of surface and subsurface flow CW	*Escherichia coli*	99.9	[32]
		Total coliforms	99.9	
		Clostridium perfringens	100	
		Faecal streptococci	97	
		Giardia cysts	100	
		Cryptosporidium oocysts	100	

5 Conclusions

For the regulation in future, the micropollutants show potential candidature based on the occurrence and their potential health effects as they belong to the unregulated pollutants due to continuous introduction into the aquatic environment. Undoubtedly, due to numerous cost advantages (both capital and operation and maintenance costs), the natural treatment system might prove to be the outstanding technological substitute. Thus, in the developing tropical countries, the low-cost, reliable, effective and sustainable technology i.e. constructed wetland exhibits the significant potential for the prevention and management of waterborne diseases through the removal of pathogens. Hence, in rural and peri-urban communities, the constructed wetlands are of distinctive importance for the treatment as well as reuse of sewages and sullages.

Acknowledgements Authors acknowledge the co-funding from Rajiv Gandhi Science and Technology Commission, Government of Maharashtra and Indian Institute of Technology Bombay for this work.

References

1. Luo, Y., Guo, W., Hao, H., Duc, L., Ibney, F., Zhang, J., Liang, S., Wang, X.C.: A review on the occurrence of micropollutants in the aquatic environment and their fate and removal during wastewater treatment. Sci. Total Environ. **473–474**, 619–641 (2014)
2. Renge, P.V.C., Khedkar, P.S.V., Bhoyar, K.S.: Micropollutant removal from waste water treatment plant—a review. Int. J. Adv. Eng. Technol. **3**, 27–30 (2012)
3. Li, Y., Zhu, G., Ng, W.J., Tan, S.K.: A review on removing pharmaceutical contaminants from wastewater by constructed wetlands: Design, performance and mechanism. Sci. Total Environ. **468–469**, 908–932 (2014)
4. Fent, K., Weston, A.A., Caminada, D.: Ecotoxicology of human pharmaceuticals. Aquat. Toxicol. **76**, 122–159 (2006)
5. Pruden, A., Pei, R., Storteboom, H., Carlson, K.H.: Antibiotic resistance genes as emerging contaminants: studies in Northern Colorado. Environ. Sci. Technol. **40**, 7445–7450 (2006)
6. Murillo-Torres, R., Durán-Álvarez, J.C., Prado, B., Jiménez-Cisneros, B.E.: Sorption and mobility of two micropollutants in three agricultural soils: a comparative analysis of their behavior in batch and column experiments. Geoderma **189–190**, 462–468 (2012)
7. Sutar, R.S., Rathod, V.K.: Ultrasound assisted enzyme catalyzed degradation of cetirizine dihydrochloride. Ultrason. Sonochem. **24**, 80–86 (2015)
8. Arceivala, S.J., Asolekar, S.R.: Environmental Studies: A Practitioner's Approach. Tata McGraw Hill Education Pvt. Ltd., New Delhi (2012)
9. Arceivala, S.J., Asolekar, S.R.: "Wastewater Treatment for Pollution Control and Reuse", (3rd edn, 11th Reprint). Tata McGraw Hill Education India Pvt. Ltd., New Delhi (2007)
10. Reddy, L.D., Kumar, D., Asolekar, S.R.: Typologies for successful operation and maintenance of horizontal sub-surface flow constructed wetlands. Int. J. Eng. Res. **3**, 147–154 (2014)
11. Kumar, D., Asolekar, S.R., Sharma, S.K.: Post-treatment and reuse of secondary effluents using natural treatment systems: the Indian practices. Environ. Monit. Assess. **187**, 612 (2015)
12. Kumar, D., Chaturvedi, M.K.M., Sharma, S.K., Asolekar, S.R.: Sewage-fed aquaculture: a sustainable approach for wastewater treatment and reuse. Environ. Monit. Assess. **187**, 656 (2015)
13. Weber, K.P., Legge, R.L.: Pathogen removal in constructed wetlands. In: aymundo E.R. (ed) Wetlands: Ecology, Conservation & Restoration, Chapter 5, pp. 1–35 (2008)
14. Makvana, K.S., Sharma, M.K.: Assessment of pathogen removal potential of root zone technology from domestic wastewater. Univers. J. Environ. Res. Technol. **3**, 401–406 (2013)
15. Wintgens, T., Nattorp, A., Elango, L., Asolekar S.R.: "Natural Water Treatment Systems for Safe and Sustainable Water Supply in the Indian Context" Saph Pani. IWA Publishing, Alliance House, 12 Caxton Street, London SW1H 0QS, UK (2016)
16. Zhang, D., Gersberg, R.M., Ng, W.J., Tan, S.K.: Removal of pharmaceuticals and personal care products in aquatic plant-based systems: a review. Environ. Pollut. **184**, 620–639 (2014)
17. Brix, H.: Functions of macrophytes in constructed wetlands. Water Sci. Technol. **29**, 71–78 (1994)
18. Brix, H.: Do macrophytes play a role in constructed treatment wetlands? Water Sci. Technol. **35**, 11–17 (1997)
19. Avelar, F.F., de Matos, A.T., de Matos, M.P., Borges, A.C.: Coliform bacteria removal from sewage in constructed wetlands planted with *Mentha aquatica*. Environ. Technol. **35**, 2095–2103 (2014)
20. Smith, E., Gordon, R., Madani, A., Stratton, G.: Pathogen removal by agricultural constructed wetlands in cold climates. J. Environ. Inform. **6**, 46–50 (2005)
21. Gouriveau, F.: Constructed farm wetlands (CFWs) designed for remediation of farmyard runoff: an evaluation of their water treatment efficiency, ecological value, costs and benefits. Ph.D thesis, The University of Edinburgh (2009)

22. Onyiri, O.: Micro-pollutants : an emerging concern for aquatic organisms. Campbellsville Rev. 51–60 (2013)
23. Jiang, J.Q., Zhou, Z., Sharma, V.K.: Occurrence, transportation, monitoring and treatment of emerging micro-pollutants in waste water—a review from global views. Microchem. J. **110**, 292–300 (2013)
24. Verlicchi, P., Galletti, A., Petrovic, M., Barcelo, D.: Hospital effluents as a source of emerging pollutants: an overview of micropollutants and sustainable treatment options. J. Hydrol. **389**, 416–428 (2010)
25. Matamoros, V., Puigagut, J., García, J., Bayona, J.M.: Behavior of selected priority organic pollutants in horizontal subsurface flow constructed wetlands: a preliminary screening. Chemosphere **69**, 1374–1380 (2007)
26. Avila, C., Pedescoll, A., Matamoros, V., Bayona, J.M., Garcia, J.: Capacity of a horizontal subsurface flow constructed wetland system for the removal of emerging pollutants: an injection experiment. Chemosphere **81**, 1137–1142 (2010)
27. Carvalho, P.N., Araújo, J.L., Mucha, A.P., Basto, M.C.P., Almeida, C.M.R.: Potential of constructed wetlands microcosms for the removal of veterinary pharmaceuticals from livestock wastewater. Bioresour. Technol. **134**, 412–416 (2013)
28. Liu, L., Liu, C., Zheng, J., Huang, X., Wang, Z., Liu, Y., et al.: Elimination of veterinary antibiotics and antibiotic resistance genes from swine wastewater in the vertical flow constructed wetlands. Chemosphere **91**, 1088–1093 (2013)
29. Matamoros, V., Arias, C., Brix, H., Bayona, J.M.: Removal of pharmaceuticals and personal care products (PPCPs) fromurban wastewater in a pilot vertical flow constructed wetland and a sand filter. Environ. Sci. Technol. **41**, 8171–8177 (2007)
30. Breitholtz, M., Näslund, M., Stråe, D., Borg, H., Grabic, R., Fick, J.: An evaluation of free water surface wetlands as tertiary sewage water treatment of micro-pollutants. Ecotox. Environ. Safe. **78**, 63–71 (2012)
31. Berglund, B., Khan, G.A., Weisner, S.E.B., Ehde, P.M., Fick, J., Lindgren, P.E.: Efficient removal of antibiotics in surface-flow constructed wetlands with no observed impact on antibiotic resistance genes. Sci. Total Environ. **476–477**, 29–37 (2014)
32. Molleda, P., Blanco, I., Ansola, G., de Luis, E.: Removal of wastewater pathogen indicators in a constructed wetland in Leon. Spain. Ecol. Eng. **33**, 252–257 (2008)
33. Chang, N.B., Xuan, Z., Daranpob, A., Wanielista, M.: A subsurface upflow wetland system for removal of nutrients and pathogens in on-site sewage treatment and disposal systems. Environ. Eng. Sci. **28**, 1–14 (2011)
34. Larsson, D.G.J., de Pedro, C., Paxeus, N.: Effluent from drug manufactures contains extremely high levels of pharmaceuticals. J. Hazard. Mater. **148**, 751–755 (2007)
35. Reyes-Contreras, C., Hijosa-Valsero, M., Sidrach-Cardona, R., Bayona, J.M., Bécares, E.: Temporal evolution in PPCP removal from urban wastewater by constructed wetlands of different configuration: a medium-term study. Chemosphere **88**, 161–167 (2012)
36. Zhang, D.Q., Gersberg, R.M., Hua, T., Zhu, J., Tuan, N.A., Tan, S.K.: Pharmaceutical removal in tropical subsurface flow constructed wetlands at varying hydraulic loading rates. Chemosphere **87**, 273–277 (2012)
37. Dordio, A., Carvalho, A.J.P., Teixeira, D.M., Dias, C.B., Pinto, A.P.: Removal of pharmaceuticals in microcosm constructed wetlands using *Typha spp.* and LECA. Bioresour. Technol. **101**, 886–892 (2010)
38. Greenway, M.: The role of constructed wetlands in secondary effluent treatment and water reuse in subtropical and arid Australia constructed wetlands in Australia. Ecol. Eng. **25**, 501–509 (2004)
39. Karimi, B., Ehrampoush, M.H., Jabary, H.: Indicator pathogens, organic matter and LAS detergent removal from wastewater by constructed subsurface wetlands. J. Environ. Health Sci. Eng. **12**(52), 1–7 (2014)
40. Mburu, N., Thumbi, G.M., Mayabi, A.O.: Removal of bacterial pathogens from domestic wastewater in a tropical subsurface horizontal flow constructed wetland. In: The 12th World Lake Conference, pp 1010–1015 (2008)

A Review on Nitritation Process

Sourabh Dixit and S. K. Patidar

Abstract Due to the growing population and increased discharge of wastewater, the nutrient concentration is increasing in the surface water bodies. The effluent standards are now more stringent to reduce nutrient load in receiving water sources. Nitrogen removal is achieved through nitrification followed by de-nitrification in wastewater treatment plants. Recently, nitrogen removal via nitrite pathway, i.e., nitritation has caught the attention of researchers as it helps in reducing aeration costs by 25% without the need to add an extra carbon source in the de-nitritation step. In the present paper, biochemistry of nitrogen removal process, microbiology of nitrification, and factors affecting nitritation are discussed. A summary of studies involving sequencing batch reactor (SBR) for nitrite accumulation is also presented.

Keywords Nitritation · Wastewater treatment · Ammonia-oxidizing bacteria
Nitrite-oxidizing bacteria

1 Introduction

Due to the population explosion and increased discharge of wastewater, the nutrient concentration has been increasing in the surface water bodies. Thus, there was an urgent need for the legislative bodies to reform and amend the nutrient disposal laws and as a result today, we have more stringent nutrient discharge standards. The wastewater treatment plants have sought new technologies to cope up with the corresponding discharge standards [1]. In the past few decades, biological nitrogen

S. Dixit (✉) · S. K. Patidar
Department of Civil Engineering, National Institute of Technology, Kurukshetra, Kurukshetra 136119, India
e-mail: sourabhgate2014@gmail.com

S. K. Patidar
e-mail: patidarsk@rediffmail.com

A. S. Kalamdhad et al. (eds.), *Advances in Waste Management*,
https://doi.org/10.1007/978-981-13-0215-2_26

removal (BNR) has been preferred over various other physiochemical methods for lessening the nutrient load on the streams and rivers and has developed new sophisticated technologies for removing nitrogen from wastewaters. Apart from conventional nitrification and de-nitrification, many other methods, viz. partial nitrification and de-nitrification, ANAMMOX, simultaneous nitrification and de-nitrification, CANNON, etc., have been developed [2]. These processes have led to the reclassification of the global nitrogen cycle according to the potential biochemical pathway [3].

The partial nitrification process, i.e., nitrification up to nitrite and then the de-nitrification of this nitrite into gaseous nitrogen [4] promises several advantages over the conventional nitrification and de-nitrification process [5–7]. The advantages include (1) 20% less carbon dioxide emission; (2) reduced electron donor requirement in the anoxic stage by 40%; (3) 25% less oxygen consumption in the aerobic stage, simultaneously saving 60% energy costs; (4) 1.5–2 times increment in the de-nitrification rates; (5) reduction in sludge production by 33–35 and 55% in nitrification and de-nitrification stages, respectively.

The partial nitrification is difficult to be achieved in municipal wastewater treatment systems and causes some filamentous sludge bulking along with low phosphorus removal. Partial nitrification is a cheap and technologically feasible process when performing the treatment of wastewater containing high ammonium nitrogen or low *C/N* ratio [8, 9]. In the present paper, biochemistry of nitrogen removal process, microbiology of nitrification, and factors affecting nitritation are discussed. A summary of studies involving SBR for nitrite accumulation is also presented.

2 Biochemistry of Conventional Nitrogen Removal Process

Nitrification is the complete biological oxidation of ammonium nitrogen to nitrate [10]. Henze [11] stated that it is a two-step process occurring in a sequence. Stoichiometrically the overall nitrification reaction can be represented as follows (EPA 1975):

$$NH_4^+ + 3/2O_2 \rightarrow 2H^+ + H_2O + NO_2^- + 58 - 84\ \text{kcal} \quad (1)$$

$$NO_2^- + 1/2O_2 \rightarrow NO_3^- + 15.4 - 20.9\ \text{kcal} \quad (2)$$

The reaction (1) and (2) are brought about by two phylogenetically unrelated groups of bacteria, namely ammonia oxidizing bacteria (AOB) and nitrite oxidizing bacteria (NOB). Oxidation of NH_4^+ to NO_2^- proceeds with hydroxylamine as an intermediate by the action of two enzymes ammonia monooxygenase (AMO) and hydroxylamine oxidoreductase (HAO). Oxidation of ammonia is endothermic while that of hydroxylamine is energy giving reaction. Suzuki et al. [12] stated that NH_3 is the actual substrate instead of NH_4^+ in the first step.

$$NH_3{}^{+} + O_2 + 2H^{+} + 2e^{-} \rightarrow NH_2OH + H_2O - 120\ \text{kJ/mol} \quad (3)$$

$$NH_2OH + 1/2O_2 \rightarrow HNO_2 + 2H^{+} + 2e^{-} - 114\ \text{kJ/mol} \quad (4)$$

$$NH_2OH + H_2O \rightarrow NO_2{}^{-} + 5H^{+} + 4e^{-} + 23\ \text{kJ/mol} \quad (5)$$

A proton motive force is generated when the two spare electrons produced in reaction (5) are passed to the terminal oxidase via an electron transport chain. Two other electrons produced in the reaction (5) are already compensated for the input of the electrons in the reaction (3) [13]. The nitrite that is formed in the reaction (1) is oxidized to NO_3^- by NOB by the help of nitrite oxidoreductase (NOR) enzyme. One important point is that the bacteria use NH_4^+ as an energy source but do not nitrify all of it in nitrification, only a part of it is utilized for cell growth and in this process, CO_2 fulfills the need of the carbon source.

$$4CO_2 + 4H_2O + NH_4{}^{+} + HCO_3^{-} \rightarrow C_5H_7O_2N + 3H_2O + 5O_2 \quad (6)$$

The overall equation for nitrification given by the EPA (1975) suggests that production of H^+ (Eq. 8) in nitrification process causes a decrement in the alkalinity per mole of NH_4^+ removed [13].

$$NH_4{}^{+} + 1.83O_2 + 1.98HCO_3{}^{-} \rightarrow 0.021C_5H_7O_2N + 0.98NO_3{}^{-} + 1.041H_2O + 1.88H_2CO_3 \quad (7)$$

$$CO_2 + H_2O \leftrightarrow H_2CO_3 \leftrightarrow H^{+} + HCO_3 \quad (8)$$

De-nitrification [10] is the reduction of nitrates into nitrogen gas with nitrite, nitric oxide, and nitrous oxide as the intermediates.

$$NO_2{}^{-} + 0.53CH_3OH + 0.67H_2CO_3 \rightarrow 0.04C_5H_7O_2N + 0.48N_2 + HCO_3{}^{-} + 1.23H_2O \quad (9)$$

$$NO_3{}^{-} + 1.08CH_3OH + 0.24H_2CO_3 \rightarrow 0.056C_5H_7O_2N + 0.47N_2 + HCO_3{}^{-} + 1.68H_2O \quad (10)$$

Reactions (9) and (10) show the de-nitrification process over nitrite and nitrate, respectively. The reactions (1) and (9) show the nitritation and de-nitritation process, respectively.

3 Microbiology of Nitrification

As discussed in the previous section, the nitrification reaction is caused by two groups of bacteria ammonia-oxidizing bacteria (AOB) and nitrite-oxidizing bacteria (NOB). Not even a single group of bacteria has been isolated that directly oxidizes ammonia to nitrate [14–16]. Studies in the past two decades using comparative 16S rRNA analyses have shown that AOBs and NOBs can be dealt as separate lineages within the *Proteobacteria* [15, 17].

3.1 Classification of Ammonia-Oxidizing Bacteria

Distinguishable by the cell morphology and multilayered gram-negative cell walls, AOBs can be classified into two phylogenetically different groups of *Proteobacteria* in two subclasses β and γ, respectively [18]. The γ subclass has only one cluster of *Nitrosococcus* genera [19] while the β subclass has four different genera of bacteria including clusters of *Nitrosovibrio*, *Nitrosospira*, *Nitrosolobus*, and *Nitrosomonas*. 25 distinct species of AOBs have been cultured till date which obtains energy for their survival from aerobic oxidation of ammonia to nitrite. It is also believed that some peculiar species of AOBs can grow in both aerobic as well as anaerobic conditions. Nitrite is the main product in aerobic condition while anaerobic condition has nitric oxide, nitrite, and nitrogen gas as the main products of oxidation [20].

3.2 Classification of Nitrite-Oxidizing Bacteria

Eight pure cultures of NOBs exist, which can be described by four different phylogenetic groups. The α subclass is assigned to the genera *Nitrococcus* while the genera *Nitrobacter* is assigned γ subclass. Also, the genera *Nitrospira* was assigned δ subclass [18]. *Nitrospira* is believed to be the more specialized and dominant NOB in WWTPs over *Nitrobacte*r as reported by [21]. *Nitrospira* has low maximum activity and a high substrate affinity; therefore, it is called K-strategists for nitrite and O_2, whereas *Nitrobacter* is called r-strategists in limiting substrate conditions. Table 1 shows various characteristics of AOBs and NOBs.

Table 1 Important characteristics of AOB and NOB [22]

Nitrifiers	Cell shape	Flagellation of motile cells	Cell size (micrometers)	Arrangement of intra-cytoplasmic membranes	μ_{max} (h^{-1})
Nitrosococcus	Spherical to ellipsoidal	Tuft of flagella	1.5–1.8 × 1.7–2.5	Peripheral or central stacks of vesicles	–
Nitrosolobus	Pleomorphic lobate	Peritrichous	1–1.5 × 1–2.5	Compartmentalizing	–
Nitrosomonas	Straight Rods	Polar to subpolar	0.7–1.5 ×1–2.4	Peripheral flattened vesicles	0.043 0.039
Nitrosospira	Tightly coiled spirals	Peritrichous	0.3–0.8 × 1–8	Invaginations	–
Nitrosovibrio	Slender curved rods	Polar to subpolar	0.3–0.4 × 1.1–3	Invaginations	–
Nitrobacter	Pear-shaped/ pleomorphic rods	Polar to lateral	0.5–0.8 × 1–2	Polar flattened vesicles	0.20 0.032 0.043
Nitrococcus	Spherical	Polar	1.5	Randomly arranged tubules	–
Nitrospina	Slender straight rods	–	0.3–0.4 × 1.7–6.6	–	–
Nitrospira	Loosely coiled spiral	–	0.3–0.4 × 0.8–1	Invaginations	–

4 Factors Affecting Achievement and Maintenance of Nitritation

Many researchers have tried to unfold the mystery of factors affecting nitritation process over the past few decades. It has been widely reported that stable nitritation can be achieved by keeping a check on pH, DO, temperature, free ammonia, and free nitrous acid concentration. Nitritation is mainly affected by DO concentration [23].

4.1 Real-Time Control for Partial Nitrification

It is an online control technique which basically aims at providing conducive growing conditions for AOBs simultaneously favoring the washout of NOBs. Many recent studies have focused on real-time control strategies to control various factors responsible for nitritation as it helps in the reduction of start-up time to a minimum, thus helping in the achievement of stable nitritation [23]. Ge et al. [22] also support the application of real-time control in partial nitrification process to determine the duration for aeration. Nitrogen concentration measurement with NO_2^--N, NH_4^+-N, and NO_3^--N probes are used for the direct online control strategy, whereas the indirect control strategy uses sophisticated pH, DO, ORP, OUR, and BF probes.

Blackburne et al. [24] achieved nitritation by applying aerobic duration control by OUR measurement. Summary of studies involving real-time control for partial nitrification is presented in Table 2.

4.2 *Dissolved Oxygen Concentration*

The activity of nitrite oxidizers decreases more rapidly in low DO concentration than ammonia oxidizers. This is attributed to the fact that there is a difference in the oxygen half-saturation constant of AOBs and NOBs, and the AOBs release more amount of energy in consuming one mole of O_2 [29]. High and low dissolved oxygen levels are reported to cause nitrite accumulation [22]. Peng and Zhu [30] have clearly stated that DO should be between 1 and 1.5 mg/L in order to achieve ammonium oxidation along with nitrite accumulation. Katsou et al. [31] successfully treated anaerobic effluent at low DO concentration of 0.7 mg/L. Kulikowska and Bernat [32] accumulated 98% nitrite by controlling DO at 0.7 mg/L.

4.3 *Free Nitrous Acid (FNA) and Free Ammonia (FA) Inhibition*

Anthonisen et al. [33] reported that 0.1–1.0 mgN/L of FA can cause NOB inhibition (as FA interferes with the metabolism of nitrite oxidoreductase found on the cell wall of NOBs) while AOB is inhibited between 10 and 150 mgN/L concentration of FA. pH plays a major role in influencing the equilibrium between the FNA and nitrite and also the equilibrium of FA; thus, pH can be regulated to achieve nitritation. Abeling and Seyfried [34] stated that pH in the range of 7.5–8.5 is beneficial for accomplishing successful nitritation. The concentration of FA and FNA can be determined by Eqs. (11) and (12), respectively, where TAN is total ammonium and TNO_2 is total nitrite.

$$[NH_3] = \frac{[TAN]10^{pH}}{e^{\frac{6344}{T+273}} + 10^{pH}} \tag{11}$$

$$[HNO_2] = \frac{[TNO_2]10^{-pH}}{e^{-2300/T+273} + 10^{-pH}} \tag{12}$$

Peng and Zhu [30] revealed the mechanism of FNA inhibition, in which a proton is donated by HNO_2 inside the cell and this donated proton directly interferes in the synthesis of ATP. Van Hulle et al. [29] concluded that FA and FNA do not kill the NOBs; they only have an inhibition effect as the NOBs get adapted to these conditions. FA inhibition is more in the pH range (>8), while FNA has more inhibition at low pH (<7.5).

Table 2 Summary of studies involving real-time control for partial nitrification process [23]

Type of wastewater	System	RCP start-up	RCP (long term)	Strategy	Performances	References
Municipal	Bench scale	DO	DO	Evaluation of O_2 concentration	98% nitrite accumulation	[25]
Municipal	Bench scale	pH	pH	Aeration duration based on NH_4 valley	95% partial nitrification	[26]
Strong nitrogenous	Pilot demonstration scale	Blower frequency	Blower frequency	Detection of NBP in blower frequency	Achieving of Shortcut in 40 days	[27]
Swine wastewater	Pilot scale	ORP, DO, pH	–	Intermittent aeration DO <2 mg/l	83% nitrite accumulation	[28]

RCP real-time control parameters

4.4 Alternate Oxic–Anoxic Phases

Compared to AOBs, NOBs have lower growth rate. Thus, NOBs are more affected in anoxic conditions and are not able to recover from the deficiency of O_2 in the aerobic phases. SBRs have been investigated for nitritation process through alternate aerobic-anaerobic phases. Kornaros et al. [35] have successfully operated a pilot-scale SBR to treat real domestic wastewater by applying alternate oxic–anoxic phases. Ge et al. [36] also successfully accumulated more than 81% nitrite in a plug flow reactor treating municipal wastewater.

4.5 Temperature

One of the main aspects of nitritation is to enhance the growth rate of AOBs and decrease the growth rate of NOBs. This can be easily accomplished by controlling the temperature above 15 °C, so as to increase the specific growth rate of AOBs [30]. Hellinga et al. [9] added that increasing the temperature unnecessarily will not result in stable nitritation, and an optimum temperature of 25 °C was suggested by him to achieve stable nitritation. It is inferred from Van Hulle et al. [29] that the activation energy of AOB is greater than that of NOB. Therefore, the metabolism of AOB will be faster than NOB and the former will outcompete the latter. Temperature also influences the free ammonia and free nitrous acid chemical equilibrium, thus affecting nitritation. In various studies, the observed temperature for maintaining activity of AOBs and NOBs was 35 and 38 °C, respectively. Gabarró et al. [37] inferred that temperature influenced the FA and FNA inhibition. At 35 °C FA and FNA, both affected the activity of AOBs while NOBs were inhibited at 25 °C.

4.6 Other Factors

According to Sinha and Annachhatre [13], various other factors that influence the accumulation of nitrite include sludge retention time, light, and various inhibitory compounds. Ge et al. [22] reported that ultrasonic treatment and substrate concentration were also the factors responsible for bringing about nitritation.

5 Studies on Nitrite Accumulation

A summary of various studies employing variety of control strategies for nitrite accumulation in SBR is presented in Table 3.

Table 3 Summary of studies employing variety of control strategies for nitrite accumulation in SBR

S. No.	Type of wastewater	Control strategies	Details of system and operating conditions	Nitrite accumulation and nitrogen removal	References
1	Domestic wastewater	Formic acid inhibition and real-time control	Vol. 10 L; MLSS 1200–1700 mg/L; SRT 8 d; pH 7.1 ± 0.3; DO 1.8–2.2 mg/L	Nitrite accumulation 80%; nitrogen removal up to 94%	[24]
2	Landfill leachate	FA inhibition and real-time control	Vol. 8 L; MLSS 1500–2500 mg/L; DO 0.5–2 mg/L; *T* 13–17.6 °C	Nitrite accumulation 90%; nitrogen removal up to 90%	[38]
3	Anaerobic co-digestion (supernatant)	Real-time control and FA-FNA inhibition	HRT 20 d; SRT >15 d; DO 1.48–1.53 mg/L; pH 7.15; *T* >15 °C	Nitrite accumulation 100%; nitrogen removal up to 85%	[39]
4	Anaerobic digestion supernatant	FA-FNA inhibition and real-time control	DO 0.83–1.55 mg/L; *T* >15 °C	Nitrite accumulation 100%; nitrogen removal up to 85%	[40]
5	Anaerobic effluent	DO	Vol. 26 L; SRT 18 d; DO 0.7–2 mg/L; *T* 22 ± 3 °C	Nitrite accumulation >97%; nitrogen removal up to 85%	[31]
6	Landfill leachate	Artificial intelligence	Vol. 24 L; SRT 20–25 d; *T* 20 ± 1 °C	Nitrite accumulation 100%; nitrogen removal >95%	[41]
7	Landfill leachate	Alternate aerobic-anaerobic modes and real-time control	Vol. 10 L; MLSS 3500 mg/L; pH 7.1–8.5; DO 0.5–1.4 mg/L; *T* 25–35 °C	Nitrogen removal >95%	[42]
8	Landfill leachate	pH, ORP	Vol. 8 L; MLSS 3000–5000 mg/L; SRT 30–40 d; pH 8.9–9.5; DO 3 mg/L; *T* 27–29 °C	Nitrite accumulation 89%; nitrogen removal 98%	[43]
9	Landfill leachate	FA inhibition and real-time control	Vol. 9 L; MLSS 1340–4400 mg/L; *T* 9–32.1 °C	Nitrite accumulation up to 90%; nitrogen removal 99.5%	[44]

(continued)

Table 3 (continued)

S. No.	Type of wastewater	Control strategies	Details of system and operating conditions	Nitrite accumulation and nitrogen removal	References
10	Landfill leachate	FA inhibition and real-time control	Vol. 10 L; T 25 °C	Nitrite accumulation up to 98%	[45]
11	Landfill leachate	FA and FNA inhibition and real-time control	Vol. 250 L; HRT 4.5–12 d; pH <8; DO <2 mg/L; T 25 or 35 °C	–	[37]
12	Landfill leachate	pH	Vol. 70 L; pH 7.76–8.91	Nitrite accumulation up to 92%; nitrogen removal 98%	[46]
13	Synthetic wastewater	Real-time control	Vol. 11 L and 7 L; SRT 16 d; DO 0.5–1 mg/L	Nitrite accumulation >95%	[47]
14	Co-digested liquor	FA and FNA inhibition	SRT 3–40 d; pH 7.8–8.2; DO 0.8 mg/L; T 25 or 30 °C	Nitrite accumulation >80%; nitrogen removal up to 96%	[48]
15	Sludge de-watering liquor	Real-time control	Vol. 4–6 L; pH > 6.2; DO 1–1.5 mg/L	Nitrite accumulation >95%; nitrogen removal around 85–90%	[49]
16	Domestic wastewater	Real-time control	SRT 5–40 d	Nitrite accumulation >88%	[50]
17	Domestic wastewater	Real-time control	Vol. 11 L; MLSS 3000–3500 mg/L; SRT 30–35 d; DO 1–2 mg/L; T 19 ± 1 °C	Nitrite accumulation >95%; nitrogen removal >97%	[26]
18	Synthetic wastewater	FNA	Vol. 8 L; HRT 1 d; SRT15 d; pH 7.2–7.5; DO 1.5–2 mg/L; T 22 ± 2 °C	Nitrite accumulation >80%	[51]
19	Anaerobic effluent	FA	Vol. 26 L; SRT 18 d; pH 7.2–8.1; DO 0.7–2 mg/L; T 22 ± 3 °C	Nitrite accumulation 100%; nitrogen removal up to 85%	[31]
20	Landfill leachate	DO and FA	Vol. 5 L; DO 0.7 ± 0.2 mg/L; T 20–22 °C	Nitrite accumulation 98–99%; nitrogen removal 65.1–78%	[32]

(continued)

Table 3 (continued)

S. No.	Type of wastewater	Control strategies	Details of system and operating conditions	Nitrite accumulation and nitrogen removal	References
21	Anaerobic effluent	FA and real-time control	Vol. 9 L; *T* 9–32.1 °C	Nitrite accumulation 95.5%; nitrogen removal 89–99%	[44]
22	Domestic wastewater	Real-time control based on blower frequency	Vol. 7 m^3; HRT 16–20 h; SRT 15–18 d; DO 2 mg/L; *T* 11–26 °C	Nitrite accumulation 95%; nitrogen removal 98%	[27]
23	Domestic wastewater	Real-time control	Vol. 10 L; HRT 8–12 h; SRT 30 d; *T* 12–25 °C	Nitrite accumulation >90%	[25]
24	Soybean mill wastewater and tap water	Real-time control	Vol. 38 L; SRT 53 d; DO 0.2–7.2 mg/L; *T* 31 ± 0.5 °C	Nitrite accumulation >96%	[28]
25	Real domestic saline sewage	Real-time control	Vol. 10 L; HRT 0.17–7 h; MLSS 2.5 g/L; SRT 14–28 d; *T* 22 ± 1 °C	Nitrite accumulation >95%; nitrogen removal >98%	[52]
26	Synthetic wastewater	Chlorate addition	Vol. 2.5 L; pH 7.8–8.2; DO >9 mg/L; *T* 25 °C	Nitrogen removal 98.93%	[53]
27	Synthetic wastewater	Hydroxylamine inhibition	Vol. 2.5 L; HRT 6–8 h; pH 7.8–8.2; DO 5.5 mg/L; *T* 25 °C	Nitrite accumulation 99.8%; nitrogen removal 81%	[54]
28	Synthetic wastewater	Sulfide addition	Vol. 1.5 L; HRT 1.5 d; pH 7.5 ± 0.2; DO <0.2 mg/L; *T* 23–28 °C	Nitrite accumulation 75%; nitrogen removal 83%	[55]
29	Domestic wastewater	Intermittent aeration	Vol. 2 m^3; SRT 16 d; pH 7.2–7.6; DO <0.2 mg/L; *T* 12.6–24 °C	Nitrite accumulation >95%	[35]
30	Synthetic wastewater	Intermittent aeration	Vol. 2 L; pH 8.2–8.5; DO <0.2 mg/L; *T* 25 °C	Nitrite accumulation 92.95%	[56]

(continued)

Table 3 (continued)

S. No.	Type of wastewater	Control strategies	Details of system and operating conditions	Nitrite accumulation and nitrogen removal	References
31	Domestic wastewater	Real-time control	Vol. 54 m^3; MLSS 2000–3000 mg/L; SRT 13 d; DO 2.5 mg/L; T 11.9–26.5 °C	Nitrite accumulation >95%; nitrogen removal >90%	[57]
32	Domestic wastewater	SRT	Vol. 60 L; HRT 12 h; SRT 8 d; DO >3 mg/L; T 13–15 °C	Nitrogen removal 72%	[58]

6 Conclusion

The nitrite accumulation is practically impossible in natural aquatic systems. Despite this fact some researchers have unlocked new avenues in the water research field and have tried to accumulate nitrite, it is an intermediate in both, nitrification and de-nitrification. The rate of formation and transformation of nitrite directly affects its accumulation. Various investigators have manipulated parameters such as DO, pH, SRT, and FA-FNA concentration to accumulate nitrite in the water systems. With the application of sophisticated molecular techniques, it has now been possible to identify the bacteria responsible for formation and transformation of nitrite (i.e., AOB and NOB). Based on the literature review, various physico-chemical parameters conducive for nitrite accumulation include pH value of 7.5–8.5, temperature ranging between 20 and 25 °C and DO concentration of 0.5–1.5 mg/L. The nitritation process offers a 25% saving in aeration costs and removes the need of adding an external carbon source in the de-nitritation process.

References

1. STOWA: One reactor system for ammonia removal via nitrite. Report No 96-01. Utrecht, the Netherlands: STOWA, ISBN 90 74476 39 (1996)
2. Verstraete, W., Philips, S.: Nitrification-denitrification processes and technologies in new contexts. Environ. Pollut. **102**(1), 717–726 (1998)
3. Ahn, Y.H.: Sustainable nitrogen elimination biotechnologies: a review. Process Biochem. **41** (8), 1709–1721 (2006)
4. Fdz-Polanco, F., Villaverde, S., Garcia, P.A.: Nitrite accumulation in submerged biofilters—combined effects. Water Sci. Technol. **34**(3), 371–378 (1996)
5. Beccari, M., Passino, R., Ramadori, R., Tandoi, V.: Kinetics of dissimilatory nitrate and nitrite reduction in suspended growth culture. J. (Water Pollut. Control Fed.) 58–64 (1983)

6. Van Kempen, R., Mulder, J.W., Uijterlinde, C.A., Loosdrecht, M.: Overview: full scale experience of the SHARON® process for treatment of rejection water of digested sludge dewatering. Water Sci. Technol. **44**(1), 145–152 (2001)
7. Turk, O., Mavinic, D.S.: Benefits of using selective inhibition to remove nitrogen from highly nitrogenous wastes. Environ. Technol. **8**(1–12), 419–426 (1987)
8. Surmacz-Górska, J., Cichon, A., Miksch, K.: Nitrogen removal from wastewater with high ammonia nitrogen concentration via shorter nitrification and denitrification. Water Sci. Technol. **36**(10), 73–78 (1997)
9. Hellinga, C., Schellen, A.A.J.C., Mulder, J.W., Van Loosdrecht, M.C.M., Heijnen, J.J.: The SHARON process: an innovative method for nitrogen removal from ammonium-rich waste water. Water Sci. Technol. **37**(9), 135–142 (1998)
10. Beylier, M.R., Balaguer, M.D., i Nàcher, C.P., Smets, B.F., Sun, S.P., Wang, R.C.: Biological nitrogen removal from domestic wastewater. In Comprehensive Biotechnology. Elsevier Science (2011)
11. Henze, M. (ed.): Biological Wastewater Treatment: Principles, Modelling and Design. IWA Publishing (2008)
12. Suzuki, I., Dular, U., Kwok, S.C.: Ammonia or ammonium ion as substrate for oxidation by *Nitrosomonas europaea* cells and extracts. J. Bacteriol. **120**(1), 556–558 (1974)
13. Sinha, B., Annachhatre, A.P.: Partial nitrification—operational parameters and microorganisms involved. Rev. Environ. Sci. Bio/Technol. **6**(4), 285–313 (2007)
14. Regan, J.M., Harrington, G.W., Noguera, D.R.: Ammonia-and nitrite-oxidizing bacterial communities in a pilot-scale chloraminated drinking water distribution system. Appl. Environ. Microbiol. **68**(1), 73–81 (2002)
15. Teske, A., Alm, E., Regan, J.M., Toze, S., Rittmann, B.E., Stahl, D.A.: Evolutionary relationships among ammonia-and nitrite-oxidizing bacteria. J. Bacteriol. **176**(21), 6623–6630 (1994)
16. Lipponen, M.T., Martikainen, P.J., Vasara, R.E., Servomaa, K., Zacheus, O., Kontro, M.H.: Occurrence of nitrifiers and diversity of ammonia-oxidizing bacteria in developing drinking water biofilms. Water Res. **38**(20), 4424–4434 (2004)
17. Head, I.M., Hiorns, W.D., Embley, T.M., McCarthy, A.J., Saunders, J.R.: The phylogeny of autotrophic ammonia-oxidizing bacteria as determined by analysis of 16S ribosomal RNA gene sequences. Microbiology **139**(6), 1147–1153 (1993)
18. Koops, H.P., Pommerening-Röser, A.: Distribution and ecophysiology of the nitrifying bacteria emphasizing cultured species. FEMS Microbiol. Ecol. **37**(1), 1–9 (2001)
19. Junier, P., Molina, V., Dorador, C., Hadas, O., Kim, O.S., Junier, T., Imhoff, J.F.: Phylogenetic and functional marker genes to study ammonia-oxidizing microorganisms (AOM) in the environment. Appl. Microbiol. Biotechnol. **85**(3), 425–440 (2010)
20. Schmidt, I., Bock, E.: Anaerobic ammonia oxidation with nitrogen dioxide by *Nitrosomonas eutropha*. Arch. Microbiol. **167**(2–3), 106–111 (1997)
21. Nogueira, R., Melo, L.F.: Competition between *Nitrospira* spp. and *Nitrobacter* spp. in nitrite-oxidizing bioreactors. Biotechnol. Bioeng. **95**(1), 169–175 (2006)
22. Ge, S., Wang, S., Yang, X., Qiu, S., Li, B., Peng, Y.: Detection of nitrifiers and evaluation of partial nitrification for wastewater treatment: a review. Chemosphere **140**, 85–98 (2015)
23. Zanetti, L., Frison, N., Nota, E., Tomizioli, M., Bolzonella, D., Fatone, F.: Progress in real-time control applied to biological nitrogen removal from wastewater. A short-review. Desalination **286**, 1–7 (2012)
24. Blackburne, R., Yuan, Z., Keller, J.: Demonstration of nitrogen removal via nitrite in a sequencing batch reactor treating domestic wastewater. Water Res. **42**(8), 2166–2176 (2008)
25. Guo, J.H., Peng, Y.Z., Wang, S.Y., Zheng, Y.N., Huang, H.J., Ge, S.J.: Effective and robust partial nitrification to nitrite by real-time aeration duration control in an SBR treating domestic wastewater. Process Biochem. **44**(9), 979–985 (2009)
26. Zeng, W., Zhang, Y., Li, L., Peng, Y.Z., Wang, S.Y.: Control and optimization of nitrifying communities for nitritation from domestic wastewater at room temperatures. Enzyme Microb. Technol. **45**(3), 226–232 (2009)

27. Gu, S., Wang, S., Yang, Q., Yang, P., Peng, Y.: Startup partial nitrification at low temperature with a real-time control strategy based on blower frequency and pH. Biores. Technol. **112**, 34–41 (2012)
28. Gao, D., Peng, Y., Li, B., Liang, H.: Shortcut nitrification–denitrification by real-time control strategies. Biores. Technol. **100**(7), 2298–2300 (2009)
29. Van Hulle, S.W., Vandeweyer, H.J., Meesschaert, B.D., Vanrolleghem, P.A., Dejans, P., Dumoulin, A.: Engineering aspects and practical application of autotrophic nitrogen removal from nitrogen rich streams. Chem. Eng. J. **162**(1), 1–20 (2010)
30. Peng, Y., Zhu, G.: Biological nitrogen removal with nitrification and denitrification via nitrite pathway. Appl. Microbiol. Biotechnol. **73**(1), 15–26 (2006)
31. Katsou, E., Malamis, S., Frison, N., Fatone, F.: Coupling the treatment of low strength anaerobic effluent with fermented biowaste for nutrient removal via nitrite. J. Environ. Manag. **149**, 108–117 (2015)
32. Kulikowska, D., Bernat, K.: Nitritation–denitritation in landfill leachate with glycerine as a carbon source. Biores. Technol. **142**, 297–303 (2013)
33. Anthonisen, A.C., Loehr, R.C., Prakasam, T.B.S., Srinath, E.G.: Inhibition of nitrification by ammonia and nitrous acid. J. (Water Pollut. Control Fed.) 835–852 (1976)
34. Abeling, U., Seyfried, C.F.: Anaerobic-aerobic treatment of high-strength ammonium wastewater-nitrogen removal via nitrite. Water Sci. Technol. **26**(5–6), 1007–1015 (1992)
35. Kornaros, M., Marazioti, C., Lyberatos, G.: A pilot scale study of a sequencing batch reactor treating municipal wastewater operated via the UP-PND process. Water Sci. Technol. **58**(2) (2008)
36. Ge, S., Peng, Y., Qiu, S., Zhu, A., Ren, N.: Complete nitrogen removal from municipal wastewater via partial nitrification by appropriately alternating anoxic/aerobic conditions in a continuous plug-flow step feed process. Water Res. **55**, 95–105 (2014)
37. Gabarró, J., Ganigué, R., Gich, F., Ruscalleda, M., Balaguer, M.D., Colprim, J.: Effect of temperature on AOB activity of a partial nitritation SBR treating landfill leachate with extremely high nitrogen concentration. Biores. Technol. **126**, 283–289 (2012)
38. Sun, H., Peng, Y., Wang, S., Ma, J.: Achieving nitritation at low temperatures using free ammonia inhibition on Nitrobacter and real-time control in an SBR treating landfill leachate. J. Environ. Sci. **30**, 157–163 (2015)
39. Frison, N., Katsou, E., Malamis, S., Bolzonella, D., Fatone, F.: Biological nutrients removal via nitrite from the supernatant of anaerobic co-digestion using a pilot-scale sequencing batch reactor operating under transient conditions. Chem. Eng. J. **230**, 595–604 (2013)
40. Frison, N., Chiumenti, A., Katsou, E., Malamis, S., Bolzonella, D., Fatone, F.: Mitigating off-gas emissions in the biological nitrogen removal via nitrite process treating anaerobic effluents. J. Clean. Prod. **93**, 126–133 (2015)
41. Spagni, A., Marsili-Libelli, S.: Nitrogen removal via nitrite in a sequencing batch reactor treating sanitary landfill leachate. Biores. Technol. **100**(2), 609–614 (2009)
42. Wang, K., Wang, S., Zhu, R., Miao, L., Peng, Y.: Advanced nitrogen removal from landfill leachate without addition of external carbon using a novel system coupling ASBR and modified SBR. Biores. Technol. **134**, 212–218 (2013)
43. Lina, W.U., Chengyao, P.E.N.G., ZHANG, S., Yongzhen, P.E.N.G.: Nitrogen removal via nitrite from municipal landfill leachate. J. Environ. Sci. **21**(11), 1480–1485 (2009)
44. Sun, H., Peng, Y., Shi, X.: Advanced treatment of landfill leachate using anaerobic–aerobic process: organic removal by simultaneous denitritation and methanogenesis and nitrogen removal via nitrite. Biores. Technol. **177**, 337–345 (2015)
45. Miao, L., Wang, S., Li, B., Cao, T., Xue, T., Peng, Y.: Advanced nitrogen removal via nitrite using stored polymers in a modified sequencing batch reactor treating landfill leachate. Biores. Technol. **192**, 354–360 (2015)
46. Yabroudi, S.C., Morita, D.M., Alem, P.: Landfill leachate treatment over nitritation/denitritation in an activated sludge sequencing batch reactor. APCBEE Procedia **5**, 163–168 (2013)

47. Zeng, W., Yang, Y., Li, L., Wang, X., Peng, Y.: Effect of nitrite from nitritation on biological phosphorus removal in a sequencing batch reactor treating domestic wastewater. Biores. Technol. **102**(12), 6657–6664 (2011)
48. Scaglione, D., Tornotti, G., Teli, A., Lorenzoni, L., Ficara, E., Canziani, R., Malpei, F.: Nitrification denitrification via nitrite in a pilot-scale SBR treating the liquid fraction of co-digested piggery/poultry manure and agro-wastes. Chem. Eng. J. **228**, 935–943 (2013)
49. Fux, C., Velten, S., Carozzi, V., Solley, D., Keller, J.: Efficient and stable nitritation and denitritation of ammonium-rich sludge dewatering liquor using an SBR with continuous loading. Water Res. **40**(14), 2765–2775 (2006)
50. Changyong, W.U., Yongzhen, P.E.N.G., Shuying, W.A.N.G., Xiaoling, L.I., Randeng, W.A. N.G.: Effect of sludge retention time on nitrite accumulation in real-time control biological nitrogen removal sequencing batch reactor. Chin. J. Chem. Eng. **19**(3), 512–517 (2011)
51. Wang, Q., Ye, L., Jiang, G., Hu, S., Yuan, Z.: Side-stream sludge treatment using free nitrous acid selectively eliminates nitrite oxidizing bacteria and achieves the nitrite pathway. Water Res. **55**, 245–255 (2014)
52. Ye, L., Peng, C.Y., Tang, B., Wang, S.Y., Zhao, K.F., Peng, Y.Z.: Determination effect of influent salinity and inhibition time on partial nitrification in a sequencing batch reactor treating saline sewage. Desalination **246**(1), 556–566 (2009)
53. Xu, G., Xu, X., Yang, F., Liu, S.: Selective inhibition of nitrite oxidation by chlorate dosing in aerobic granules. J. Hazard. Mater. **185**(1), 249–254 (2011)
54. Xu, G., Xu, X., Yang, F., Liu, S., Gao, Y.: Partial nitrification adjusted by hydroxylamine in aerobic granules under high DO and ambient temperature and subsequent Anammox for low C/N wastewater treatment. Chem. Eng. J. **213**, 338–345 (2012)
55. Erguder, T.H., Boon, N., Vlaeminck, S.E., Verstraete, W.: Partial nitrification achieved by pulse sulfide doses in a sequential batch reactor. Environ. Sci. Technol. **42**(23), 8715–8720 (2008)
56. Li, J., Elliott, D., Nielsen, M., Healy, M.G., Zhan, X.: Long-term partial nitrification in an intermittently aerated sequencing batch reactor (SBR) treating ammonium-rich wastewater under controlled oxygen-limited conditions. Biochem. Eng. J. **55**(3), 215–222 (2011)
57. Yang, Q., Peng, Y., Liu, X., Zeng, W., Mino, T., Satoh, H.: Nitrogen removal via nitrite from municipal wastewater at low temperatures using real-time control to optimize nitrifying communities. Environ. Sci. Technol. **41**(23), 8159–8164 (2007)
58. Yuan, Q., Oleszkiewicz, J.: Low temperature biological phosphorus removal and partial nitrification in a pilot sequencing batch reactor system. Water Sci. Technol. **63**(12) (2011)

Agricultural Waste Utilization in Sustainable and Resilient Construction

Poonam Shekhawat, Sandeep Shrivastava and Neha Shrivastava

Abstract Today, the world is suffering from some degree of climate change that is causing extreme weather events including frequent flooding followed by drought, cold wave, heat wave, landslides, cyclones, etc. This scenario is imposing a need to adopt sustainable and resilient construction to mitigate the cause of climate change and to provide sustainable habitat. The present work is a study on achieving resilient construction with locally available materials. It uses agricultural wastes as fibers in improving the resiliency of earthen structures prone to flood damages. Straw bale and jute coir wastes were used to reinforce earthen soil. The initial properties and a simplified flood resilient test as adopted from the literature were observed. The unreinforced sample failed after first flooding, while the straw and coir-reinforced sample stood for the entire test duration without any failure. The maximum vertical displacement, which indicates swelling, of approximately 2.7 mm in the case of the unreinforced sample, 2.3 mm for the straw-reinforced sample, and 1.95 mm for the coir-reinforced sample was recorded. It was concluded that both natural fibers could improve the resilient properties of earthen construction.

Keywords Agricultural fiber · Construction · Recycled materials
Flood resilient · Sustainable · Waste

P. Shekhawat · S. Shrivastava (✉)
Malaviya National Institute of Technology Jaipur, Jaipur, India
e-mail: san.civil@gmail.com

N. Shrivastava
Swami Keshvanand Institute of Technology Management & Gramothan Jaipur, Jaipur, India

A. S. Kalamdhad et al. (eds.), *Advances in Waste Management*,
https://doi.org/10.1007/978-981-13-0215-2_27

1 Introduction

Climate change is causing extreme weather events imposing the need to adopt resilient construction to mitigate the cause of climate change and to provide sustainable and resilient habitat. Sustainable building construction includes eco-friendly and economical materials and methods for the duration of a building's life cycle from outline to development, operation, maintenance, re-design, and deconstruction [1]. In the context of climate change, resiliency is defined as the ability of a system to absorb impacts of a climatic event without failing or changing its state [2]. Earthen construction is a sustainable method of construction, which is still prevalent in rural areas in developing countries like India. Earthen construction is environmentally friendly and economical method of construction and involves low carbon content, low initial cost, and embodied energy [3]. It also involves locally available material and easier construction techniques. The common types of earth construction are cob, rammed earth, adobe, and cut blocks, etc. Cob is a traditional mixture of clay, sand, gravel, straw, and water mixed by hand or machinery. Cob structures are low height, minor load-bearing structures. The walls are about 30–60 cm thick and are built in layers 20–35 cm high [4]. Cob is a type of soil reinforcement technique. Soil reinforcement is a technique consisting of discrete randomly distributed materials in soil. These materials can be in the form of powder, fibers, and chips, which provide an improvement in the mechanical behavior of the soil composite [5]. For the present study, different fibers from the waste of agricultural products were used as soil reinforcement.

2 Materials and Methods

Clay, coarse sand, fine sand, gravel, and fibers (straws, jute coir each at a time) were used in the present study (Fig. 1). Straw and coir fiber were used as agricultural waste fibers. Straw is largely produced in rural areas and used as reinforcement in composite soil blocks. In a study [6], it was observed that with the addition of straw as soil reinforcement, compressive strength is improved and shrinkage is reduced. In another study [7], an increment in shear strength was observed with the inclusion of 1% optimum straw content as soil reinforcement.

Coir is a fibrous material, which is outer part of an aged coconut. The fibers are normally 50–350 mm long. Coir fibers have good tensile strength, and this property remains almost the same even in the wet condition. Elongation in coir fiber without failure is much higher though it has less tenacity [8]. Coir grids, geotextiles are nowadays available in various categories, which are economical and environment-friendly [9]. For instance, in a study [10], it was observed that coir has shown greater improvement (47.50%) than that of synthetic fibers (40.0%) in resilient modulus or strength of the soil.

Fig. 1 Materials used in the preparation of samples

For the preparation of samples, a basic composition of cob construction was used and the materials were blended together homogenously. Tables 1 and 2 show percentage required for different components and the properties of samples prepared, respectively.

In the first stage, some characterization tests, viz. liquid limit, plastic limit, and specific gravity test were performed. In the second stage, compaction test was performed with the aim to investigate the optimum moisture content at which maximum dry density of different samples (unreinforced, straw-reinforced, and coir reinforced) was achieved.

Standard proctor test method was used in the investigation as per the IS-2720-PART-7-1980 [12]. In this method, the damp soil in the mold is compacted in three equivalent layers. Every layer was given 25 blows with a 2.6 kg rammer dropped from the height of 310 mm over the soil. The soil is then expelled from the

Table 1 Particle size distribution and details of preparation (data from [11])

Material	Percentage required (% by weight)	Details of preparation
Clay	20.6	Dried in oven at 105 °C for 24 h
Fine sand	32.5	Dried in oven at 105 °C for 24 h
Coarse sand	19.7	Dried in oven at 105 °C for 24 h
Gravel	24.4	–
Fiber	1.25	Cut into approximate length of 15–50 mm
Water	1.55	–

Table 2 Results obtained from characterization tests

Properties of soil samples	Value
Liquid limit	17.2%
Plasticity index	4%
Specific gravity (without reinforcement)	2.56
Specific gravity (with straw)	2.4
Specific gravity (with coir)	2.38

mold, and a representative soil sample is acquired for water content determination. Figure 2 shows the comparison of optimum moisture content (OMC) and maximum dry density (MDD) for different samples.

By comparing the proctor test results of the three samples, it was observed that adding fibers in soil sample led to increase in OMC and decrease in MDD. Among all samples, coir-reinforced sample was with the highest OMC (9%) and the lowest MDD (2.047 gm/mL). The results demonstrate that the addition of fiber influences the amount of water that ought to be added to the blend to acquire a denser soil structure.

Further, the optimum moisture content obtained was used in the preparation of samples for unconfined compression and flooding simulation tests. Cylindrical samples (Fig. 3) were prepared using a mold of 75 mm dia. and 150 mm height. The samples were prepared in four equal soil layers. Each layer was rammed with 60 blows as adopted from the literature [3].

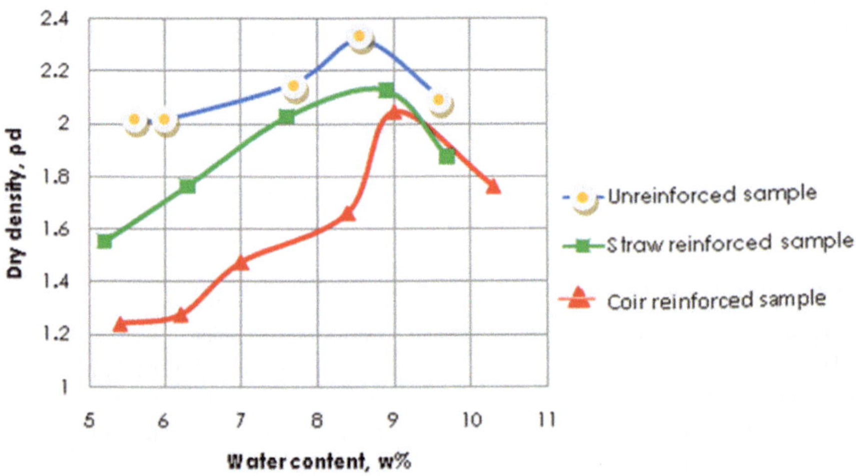

Fig. 2 Optimum moisture content (OMC) and maximum dry density (MDD) comparison for different samples

Fig. 3 Unreinforced and reinforced samples

2.1 Unconfined Compression Test

Unconfined compression tests were performed to explore the behavior of the specimens under compression (Fig. 4a); a comparable circumstance would be seen in the field when a vertical load is applied on the top of the earthen wall [3].

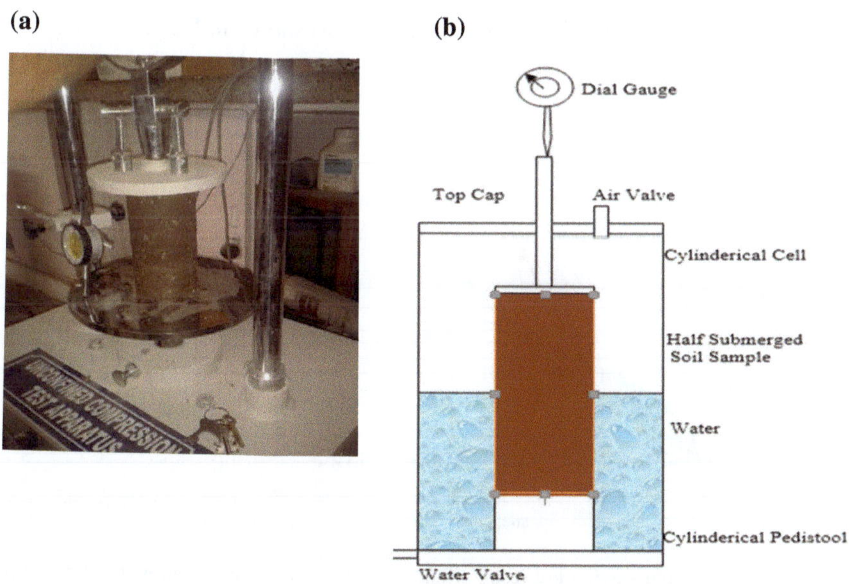

Fig. 4 **a** Unconfined compression setup, **b** flooding simulation test configuration (based on [13])

2.2 *Flooding Simulation Test*

The test setup was adopted from [13] to explore the behavior of the specimens under three cycles of flooding followed by drying periods for a time around 15 days. Figure 4 illustrates the experimental setup. The cylindrical cell of 300 mm × 600 mm was assembled with water valve and dial gauge for vertical displacement measurement (Fig. 4b). Forster et al. [13] called the proposed test a flooding simulation test (FST).

A series of FSTs was performed for the specimens compacted with 60 blows every layer without reinforcement and with reinforcement (straw fibers and coir fibers). The specimens were exposed to water for a controlled time of 24, 60, and 120 h, followed by a drained environment for 72, 60, and 12 h, respectively. The half portions of the samples were exposed to water.

3 Results and Discussion

3.1 *Unconfined Compression Test*

By comparing the results of an unconfined compression test, the peak strength for the unreinforced sample was 63.9 kPa, whereas it was 87.23 kPa for straw-reinforced sample and 119.15 kPa for coir-reinforced sample as shown in Fig. 5.

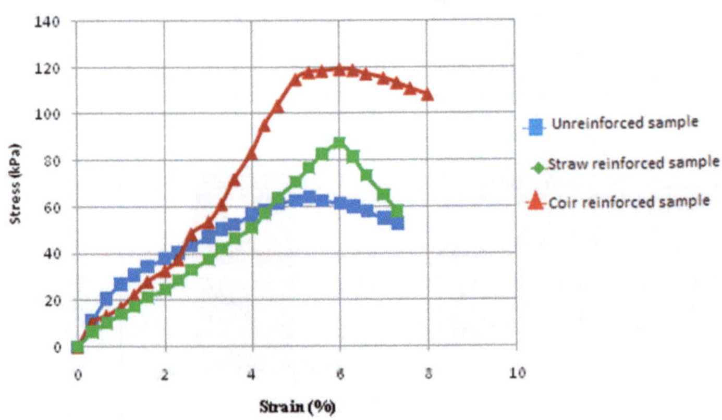

Fig. 5 Unconfined compression test results comparison for samples prepared with 60 blows per layer

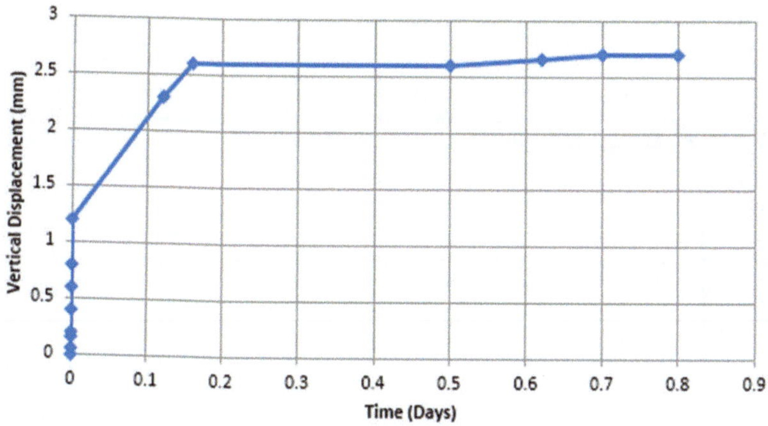

Fig. 6 Flooding simulation test (FST) results for sample prepared without reinforcement

3.2 *Flooding Simulation Test*

The unreinforced sample failed after first flooding. The sample showed considerable swelling with the maximum vertical displacement of 2.7 mm (Fig. 6). For straw-reinforced sample, the first flooding event caused an increase in the vertical height equivalent to 3% of the exposed height under water. The maximum vertical displacement of 2.3 mm was detected in the case of straw-reinforced sample (Fig. 7). For the coir-reinforced sample, the first flooding event caused an increase in the vertical height equivalent to 2.5% of the exposed height under water, which is lesser than straw-reinforced sample. Coir-reinforced sample showed a maximum

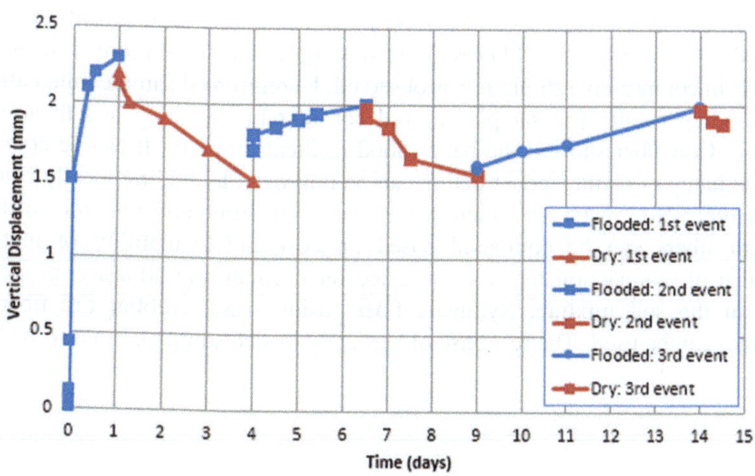

Fig. 7 Flooding simulation test (FST) results for straw-reinforced sample

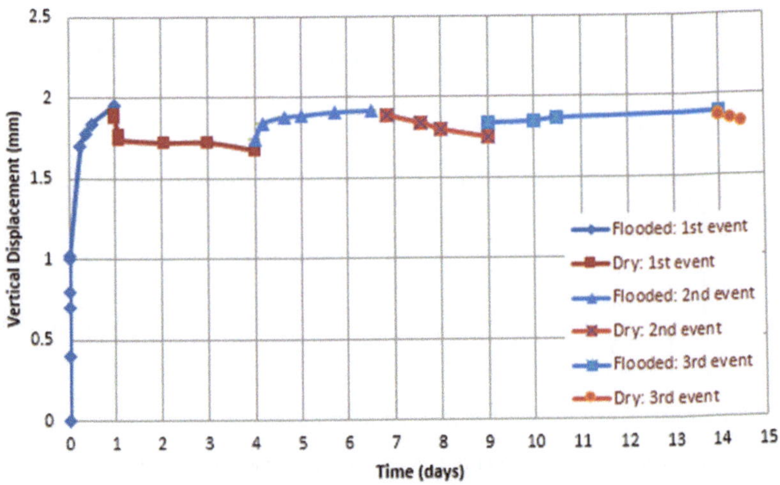

Fig. 8 Flooding simulation test (FST) results for sample reinforced with coir fibers

displacement of 1.95 mm, which is much lesser than other samples (Fig. 8). The straw- and coir-reinforced sample stood for the entire test duration without any failure though a little neck formation was detected.

4 Conclusions

In the present study, compaction, unconfined compressive strength, and flood resilient properties of unreinforced and two-reinforced soil samples with the inclusion of 1.25% fibers were tested. Straw and coir were used as soil-reinforcing materials. Comparing coir fiber-reinforced sample with unreinforced sample, an increase in compressive strength was observed. Unreinforced sample failed after the first flooding event. The sample showed considerable swelling and then sudden collapse. Coir fiber showed the better flood resilient property. It can be concluded that resiliency of earthen construction can be improved by adding locally available agricultural waste materials which in this case were straw and jute coir. In future, different fibers can be compared based on cost and availability of materials. Influence of compaction rate can be tested on strength and adhesion property of fibers in the soil mixture. Synthetic fibers using waste (rubber tire fibers, for example) can be used. The response of vertically loaded samples can also be tested.

References

1. United State Environmental Protection Agency, http://www3.epa.gov/
2. Larsen, L., Rajkovich, N., Leighton, C., McCoy, K., Calhoun, K., Mallen, E., Bush, K., Enriquez, J., Pyke, C., McMahon, S., Kwok, A.: Green Building and Climate Resilience: Understanding Impacts and Preparing for Changing Conditions. Technical Report, University of Michigan; U. S. Green Building Council (2011)
3. Medero, G.M., Kennedy, J.H., Woodward, P.K., Banimahd, M.: Flooding effect on earth walls. Sustainability **3**, 69–81 (2010)
4. Mrinmaye: General Advice for Mud Block. Earth Construction in India J14 (2014)
5. Li, C.: Mechanical Response of Fiber-Reinforced Soil. Ph.D. Thesis, University of Texas at Austin (2005)
6. Bouhicha, M., Aouissi, F., Kenai, S.: Performance of composite soil reinforced with barley straw. Cement Concr. Compos. 17–21 (2005)
7. Abtahi, M., Okhovat, N., Pourhosseini, R., Hejazi, M.: Improvement of soil strength by natural fibers. From res to des Europprac, Bratislava, Slovak Republic (2010)
8. Sivkumar Babu, G.L., Vasudevan, A.K.: Strength and stiffness response of coir fiber-reinforced tropical soil. J. Mater. Civ. Eng. **20**, 571–577 (2008)
9. Subaida, A., Chandrakaran, E., Sankar, N.: Laboratory performance of unpaved roads reinforced with woven coir. Geotext. Geomembr. **27**(3), 204–210 (2009)
10. Chauhan, S., Mittal, S., Mohanty, B.: Performance evaluation of silty sand subgrade reinforced with fly ash and fiber. Geotext. Geomembr. **26**, 429–435 (2008)
11. Williams-Ellis, C., Eastwick-Field, C., Eastwick-Field, E.: Building in Cob, Pise, and Stabilised Earth, Reprinted edn. Donhead Publishing, Shaftesbury (1999)
12. IS: 2720-7: Methods of Test for Soils—Part VII Determination of Water Content—Dry Density Relation Using Light Compaction. Indian Standard (1980)
13. Forster, A.M., Medero, G.M., Morton, T., Buckman, J.: Traditional cob wall: response to flooding. Struct. Surv. **26**(4), 302–321 (2008)

Improving the Hydrophilicity of Polysulfone Membrane by the Addition of Imidazol with Polyvinyl Pyrrolidone for Crystal Violet Dye Removal

Nilay Sharma and Mihir Kumar Purkait

Abstract Asymmetric polysulfone (PSn) ultrafiltration membranes were prepared using polyvinyl pyrrolidone (PVP) of three different molecular weights with imidazol (IMZ). Phase inversion technique was adopted for fabricating the PVP and IMZ blended PSn-based membrane. Morphological analysis of membranes was done by field emission scanning electron microscope (FESEM) and atomic force microscopy (AFM). Liquid–liquid displacement porosimetry (LLDP) method was used for determining the average pore size, number of pores, and permeability of membranes. Hydrophilicity was found to be increased with increasing molecular weight of PVP from 24,000 to 360,000 Da with IMZ, as compared to the plain membrane. Anionic surfactant sodium dodecyl sulfate (SDS) was used in aqueous media for investigating the permeation and rejection performance of fabricated membrane for crystal violet dye (CVD).

Keywords Polysulfone membrane · Hydrophilicity · Dye removal
Contact angle · Imidazol

1 Introduction

Dye consumption is very high in various production units such as paints, textile, and paper. Effluent released from these units is polluted with different coloring agents, i.e., dye, which is harmful for both human being and vegetation. It is needed to properly treat the colored wastewater before its disposal. Trivial amount of dye present in the water can hamper the path of sunlight which affects the cleanness of water and also the aquatic life [1]. Dye effluent has complex nature which does not

N. Sharma (✉) · M. K. Purkait
Department of Chemical Engineering, Indian Institute of Technology Guwahati, Guwahati 781039, Assam, India
e-mail: s.nilay@iitg.ernet.in

A. S. Kalamdhad et al. (eds.), *Advances in Waste Management*,
https://doi.org/10.1007/978-981-13-0215-2_28

allow the satisfactory treatment of it. Many decolorization techniques including physical [2, 3], chemical, and biological processes [4, 5] have been reported in last decades. However, performance of combination of two methods is proven to be more satisfactory than that of single one.

Removal of organic dyes from tinted wastewater can be treated by micellar-enhanced ultrafiltration (MEUF). It is one of the promising techniques for dye removal from aqueous media. Higher removal efficiency with elevated permeate flux and lower operating cost is the major advantages of MEUF over other membrane-based separation processes [6]. In this process, micelle formation occurs by the addition of surfactant in aqueous medium at the same or higher critical micelle concentration (CMC) level. Then, the oppositely charged solute molecules are solubilized over the micellar surface. Thus, by the increment of size the solutes separated after trapping with the micelles [7]. Recovery of surfactant in MEUF is a well-established technique; Purkait et al. have done work on aromatic alcohols separation and dye removal from water using MEUF and further recovery of surfactant [8, 9].

PSn-based asymmetric ultrafiltration (UF) membranes are widely used in water treatment, food processing, and biotechnology [10, 11]. This is one of the most promising clean energy technologies. PSn has certain potential advantages, including its excellent acidic and basic resistance, good thermal stability, wide pH range, and film forming ability [12–15]. Major disadvantages of PSn membranes include their low infiltration capacity and severe membrane fouling because of their hydrophobic nature [16]. It is very necessary to overcome these problems for increasing the permeability and life span of the membrane. Surface characteristics of membrane such as morphology, porosity, pore size, and hydrophilicity affect the membrane fouling [17]. Flux-declining and membrane fouling can be reduced by improving the hydrophilicity of the membrane. Many researchers have done work by altering the composition of the membrane for finding an additional hydrophilic material. So, membrane fouling can be remarkably reduced by increasing the hydrophilicity of the membrane surface as well as porous surfaces [18, 19].

Hydrophilic properties of the PSn membrane can be improved by the addition of different organic additives with it. Surface modification done by the blending of additives is resulted in increased hydrophilicity of PSn-based membrane [20–22]. Yan et al. [23] prepared imidazolium-functionalized polysulfone hydroxide exchange membranes for potential applications in alkaline membrane direct alcohol fuel cells. Kumari et al. [24] have done a review on imidazol and its biological activities. They resulted with the fact that being a polar and ionizable aromatic compound, it improves pharmacokinetic characteristics of lead molecules and thus used as a remedy to optimize solubility and bioavailability parameters of proposed poorly soluble lead molecules. Sun et al. [25] have identified a number of liquid proton-conducting electrolytes, based on imidazol (IMZ) compounds and/or organic molten salts, as candidates for impregnation into polymer membranes. Hydrophilic group containing polymeric membranes shows high porosity. These membranes are resistant to attack by acids as well as have high-temperature stability. Oxidizing and reducing atmospheres were also found in operating fuel cells.

A critical review of above the literature envisages that lots of works have been reported on the addition of different IMZ-based materials to polymeric membrane, but effect of direct addition of IMZ with PVP on polymeric membranes was not studied yet. Therefore, in the present work a detailed study was done for finding the effects of addition of IMZ with three different molecular weights of PVP into the casting solution of PSn membrane; in order to increase the hydrophilicity and crystal violet dye (CVD) rejection performance of PSn membrane, PVP (24,000, 40,000 and 360,000 Da) was taken as pore former in this study and blended with IMZ in PSn membrane casting solution. Effects of IMZ on the hydrophilicity, water flux as well as permeation and rejection behavior of CVD under different solution conditions was studied. For characterizing the prepared membranes, morphological parameters were examined. Hydrophilicity of membranes was evaluated by contact angle measurement. All the parameters were examined and explained well.

2 Materials and Methods

Polysulfone (average molecular weight 30,000 Da) was procured by Sigma-Aldrich Co. USA, and used as base polymer. *N*-methyl pyrrolidone (NMP) (supplied by LOBA Chemie, India) was used as solvent for membrane preparation. Otto Chemie Private Limited, India, supplied the PVP (average molecular weight 24,000, 40,000, and 360,000 Da which was used as pore-forming additive. CDH Laboratory, India, supplied the CVD with molecular weight of 407.9 Da and Sisco Research Laboratories, India, provided the surfactant SDS (MW 288.38). IMZ was supplied by Otto Chemie Private Limited, India. Detailed procedure for membrane preparation is discussed in the literature (Sharma et al.). Membranes with altering compositions were defined as m1, m2, m3, and m4. Composition of membranes is given in Table 1.

Table 1 Composition of PSn casting solutions of the membranes

S. No.	Membrane	PSn (wt%)	PVP-24,000 (wt%)	PVP-40,000 (wt%)	PVP-360,000 (wt%)	NMP (wt%)	IMZ (wt%)
1	m1	14	0	0	0	86	0
2	m2	14	4.8	0	0	78	3.2
3	m3	14	0	4.8	0	78	3.2
4	m4	14	0	0	4.8	78	3.2

3 Results and Discussion

3.1 Morphological Study

For knowing the surface roughness parameters, AFM images were used. High-resolution FESEM was used for examining surface and cross-sectional morphology of the membranes, i.e., size and shape of the pores.

3.1.1 Image Analysis by FESEM

Cross-sectional view of the fabricated membranes is shown in Fig. 1. Asymmetric porous structure can be clearly shown in these FESEM images. The figure also depicted dense top layer and a porous sub-layer, except for membrane m1. The porosity of the fabricated membranes was found to be increased by the addition of higher molecular weight of PVP with IMZ. This information may be explained by two facts, viz. (i) interactions between polymer chains got reduced as PVP-IMZ molecules and polymer came into contact and interacted with each other. (ii) Solvent (NMP) outflow decreased and non-solvent (water) inflow increased due to the hydrophilicity of added IMZ to the casting solution [26]. Further, ratio of

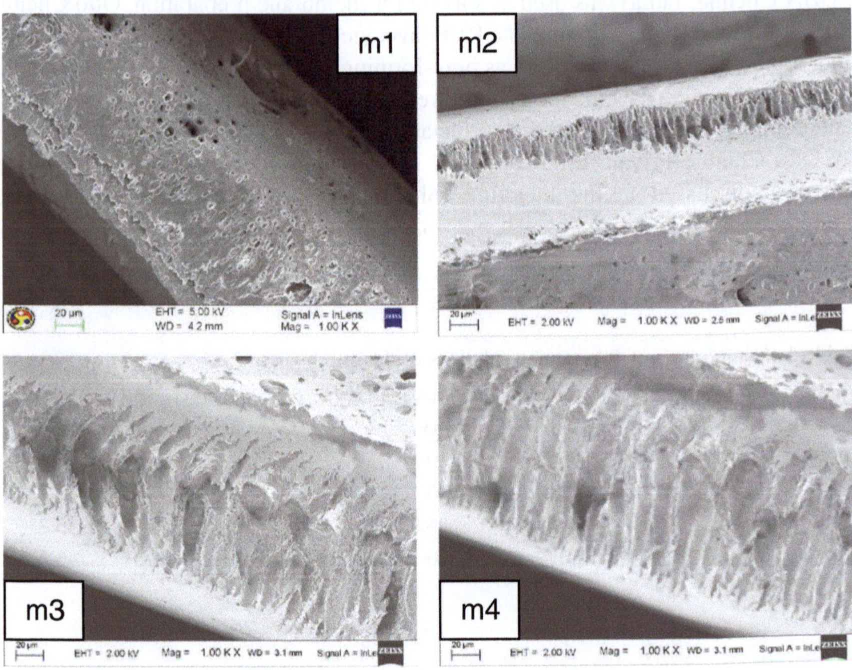

Fig. 1 Cross-sectional FESEM images of membrane m1, m2, m3, and m4

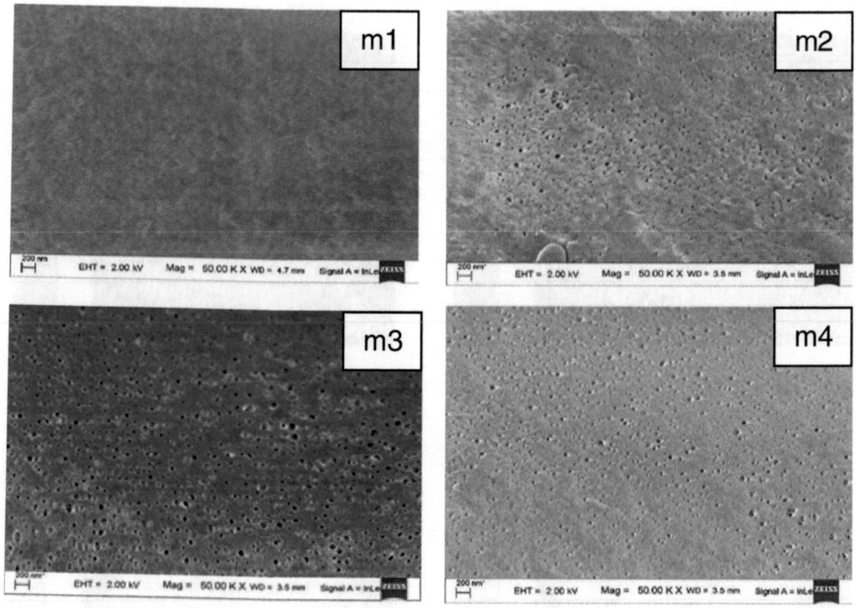

Fig. 2 FESEM images of top surface of the membrane m1, m2, m3, and m4

non-solvent inflow to solvent outflow increases by increasing the viscosity which resulted in a more porous membrane [27].

Figure 2 depicts the top surface images of m1, m2, m3, and m4 membranes captured by FESEM. Spinodal demixing may be the cause of the development of dense top surface. This may be because of the fact that the diffusion process was so fast to become extremely unstable and cross the spinodal curve during formation of top layer [28, 29].

3.1.2 AFM Analysis

Atomic force microscopy (AFM) was used for analyzing the surface morphology and roughness of the membranes. Small squares of membranes (Approximately 1.5 cm^2) were taken for analysis. Figure 3 depicts the AFM images of all the four membranes. Root mean square (RMS) roughness was measured using the 3D AFM images. RMS roughness was found to be increased with the addition of PVP and IMZ. Further, it was also observed that higher molecular weight of PVP containing membranes shows more RMS roughness than plain and lower molecular weight of PVP containing membranes. It may be attributed to the fact that porosity was increasing since number of pores on the surface was increasing by the addition of PVP and IMZ [20]. RMS roughness for m1, m2, m3, and m4 was measured as 3.6, 5.8, 6.9, and 8.1 nm, respectively.

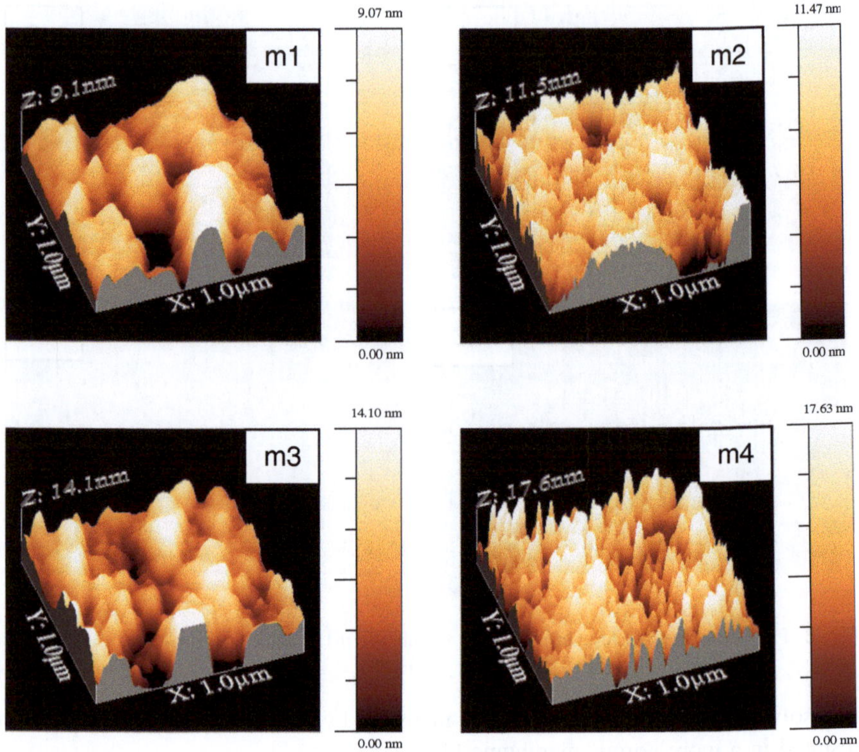

Fig. 3 AFM images of membrane m1, m2, m3, and m4

3.2 Determination of Molecular Weight Cut-off (MWCO)

MWCO values are the measures to determine qualitative pore sizes of the membranes while the pure water permeability is related to porosity (pore number density) of the membranes [30]. Measurement of MWCO values was done by passing aqueous solutions of polyvinyl pyrrolidone (PVP) of two different molecular weights (i.e., PVP 24,000 Da and PVP 40,000 Da) and a protein bovine serum albumin (molecular weight 68,000 Da). Standard technique [31] was opted for finding the concentration of PVP in aqueous solution, whereas protein concentration was measured by UV–vis spectrometer. Rejection (% R) of polymer and protein solutes was used for determining the MWCO of the membranes. Figure 4 depicts the plot of rejection versus molecular weight for all the three membranes. The value of molecular weight at rejection 90% was taken as MWCO of membrane assuming that there is a linear correlation between the rejection and molecular weight of solute. The MWCO value was found to be about 108, 97, 85, and 68.5 kDa for the membrane m1, m2, m3, and m4, respectively. MWCO value was lowest for membrane m4 among all the four membranes. Figure 4 shows the rejection (%) versus molecular weight for finding the MWCO of the membranes.

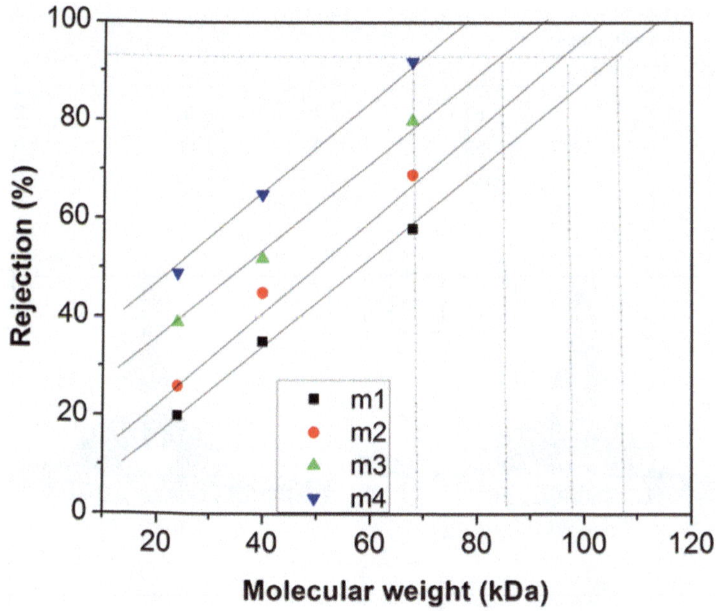

Fig. 4 Rejection (%) versus molecular weight for determining MWCO value of the membranes

3.3 *Permeation Studies*

Effect of the addition of PVP with IMZ on fabricated membranes was checked by permeation behavior. Effects of both the additives were studied on equilibrium water content (EWC) and porosity. Membranes were tested for flux and rejection (%) of CVD with and without SDS. Lastly, at different pHs the membrane performance was checked for its rejection and permeate flux performance with CVD. Outcomes of the research have been discussed in subsequent sections.

3.3.1 Hydrophilicity, EWC (%), and Porosity Analysis

Role of hydrophilicity and porosity of the membrane is undoubtedly significant in membrane permeation process. These are based on the morphology and subsequently affect the pure water flux through membrane. Surface hydrophilicity is principally described by the contact angle measurement [32]. Generally, it is a common phenomenon that as the value of contact angle decreases the hydrophilicity increases. Porosity and EWC (%) of the membranes were measured using standard method [20, 21]. Values of porosity and EWC (%) of membranes are shown in Table 2. Figure 5 depicts the images of contact angle. Contact angle decreased and porosity increased with the addition PVP with IMZ. Hydrophilicity increases by increasing the molecular weight of PVP; it was highest for membrane

Table 2 Values of some characterization parameters of all the four membranes

Membrane	EWC (%)	Pore number $N_t \times 10^{-8}$ (m^{-2})	Porosity
m1	36.0	1.21	0.17
m2	41.1	1.46	0.22
m3	46.3	3.11	0.28
m4	54.2	4.60	0.43

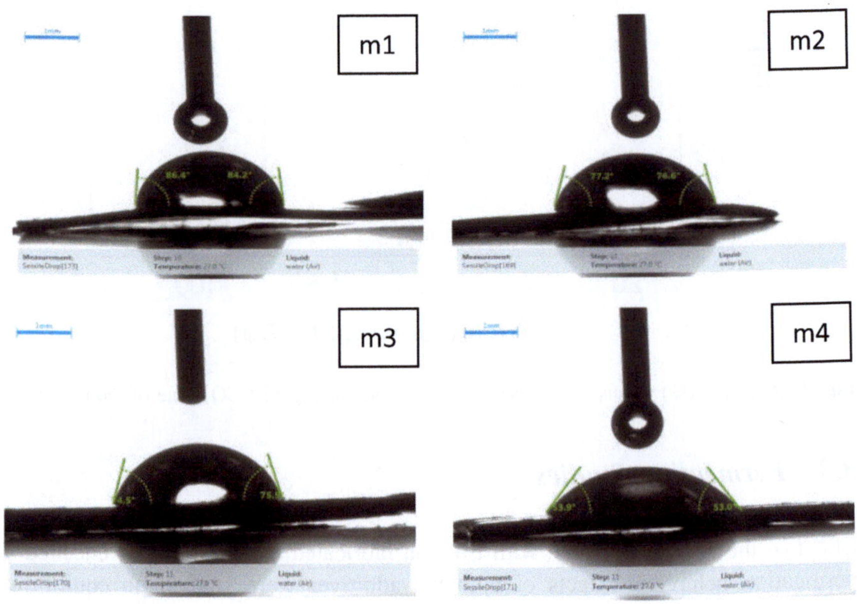

Fig. 5 Images of contact angles for membrane m1, m2, m3, and m4

m4 among all the four membranes. It may be due to the increased hydrophilic group on the membrane surface by the addition of PVP and IMZ. The contact angle for m1, m2, m3, and m4 was 85.5°, 76.9°, 75°, and 53.4°, respectively. Similarly, porosity for m1, m2, m3, and m4 was 0.17, 0.22, 0.28, and 0.43, respectively. The change in porosity was due the difference in viscosity of the casting solution.

3.3.2 Analysis of Liquid–Liquid Displacement Porosimetry Results

Radius of the pore, permeability of the membranes, and pore number per unit area for each membrane were calculated using standard equations [33, 34]. Pore size distribution of the membranes observed by LLDP is shown in Fig. 6. Around 18.5% of the pores were in the size of 2–3 nm for m4. For m1, m2, and m3, these numbers are 17, 17.1, and 17.3%, respectively. It can be seen from Fig. 6 that the

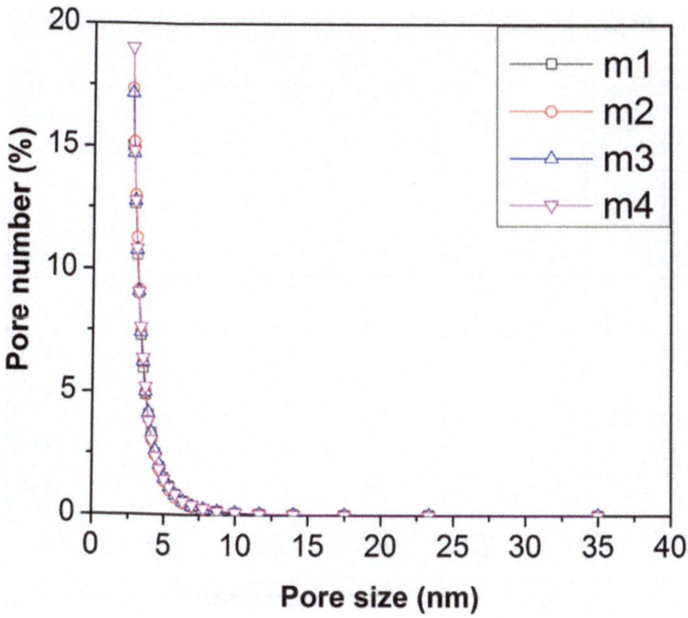

Fig. 6 Pore size distribution of all the membranes by LLDP method

pores approximately 90% for all the five membranes are in the range of 3–5 nm which clearly brought them under UF range. Figure 7 depicts the variation in cumulative permeability (%) with pore size (nm). Though, larger pores play major role in overall permeability; as contribution from the small number of larger pores (>50 nm) can be quite highly compared to the smaller size pores (<5 nm), they are majority in number [22]. Hagen–Poiseuille equation explains that the enlarged pore radius can be responsible for increase in flux. Calvo et al. [35] observed the same results for UF membranes formed by track etched method. However, it is difficult and challenging to measure the accurate part of the larger pores (>50 nm) and smaller pores (<2 nm) by the LLDP method. Results obtained by LLDP method are placed in Table 2. It is observed that by the addition of PVP with IMZ the number of pores for all the membranes increases, especially addition of higher molecular weight of PVP resulted in more porous membranes. Pores were increased from 1.2×10^8 m^{-2} to 4.6×10^8 m^{-2} for membrane m1 and m4, respectively. The mean pore size r_m for all the five membranes was slightly decreased by the addition of PVP and IMZ. It was calculated as 2.6, 2.4, 2.21, and 1.9 nm for m1, m2, m3, and m4, respectively.

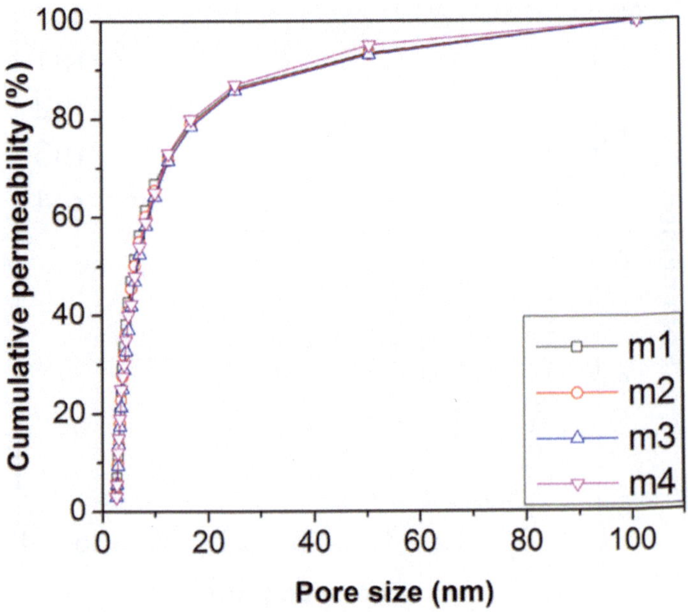

Fig. 7 Variation in cumulative permeability (%) with pore size (nm)

3.4 *Ultrafiltration of CVD*

Transmembrane pressure is a key factor for finding the characteristic of flux other than that the flux description and rejection of solute through the membranes mostly assessed by the morphology of the membrane and the properties of the feed solution, specially its pH. So, fabricated UF membranes were also tested by measuring flux at different pH of CVD solution (20 mg L^{-1}).

3.4.1 Flux and Rejection of CVD at pH 7: With and Without Anionic Surfactant Sodium Dodecyl Sulfate (SDS)

In the present study, SDS is used for the micellar-enhanced ultrafiltration (MEUF). In this process, micelles of surfactant help to solubilize organic or inorganic contaminants from aqueous stream. Effect of the addition of SDS in CVD removal is depicted in Fig. 8. SDS was taken at a concentration of 8.2 mM which is the critical micellar concentration (CMC) of SDS [36], and CVD concentration was taken as 20 mg L^{-1}. Surfactant–dye interaction in aqueous solutions has been studied in the different literature [37–40]. Both hydrophobic and electrostatic forces play significant role in association between ionic dyes and ionic surfactants [41]. Figure 8 clearly depicts that in the absence of surfactant rejection of CVD is lower. It is due

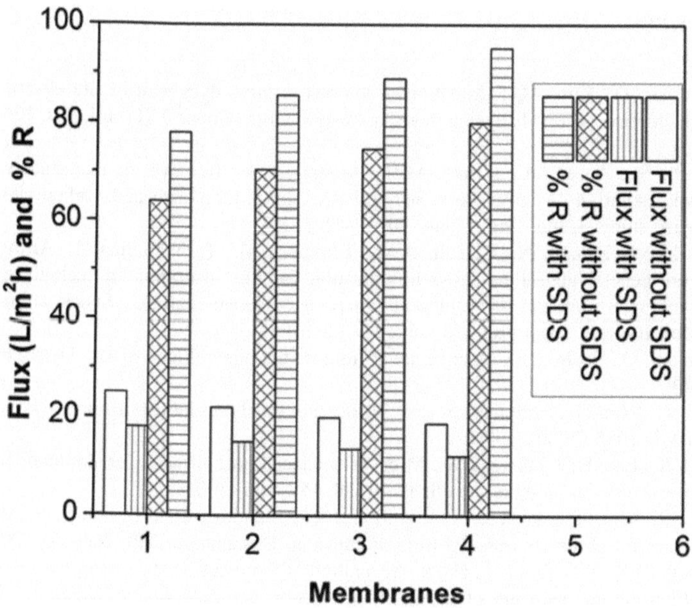

Fig. 8 Flux and % rejection after addition of SDS in dye solution

the fact that the size of pores of membrane is bigger than the CVD molecules. Conversely, by the addition of SDS in dye solution, the rejection of CVD enhanced up to 95.5%. Therefore, it is clear that the dye molecules were retained by the UF membrane after solubilization on the surfactant micelles [41].

Figure 8 also depicts the change in the permeate flux with and without SDS. Without SDS, flux was found more. However, by the addition of SDS flux was decreased. Concentration polarization on the membrane is responsible for reduced permeate flux [41]. Since the micelles were accumulated over the membrane surface. Purkait et al. also found similar observations [42].

4 Conclusions

Major findings of the work are:

I. PVP (average molecular weight of 360,000 Da) with IMZ depicted best performance among all the membranes for every performance parameter.
II. Porosity was found to increase from 0.17 to 0.43 for m1 and m4, respectively.
III. CVD rejection (%) through the membranes was improved from 77.5% for plain membrane m1 to 95.5% for m4 membrane with SDS at normal pH 7.

References

1. Rajkumar, D., Kim, J.G.: Oxidation of various reactive dyes with in situ electro-generated active chlorine for textile dyeing industry wastewater treatment. J. Hazard. Mat. **136**, 203–212 (2006)
2. Ghaedi, M., Ansari, A., Habibi, M.H., Asghari, A.R.: Removal of malachite green from aqueous solution by zinc oxide nanoparticles loaded on activated carbon: kinetics and isotherm study. J. Ind. Eng. Chem. **20**, 17–28 (2014)
3. Ghaedi, M., Zeinali, N., Ghaedi, A.M., Teimuori, M., Tashkhourian, J.: Artificial neural network-genetic algorithm based optimization for the adsorption of methylene blue and brilliant green from aqueous solution by graphite oxide nanoparticles. Spectrochim. Acta Part A: Mol. Biomol. Spectrosc. **125**, 264–277 (2014)
4. Marmara, O., Poste, C.: Colour removal from textile plant effluents. Am. Dyest. Rep. 15–21 (1996)
5. Crini, G.: Non-conventional low-cost adsorbents for dye removal: a review. Bior. Technol. **97**, 1061–1085 (2006)
6. Baek, K., Lee, H.H., Yang, J.W.: Micellar-enhanced ultrafiltration for simultaneous removal of ferricyanide and nitrate. Desalination **158**, 157–166 (2003)
7. Bade, R.: Micellar enhanced ultrafiltration (MEUF) and activated carbon fibre (ACF) hybrid processes for chromate removal from wastewater. Desalination **229**, 264–278 (2008)
8. Purkait, M.K., DasGupta, S., De, S.: Separation of aromatic alcohols using micellar-enhanced ultrafiltration and recovery of surfactant. J. Membr. Sci. **250**, 47–59 (2005)
9. Purkait, M.K., DasGupta, S., De, S.: Removal of dye from wastewater using micellar-enhanced ultrafiltration and recovery of surfactant. Sep. Purif. Technol. **37**, 81–92 (2004)
10. Mulder, M.: Basic Principles of Membrane Technology, vol. 56, pp. 49–51. Kluwer Academic Publishers, Netherlands (1996)
11. Cheryan, M.: Ultrafiltration and Microfiltration Handbook, vol. 345, pp. 49–51. Technomic Publishing Company Inc., USA (1998)
12. Kim, J.H., Lee, K.H.: Effect of PEG additive on membrane formation by phase inversion. J. Membr. Sci. **138**, 153–163 (1998)
13. Bhadra, S., Singha, N.K., Khastgir, D.: Effect of aromatic substitution in aniline on the properties of polyaniline. Europ. Polym. J. **44**, 1763–1770 (2008)
14. Geise, G.M., Lee, H.S., Miller, D.J., Freeman, B.D., McGrath, J.E., Paul, D.R.: Water purification by membranes: the role of polymer science. J. Polym. Sci. Part B: Polym. Phys. **48**, 1685–1718 (2010)
15. Tweddle, T.A., Kutowy, O., Thayer, W.L., Sourirajan, S.: Polysulfone ultrafiltration membranes. Ind. Eng. Chem. Prod. Res. Dev. **22**, 320–326 (1983)
16. Zhang, Y., Shana, L., Tua, Z., Zhang, Y.: Preparation and characterization of novel Ce doped non stoichiometric nano silica/polysulfone composite membranes. Sep. Purif. Technol. **63**, 207–212 (2008)
17. Vatanpour, V., Madaeni, S.S., Moradian, R., Zinadini, S., Astinchap, B.: Novel antifouling nanofiltration polyether sulfone membrane fabricated from embedding TiO_2 coated multi-walled carbon nanotubes. Sep. Purif. Technol. **90**, 69–82 (2012)
18. Li, Y.S., Yan, L., Xiang, C.B., Hong, L.J.: Treatment of oily wastewater by organic–inorganic composite tubular ultrafiltration (UF) membranes. Desalination **196**, 76–83 (2006)
19. Ahmad, A.L., Abdulkarim, A.A., Ooi, B.S., Ismail, S.: Recent development in additives modifications of polyethersulfone membrane for flux enhancement. Chem. Eng. J. **223**, 246–267 (2013)
20. Sharma, N., Purkait, M.K.: Preparation of hydrophilic polysulfone membrane using polyacrylic acid with polyvinyl pyrrolidone. J. Appl. Polym. Sci. (2015). https://doi.org/10.1002/app.41964

21. Sinha, M.K., Purkait, M.K.: Increase in hydrophilicity of polysulfone membrane using polyethylene glycol methyl ether. J. Membr. Sci. **437**, 7–16 (2013)
22. Chakrabarty, B., Ghoshal, A.K., Purkait, M.K.: Preparation, characterization and performance studies of polysulfone membranes using PVP as an additive. J. Membr. Sci. **315**, 36–47 (2008)
23. Yan, X., He, G., Gu, S., Wua, X., Dua, L., Wang, Y.: Imidazolium-functionalized polysulfone hydroxide exchange membranes for potential applications in alkaline membrane direct alcohol fuel cells. Int. J. Hydrogen Energy **37**, 5216–5224 (2012)
24. Kumari, S., Sharma, P.K., Kumar, N.: Imidazole and its biological activities: a review. Pelagia Res. Libr. **1**(3), 36–40 (2010)
25. Sun, J., Jordan, L.R., Forsyth, M., MacFarlane, D.R.: Acid-organic base swollen polymer membranes. Electrochim. Acta **46**, 1703–1710 (2001)
26. Gendi, A.E.L., Abdalla, H., Ali, S.: Construction of ternary phase diagram and membrane morphology evaluation for polyamide/formic acid/water system. Aust. J. Basic Appl. Sci. **6** (5), 62–68 (2012)
27. Young, T.H., Chen, L.W.: A diffusion-controlled model for wet casting membrane formation. J. Membr. Sci. **59**, 169–181 (1991)
28. Kimmerle, K., Strathmann, H.: Analysis of the structure-determining process of phase inversion membranes. Desalination **79**, 283–302 (1990)
29. Reuvers, J., Smolders, C.A.: Formation of membranes by means of immersion precipitation. Part II. The mechanism of formation of membranes prepared from the system cellulose acetate–acetone–water. J. Membr. Sci. **34**, 67–86 (1987)
30. Maurya, S.K., Parashuram, K., Singh, P.S., Ray, P., Reddy, A.V.R.: Preparation of polysulfone–polyamide thin film composite hollow fiber nanofiltration membranes and their performance in the treatment of aqueous dye solutions. Desalination **304**, 11–19 (2012)
31. Jones, S.A., Martin, G.P., Brown, M.B.: Determination of polyvinyl pyrrolidone using high-performance liquid chromatography. J. Pharm. Biomed. Anal. **35**, 621–624 (2004)
32. Farbod, M., Rezaian, S.: An investigation of super-hydrophilic properties of TiO_2/SnO_2 nano-composite thin films. Thin Solid Films **520**, 1954–1958 (2012)
33. Abaticchio, P., Bottino, A., Capannelli, G., Munari, S.: Characterization of ultrafiltration polymeric membranes. Desalination **78**, 235–255 (1990)
34. Munari, S., Bottino, A., Capannelli, G., Moretti, P.: Membrane morphology and transport properties. Desalination **53**, 11–23 (1985)
35. Calvo, J., Bottino, A., Capannelli, G., Hernández, A.: Comparison of liquid–liquid displacement porosimetry and scanning electron microscopy image analysis to characterise ultrafiltration track-etched membranes. J. Membr. Sci. **239**(2), 189–197 (2004)
36. Elworthy, P.H., Mysels, K.J.: The surface tension of sodium dodecyl sulfate solutions and the phase separation model of micelle formation. J. Colloid Interface Sci. **21**, 331–347 (1966)
37. Sarkar, M., Poddar, S.: Studies on the interaction of surfactants with cationic dye by absorption spectroscopy. J. Colloid Interface Sci. **221**, 181–185 (2000)
38. Sabaté, R., Gallardo, M., de la Maza, A., Estelrich, J.: A spectroscopy study of the interaction of pinacyanol with n-dodecyltrimethyl ammonium bromidemicelles. Langmuir **17**, 6433–6437 (2001)
39. Göktürk, S., Tuncay, M.: Spectral studies of safranin-O in different surfactant solutions, Spectrochim. Acta Part A: Mol. Biomol. Spectrosc. **59**, 1857–1866 (2003)
40. Sharma, N., Purkait, M.K.: Enantiomeric and racemic effect of tartaric acid on polysulfone membrane during crystal violet dye removal by MEUF process. J. water Process Eng. **10**, 104–112 (2016)
41. Zaghbani, N., Hafiane, A., Dhahbi, M.: Separation of methylene blue from aqueous solution by micellar enhanced ultrafiltration. Sep. Purif. Technol. **55**, 117–124 (2007)
42. Purkait, M.K., DasGupta, S., De, S.: Micellar enhanced ultrafiltration of phenolic derivatives from their mixture. J. Colloid Interface Sci. **285**, 395–402 (2005)

Performance Evaluation of Low-Cost Microbial Fuel Cell Employing Clayware Separator of Different Thickness

Sudhansu Behera and Manaswini Behera

Abstract The performance of low-cost microbial fuel cell (MFC) employing earthenware separator of different thickness is studied. The earthenware separator composed of red soil blended with montmorillonite; a soil mineral having high cation exchange capacity was used for the study. A comparative study of MFC employing 4 mm (MFC-1) and 5 mm (MFC-2) separator thickness was established where MFC-1 outperformed MFC-2. The dual-chambered MFC with inner anodic chamber and concentric outer cathode chamber showed a maximum power density of 47.77 and 33.44 mW/m^2 in MFC-1 and MFC-2, respectively, while fed with synthetic dairy wastewater of chemical oxygen demand (COD) 1920 $\pm$ 20 mg/L. A comparatively lesser internal resistance of 71 Ω was observed in MFC-1 while 105 Ω in MFC-2 at maximum COD removal efficiency of 89%. MFC-1 and MFC-2 depicted an open-circuit voltage of 822 and 490 mV, respectively, with 150 Ω external resistance.

Keywords Microbial fuel cell (MFC) · Earthenware separator · Chemical oxygen demand (COD) · Power density · Internal resistance

1 Introduction

With the increasing rate of production of domestic and industrial wastewater containing enormous amount of nutrients and insufficient availability of on-site treatment plants, the demand has been a primary concern for the society. For instance, the effluents released to the water body thereby reduce the dissolved oxygen (DO) level in freshwater, and the existence of aquatic animals and herbs in the toxic environment thus becomes impossible. In developing countries like India,

S. Behera (✉) · M. Behera
School of Infrastructure, Indian Institute of Technology Bhubaneswar, Bhubaneswar 751013, India
e-mail: sb33@iitbbs.ac.in

A. S. Kalamdhad et al. (eds.), *Advances in Waste Management*,
https://doi.org/10.1007/978-981-13-0215-2_29

this scenario is very common and there is a need for on-site wastewater treatment facility for removal of excess organic matter and nutrients. The concept pertaining to generation of electric current through bacteria culture was first reported by Potter (1911) [1]. The microbial fuel cell (MFC) is a bioelectrochemical device that continuously produces electricity by the exoelectrogens (anaerobes) present in wastewater. The conventional MFC consists of two chambers, namely anaerobic anode chamber and aerobic cathode chamber, where wastewater is supplied at anode. Two electrodes one at anode and other at cathode are connected through an external electric circuit and a load (resistor) in between. The anaerobic degradation of wastewater and the metabolic activity of bacteria produce electrons and protons in anode chamber where the electrons are transferred to the cathode through external electric circuit, while protons are migrated to cathode through the cation exchange membrane (CEM). At cathode, the electrons, protons, and oxygen all combine to form water as by-product. Therefore, there is a potential difference in anode and cathode chamber which produces electricity. A continuous supply of atmospheric oxygen is thus ensured through active aeration in cathode chamber. The results thus obtained from the current study depicted a significant improvement over conventional MFCs employed with high cost membranes for proton mass transfer. The unique thermal, chemical, and mechanical characteristics of ceramic membrane separators offer greater advantages over polymeric membranes in fuel cells and other applications [2].

However, for the scale-up of power and practical application, still more comprehensive knowledge on electrochemical, biological reactions and microbiological study is required. The use of high-cost materials in MFC also effects the efficacy and its sustainability. The separator, i.e., CEM, has more significance for effective proton permeability from anode to cathode and for inhibiting oxygen transfer and electrolyte crossover. The most commonly used separators are Nafion®; a sulfonated group possess high proton conductivity. However, it is very expensive ($1733/m^2) and its cost has a significant impact on the overall manufacturing cost of an MFC [3]. So, the current research is based on developement of a low cost eco-friendly CEM made up of locally available clayey soil which can sustain the worst environment along with its behaviour of different thickness.

2 Materials and Methods

2.1 Earthenware Separator

Earthenware separators were prepared by blending 80% of red soil (laterite) and 20% of montmorillonite (cation exchanger) and adding small amount of water. Wet clay after achieving certain plasticity was filled in two molds of size 7 cm × 9 cm × 0.5 cm and 7 cm × 9 cm × 0.4 cm, respectively. They are dried in open air for 5 days, followed by drying in hot air oven at 150 °C for 2 h, and finally kept in muffle furnace at 650 °C for 30 min. Energy Dispersive X-ray Spectroscopy (EDS) analysis was carried out to study particle characterization of earthenware separator. The separator possess SiO_2—49.23%, Al_2O_3—21.6%, Fe—21.5%, K—2.15%, MgO—1.11%, Ti—1.45%, Na—0.95%, Cu—2%. The earthenware separators were placed in the MFC by Araldite, an adhesive material.

2.2 Electrodes

Stainless steel (SS) mesh was used as anode electrode with total projected surface area 502.4 cm^2. Similarly, graphite plates of size 6 cm × 8 cm having thickness of 0.5 cm were used as cathode electrode for both the reactors. The anode and cathode electrodes were placed very close to earthenware separator at a distance of 0.7 cm apart. Thin copper wire with full insulation was used in the external electric circuit with 150 Ω external resistance.

2.3 Fabrication of MFC and Its Operation

Two dual-chambered MFC reactors were fabricated each having outer cathode chamber and concentric inner anode chamber. Hereafter, the MFC reactor employing 4 mm is noted as MFC-1 while that of MFC employing 5 mm is noted as MFC-2. The dual-chambered MFC (Fig. 1) was fabricated using transparent poly acrylic material. The anode chamber has working volume of 780 and 800 mL in MFC-1 and MFC-2, respectively, while that of cathode it is 6.6 L for both the reactors. Four openings each having size of 6 cm × 8 cm were made at four faces of anode chamber for placement of earthenware separators. Wastewater was fed to the anode chamber from the bottom. Two openings were made at the top and bottom of anode chamber for passage of effluent from anode chamber to cathode chamber.

The dual-chambered MFC was operated with continuous supply of synthetic dairy wastewater having chemical oxygen demand (COD) of 1920 ± 20 mg/L. The substrate was supplied at a flow rate of 2.1 mL/min (HRT = 6 h) by using a peristaltic pump. The constituents of synthetic dairy wastewater are milk powder, 1.6 g/L; $FeCl_3{\cdot}6H_2O$, 15 mg/L; $MgSO_4{\cdot}7H_2O$, 70 mg/L; $CaCl_2{\cdot}H_2O$, 30 mg/L; $MnSO_4{\cdot}7H_2O$, 15 mg/L; KH_2PO_4, 60 mg/L; NH_4Cl, 120 mg/L; $NaHCO_3$, 1 g/L [4, 5]. The substrate was given anaerobic treatment in the anode chamber, and the effluent of anode chamber was given aerobic treatment in the cathode chamber with the help of active aeration. The anode chamber was inoculated with bottom sludge of domestic sewer having volatile suspended solid (VSS) of 32.5 g/L. The experimental setup is illustrated in Fig. 1.

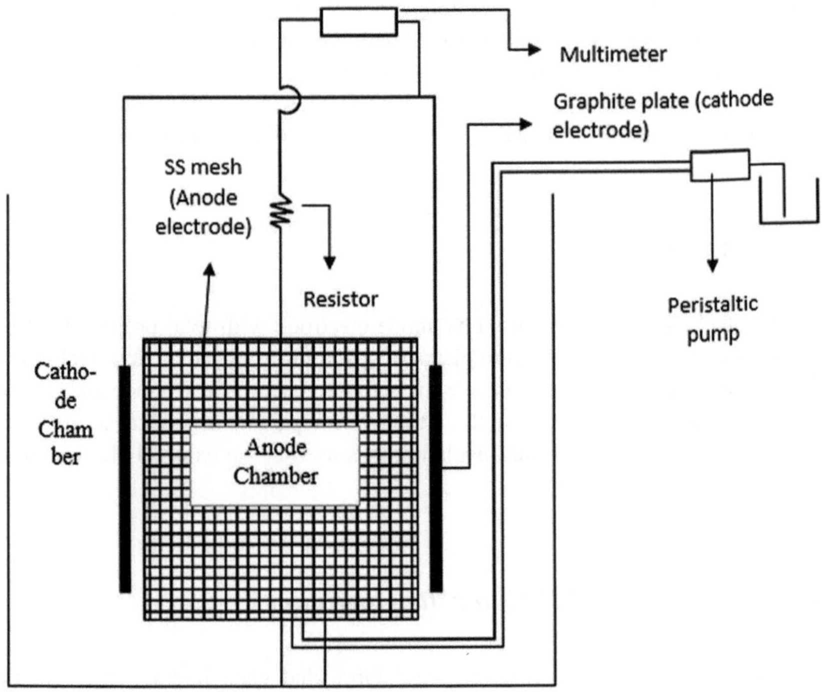

Fig. 1 MFC experimental setup

2.4 Analyses and Calculation

The COD concentrations of influent and effluent, DO, pH, volatile suspended solid (VSS), total solid (TS) were measured according to APHA standard procedure [6]. The current (*I*) and voltage (*V*) were measured by a true RMS professional multimeter (Extech, India). The power can thus be calculated by $P = V * I$. The open-circuit voltage (OCV) and working voltage (external resistance of 150 Ω) were measured every day throughout the working period of 45 days. The power density and volumetric power were calculated normalized to anodic electrode area and net liquid volume of anode chamber, respectively. The anode and cathode cell potentials were measured by KCl reference electrode. Polarization study was conducted after achieving a stable performance by varying external resistances from 50 to 4500 Ω. The internal resistance associated with MFC was calculated from the slope of line plotted between current and voltage.

The coulombic efficiency (CE) is defined as the ratio of total coulombs recovered during the operation per total coulombs present in the substrate [3]. The CE of MFC operated in continuous mode of substrate supply can be found out from the formula:

$$\mathrm{CE} = \frac{8I}{Fq\Delta\mathrm{COD}} \tag{1}$$

where I is the measured electric current; 8 is a constant used for substrate measured in terms of COD, based on molecular weight of O_2 as 32 g/mol; F is the Faraday's constant = 96,485 C/mol; q is the flow rate of substrate in L/day; and ΔCOD is the change in COD value of the influent and effluent in mg/L. The morphological characteristics of earthenware separator were studied by Energy dispersive X-ray Spectroscopy (EDS) analysis.

2.5 *Proton Mass Transfer Coefficients of Earthenware Separator*

The proton mass transfer coefficients (k_H, cm/s) for both 4 and 5 mm separators were determined abiotically with one chamber filled with deionized water (pH ≈ 6.8) and using 0.1 M NaOH solution with pH 10.5 in the other chamber. The change in pH was observed using a pH probe (Systronics, digital pH meter 335) continuously at intervals of 15 min. The mass transfer coefficient was then calculated as per Eq. (2) [7].

$$k_H = -\frac{v}{2At}\ln\left(\frac{C_a + C_c - 2C_p}{C_a}\right) \tag{2}$$

where C_a and C_c are the initial proton concentration in anodic and cathodic chamber, respectively, and C_p is the proton concentration in the cathodic chamber at time t. Similarly, the proton diffusion coefficient (D_H, cm^2/s) was calculated from average thickness of ceramic separator (L_{th}) as per Eq. (3)

$$D_H = k_o L_{th} \tag{3}$$

3 Results and Discussion

3.1 *Physical Characterization of Separator*

3.1.1 EDS Analysis

The microscopic analysis of earthenware PEM has shown an appreciable amount of silica and aluminum as the key inorganic element in a typical soil sample. The chemical constituents of montmorillonite and the composite soil membrane are listed in Table 1. The interpretation for the elements present in the composite

Table 1 Composition of soil with their weight percentages

Elements	Montmorillonite (%)	Composite (montmorillonite + red soil) (%)
Al_2O_3	19.68	25.64
SiO_2	51.42	45.78
K (feldspar)	1.89	2.2
Fe	17.76	22.94
Cu	2.44	3.43
Ti	2.81	–
Na (albite)	1.19	–
MgO	1.95	–
Ca (wollastonite)	0.86	–

membrane shows high amount of oxides of silica which represents a potential membrane for cation exchange. The composite membrane contains huge amount of Silicon ion creates negative charge when combines with silicon-oxygen tetrahedron. Thus, the presence of silicon ions helps in retaining the positive ions, i.e., protons (H^+) ions on the surface of earthenware separator. Montmorillonite is found to have highest cation exchange capacity (CEC) ranging about 70–100 meq/100 g among all minerals present in soil.

The CEC is the number of exchangeable cations per dry weight that a soil is able to hold. Therefore, a blend of bentonite clay and montmorillonite will certainly exhibit high cation exchange and could possibly overcome major difficulties associated with MFC. A separate study also conducted by Ghagde and Ghangrekar [8] on earthenware separator and concluded that MFC show highest performance by blending 20% of montmorillonite with soil in the composite membrane. Further, the clay minerals are very fine and by blending with soil having higher particle size could possibly reduce the voids. Thus, montmorillonite has the potential to act as proton exchanger, thereby enhancing the efficacy of MFC.

3.1.2 Proton Mass Transfer

The proton mass transfer coefficient (k_H) and diffusion coefficient (D_H) for the earthenware separator were evaluated in un-inoculated MFCs at room temperature. The pH was monitored throughout the period. During the experiment, MFC-1 recorded proton mass transfer coefficient of 8.34×10^{-4} cm/s which is 30% higher than MFC-2 of value 6.35×10^{-4} cm/s. Similarly, the proton diffusion coefficients were calculated as 3.32×10^{-4} cm^2/s and 3.17×10^{-4} cm^2/s for MFC-1 and MFC-2, respectively. The D_H value is higher than the value reported by Ghadge and Ghangrekar (2009) which is 3.25×10^{-6} cm^2/s. The high proton diffusion property is due to the presence of different functional groups [2]. The use of thinner earthenware separator supports more proton flux diffusion at cathode which could be due to decrease in the resistance for proton transfer through the thinner separator in comparision with higher wall thicknesses.

3.2 Electricity Generation in MFC

The direct electricity obtained in MFC from the oxidation of organic substrates is a function of strength (COD) of wastewater. The more is the oxidation of organic substrates during stable period, the more would be the power production. The MFC-1 and MFC-2 showed maximum open-circuit voltage (OCV) of 0.822 and 0.490 V at 60th and 14th day of operation, respectively. A delay in obtaining stable and optimum performance was observed in MFC-1 because of seasonal variation of temperature in operation as MFC-2 and MFC-1 were operated in summer and winter season, respectively. The rate of metabolism and substrate degradation is generally higher in summer; therefore, it took less time for MFC-2 than MFC-1 to obtain stable performance. The maximum and average power production values are given in Table 2. The power was calculated by the formula $P = V * I$, where P is the power generated in watt, V is the working voltage, and I is the measured current with a load of 150 Ω. The volumetric power obtained is much higher than the earlier studies based on ceramic separators; thus, the current composite separator having thickness of 4 mm has shown appreciable performance for power production along with wastewater treatment. The earthenware separator also sustained for such a hydraulic load and can sustain for 14 months as reported by Ghadge and Ghangrekar [8].

3.2.1 Coulombic Efficiency

The average CE during the period of stable performance was 5.25% and 4.16% in MFC-1 and MFC-2, respectively. The low CE shown by the MFCs could possibly due to utilization of large percentage of substrate by anaerobic microbes such as methanogenic archaea or sulfate-reducing bacteria other than anode-respiring bacteria (He et al. 2005) [9]. The low coulombic efficiency could also be due to oxygen and substrate diffusion through the separator as reported by earlier studies.

Table 2 Power production values

	Working voltage (mV)	Power density (mW/m^2)	Volumetric power (W/m^3)	OCV (mV)
MFC-1	600 (max)	47.77	3	822
	466 (average)	30.23	1.89	686
MFC-2	470 (max)	29.31	1.88	490
	422 (average)	23.71	1.52	440

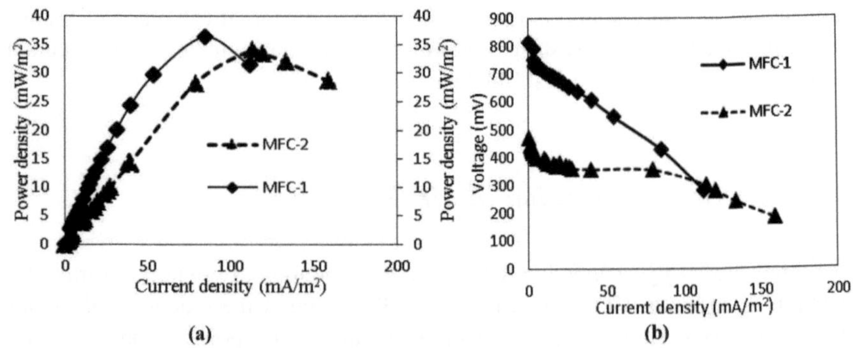

Fig. 2 Polarization study of MFC-1 and MFC-2

3.2.2 Polarization Study and Internal Resistance

The polarization study is mainly conducted to know the effect of different external resistances on the performance of the system. Figure 2 shows the polarization curve as a function of current density, voltage, and power density measured at variable external circuit load (50–4500 Ω). The voltage and current were measured in the multimeter after achieving a stable reading at an interval of 5 min. During polarization, current generation is decreased with increase in external resistance, which indicated typical fuel cell behavior. The power density curve showed a maximum power density of 36.29 and 33.44 mW/m^2 in MFC-1 and MFC-2, respectively, at an external resistance of 100 and 200 Ω during polarization.

The internal resistance associated with both the reactors enumerated from the slope of the curve between voltage and current density as shown in Fig. 3. The ohmic resistance mentioned is the slope of curve between voltage and current density during stable period obtained from the mid-straight portion. The internal resistance of MFC-1 was 71 Ω which is less than the value 105 Ω obtained in MFC-2. This is due to thinner earthenware separator facilitated a decrease in the resistance to mass transfer of protons and increase the flux which enhanced the reduction reaction on the cathode surface [10]. Therefore, the use of earthenware separator of lesser thickness has significant improvement over internal resistance.

3.3 *COD Removal*

The overall COD removal efficiency of MFC-1 and MFC-2 was 85.25% and 83.3%, respectively, during the stable operational period. There has been a consistent increase in the COD removal efficiency in MFC-2 after an acclimatization period of 12 days of operation, while MFC-1 took 18 days to attain stable performance because of low opearating temperature. The COD removal efficiency in anode chamber was 61.44% and 69.7%, while the COD removal efficiency in

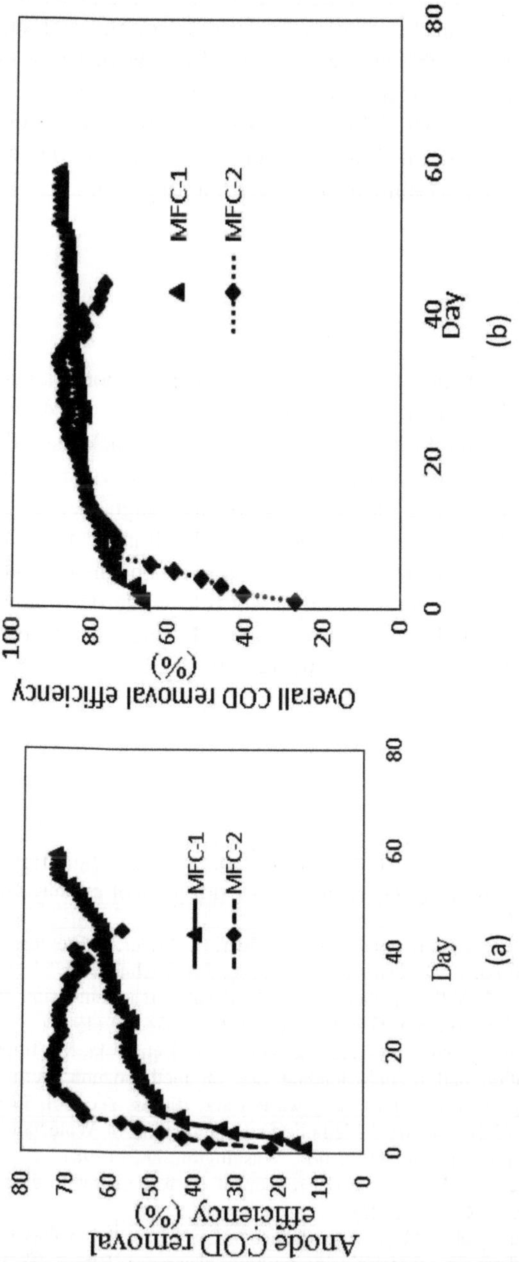

Fig. 3 a, b Anodic COD removal and overall COD removal efficiency cathode chamber was 61.74% and 45.2% in MFC-1 and MFC-2, respectively. A marginal improvement on COD removal was observed in MFC-1 as compared to MFC-2, but most importantly the difference was observed in electricity production in both the reactors. From this experiment, it was observed that summer temperature supports anaerobic substrate degradation. The anaerobically treated effluents from anode chamber were again supplied to cathode chamber to undergo aerobic treatment. The dual-chambered MFC was found to be efficient to treat synthetic dairy wastewater. The introduction of biocathode also helped to reduce the strength of organic matter of anode effluent by continuous aeration.

4 Conclusion

The study investigated the effect of earthenware separator thickness on the performance of MFC. The power output in MFC-1 employed with 4 mm separator was significantly higher than MFC-2 having 5 mm wall thickness. The earthenware separator can be an alternative for high-cost polymer separators during operation. Therefore the use of thinner cation exchange membranes as provided in MFC-1 will certainly overcome the major hindrances in obtaining elevated performance in conventional MFCs. The improvement in internal resistance and proton mass transfer also plays an important role while designing stacked MFC. The MFC employed with low-cost separators is thus sustainable for providing renewable energy during wastewater treatment.

References

1. Potter, M.C.: Electrical effects accompanying the decomposition of organic compounds. Proceedings of the Royal Society B: Biol. Sci. **84**(571), 260–276 (1911)
2. Sondhi, R., Bhave, R., Jung, G.: Applications and benefits of ceramic membranes. Membr. Technol. **11**, 5 (2003)
3. Logan, B.E., Hamelers, B., Rozendal, R., Schröder, U., Keller, J., Freguia, S., et al.: Microbial fuel cells: methodology and technology. Environ. Sci. Technol. **40**, 5181–5192 (2006)
4. Raj, S.A., Murthy, D.V.S.: Synthetic dairy wastewater treatment using cross flow medium trickling filter. J. Environ. Sci. Health. Part A **34**(2), 357–369 (1999)
5. Zielińska, M., Cydzik-Kwiatkowska, A., Zieliński, M., Dębowski, M.: Impact of temperature, microwave radiation and organic loading rate on methanogenic community and biogas production during fermentation of dairy wastewater. Biores. Technol. **129**, 308–314 (2013)
6. APHA, Awwa, WPCF: Standard Methods for Examination of Water and Wastewater, 20th edn. American Public Health Association, Washington, DC (1998)
7. Lovley, D.R: The microbe electric: conversion of organic matter to electricity. Curr. Opin. Biotechnol. **19**(6), 564–571 (2008)
8. Ghadge, A.N., Ghangrekar, M.M.: Performance of low cost scalable air–cathode microbial fuel cell made from clayware separator using multiple electrodes. Biores. Technol. **182**, 373–377 (2015)

9. He Z., Minteer, S.D., Angenent, L.T.: Electricity generation from artificial wastewater using an upflow microbial fuel cell. Environ. Sci. Technol. **39**(14), 5262–5267 (2005)
10. Behera, M., Ghangrekar, M.M.: Electricity generation in low cost microbial fuel cell made up of earthenware of different thickness. Water Sci. Technol. **64**(12), 2468–2473 (2011)

Characterization and Disposal of Ashes from Biomedical Waste Incinerator

J. M. Vivek, Richa Singh, Rahul S. Sutar and Shyam R. Asolekar

Abstract Incineration happens to be the most commonly practiced method of disposal of biomedical wastes (BMWs) in India. This method has some important advantages such as pathogen destruction and volume reduction. The ashes generated are sent to engineered hazardous wastes' landfills. There are many issues of concern of ash disposal due to the potential for dislocation of pollution through the presence of heavy metals, inorganics and other refractory substances. The bottom ash samples characterized in this study were collected from a biomedical wastes' incinerator having a capacity of 50 kg BMW/h. The particle size distribution, morphology, mineralogy and chemical composition were determined using analytical techniques like ICP-AES, XRD and SEM. The analysis by ICP-AES and SEM indicated the presence of heavy metals such as Fe, Pb, As, Cr in all the samples. The presence of chlorine was also detected which indicated the prevalence of burning of polyvinylchloride compounds in the incinerator. Bottom ash was enriched in Mg, Zn, Al, Ca and Ti. Powder-XRD data showed bottom ash to be an amorphous mass. Some rare earth elements were also detected. A set of alternate solutions for keeping ash out of hazardous waste landfills and sustainable management methods are also discussed.

Keywords Incinerator · Biomedical waste · Bottom ash · Heavy metals
Hazardous waste landfill

J. M. Vivek · R. Singh · R. S. Sutar · S. R. Asolekar (✉)
Centre for Environmental Science and Engineering, Indian Institute of Technology Bombay, Powai, Mumbai 400076, India
e-mail: asolekar@iitb.ac.in

J. M. Vivek
e-mail: vivekjm@iitb.ac.in

R. Singh
e-mail: richasingh@iitb.ac.in

R. S. Sutar
e-mail: rahulsutar@iitb.ac.in

A. S. Kalamdhad et al. (eds.), *Advances in Waste Management*,
https://doi.org/10.1007/978-981-13-0215-2_30

1 Introduction

The wastes generated from human activities generally include the heterogeneous disposals from both municipal areas as well as the rural settings. It is generally characterized as municipal wastes, hazardous wastes, biomedical wastes or hospital wastes. The rules governing each category point out at the explicit features leading to specific rules, regulations and protocols for proper treatment, storage and disposal.

The growth of population and proportionate massive outburst in diseases and accidents generate more and more infectious and non-infectious wastes from healthcare facilities. Biomedical wastes (popularly referred by various names as hospital wastes, clinical wastes, healthcare wastes) are produced while the diagnosis, treatment or immunization of humans (or animals) is carried out or when research activities are conducted in these areas. The generated biomedical wastes can be hazardous as well as infectious and contain toxic chemicals and pathogenic organisms which pose contamination risks to both people and the environment. There is a huge irony that in the process of offering healthcare facilities, waste products are generated that are potentially more harmful. The wastes generated from healthcare facilities, research centres and laboratories can be broadly classified into "hazardous-biomedical" wastes which comprise around 25% of the total waste fraction (sharps, infectious, pathological, pharmaceutical, cytotoxic, chemical and radioactive wastes) and 75% "non-hazardous-biomedical" wastes that mainly comprise of domestic waste and garbage [1]. Due to lack of awareness and improper processes of segregation, the generated domestic waste also gets contaminated with the infectious wastes, leading to the generation of huge quantity of potentially infectious wastes. The important parameters of concern for any waste treatment disposal system facility should be the quantity and quality of wastes, the competency of the dedicated disposal facility to manage that much quantity, the treatment efficiency observed, available space for the equipment, effluents and gaseous emissions from the facility, the economy of the operation and most importantly the occupational health and safety mainly for the workers who are continuously exposed in the facility [2]. In a study conducted for biomedical waste management in a hospital in Belgaum, India, the care ensured by personnel for handling the biomedical wastes elucidates the need of systematic handling of infectious and non-infectious wastes [3].

Generally, four methods are employed to treat the hazardous wastes from the healthcare sector. These are: thermal (autoclaving, using microwaving, frictional heat treatment system, dry heat technologies, incinerators), chemical (using either dissolved chlorine dioxide or sodium hypochlorite, glutaraldehyde/quaternary ammonium compound-based treatment, lime slurry or calcium oxide treatment, ozone treatment), irradiative (ionizing radiations, UV-C irradiation) and biological methods (using enzymes to treat organic wastes) [4]. But out of all the methods, the most viable system for the disposal is burning in incinerators [5, 6]. In thermal treatment and disposal of biomedical wastes, the wastes are burned in a supervised condition which converts the waste into ash and gases. However, the process is energy-intensive. Biomedical waste incinerators are operated either using oil or electricity or the combination of both. Multiple hearth type, rotary kiln and controlled air types are the commonly used types of incinerators for disposal of healthcare/infectious wastes. The commonly used incinerators have a refractory lining with primary and secondary chambers for ensuring optimal combustion. The primary chamber has pyrolytic conditions where temperature builds up for a range of about 800 ± 50 °C followed by the secondary chamber that is usually operated in excess air conditions and temperature is around 1050 ± 50 °C as per schedule V of the Bio-Medical Waste (Management and Handling) Rules, 1998 [7]. The volatile compounds are liberated in the first chamber, followed by complete destruction in the second chamber. As per the Bio-Medical Waste (Management and Handling) Rules, it has been suggested to dispose cytotoxic drugs, discarded medicines, human anatomical wastes, animal wastes and other soiled wastes by incineration. As per the report on status and issues on implementation of Bio-Medical Waste (Management an Handling) Rules, Indian hazardous and biomedical waste management system employs around 200 hazardous waste and biomedical waste incinerators, with a major fraction employed in the biomedical waste sector only [8].

In the incineration process, certain wastes must be handled in a careful manner. These complex wastes include pressurized gas vessels, halogenated plastics (e.g. PVC), radioactive wastes, huge quantities of reactive chemical wastes, mercury-containing substances, silver salts, etc. As the process of segregation of wastes is improper, a huge heterogeneous mixture is usually being fed into the incinerator. Even though incineration is well suited for chemical, infectious and pharmaceutical wastes and helps in drastic reduction of weight and volume of wastes; production of ash is a big nuisance. The incineration of biomedical wastes also releases gases like CO, CO_2, NO_2, dioxins and furans into the environment. The solid residues generated in the system, commonly referred as ash, may contaminate soil as well as groundwater due to the occurrence of toxic metals, organic and inorganic compounds [9]. The bottom ash disposal in a hazardous waste landfill void of any systematic treatment strategies may lead to the contamination of the soil and water due to the production of leachate in due course of time. Fly ash which is very fine and light in weight settles on post-burner devices like the scrubbers. A comprehensive research has been done on the characteristics of fly ash, and it has been added in the list of hazardous waste materials by the European Union with code 190,103 [10]. Bottom ash was added recently in the list, in 2003. In India, special rules and regulations are framed for the proper disposal and management of fly ash. Fly ash is believed to have high content of heavy metals and hence demands for proper environmental management plan [11]. There is also a dire need to evaluate the bottom ash produced in the country. Hazardous Waste Management Report 2008 released by Central Pollution Control Board (CPCB), India, highlights that approximately 0.82 lakh MTA (metric tons per annum) of ash is produced as a result of incinerating 3.28 lakh MTA of hazardous waste [12]. As the population increases and land footprint decreases, the voluminous ash generated is an issue of concern as its disposal pathways are concerned.

This paper focuses on the bottom ash characteristics mainly its particle size distribution, elemental and mineralogical composition. Very limited studies have been done on bottom ash from BMW incinerators. It is inevitable that more research is needed to understand the specific characteristics of bottom ash so that it will aid in developing a secondary raw material for manufacturing building materials. The characterization and evaluation of properties of bottom ash will also help in deciding whether it is to be deemed hazardous, non-hazardous or inert. This will help in ensuring a sustainable environmental management solution for bottom ash disposal.

2 Materials and Methods

Bottom ash generated from BMW management facility in Maharashtra, India, was studied (ash collected at bottom of incinerator as shown in Fig. 1). The incinerator at site is a Thermex model PY-75 installed in 2003. The fuel used in the facility is diesel with waste feed rate of 50 kg/h. The facility handles around 1400 healthcare establishments with 4802 beds. Around 2900 kg/day is supposed to be handled by the facility.

Around 50% of the total waste reaching the facility is supposed to be incinerable in nature. The collection periods of bottom ash ranged over a time span of three months from the facility and were given names as BAS1, BAS2 and BAS3 (sampling was done at an interval of 29–30 days). The type of the waste received at the facility is presented in Table 1. The collection system ensures that the infectious and non-infectious wastes are segregated in different coloured bags. The yellow- and red-coloured bags are fed directly into the incinerator without opening the bags. If mixing of wastes is suspected in other bags (blue and black), then the suspicious bags are fed into the incinerator. The blue bag components are disinfected using autoclaving and hypochlorite solution treatment, and recyclables are sent for

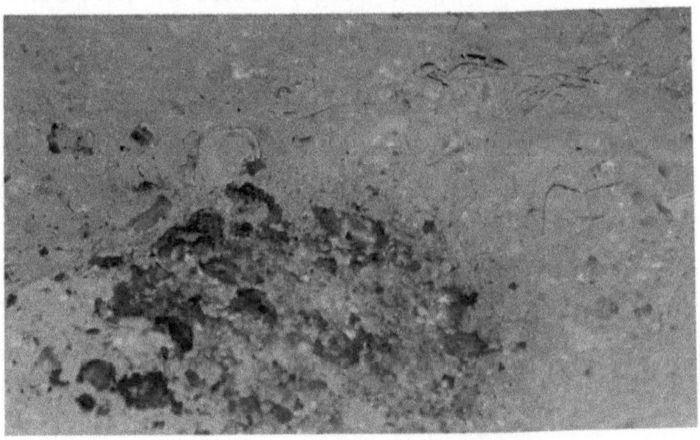

Fig. 1 Residues collected at the bottom of biomedical waste incinerator

Table 1 Categories of waste received at the study facility

Bag for transport of waste	Waste category
Yellow plastic bags	Human body parts, dead foetus, placenta, pathology waste
Red plastic bags	Microbiology and biotechnology waste, soiled waste, gloves, infected plastic wastes
Blue translucent plastic bag	Syringe body, ampules, tubes, dialysis membranes
Black plastic bag	Incineration ash, chemical wastes, discarded medicines

recovery. The incinerator ash is packed and sent to hazardous waste landfill site, and the other domestic garbage items like waste from cafeteria are handed over to the municipal waste disposal system.

The collected samples of bottom ash weighed about 3 kg was dried at room temperature, and then packed and stored properly. For the study of bottom ash, the following analyses were performed:

1. Bottom ash particle size gradation—Sieve analysis
2. Elemental identification and quantitative compositional information—scanning electron microscope (SEM)
3. Detection of elements—inductively coupled plasma atomic emission spectrometry (ICP-AES)
4. Crystalline study for mineral identification—X-ray powder diffraction

2.1 *Bottom Ash Particle Size Gradation—Sieve Analysis*

The ash samples for grading were obtained from the incinerator and allowed to get cooled to room temperature. Visible, unburned residues including unshaped metallic objects, melted glass were segregated prior to the gradation study. The samples were oven dried at 105 °C for 24 h and thereafter cooled down to room temperature. 1 kg of each sample was mechanically sieved through set of sieves starting with 4.75 mm sieve at top, 2.36 mm, 1.18 mm, 600 μm, 300 μm, 150 μm, 90 μm and pan at bottom.

2.2 *Elemental Identification and Quantitative Compositional Information—Scanning Electron Microscope (SEM)*

The sample of bottom ash collected from the incinerator was finely ground, prepared and coated with platinum for around 200 s. The powder was examined in a scanning electron microscope (SEM) fitted with an energy-dispersive X-ray (EDX) prime energy-dispersive analysis system. In the EDX technique, the energy of X-rays produced is collected and measured. It is displayed as a graph of X-ray energy (in keV) versus the frequency of occurrence. The spectrum produced shows a number of characteristic X-ray peaks, associated with the elements present. The instrument used in the study is of Make FEI Model: QUANTA 200 with specifications of operating voltages between 0.7 and 30 kV, with emission current of 100 μA, pressure range in low vacuum and ESEM mode between 0.1 and 40 torr having a resolution of 3.5 nm and magnification range between low 20× to high 50,000×.

2.3 Detection of Elements—Inductively Coupled Plasma Atomic Emission Spectrometry (ICP-AES)

One gram of each ash sample was taken in a beaker. 7.0 mL of HNO_3 and 3.0 mL of HF were added. The mixture was gently swirled and kept for approximately 10 min before closing the digestion vessel. The excess HF present in the solution was complexed with a saturated H_3BO_3 solution to avoid damaging any glassware used during analysis. The sample digestion was carried for 40–45 min in the digester at a maximum temperature of 180 °C. After cooling the samples, the sample was diluted with distilled water and made up to 25 mL. Solution of every sample was filtered to discard any remains as the impurities interfere with the detection process. Metal concentrations in the filtered solutions were subjected to analysis by ICP-AES (Make: SPECTRO Analytical Instruments GmbH, Germany; Model: ARCOS). The ICP spectrometer with radial plasma can analyse the aqueous solutions having high dissolved solid content even up to 30 wt%.

2.4 Crystalline Study for Mineral Identification—X-Ray Powder Diffraction

X-ray powder diffraction (XRD) study helps to analyse the structure of crystalline materials encompassed in the bottom ash matrix. The identification and characterization of the compounds are based on their diffraction pattern. A diffractometer of type X'Pert MPD with a diffraction angle of between 5° and 90° (2θ), 4°/min as rate of scanning, was employed for analysis. The monitoring of crystalline phase and structure was performed using software, Xpert high score.

3 Results and Discussion

3.1 *Bottom Ash Particle Size Gradation*

The distribution of bottom ash particle size ranges from fine gravel to fine sand with low percentages of gravel and silt–clay-sized particles. Bottom ash on analysis was found to be mainly sand fraction with around 70–90% passing a 4.75 mm sieve and 4–10% passing a 0.075 mm sieve as shown in Fig. 2. Bottom ash is found to be well-graded material although variations may be encountered. In this study, the particle size distribution obtained was in close agreement with studies conducted on ash from a healthcare waste incinerator in Greece [13]. The bottom ash generated in municipal solid waste incinerators from Taiwan distinctively resembles the ash grading in the present study [14].

3.2 *Elemental Identification and Quantitative Compositional Information*

The ash samples appeared to be amorphic as shown in Fig. 3.

Subsequent investigation of bottom ash has shown that the particles are composed of Si, Ca, Al, Na, Fe, Mg, Ti, Cu, Zn along with Cl being detected in the samples. This Cl content reveals the burning of polyvinylchloride (PVC) compounds and other chlorinated organics in the incinerator. The presence of elements like Ti associated with others reveals the use of alloys in metallic instruments which are fed as biomedical wastes. The point analysis studies in Greece have also indicated a higher composition of heavy metals like Fe, Cu, Zn and Cr [15].

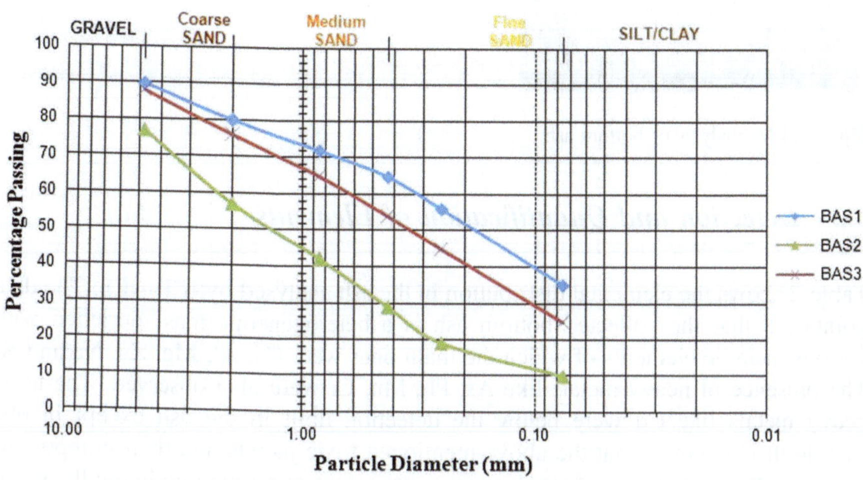

Fig. 2 Sieve analysis of bottom ash

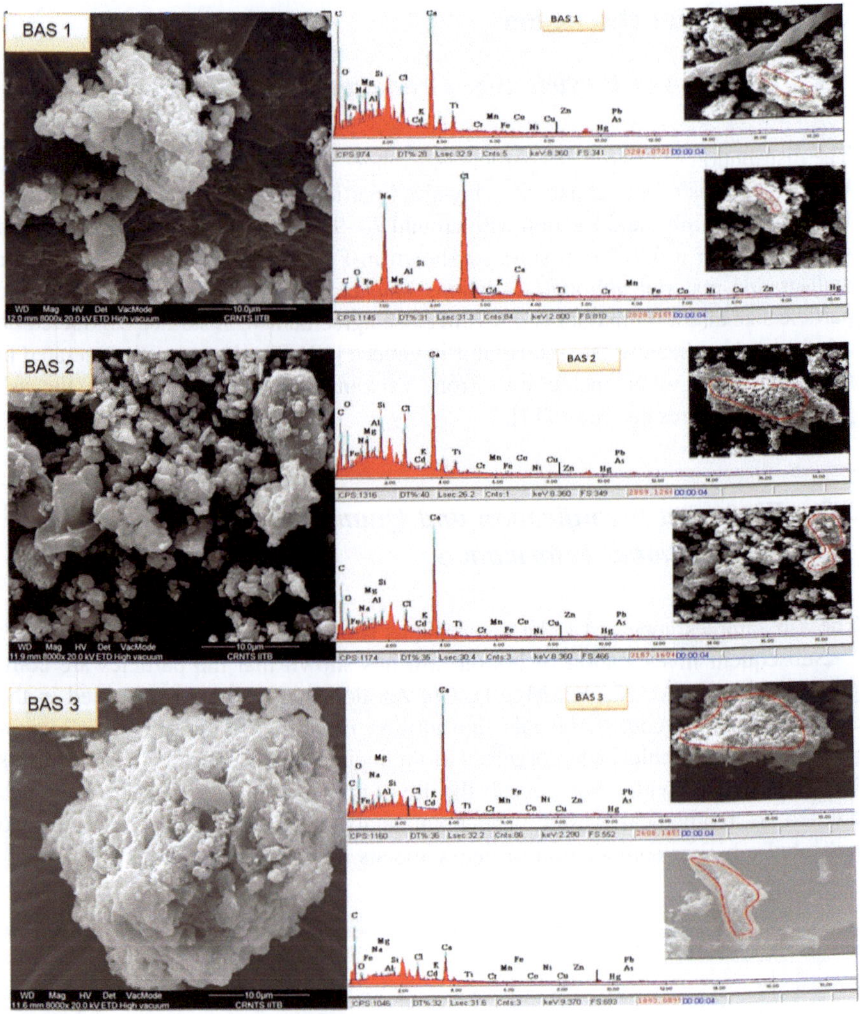

Fig. 3 SEM analysis of bottom ash

3.3 Detection and Quantification of Elements

Table 2 shows the elemental distribution in the ash analysed by ICP-AES. The data points out that the collected bottom ash is a heterogeneous mass enriched with various metallic elements of which the main ones were Ca, Al, Mg, Zn, Na and K. The presence of heavy metals like As, Pb, Mn, Cr were also observed. The toxic heavy metals like Cd were below the detection limit in the ash except in one sample. It is believed that the above-mentioned toxic metals and their compounds may have found their way through gaseous emissions and ended up in the fly ash as

Table 2 Elemental composition of bottom ash (g/kg)[a]

Elements	Ti	Na	K	Al	Ca	Co	Cr	Cu	Fe
BAS1	9.95	0.33	0.02	6.25	12.44	0.01	0.09	0.08	1.99
BAS2	9.83	0.28	0.02	7.40	12.68	0.01	0.10	0.23	4.74
BAS3	11.41	0.33	0.02	3.96	13.50	0.01	0.12	0.28	4.00
Elements	Mg	Mn	Ni	Pb	Sr	Zn	As	Cd	Hg
BAS1	2.54	0.13	0.01	0.47	0.06	3.92	0.50	ND	ND
BAS2	2.65	0.19	0.03	0.47	0.05	3.63	0.44	0.002	ND
BAS3	3.97	0.18	0.03	0.41	0.06	4.44	0.48	ND	ND

[a]All samples analysed in triplicates, ND—not detectable

these are easily volatile in nature. A comprehensive study involving fly ash and bottom ash is required to validate this belief. The bottom ash was found to contain much higher amounts of Zn and Ti. Generally medical instruments made of metal alloys are made of these elements. Hence the presence of these two elements is justifiable [16]. Moreover, Ti has a higher boiling point and has higher chances of ending up in the ash collected at the bottom of BMW incinerator. Also, the higher presence of Ca and Mg are not surprising as the biomedical waste incinerator is also fed with amputated human body parts, tissues, carcasses of dead foetuses. The presence of mercury is found to be very much low and was below detectable limit. This may be an indication that the new regulations [17] laid out by Central Pollution Control Board (CPCB), India, for management of mercury-containing devices in healthcare facilities (HCFs) is in a good pace for implementation. The variations of elements for the three samples are represented in Fig. 4.

3.4 *Crystalline Study for Mineral Identification*

The diffraction study result as given in Fig. 5 reveals that the bottom ash exhibited a highly complex crystalline matrix. The main minerals detected in the study were calcite ($CaCO_3$), halite (NaCl) and anhydrite ($CaSO_4$). The other distinguishing fact is the higher percentage of amorphous mass which is reflected upon in the large background signal and distortions. The bottom ash generated in municipal solid waste incinerators in Taiwan [14] also has the same crystal structures as identified in this study. In the study conducted for the environmentally sound disposal of ash from biomedical waste incinerators by stabilization/solidification, the XRD results of bottom ash have shown a good similarity to the current study [18]. The other important parameter of concern is that there is a huge variation at the mineralogical composition for varying samples. This points out to the fact that the mixture fed into the incinerator varies a lot in its composition in due course of time. There is the presence of rare earth elements detected in the samples, which conveys that a future scope of research lies in identifying and characterizing these rare earth elements. This is in total agreement with some previous studies conducted in China [16].

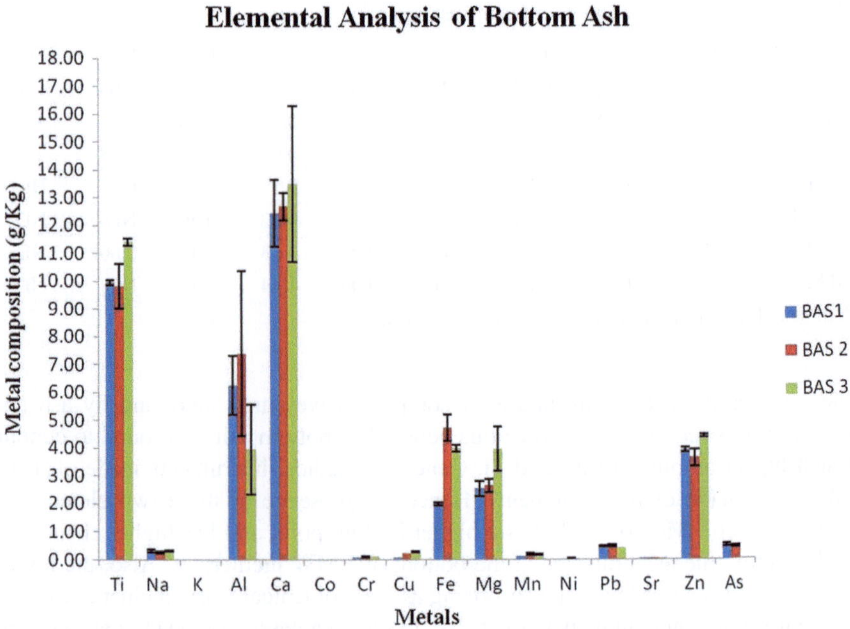

Fig. 4 ICP-AES analysis of collected ash

4 Proposed Alternatives for Sustainable Ash Management

The prevalent method of dumping of bottom ash in landfills is not sustainable. BMW incineration in Italy produced ash with potentially toxic concentrations of Cd, Cr, Cu, Ni, Pb [19]. In the study, it was found that absolute concentrations of these metals in the ash were not sufficient to prove the toxicity. However, the elution/leaching tests pointed out that sufficient care has to be taken before disposing the ash into a landfill.

Thus, to sustainably manage the wastes ash in a landfill site, following methods (either by solidification/stabilization, vitrification or by reusing the waste ash as a secondary construction material) can be practiced:

4.1 Solidification/Stabilization (S/S)

The process of solidification/stabilization (s/s) refers to the technique of using additives or binders such as lime, cement natural or synthetic polymers to immobilize the hazardous substances present in wastes by chemical and/or physical methods. This will ensure that the whole matrix remains intact [17]. This will enable proper encapsulation of the waste material thereby reducing the leachability.

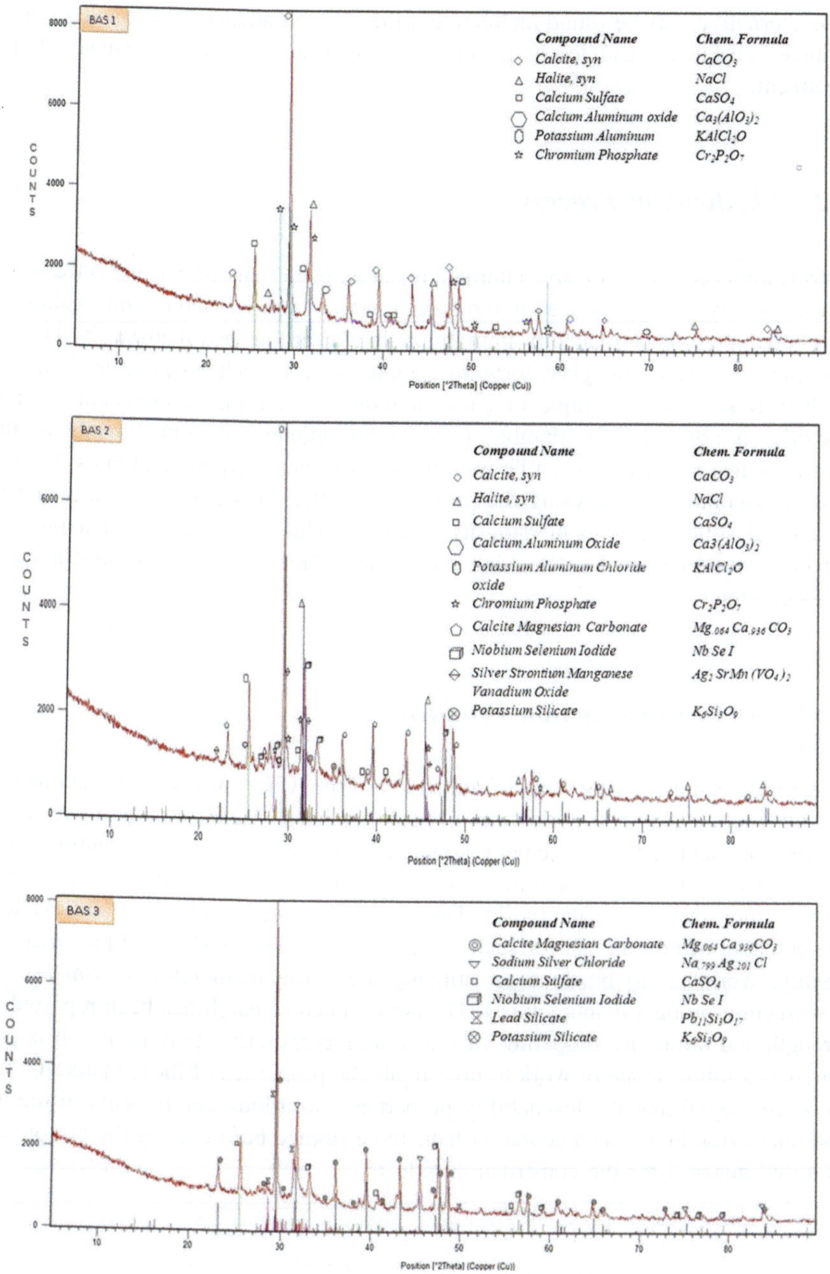

Fig. 5 XRD analysis of bottom ash

The alternative may be found ineffective if the selected additives react and produce soluble substances, and leaching behaviour is observed in the aftermath of the treatment.

4.2 *Vitrification Process*

Vitrification process is a viable alternative that helps to transform a substance into a glass. Usually, being practiced in radioactive waste disposal method, this technique can be adopted to permanently immobilize the hazardous ash matrix. The bottom ash can be mixed with glass-forming chemicals in a smelter device to form the molten glass. This technique of encapsulation of hazardous components of ash provides a non-leaching, durable material entrapping the contaminants in the matrix. Thus, vitrified ash can be used to produce glass-ceramic materials. Studies had been conducted analysing the properties of glass and glass-ceramic made of municipal solid waste incinerator fly ash [20]. This study presented a positive direction pointing out less leaching out of heavy metal ions from the glass and glass-ceramics.

4.3 *Use in Construction Products*

There has been a growing trend of utilization of the waste materials in manufacture of construction material worldwide. Fly ash has been in use as an additive in the process of manufacture of cement, concrete and other various construction products. However, the use of bottom ash for the production of construction materials is not common throughout the world. Thus, there is a need to examine and evaluate the potential of bottom ash for such uses. To curtail the ash disposal problem into landfill, work has to be done on utilizing the waste incinerator bottom ash for construction product manufacture [21]. As not much research has been reported on strength and durability properties of materials incorporating bottom ash, it is proposed as a future scope of work to investigate the properties of these materials. It is to be envisaged that the leachability properties, hazardous components in the ash and interaction in wholesome matrix have to be studied before using the bottom ash as a raw material for the construction industry.

4.4 *Change in Policy Strategies*

As observed in the study, chlorine and associated compounds are a significant component in incinerator residues. The predominant source of chlorine is PVC plastic, which enters the incinerator system en route packaging and from many disposed medical products. The release of chlorinated waste products released from the BMWs is a significant issue. Hospitals and clinics in Austria, Germany and Denmark are reducing the quantity of wastes by switching to reusables, which can be sterilized and used again.

The first step in the proper disposal of wastes in a sustainable manner with less ecological burden is waste minimization and systematic scientific segregation. This methodology outweighs other disposal alternatives in all aspects and particularly from an economic standpoint. All sustainable strategies must include:

- Reduction at source: This can be initiated by imparting scientific education to the healthcare workers, nurses, doctors, adopting methods like good operating and housekeeping practices, inventory control and necessary technology changes.
- Segregation: This process of separating the wastes at the point of generation helps in adopting suitable recovery, recycling and disposal techniques. Often it is observed that the domestic or garbage wastes get mixed up with infectious matrix. If not segregated, all healthcare wastes are considered as potentially infectious. Thus, this strategy enables to treat less quanta of infectious waste which usually requires special methods of treatment.
- Source recovery and recycling: In the well-known hierarchical steps of waste management, recovery and reuse of materials play an important role. If the segregation of infectious and non-infectious wastes is proper, recovery of recyclables can be performed easily.

5 Conclusions

The BMWs generated in an exponential way have a major impact on the health of humans if the wastes remain unattended scientifically. It is essential that an environmentally sound waste management strategy with proper education and awareness for the operators should be adopted. Out of all the disposal methods for complete destruction of wastes and reduction of volume, incineration is widely preferred. The major drawbacks of incineration include generating a new waste stream of ashes (fly ash and bottom ash), furans, dioxins and poisonous emissions. This study reveals some potential insight of contaminants found in bottom ash from BMW incinerator operated in Mumbai, India. The unrestrained and less scientific methods of dumping of the ashes may contaminate the soil as well as surface and underground water.

Elemental analysis indicated low levels of mercury and cobalt. However, the presence of higher levels of toxic metals such as arsenic, chromium and lead is alarming. The recovery of rare earth elements requires a more scientific study incorporating the time and seasonal variation aspects of bottom ash characteristics in India. It should also cover more BMW disposal facilities thus providing a thorough analysis, helping to develop data-driven policies for ash management. A meticulous investigation of physical and chemical properties of bottom ash is required before it is widely accepted as a raw material to be used in production of construction materials or in glass industry. Further studies are needed to analyse the potential of the ash for use as per the alternatives recommended in the study. For the proposed suggestive measures to be effective, a thorough systematic, scientific study on their properties, particularly leachability of the toxic metals and their subsequent long-term bioaccumulation in ecosystem has to be carried out.

Acknowledgements The authors acknowledge the funding from Indian Institute of Technology, Bombay, for this work.

References

1. Rushbrook, P., Zghondi, R.: Better health care waste management: an integral component of health investment. World Health Organization (2005)
2. International Committee of The Redcross: Medical waste management. International Committee of the Red Cross, Geneva, Switzerland (2011). https://www.icrc.org/eng/assets/files/publications/icrc-002-4032.pdf
3. Patil, G.V., Pokhrel, K.: Biomedical solid waste management in an Indian hospital: a case study. Waste Manage. **25**(6), 592–599 (2005)
4. UNEP.: Compendium of technologies for treatment/destruction of healthcare waste, united nations environment programme, division of technology industry and economics, international environmental technology centre, Osaka, Japan (2012)
5. Altin, S., Altin, A., Elevli, B., Cerit, O.: Determination of hospital waste composition and disposal methods: a case study. Pol. J. Environ. Stud. **12**(2), 251–255 (2003)
6. Jang, Y.C., Lee, C., Yoon, O.S., Kim, H.: Medical waste management in Korea. J. Environ. Manage. **80**, 1–9 (2005)
7. The Gazette of India: Biomedical wastes (management and handling) rules, extraordinary part II section 3-sub section (ii), pp. 10–20. India: Ministry of Environment and Forests, Government of India. Notification dated 20th July (1998)
8. CPCB: Status and issues on implementation of bio-medical waste (management and handling) rules, 1998 and as amended—final and updated annual report—2013. Hazardous Waste Management Division, Central Pollution Control Board, India (2015)
9. Tufail, M., Khalid, S.: Heavy metal pollution from medical waste incineration at Islamabad and Rawalpindi Pakistan. Microchem. J. **90**(1), 77–81 (2008)
10. Directive, Council: Council Directive 94/904/EC of 22 December 1994: Council Decision of 22, establishing a list of hazardous waste pursuant to Article 1 (4) of Council Directive 91/689/EEC on hazardous waste. Off. J. EC L **356**(31/12), 0014–0022 (1994)
11. Alba, N., Gasso, S., Lacorte, T., Baldasano, J.M.: Characterization of municipal solid waste incineration residues from facilities with different air pollution control systems. J. Air Waste Manage. **47**(11), 1170–1179 (1997)
12. CPCB: Hazardous waste management, Chapter 9. Highlights 2008. Central Pollution Control Board, India http://www.cpcb.nic.in/Highlights/2008/chapter-9.pdf
13. Gidarakos, E., Petrantonaki, M., Anastasiadou, K., Schramm, K.W.: Characterization and hazard evaluation of bottom ash produced from incinerated hospital waste. J. Hazard. Mater. **172**(2), 935–942 (2009)
14. Chang, F.Y., Wey, M.Y.: Comparison of the characteristics of bottom and fly ashes generated from various incineration processes. J. Hazard. Mater. **138**(3), 594–603 (2006)

15. Kougemitrou, I., Godelitsas, A., Tsabaris, C., Stathopoulos, V., Papandreou, A., Gamaletsos, P., Economou, G., Papadopoulos, D.: Characterisation and management of ash produced in the hospital waste incinerator of Athens Greece. J. Hazard. Mater. **187**(1), 421–432 (2011)
16. Zhao, L., Zhang, F.S., Zhang, J.: Chemical properties of rare earth elements in typical medical waste incinerator ashes in China. J. Hazard. Mater. **158**(2), 465–470 (2008)
17. CPCB: Environmentally sound management of mercury waste in health care facilities. Central Pollution Control Board, India (2010)
18. Anastasiadou, K., Christopoulos, K., Mousios, E., Gidarakos, E.: Solidification/ stabilization of fly and bottom ash from medical waste incineration facility. J. Hazard. Mater. **207**, 165–170 (2012)
19. Santarsiero, A., Ottaviani, M.: Evaluation of heavy metals in slags from medical waste incinerator. Microchem. J. **51**(1–2), 166–169 (1995)
20. Park, Y.J., Heo, J.: Conversion to glass-ceramics from glasses made by MSW incinerator fly ash for recycling. Ceram. Int. **28**(6), 689–694 (2002)
21. Carbone, L.G., Gutenmann, W.H., Lisk, D.J.: Element immobilization in refuse incinerator ashes by solidification in glass, ceramic or cement. Chemosphere **19**(12), 1951–1958 (1989)

Emissions Associated with the Transportation of Municipal Solid Waste in Mumbai

Bhupendra K. Sharma and Munish K. Chandel

Abstract Mumbai is the largest metropolitan city in India with an estimated population of 12.4 million inhabitants in 2011. The city generates ~9,000 metric tonnes of municipal solid waste (MSW) daily with the average waste generation rate of 0.45 kg per capita per day. The generated waste is collected and transported to the disposal sites. In addition to the greenhouse gas (GHG) released from a disposal site, emissions are also released while transportation of MSW by using diesel-fuelled vehicles. Hence, there is a need to determine the amount of emissions generated by these diesel vehicles into the environment if we want to curb these emissions. Objective of this work is to evaluate emissions from transportation of MSW for Mumbai city. An analysis was done to ascertain the emissions produced by the diesel vehicles while transporting MSW. The pollutants (CO_2, CO, HC, NO_x and PM) were evaluated based on the vehicle kilometres travelled method. The results show that CO_2, CO, HC, NO_x and PM were 67,401.38 tonnes/year, 579.11 tonnes/year, 107.36 tonnes/year, 613.51 tonnes/year and 99.94 tonnes/year, respectively. The results show that the transportation of MSW also plays an important role in the contribution of pollutants especially GHGs to the atmosphere. The total global warming potential (GWP) from the transportation of MSW is obtained as 0.07 million tonnes CO_2 eq. per year which is approximately 2.8% of the total GWP resulted from the present MSW management system in Mumbai.

Keywords Municipal solid waste · Municipal solid waste transport
Diesel vehicles · Gas emissions · Environmental impact

B. K. Sharma · M. K. Chandel (✉)
Centre for Environmental Science and Engineering,
Indian Institute of Technology Bombay, Powai, Mumbai, India
e-mail: munish.chandel@iitb.ac.in

B. K. Sharma
e-mail: bhupendrasharma@iitb.ac.in

A. S. Kalamdhad et al. (eds.), *Advances in Waste Management*,
https://doi.org/10.1007/978-981-13-0215-2_31

1 Introduction

The management of MSW has become a severe problem in India due to increase in waste generated because of rapid population growth and economic development. In recent years, many municipalities have been compelled to evaluate their MSW management system due to the environmental and health concerns. The city of Mumbai is the largest metropolitan city in India with an estimated population of 12.4 million [1]. Such a huge population generates ~9,000 metric tonnes of MSW daily [2] with the average waste generation rate of 0.45 kg per capita per day [3]. From the bins, the waste is collected by Municipal Corporation of Greater Mumbai (MCGM) and transported to the disposal sites at Deonar, Mulund and Kanjur. The waste disposal in Mumbai is done in two open dump sites (Deonar and Mulund) and one bioreactor landfill (Kanjur). The capacity of these three sites is 3,200 tonnes/day, 3,000 tonnes/day [3] and 2,800 tonnes/day [4], respectively.

In addition to the GHG and other emissions released from a landfill site, emissions are also released while transportation of MSW. The GHG emissions from transport sector are significant as 15.71 lakh tonnes of carbon dioxide equivalent GHG emissions was released for Mumbai city in 2011 [5]. The GHG emissions as well as other pollutants from heavy diesel vehicles used for transportation of MSW cannot be ignored as they may have considerable impact on the environment. Emissions from the road transport depend on the combinations of parameters such as the vehicle type, fuel used and driving patterns [6].

The MSW vehicles deployed by MCGM for the transportation of MSW generously contribute pollutants to atmosphere. Hence, there is a need to determine the amount of emissions generated by these diesel vehicles into the environment. There are only limited studies carried out on the emission from MSW vehicles in India. The emissions of various gases into the environment from the vehicles used for transportation of MSW in Mumbai have been focused in the present work. In this study, the emissions associated with the current MSW management practice in Mumbai are compared with the emissions generated from the transportation of MSW to the site.

2 Methodology

The number of registered motor vehicles, distance travelled by the vehicles and the number of trips done by the vehicles in Mumbai were collected from the MCGM Office through the personal interview. Municipal and private contract vehicles both are operated by MCGM for collecting and transporting the waste. The vehicles used for transportation of the waste generated in Mumbai make 7,092 trips each day. The total number of municipal vehicles is 155 while the total number of private vehicles is 989 (Table 1).

Table 1 Vehicles used by MCGM for transportation of waste [4]

Type of vehicles	Capacity (MT)	Distance travelled (km/trip)	Trips/day	Municipal	Private	Total
Large compactor	6	80	3	117	452	569
Mini compactor	2.5	80	6	38	322	360
Small closed vehicles	0.6	20	15	0	215	215

Table 2 ARAI emission factors in g/km for diesel vehicles [7]

Parameters	Large compactor	Mini compactor	Small closed vehicles
CO_2	762.39	401.25	173.85
CO	6	3.66	2.09
HC	0.37	1.35	0.16
NO_x	9.3	2.12	0.69
PM	1.24	0.475	0.347

The emissions from three types of vehicles are estimated based on the emission factors provided by Automotive Research Association of India (ARAI) for diesel vehicles (Table 2). The pollutants (CO_2, CO, HC, NO_x and PM) were evaluated based on the vehicle kilometres travelled (VKT) method. The emission load, *E*, is characterized as the product of the number of vehicles, distance travelled by vehicle and the emission factor for the vehicle and contamination of concern, as shown in Eq. 1.

$$\mathrm{E} = \text{Total number of vehicles} \times \text{Distance travelled per day(km)} \times \text{Emission factor} \quad (1)$$

2.1 *Estimation of Emissions from the MSW Transportation Vehicles*

2.1.1 Estimation of Emissions from Large Compactors

There are approximately 569 large compactors operating in Mumbai city. Each large compactor makes three trips in each day, and the distance covered by a single large compactor is about 240 km. On an average, the total distance covered by 569 compactors is 136,560 km in a day. Table 3 shows the calculated emissions load from the large compactor.

Table 3 Emissions from large compactor

Parameters	g/day	tonnes/year
CO_2	104,111,978.4	38,000.87
CO	819,360	299.07
HC	50,527.2	18.44
NO_x	1,270,008	463.55
PM	169,334.4	61.81

2.1.2 Estimation of Emissions from Mini Compactors

There are approximately 360 mini compactors operating in Mumbai city. Each mini compactor makes six trips in each day, and the distance covered by a single mini compactor is about 480 km. On an average, the total distance covered by 360 compactors is 172,800 km in a day. Table 4 shows the calculated emissions load from the mini compactor.

2.1.3 Estimation of Emissions from Small Closed Vehicles

There are approximately 210 small closed vehicles operating in Mumbai city. Each small closed vehicle makes fifteen trips in each day, and the distance covered by a single small closed vehicle is about 300 km. On an average, the total distance covered by 210 small closed vehicles is 64,500 km in a day. Table 5 shows the calculated emissions load from the small closed vehicles.

Table 4 Emissions from mini compactor

Parameters	g/day	tonnes/year
CO_2	69,336,000	25,307.64
CO	632,448	230.84
HC	233,280	85.15
NO_x	366,336	133.71
PM	82,080	29.96

Table 5 Emissions from small closed vehicles

Parameters	g/day	tonnes/year
CO_2	11,213,325	4,092.86
CO	134,805	49.20
HC	10,320	3.77
NO_x	44,505	16.24
PM	22,381.5	8.17

2.2 *Estimation of Emissions from the Waste Disposal*

Out of the total MSW (9,000 tonnes/day) generated in Mumbai, approximately 31% of the waste is treated through bioreactor landfill (Kanjur) and the rest (69%) of the waste is dumped in two open dumps: Deonar and Mulund. In this study, IPCC Tier I methodology has been used to estimate CH_4 from open dumps, which is based on a mass balance approach [8]. The GHG emissions from landfill were calculated using the following equation [8, 9] in which MSW_T is the total MSW generated and DOC is the degradable organic carbon.

$$CH_4\ \text{emissions(kg/tonne of MSW)} = (MSW_T \times MSW_F) \times CF \times DOC \times DOC_F \times F \times (16/12{-}R) \times (1 - OX) \quad (2)$$

$$CO_2\ \text{emissions(kg/tonne of MSW)} = CH_4\ \text{emission(kg/ton MSW)} \times ((1 - F/F) + OX)) \times 44/16 \quad (3)$$

Default values for empirical constants like methane correction factor (MCF = 0.6), fraction of DOC dissimilated ($DOC_F = 0.77$), oxidation factor (OX = 0), fraction of MSW disposed at landfill (MSW_F = 100%), CH_4 fraction by volume in landfill gas (F = 50%) and recovered methane (R = 0) are used for computation as per the IPCC Tier I method [8, 10]. Direct CO_2 emissions from the open dumping are of biogenic origin and were not accounted towards GHG emissions, i.e. biogenic CO_2 emissions are considered to be "carbon-neutral" [11, 12].

The air emissions generated from the bioreactor landfill are assumed same as the emissions generated in case of sanitary landfill [13]. The default methodology has been used for bioreactor landfill in the estimation of CH_4, similar to open dumping. The value of MCF for the bioreactor landfill was taken as 1.0, default value from IPCC report [8]. It has been assumed that 50% of the landfill gas is collected [14] and flared and rest is released to the atmosphere and contributes towards GHG emissions.

3 Results and Discussions

The emissions of CH_4 and CO_2 have significant effects on global warming impact category. The GWP for GHGs over a time horizon of 100 years is 1 for CO_2 and 23 for CH_4. The GWP for MSW characteristics of Mumbai under the current scenario which is open dumping plus bioreactor landfill is calculated as 2.4 million tonnes CO_2 eq. per year due to the high emission of CH_4 during the biological degradation of organic waste.

The total amount of CO_2, CO, HC, NO_x and PM emitted from the transportation of MSW were obtained as 67,401 tonnes/year, 579.11 tonnes/year, 107.36 tonnes/year, 613.51 tonnes/year and 99.94 tonnes/year, respectively (Fig. 1). Maximum contribution was obtained from CO_2 which is a major greenhouse gas followed by NO_x. The total GWP from the transportation of MSW is obtained as 67,980 tonnes CO_2 eq. per year which is approximately 2.8% of the total GWP resulted from the present MSW management system in Mumbai, as shown in Fig. 2. The GHG emissions as well as other pollutants from heavy diesel vehicles used for transportation of MSW cannot be ignored as they have considerable impact on the environment. The results show that the transportation of MSW also plays an important role in the contribution of pollutants especially GHGs to the atmosphere.

This study helps decision-makers to understand the GHG and other pollutants released due to the vehicle pollution in MSW management sector. In order to reduce the emission of these pollutants, 3R policies (Reduction, Reuse and Recycling) for the waste can be implemented in the waste management sector which reduces the transportation of MSW resulting in the reduction of vehicle pollution. The decentralized facility can also be implemented within the different wards of the city so that the waste generated can be processed in the ward itself instead of being sent to the overburdened dumping sites at Deonar, Mulund and Kanjur which can also help in reducing the vehicle emissions as well as the transportation cost.

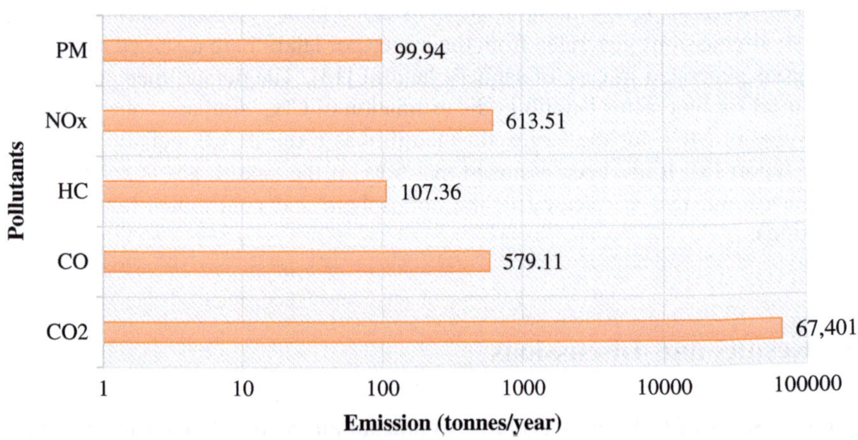

Fig. 1 Emissions from the transportation of MSW in Mumbai

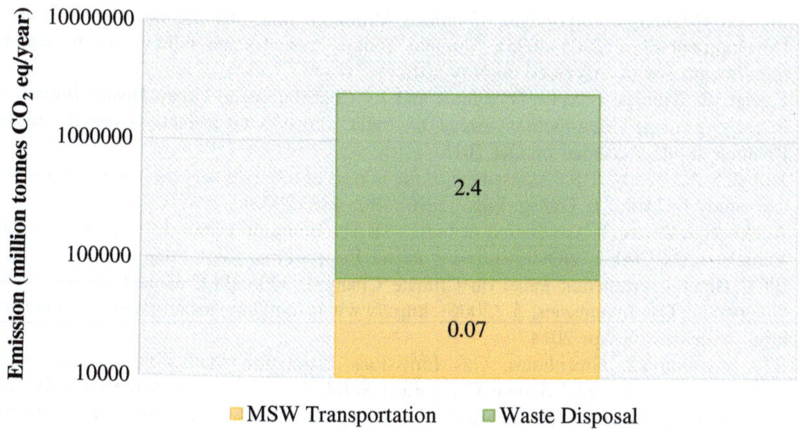

Fig. 2 Comparison of CO_2 emission from MSW transportation and current waste disposal scenario

4 Conclusion

The comparison between GHG emissions from MSW transportation and current MSW management practice (open dumping and partly bioreactor landfill) for Mumbai city was conducted. The emissions of CO_2, CO, HC, NO_x and PM generated from the transportation of MSW were obtained as 67,401 tonnes/year, 579.11 tonnes/year, 107.36 tonnes/year, 613.51 tonnes/year and 99.94 tonnes/year, whereas the CO_2 emission has the maximum contribution followed by NO_x. The net GHG emissions generated from the diesel-based vehicles deployed for transportation of MSW in Mumbai are in considerable amount. The total GWP from the transportation of MSW is obtained as 0.07 million tonnes CO_2 eq. per year which is approximately 2.8% of the total GWP resulted from the present MSW management system in Mumbai. The results show that the transportation of MSW also plays an important role in the entire life cycle impacts of MSW.

References

1. Census of India: Provisional Population Totals: India: Census (2011). http://censusindia.gov.in/2011-prov-results/indiaatglance.html. Accessed on Aug 2014
2. MPCB. Maharashtra Pollution Control Board Annual Report on Implementation of Municipal solid waste (Management and Handling) Rules, 2000. (2013) http://mpcb.gov.in/muncipal/pdf/AR_MSW_12-13.pdf. Accessed on Oct 2014
3. Pendse, S.S.: Solid waste management in India—a study of Mumbai. Online Int. Interdisc. Res. J. {Bi-Monthly} **II**(IV) (2012). ISSN2249-9598

4. MCGM (Municipal corporation of greater Mumbai), Index for the Greater Mumbai City Development Plan (2005–2025). Mumbai Today: Present status-solid waste management. http://mcgm.gov.in. Accessed on Nov 2015
5. Kumar, R. Transport Sector Dynamics and Its Contribution to Urban Health Burden in a Rapidly growing Metropolitan Areas of India. https://www.theicct.org/sites/default/files/RKumar_0.pdf. Accessed on Oct 2015
6. Jalihal, S.A., Reddy, T.S.: Assessment of the impact of improvement measures on air quality: case study of Delhi. J. Transp. Eng. 132(6), 482–488 (2006)
7. Automotive Research Association of India (ARAI), Emission Factor development for Indian Vehicles (AFL/2006-07/IOCL/Emission Factor Project/Final Rep), Pune (2008)
8. IPCC (Intergovernmental Panel on Climate Change). 2006 IPCC Guidelines for National Greenhouse Gas Inventories, 5. (2006). http://www.ipcc-nggip.iges.or.jp/public/2006gl/vol5.html. Accessed on Apr 2014
9. RTI International. Greenhouse Gas Emissions Estimation Methodologies for Biogenic Emissions from Selected Source Categories: Solid Waste Disposal Wastewater Treatment Ethanol Fermentation. (2010). https://www3.epa.gov/ttn/chief/efpac/ghg/GHG_Biogenic_Report_draft_Dec1410.pdf. Accessed on Oct 2015
10. Kumar, S., Gaikwad, S.A., Shekdar, A.V., Kshirsagar, P.S., Singh, R.N.: Estimation method for national methane emission from solid waste landfills. Atmos. Environ. 38(21), 3481–3487 (2004)
11. Chandel, M.K., Kwok, G., Jackson, R.B., Pratson, L.F.: The potential of waste-to-energy in reducing GHG emissions. Carbon Management 3(2), 133–144 (2012)
12. Zhao, W., van der Voet, E., Zhang, Y., Huppes, G.: Life cycle assessment of municipal solid waste management with regard to greenhouse gas emissions: case study of Tianjin China. Sci. Total Environ. 407(5), 1517–1526 (2009)
13. Babu, G.S., Lakshmikanthan, P., Santhosh, L. G.: Life cycle analysis of municipal solid waste (MSW) land disposal options in Bangalore City. In ICSI 2014: Creating Infrastructure for a Sustainable World, pp. 795-806 (2014)
14. TTI (Texas Transportation Institute), TTI Pre-feasibility analysis for the conversion of landfill gas to liquefied natural gas to fuel refuse trucks in India. US EPA Methane to Markets Partnership. The United States Environmental Protection Agency, Washington, DC (2009)

Prospects of Utilization of Liquid Fraction of Biogas Digestate as Substrate Supplement for Mushroom Cultivation

Sampriti Kataki, Gunajit Dev Sarma, Dipam Patowary and D. C. Baruah

Abstract Biogas digestate is the by-product of 'anaerobic digestion' process that produces biogas as the primary form of energy. Since digestates are the inevitable commodities of the conversion process, means of their appropriate management is essentially required for sustainability of bioenergy system. In developing countries, with faster adoption of biogas technology, a simultaneous increase in biogas residue is also expected drawing attention for their immediate and effective utilization. Since biogas digestate retains nutrients from input feedstock with enhanced bioavailability and minimum variation, these could be potentially used as growth supplement for mushroom. Present study discusses the prospects of utilization of the liquid fraction of biogas digestate as a substrate along with rice straw for enhancing the production of white oyster mushroom (*Pleurotus florida*) under laboratory conditions. The study investigates the effect of liquid fraction of biogas digestate on yield and growth parameters (stem height, cap diameter, mean weight) of mushroom. An enhancement of yield by 38% was observed under liquid digestate application compared to control using only rice straw as substrate. In addition, cost dynamics analysis also showed feasibility of production of mushroom

S. Kataki · G. D. Sarma · D. Patowary · D. C. Baruah (✉)
Energy Conservation Laboratory, Department of Energy, Tezpur University, Assam, India
e-mail: baruahd@tezu.ernet.in

S. Kataki
e-mail: sam_kat@tezu.ernet.in

G. D. Sarma
e-mail: gunajitdev@gmail.com

D. Patowary
e-mail: dipam189@gmail.com

A. S. Kalamdhad et al. (eds.), *Advances in Waste Management*,
https://doi.org/10.1007/978-981-13-0215-2_32

at commercial scale. Integrated development of biogas technology with subsequent valorization of digestate through mushroom cultivation may help in increasing overall profitability of biogas system.

Keywords *Pleurotus florida* · Digestate · Biogas residue · Anaerobic digestion

1 Introduction

Bioenergy by-products such as digestate from biogas technology and char (solid carbon-rich residue) from gasification technology are inevitable commodities of bioenergy production process which are generated along with the desired form of energy [1, 2]. Suitable utilization of these residues often determines the success and acceptability of bioenergy conversion technology. For comprehensive assessment and sustainability of bioenergy technology, there is need of management of these residues along with the main energy output [1–3]. With faster penetration of renewable energy systems in developing countries, it is expected that there will be simultaneous increase in by-product. Also, to deal with the adverse effect of climate change and limited availability of conventional energy resources, there is emphasis on adoption of renewable energy technologies, which again will indicate the need of managerial action to be taken for the by-products generated. Biogas residue or biogas digestate is such a by-product of 'anaerobic digestion' process that produces biogas as the main energy. About 5–80% of input feedstock is generated as by-product depending upon of feedstock [4] leaving a major portion of residue to be managed for economic and environmental benefits of the biogas production process. Presently, due to improper method of application and storage through unlined earthen drain and pit, there is possible paths for nutrient loss [5], leading to non-optimized utilization of digestate.

In general, biogas digestates are rich in plant nutrients as they retain nutrients from input raw material [10]. The presence of nitrogen in mineralized form (ammonia-nitrogen) and increased bioavailability of plant nutrients make these a suitable consideration as soil applicant. Therefore, till now the most common application of digestate is through their recycling as soil applicant, which can promote the closing of nutrient cycle [4, 11]. However, direct application of digestates has some limitations such as introduction of contaminants originating from feedstock [12], handling and transportation problem due to its bulky nature. This indicates the need of research to search for multiple end uses and other valorization options of digestates.

In this context, use of digestates as nutrient supplement in mushroom cultivation can be considered favourable. Mushroom is an excellent source of dietary components such as protein, vitamins, minerals, folic acid, low in fat and carbohydrates. Mushroom industry is gaining momentum as efficient and economically viable process for converting lignocelluloses residue into high-quality protein food [13]. Worldwide production of mushrooms is seen increasing, with maximum contributions from China, India and Vietnam [13]. During 2014–15, India produced 51,000 tonnes of mushroom, among which major share (28%) being contributed by Kerala [14]. India shows an increasing quantity of exported mushroom spawn, with an estimated 5.48 tonnes of spawn exported during the year 2014–15 [15]. It is to be mentioned that mycelium from a mushroom culture is usually collected onto grain. This grain/mycelium mixture is called spawn, which is the material used to transfer mycelium onto any material (substrate) from which mushrooms will grow. The state-of-the-art research focus in global mushroom industry includes technological developments through increased production, innovations in cultivation technologies, valorization of final product and utilization of mushrooms' natural qualities for environmental benefits [13].

Availability of carbon-rich substrate is the most essential component for growing mushroom. In mushroom cultivation, agricultural residues are the most widely used substrates that provide the necessary nutrients and vitamins for mushroom growth. Agricultural residue acts as the reservoir of cellulose, hemicellulose and lignin, which is utilized during the growth of spawn and during fructification [16]. Table 1 summarizes different substrate used for cultivating of oyster mushrooms and their

Table 1 Effect of different substrate on growth characteristics of oyster mushroom

S. No.	Substrate	Mushroom species	Yield, g of mushroom/g of substrate	References
1	Saw dust (50%) + wheat straw (50%)	*Pleurotus ostreatus*	436/1000	[6]
2	Saw dust (75%) + leaves (25%)		621/1000	
3	Leaves		211/1000	
4	Wheat straw (50%) + leaves (50%)		434/1000	
5	Wheat straw		447/1000	
6	Saw dust		647/1000	
7	Wheat straw	*Pleurotus ostreatus*	48.3/1500	[7]
8	Paddy straw		44.7/1500	
9	Chickpea straw		62.7/1500	
10	Sugar cane bagasse		37.0/1500	
11	Corn cobs		34.0/1500	
12	Cotton waste		37.0/1500	
13	Sunflower heads		40.7/1500	
14	Rice straw	–	987/250	[8]
15	Sugar cane bagasse		1029/250	
16	Mustard straw		945/250	
17	Banana leaves	*Pleurotus florida*	124/500	[9]
18	Rice straw		416/500	
19	Banana leaves + Rice straw		156/500	
20	Saw		199/500	

effect on yields. Some of the substrate materials used are rice straw, sugar cane bagasse, wheat straw, banana leaves, sawdust, tea waste, mustard straw, cotton waste, corn cobs, etc. C:N ratio plays an important role in mushroom spawn running [17]. Carbon in the form of cellulose and hemicellulose present in the agricultural wastes are partially available, but bound organic nitrogen and recalcitrant carbon (lignin) are not available until it is enzymatically released [18]. Improving the C:N ratio in order to accelerate the degradation process for release of nutrients needs supplementation of additional nitrogen source along with the substrate. Therefore, addition of inorganic nitrogen-rich source is an important practice of mushroom cultivation [19]. The literature shows supplementation with protein-rich materials such as soybean meal, cottonseed meal, brewer's grain and commercial nitrogen fertilizers for successful mushroom production [17].

A few studies have shown the prospects of biogas digestate supplement in mushroom substrate [16, 20–26]. These studies reported significant enhancement of mushroom yield after digestate supplement compared to basal substrate, suggesting their potential amendment in mushroom substrate as nutrient supplement. However, study on use of separated liquid fraction of the digestate in mushroom production is not widely reported except a few studies [26]. Considering the predominance of mineralized nitrogen in liquid fraction of biogas digestate, it could serve as a readily available nitrogen source required to maintain appropriate C:N ratio in mushroom production. In view with the above discussion, the objectives of the present work were set to investigate the prospects of using separated liquid biogas digestate as a substrate supplement for oyster mushroom cultivation, considering rice straw as a basal control substrate. The motive behind this work is to valorize the separated liquid fraction of biogas digestate by using it for mushroom cultivation in an ecological and economically sustainable way for enhancing the overall profitability of biogas system. The paper also discusses findings of the published literature to investigate the prospects of use of biogas digestate in mushroom cultivation. It is to be noted that the separated solid fraction of digestate can be used for compost production or can be directly used as organic amendments as shown in the previous literature [27–29].

2 Materials and Methods

To investigate the effects of liquid digestate supplement in mushroom growth, mushroom cultivation trials were carried out in the Department of Energy, Tezpur University (26° 69″ North, 92° 83″ East), during December to February, 2015–16. The site had climatic conditions typical of the cool wet winters and dry summers. During the experimental period, the temperature range was in the range of 8–29 °C and relative humidity range was 58–93%. Weather data during the experimental period were obtained from the Solar Radiation Resource Assessment (SRRA) Stations implemented by National Institute of Wind energy (NIWE), Chennai in the Tezpur University campus.

2.1 Mushroom Species

Oyster mushroom (*Pleurotus* spp.) belonging to class Basidiomycetes and family Agaricaceae is the most widely cultivated mushroom in India owing to its simple cultivation technology, high nutritional value, short growth cycle, low production cost and its adaptability to a wide range of substrates. It possesses high bioconversion ability, i.e. more than 60%. The best temperature range for oyster mushroom cultivation is 20–30 °C along with a humidity range of 80–85% [30]. Oyster species can utilize wide range of crop residues or substrate materials such as agricultural, forests and agro-industrial by-products such as wheat straws, rice straws, banana leaves, corn cobs, sawdust and bean straws due to their great adaptability (Table 1). Oyster mushroom looks like a white disc, growing on a thick stipe with decurrent gills extending to the base of the stipe.

For the experimental work, *Pleurotus* (*P. florida*) spawns grown of wheat grains were collected from Defence Research Laboratory, Tezpur, that works under Department of Defence Research and Development of Ministry of Defence. The research laboratory is en established laboratory that performs independent R&D work in mushroom cultivation and provides mushroom spawn for commercial cultivation and research work.

2.2 Substrate Collection, Processing and Analysis

Anaerobic digestate was collected from the digestate storage tank of a household biogas plant situated in the Amolapaam area near to Tezpur University campus. The plant was a cowdung-fed plant that was operated with 20–25 kg of input cowdung per day. Collected digestate was characterized for pH and electrical conductivity (EC) in a homogenized water suspension with water in the ratio of 1:10 (dry weight/volume) after agitating for 15 min using pH (Systronics digital pH meter 802) and electrical conductivity meters, respectively. Total solid (% fresh matter) was determined by drying at 105 °C for 24 h. Organic matter (total organic carbon) was determined by loss on ignition at 550 °C for 10 h. Plant nutrients (N, P, K, Cu, Fe, Mn, Zn, Ca, Mg, S) and heavy metals (Pb, As, Cd) were determined using ICP analysis.

Digestate was separated by manual pressing using 150-micron filter bags to get the liquid fraction of the digestate. The separated liquid digestate used in the experiment was stored at 4 °C until application. The separated digestates were applied directly without any further treatment, since report on the presence of pathogen or harmful microbes could not be found in manure-based digestates. Digestates generated from human excreta and sewage sludge were reported to contain harmful microbes of health concern such as Klebsiella, E. coli, Salmonella [31, 32]. Therefore, possible microbial contamination from cowdung digestate application was not expected.

Rice straw was used as a base substrate material in the experiments, which was collected from local farmers place after harvesting. For preparation of mushroom substrate, rice straw was first pasteurized by dipping in hot water (80–90 °C) for 2 h. The pasteurized rice straw was soaked overnight for moistening. Then, the excess water was drained off, and the moisture in substrates was adjusted to 60–70% by air drying the rice straw.

2.3 Experimental Procedure

To carry out the experiment, processed rice straw was compactly packed into transparent polythene bags of size height 58 cm, diameter 38 cm. On wet weight basis, 800 g dry rice straw was put in each bag to 3/4th of its capacity. Each bag was then spawned at the rate of 5% (wet weight basis) following 'through spawning' method. Packed substrate bags were perforated with holes all over the surface to allow free exchange of gases. To place the packed bags, one iron frame was prepared (108 cm × 76 cm × 80 cm) which was covered with jute bags from all sides. The bags were incubated in the chamber which was maintained at room temperature and under diffused light condition. The bags were tied on the top and hung in the chamber in a completely randomized block design. The humidity inside the room was maintained between 70–80% by moistening the jute bags by regular water sprinkling.

For the liquid digestate treatment, 80 mL (10:1: substrate: liquid digestate) liquid digestate was sprinkled daily in the bags after pinhead formation till third flush. It has been reported that liquid digestates are rich in available nitrogen. Considering the fact that soaking of rice straw prior to beginning of mushroom cultivation might lead to loss of nutrients particularly nitrogen through gaseous emission, daily application of liquid digestates was done, which would reduce the nitrogen loss making them available for ready uptake.

Rice straw substrates were used as control, and same quantity of water was sprayed daily to maintain similar moisture level in these substrates. For each treatment, three replications were used.

Observations on production of fruiting bodies were recorded up to third flush. Mushrooms were harvested at full biological maturity considering an average pileus diameter of 10–12 cm. At this stage, the in-rolled margins of the basidiomata begin to flatten and pileus becomes hard and curved outwards on the edge. Total yield was calculated as the fresh weight of mushrooms harvested per 800 g of dry substrate. Biological efficiency (%) was determined by the ratio of total fresh weight of mushrooms to dry weight of substrate per bag. Mean weight of mushroom weight per treatment was calculated as total weight of fresh mushrooms harvested divided by the number of mushrooms per plastic bag.

Data on mushroom yield and growth parameters were subjected to statistical analysis. Standard deviations were calculated for all parameters. Pearson bivariate correlation coefficients were calculated using SPSS to show the relationship between the parameters at two levels, i.e. $p < 0.05$ and $p < 0.01$. We used Z test to compare means for any statistically significant difference.

2.4 *Economics of Mushroom Production*

It is attempted to encompass the cost economics of the entire mushroom production cycle. To estimate the cost, digestate availability of 1 tonne per annum has been considered. Cost economics were investigated under polyhouse condition considering two scenarios, viz. mushroom production with rice straw as substrate and mushroom production with rice straw amended with digestate as substrate.

Again, the economic analysis of production has been carried out taking into consideration two subcost parameter, i.e. recurring and non-recurring cost. Specific key components to develop a production unit are identified under each cost parameter. The various components of non-recurring cost are (i) building, (ii) equipment (cooler/humidity/temperature meter/rack). While the various components of variable cost are (i) chemicals/spawn//bags/tray, (ii) interest on investment, (iii) insurance, (iv) depreciation, (v) manpower.

3 Results and Discussions

3.1 *Biogas Digestate: A Prospective Mushroom Growth Supplement*

Biogas digestate has been shown to be a prospective nutrient supplement in previous studies [16, 24, 26]. Biogas digestate rich in mineralized nitrogen is considered for supplementation mainly for balancing the nitrogen requirement in substrate. However, their phosphorus, calcium, etc., add to substrate nutrient concentration enhancing mushroom yield and nutrient content [24], as nutrient content of substrates has been reported to affect the growth and formation of mushroom fruit bodies of [33]. Digestate application also helps in maintaining appropriate moisture level for mushroom growing [16]. Table 2 summarizes effect of digestate (both unseparated and liquid digestate) supplement on mushroom growth, mushroom species tested, origin of digestate, substrate used as reported in the literature.

Yield of *P. florida* was seen increasing by 59% after rice straw supplementation with anaerobic digestates compared to using only rice straw as substrates [16]. Biogas digestate application in agro-residue substrate also enhanced the

Table 2 Effect of digestate supplement on mushroom growth

	Species	Feedstock for biogas digestate	Substrate	Effect on mushroom	References
1	*Pleurotus flabellatus*	*Euphorbia notoptera* (50%), *Synedrella nodiflora* (30%) (weed)	Paddy straw	• Intensity of fruit body and mushroom production are higher in digestate supplemented substrate • Significant increase in simple protein fraction and reduction in complex protein in substrate mixture	[20]
2	*Volvariella volvacea*	Cowdung, poultry litter, municipal solid waste, jute caddis	Rice straw	• Increased mineral content of mushroom after digestate application (P, K, Ca, Cu, Fe, Mn, Zn) • Supplementation of cowdung digestate with rice straw in the ratio 1:1 gave better yield compared to 2:1 and 1:2 ratio • Cowdung digestate supplementation gave highest yield (78%) followed by jute caddis digestate (33%), municipality solid waste (10%) and poultry litter (13%), respectively, compared to control • Protein content increased after digestate application, while carbohydrate content decreased	[21]
3	*Pleurotus sajor-caju*	Cattle dung, poultry litter, jute caddis or municipal solid waste	Rice straw	• Improved nutritional quality of the mushroom • Protein content was increased, while carbohydrate content was reduced • Digestate:straw in the ratio 1:1 was most effective compared to1:2 and 2:1 (high digestate may cause accumulation of excess moister restricting mycelium development)	[22]

(continued)

Table 2 (continued)

	Species	Feedstock for biogas digestate	Substrate	Effect on mushroom	References
				• Productivity maximum in rice straw was supplemented with digestate from poultry litter followed by jute caddis, municipal solid waste and cattle dung	
4	*Agrocybe aegerita*	Poultry litter	Dry solid digestate and wheat straw millet	• Substrates composed of 10–20% dry solid digestate, 70–80% wheat straw and 10% millet were found to produce the highest mushroom yield • Digestate supplementation increased crude protein in mushrooms • Addition of solid digestate above 50% reduced mushroom yield could be attributed to the low C:N ratio or high lignin content of digestate, making substrate degradation difficult • Mushroom growth degraded 0.6–22% lignin, 33–55%, cellulose and 14–54% hemicellulose	[24]
5	*Pleurotus florida*	Banana leaf	Coir pith, paddy straw	• Significant reduction of lignin, C: N ratio (maximum C:N, lignin reduction in paddy straw compared to coir pith) • Maximum mushroom yield of 2.3 kg/kg of substrates was observed in paddy straw supplemented with digestates compared to only paddy straw	[16]

(continued)

Table 2 (continued)

	Species	Feedstock for biogas digestate	Substrate	Effect on mushroom	References
6	*Agaricus bisporus* and *Agaricus subrufescens*	Source-separated food waste	Wheat straw	• No significant differences between straw + digestate and control in terms of mushroom dry matter, size, nitrogen or ash content • High digestate content (up to 500 g kg^{-1} by dry weight) may be used without influencing yield of A. bisporus • For A. subrufescens, less digestate should be used, preferably less than 240 g kg^{-1} by dry weight • High yields of A. subrufescens (200gkg^{-1}) were related to drier composts of lower digestate content (more straw) and lower pH values at inoculation	[25]
7	*Pleurotus florida* and *Pleurotus flabellatus*	Banana leaves	Paddy straw and coir pith	• Highest productivity in paddy straw along with digestate for *P. florida* (2.32 kg/kg of substrate) • Digestate supplement in straw resulted 37% increase in yield for *P. florida* and 24% for *P. flabellatus* • Digestate supplement with coir pith increased mushroom yields by 18.2% for *P. florida* and 18.6% for *P. flabellatus* • Addition of digestate enhances the uptake of N, P and C by mushroom • 30% substitution of straw with digestate gave better yield • Time taken for pinhead formation (by 3 days) was reduced after digestate application	[26]

productivity of *P. florida* with a biological efficiency of 232% compared to using sole agro-residue as substrate in which biological efficiency was found to be 145% [16]. Chanakya et al. reported a higher yield (20–30%) in digestate amended substrate compared to non-amended substrate [26]. However, the improved yield of mushroom from digestate addition is found to be species specific. In case of *P. florida*, digestate amended rice straw resulted in 37% yield enhancement, whereas for *P. flabellatus*, 24% yield increase was observed in a study by Chanakya et al. [26]. Mushroom growth is also influenced by digestate originated from different feedstock [21]. Use of cowdung digestate in agro-residue substrate was reported to give highest yield (78%) followed by digestates from jute caddis (33%), municipality solid waste (10%) and poultry litter (13%) compared to control [21, 22].

Proportion of biogas digestate to basal substrate material also needs to be optimized for optimum mushroom production. A ratio of 1:1 of digestate to basal substrate was recommended in order to achieve the optimum mushroom in case of *P. sajor-caju* [21]. Chanakya and Ganguly, 1994, observed 3:7 (digestate: straw) as the most effective ratio. The yield was observed decreasing when the ratio was changed to 1:2 and 2:1 [20]. Increased digestate proportion in lingo cellulosic substrate would lead to increased lignin content in substrate, contributing to higher recalcitrance nature of substrate. This would further reduce the decomposable content in substrate by mushroom mycelium and hence mineralization rate of the substrate. This is the reason (high lignin) that growing of mushroom solely on biogas digestate led to minimum yield than when digestate was used along with agro-waste [32]. High digestate content may also cause accumulation of excess moister restricting mycelium development. Isikhuemhen et al. 2009 reported a substrates composition of 10–20% dry solid digestate, 70–80% wheat straw and 10% millet to produce the highest mushroom yield [24].

Digestate amendments also improve the nutritional value of mushroom. Protein content was increased on digestate grown mushroom compared to control substrate, while carbohydrate content was reduced [24]. Digestate being rich in mineralized nutrients increases the mineral content of mushroom such as Na, K, Ca, Fe, Mn, Cu, Zn and P [21]. Plant nutrient (N, P and C) uptake was higher when digestate was supplemented to rice straw than that of only rice straw grown mushroom [26]. Digested material provides mushroom with a sufficient amount of easily metabolizable organic matter and non-assimilated carbohydrates that favours the growth and yield of the mushroom [16].

Mushroom cultivation in lignocellulosic substrates helps faster decomposition of such less decomposable material as mushroom hyphae secrete lingo cellulolytic enzymes. The extent of this lignin degradation was reported to be higher when digestate is added to such substrates [16]. It is clearly evident that supplementation of biogas digestate to mushroom growth substrate clearly improves the overall efficiency of the production process in comparison to only agricultural residue when used as substrate.

3.2 Characteristics of Biogas Digestate

Table 3 shows the composition of whole biogas digestate used in the present work. pH of the digestate was found to be basic side because of volatile fatty acid degradation and production of ammonia through organic N mineralization during digestion [34]. Electrical conductivity which is an indication of mineralization of organically bound nutrients is found to be similar to that of manure-based digestates. TOC, TN, P, K and other micronutrients were also in the range of those reported for manure-based digestates. Heavy metal contents of the digestates were found to be lower than those reported for other digestates such municipal sludge [21]. From the characterization status, mainly its high N, P and K content and low heavy metal content, digestate appears to be a suitable supplement in mushroom production.

Though the liquid fraction after separation has not been separately analysed in the present work, from our earlier publication [29], it has been found that separation of whole digestate into solid–liquid fraction generally results in a liquid fraction with higher pH, higher electrical conductivity than whole digestate. Higher pH and EC of liquid digestates mean the presence of mineralized nutrient fraction in liquid fraction. Overall, separation concentrates plant nutrients in solid digestates. Total organic carbon of liquid digestate remains minimum (0.5–2%). However, 60–90% of the ammonia-N content of the whole digestate ends up in the liquid fraction which indicates its suitable use as readily available nitrogen source in mushroom cultivation.

Table 3 Characterization of collected raw digestate (values expressed on dry matter basis)

Parameter	Value
pH	7.8
EC, mmho cm^{-1}	4.1
TOC, % dw	26
TN, g kg^{-1}	21.2
P, g kg^{-1}	14.46
K, g kg^{-1}	10.05
Ca, g kg^{-1}	7.26
Mg, g kg^{-1}	7.37
S, g kg^{-1}	1.50
Cu, mg kg^{-1}	15.13
Fe, mg kg^{-1}	1026
Mn, mg kg^{-1}	915
Zn, mg kg^{-1}	146
Pb, mg kg^{-1}	1.74
As, mg kg^{-1}	0.62
Cd, mg kg^{-1}	0.62

3.3 *Effect of Liquid Biogas Digestate on Mushroom in Laboratory-Based Experiment*

In Assam, optimum temperature of *Pleurotus* spp. is 23–25 °C and best period to grow is October to April [35]. Abrupt fluctuation in temperature may result in arrested mycelial growth, and production may be reduced [32]. Though the average time of pin head emergence remains in between 15 and 28 days for oyster mushroom, pin head formation was observed to be delayed in both the treatments in the present study. Pinhead formation occurred on 23–25 days after spawning. However, the higher time interval needed for pinhead formation and fruiting body formation could not be attributed to difference in treatments, as in the present study, application of the digestate was started only after pin head formation. Therefore, late emergence of pin heads could be attributed to decline in temperature up to 11 °C during the month of December hampering mycelium fruiting ability. Similar findings of reduced mushroom yield due to decline in temperature were reported by Sarmah et al. [35].

Comparative effect of straw and digestate amended straw on mushroom yield and growth parameters is shown in Fig. 1. It is seen that application of digestate gave better yield over basal substrate rice straw. In digestate treatment, a total of 615 g/800 g of mushroom was produced, which was higher than rice straw alone treatment (447 g/800 g). In the present study, we observed a percentage increase of 38% when substrate is supplemented with liquid fraction of the biogas digestate. The results of present study are in agreement with those that reported enhancement of mushroom yield after digestate application. Chanakya et al. [26] also reported a similar enhancement of yield (37%) in case of *P. florida*, species in digestate amended rice straw compared to rice straw with no digestate supplement. It has been reported that digested material provides mushroom with a sufficient amount of easily metabolizable organic matter and non-assimilated carbohydrates which finally favours the growth and yield of the mushroom [16].

Biological efficiency of digestate treatment was also found to be higher (76%) compared to biological efficiency of mushroom grown on rice straw alone (55%).

Fig. 1 Comparative effect of straw and digestate amended straw on mushroom yield parameters

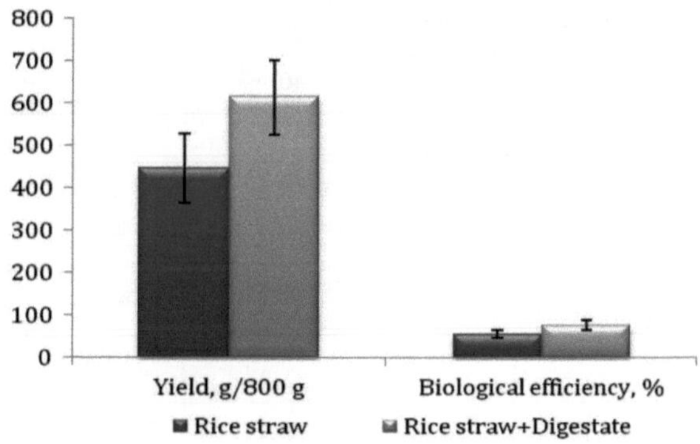

However, the yield and biological efficiency was found to be lower than those reported in these studies.

While analysing whether the yield and biological efficacy observed under both the treatments are significantly different or not, we found a significant difference between the two at $p < 0.05$. Since under the digestate supplement we found an enhanced yield compared to the digestate unamended production, this indicates overall remarkable impact of digestate on mushroom yield.

Comparative effect of straw and digestate amended straw on mushroom growth is shown in Fig. 2. A higher number of fruiting bodies (18) were observed on digestate treatment, whereas for rice straw treatment, a less (10) numbers of fruiting bodies were observed.

Though the total number of mushroom was higher in digestate amended substrate, mean weight of mushroom was observed to be less in case of substrate with digestate application. Digestate application produced mushroom with mean weight 36.17 g, but for rice straw alone treatment, it was found to be 44.72 g (Fig. 2). The mean weight observed under both the treatments was found to be significantly different at $p < 0.05$ as determined by Z test.

Higher number of mushroom in digestate treatment also seemed to decrease the average cap diameter (9.47 cm) of the digestate grown mushroom to that of rice straw alone substrate (9.73 cm). However, there was not any statistically significant difference between the two treatments.

Stem height of the digestate treatment was lower than that of the rice straw alone treatment as shown in Fig. 2. As cap is the preferred part of the mushroom, therefore, mushrooms with high cap weights and low stalk heights are desirable. Similar findings were reported by Stoknes et al. 2013 [25] who found no significant difference in mushroom size (species *Agaricus bisporus* and *Agaricus subrufescens*) grown in digestate treated straw than control using anaerobically digested food waste. Insignificant difference of stem height under both treatments is supported by Z test.

We did not find any statistically significant correlation among the parameters under study.

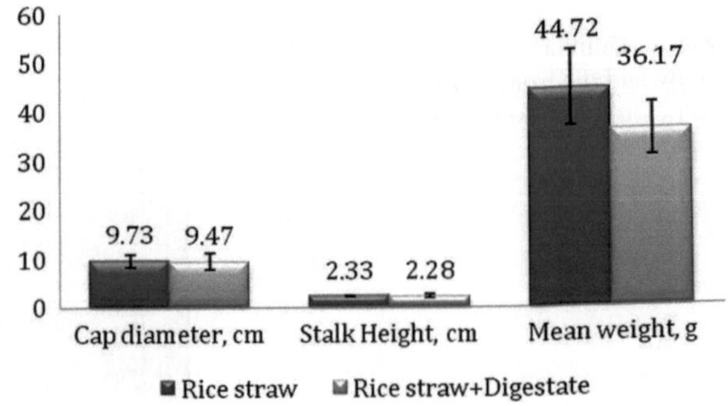

Fig. 2 Comparative effect of straw and digestate amended straw on mushroom growth

The reason for higher yield in digestate amended straw could be attributed to availability of plant nutrients such as P, K and microelements such as Mn, Mg, Fe and Cu as seen from characterization study. In general, digestates are reported to contain nutrients with enhanced bioavailability (60–80% of total nitrogen in mineralized form along with bioavailable P and K and other micro nutrients), making these a suitable consideration as plant nutrient source [36, 37]. Similar positive effect of micronutrients, viz. Mn, Mg, Fe and Cu present in biogas slurry on mushroom fruit induction and development, was reported by Navathe and Kadam [38]. They attributed increased crown width, stalk length, fruit bodies' weight of *Pleurotus tubergium* mushroom to the balanced nutrient contents of the organic media such as digestate. Figures 3, 4, and 5 show the photographs of mycelium run, growing fruiting body and matured fruiting body grown during the laboratory experiments.

It is to be noted that nutrient concentration of mushroom produced using digestate supplement is an important aspect of study. It is the source of digestate that will affect the mushroom nutritional quality as well as its heavy metal uptake. Though there is no particular standard available for digestate application as crop nutrient in Indian context, however, the cowdung digestate used in the current work showed a heavy metal concentration lower than that reported for cowdung digestate in the other relevant literature. Therefore, it is expected that the heavy metal contamination in mushroom will not be a matter of concern, though a thorough investigation would be required to conform the same.

Fig. 3 Mycelium run

Fig. 4 Formation of fruiting body

Fig. 5 Fruiting body ready to be harvested

3.4 *Economics of Mushroom Production*

Table 4 shows the cost related to mushroom production under two subcost parameters, viz. recurring and non-recurring. Considering a polyhouse-type mushroom production unit, comparison of mushroom economics has been made with 1 tonne per annum digestate utilization. As observed in the experimentation, we have considered 55 and 76% biological efficiency for mushroom production under straw treatment and straw + digestate treatment, respectively. In addition, we also considered that the digestate would be from household biogas plant, which is produced as residue of the plant and hence would be available to the grower without incurring any cost. Therefore, for economic analysis, no cost is added to it, only that

Table 4 Economics of mushroom production

Particulars	Units	Unit price (INR)	With digestate supplementation		Without digestate supplementation	
			Quantity	Total price (INR)	Quantity	Total price (INR)
Non-recurring cost						
Cost of polyhouse	m^2	700	60	42,000	60	42,000
Steel rack (for keeping bags)	No.	12,000	20	240,000	20	240,000
Thermometer cum humidity meter		500	1	500	1	500
Sprayer (for spraying digestate and liquid)		2000	2	4000	1	2000
Drum (500 l) (for storing liquid digestate)		3000	2	6000	0	0
LPG burner (for preparing straw)		3000	1	3000	1	3000
Container (for storage of miscellaneous)		500	10	5000	10	5000
Cooler (1.5 tons) (for maintaining temperature)		25,000	1	25,000	1	25,000
Cold storage (300 L deep fridge) (for storage of mushroom)		25,000	1	25,000	1	25,000
Recurring cost						
Polythene bag	kg	180	55	9900	55	9900
Spawn	kg	33	330	10,890	330	10,890
Rice straw	ton	1500	10	15,000	10	15,000
Nylon rope	kg	60	10	600	10	600
Electricity	kWh	7	540	3780	540	3780
Miscellaneous (pesticides, packaging)				5000		5000
Labour	Man-days	250	90	22,500	45	11,250
Net total			418,170		398,920	
Depreciation (@ 5% per annum)			17,525		17,125	
Interest (@ 7%)			24,535		23,975	
Insurance and taxes @15% of sum insured			60,000		60,000	
Gross expenditure, Rs.			520,230		500,020	
Total mushroom production, tons			7.6		5.5	
Revenue earned (@ Rs. 100/kg), Rs.			760,000		550,000	
Annual profit (first year), Rs.			239,770		49,980	

the grower has to pay for the processing cost. An additional manpower of 90 man-days was taken into account for processing and application of the digestate part under 'straw + digestate production scenario', whereas 45 man-days were considered for the 'only straw' production scenario.

From the table, it is found that using digestate as substrate, supplement which could increase the mushroom yield by about 38% in the laboratory experiment compared to control condition) could generate an profit of Rs. 2.39 lacs in the first year in comparison to the profit of Rs. 49,980 made under only rice straw as substrate. An amount of 1.89 lacs was estimated as the additional profit under straw + digestate production scenario', compared to 'only straw' production scenario.

The economic estimation in the present study is made considering the digestate availability from a typical household biogas plant of size 1–2 m^3. Accordingly, handling and processing of digestates could be managed without substantial technological input. However, for large-scale application, use of mechanical separation machines like screw press, belt press can be employed. In that case, there is a requirement to find out the economics considering digestate processing technique, storage for economic viability of the whole production process.

4 Future Prospects

Mushroom cultivation can lead to many associated benefits requiring further research focus. The efficient lignocellulosic degrading capacity (lignocellulolytic nature) of Pleurotus species indicates their potential use in the conversion of agricultural wastes into commercial commodity like mushroom [16]. However, there are number of aspects to be optimized for successful commercial exploitation of the process through further research considering higher number of replicates, comparison with other conventional supplements used, nutrient quality of mushroom, cost of manpower/machinery engagement in the entire process in a holistic way. Further, biogas liquid application in mushroom substrate can lead to production of leachate rich in lignocellulolytic enzymes, viz. peroxidase and laccase and low in cellulose, potential application of which has been shown for softening of biomass for higher biogas production (by 29%) and bioethanol production [39]. Spent mushroom substrate remaining after mushroom cultivation can be used as biofertilizer for soil health improvement for its suitable C:N ratio.

5 Conclusion

Liquid fraction of biogas digestate could be an alternative nutrient-rich substrate supplement for mushroom growth. Considering the growing market of mushroom in India, the benefits highlighted from biogas plant could be extended by value addition of residual digestate through its utilization in mushroom production. Digestate use in mushroom cultivation will offer the opportunity of recycling digestate nutrients in the production of edible, protein-rich food. Integrated development of biogas and mushroom cultivation system may contribute to achieving food security along with promotion of bioenergy. Analysis of cost dynamics shows feasibility of digestate utilization in mushroom production; however, low-cost region specific production unit using variety of mushroom for round the year commercial cultivation would help in promoting such systems.

Acknowledgements This work was supported by the following grant award: J000361/1 DST-Rural Hybrid Energy Enterprise Systems. We thank Professor H. N. Chanakya, Centre for sustainable Technologies, Indian Institute of Science, Bangalore, India, for his insightful comment to initiate the work.

References

1. Taheripour, F., Hertel, T.W., Tyner, W.E., Beckman, J.F., Birur, D.K.: Biofuels and their by-products: global economic and environmental implications. Biomass Bioenergy **34**, 278 (2010)
2. Galvez, A., Sinicco, T., Cayuela, M.L., Mingoranceb, M.D., Fornasier, F., Mondinia, C.: Short term effects of bioenergy by-products on soil C and N dynamics, nutrient availability and biochemical properties. Agric. Ecosyst. Environ. **160**, 3–14 (2012)
3. Wang, M., Huo, H., Arora, S.: Methods of dealing with co-products of biofuels in life-cycle analysis and consequent results within the U.S. context. Energy Policy **39**, 5726 (2011)
4. Moller, K., Müller, T.: Effects of anaerobic digestion on digestate nutrient availability and crop growth: a review. Eng. Life Sci. **12**, 242–257 (2012)
5. Sharpley, A., Moyer, B.: Phosphorus forms in manure and compost and their release during simulated rainfall. J. Environ. Qual. **29**, 1462–1469 (2000)
6. Shah, Z.A., Ashraf, M., Ishtiaq, C.M.: Comparative study on cultivation and yield performance of oyster mushroom (Pleurotus ostreatus) on different substrates (wheat straw, leaves, saw dust). Pak. J. Nutr. **3**(3), 158–160 (2004)
7. Iqbal, S.M., Rauf, C.A., Sheikh, M.I.: Yield performance of oyster mushroom on different substrates. Int. J. Agri. Biol **7**(6), 900–903 (2005)
8. Dey, R.C., Nasiruddin, K.M., Haque, M.S., Al Munsur, M.A.Z.: Production of oyster mushroom on different substrates using cylindrical block system. Progr. Agric. **19**(1), 7–12 (2013)
9. Mondal, S.R., Rehana, J., Noman, M.S., Adhikary, S.K.: Comparative study on growth and yield performance of oyster mushroom (*Pleurotus florida*) on different substrates. J. Bangladesh Agric. Univ. **8**(2), 213–220 (2010)
10. Gell, K., van Groenigen, J.W., Cayuela, M.L.: Residues of bioenergy production chains as soil amendments: Immediate and temporal phytotoxicity. J. Hazard. Mater. **186**, 2017–2025 (2011)
11. Gunnarsson, A., Bengtsson, F., Caspersen, S.: Use efficiency of nitrogen from biodigested plant material by ryegrass. J. Plant Nutr. Soil Sci. **173**, 113–119 (2010)
12. WRAP, Anaerobic digestate, Partial Financial Impact Assessment of the introduction of a Quality Protocol for the production and use of anaerobic digestate, Waste Protocols Project (2009)
13. Chang, S.T.: The world mushroom industry-trends and technological development. Intl. J. Med. Mushrooms. **8**, 297–314 (2006)

14. National Horticultural Board, Available at: http://nhb.gov.in/PDFViwer.aspx?enc=3ZOO8K5CzcdC/Yq6HcdIxPt1uDlARRr3zvrry1RS+gI=
15. DGCIS annual report, Annual report, Directorate General of Commercial Intelligence and Statistics (2014)
16. Udayasimha, L., Vijayalakshmi, Y.C.: Sustainable Waste Managementby growing mushroom (*Pleurotus florida*) on anaerobically digested waste and agro residues, Int. J. Eng. Res. Technol. 1 (2012)
17. Naraian, R., Sahu, R.K., Kumar, S., Garg, S.K., Singh, C.S., Kanaujia, R.S.: Influence of different nitrogen rich supplements during cultivation of *Pleurotus florida* on corn cob substrate. Environmentalist **29**, 1 (2009)
18. Nayak, B.K., Bathmarajan, V., Nanda, A.: Effect of substrate and environmental parameters on the production of oyster mushrom in Pondichery. Der Pharmacia Letre **7**, 74 (2015)
19. Demirer, T., Rock-Okuyucu, R., Ozer, I.: Effect of different types and doses of nitrogen fertilizers on yield and quality characteristics of mushrooms (Agaricus bisporus (Lange) Sing) cultivated on wheat straw compost. Journal of Agriculture and Rural Development in the Tropics and Subtropics **106**, 71 (2005)
20. Ganguly, N.K., Chanakya, H.N.: Mushroom cultivation on spent biomass from biogas plants. Curr. Sci. **66**, 70 (1994)
21. Banik, S., Nandi, R.: Effect of supplementation of rice straw with biogas residual slurry manure on the yield, protein and mineral contents of *Volvariella volvacea* mushroom. J. scientific & industrial research **59**, 407 (2000)
22. Banik, S., Nandi, R.: Effect of supplementation of rice straw with biogas residual slurry manure on the yield, protein and mineral contents of oyster mushroom. Ind. Crops Prod. **20** (2004), 311 (2004)
23. Noble, R., Hobbs, P.J., Mead, A., Dobrovin-Pennington, A.: Influence of straw types and nitrogen sources on mushroom composting emissions and compost productivity. J. Ind. Microbiol. Biotechnol. **29**, 99 (2002)
24. Isikhuemhen, O.S., Mikiashvili, N.A., Kelkar, V.: Application of solid waste from anaerobic digestion of poultry litter in Agrocybe aegerita cultivation: mushroom production, lignocellulolytic enzymes activity and substrate utilization. Biodegradation **20**, 351 (2009)
25. Stoknes, K., Beyerb, D.M., Norgaard, E.: Anaerobically digested food waste in compost for Agaricus bisporus and Agaricus subrufescens and its effect on mushroom productivity. J. Sci. Food Agric. **93**, 2188 (2013)
26. Chanakya, H.N., Malayil, S., Vijayalakshmi, C.: Cultivation of Pleurotus spp. on a combination of anaerobically digested, plant material and various agro-residues. Energy. Sustain. Dev. **27**, 84 (2015)
27. Torres-Climent, A., Martin-Mata, J., Marhuenda-Egea, F., Moral, R., Barber, X., Perez-Murcia, M.D., Paredes, C.: Composting of the solid phase of digestate from biogas production: optimization of the moisture, C/N ratio, and pH conditions. Commun. Soil Sci. Plant Anal. **46**, 197–207 (2015)
28. Hanc and Vasak: Processing separated digestate by vermicomposting technology using earthworms of the genus Eisenia. Int. J. Environ. Sci. Technol. **12**, 1183–1190 (2015)
29. Kataki, S., Hazarika, S., Baruah, D.C.: Assessment of by-products of bioenergy systems (anaerobic digestion and gasification) as potential crop nutrient. Waste Manag. **59**, 102–117 (2017)
30. Borah and Rahman: Mushrooms in biodiversity and food security of Sikkim. http://sikenvis.nic.in/writereaddata/chapter3.pdf (2011)
31. Chang, S.T., Chiu, S.W.: Mushroom production—an economic measure in maintenance of food security. In: DaSilva, E.J., Ratledge, C., Sasson, A. (eds.) Biotechnology: economic and social aspects, pp. 110–141. Cambridge University Press, New York (1992)
32. Santi, G., Muzzini, V.G., Galli, E., Proietti, S., Moscatello, S., Battistelli, A.: Mycelial growth and enzymatic activities of white-rot fungi on anaerobic digestates from industrial biogas plants. Environ. Eng. Manage. J. **14**, 1713–1719 (2015)
33. Chang, S.T., Miles, P.G.: Mushroom biology—a new discipline. Mycologist. **6**, 64–65 (1992)

34. Nkoa, R.: Agricultural benefits and environmental risks of soil fertilization with anaerobic digestates: a review. Agron. Sustain. Dev. **34**, 473–492 (2014)
35. Sarmah, L.M., Gogoi, R., Rathaiah, A.: Possibility of milky mushroom cultivation in Assam and use of moss as a casing material. Ann. Agric. Res. New Series **27**, 37 (2006)
36. Tambone, F., Scaglia, B., D'Imporzano, G., Schievano, A., Orzi, V., Salati, S., Adani, F.: Assessing amendment and fertilizing properties of digestates from anaerobic digestion through a comparative study with digested sludge and compost. Chemosphere 81, 577 (2010)
37. Garfi, M., Gelman, P., Comas, J., Carrasco, W., Fermer, I.: Agricultural reuse of the digestate from low- cost tubular digestates in rural Andean communities. Waste Manage. **31**, 2584–2589 (2011)
38. Navathe, S., Borkar, P.G., Kadam, J.J.: Cultivation of Calocybe indica (P & C) in Konkan Region of Maharashtra. India, World J Agric. Res. 2, 187 (2014)
39. Malayil, S., Chanakya, H.N.: Fungal enzyme cocktail treatment of biomass for higher biogas production from leaf litter. Resource Efficient Waste Management: Proceedings of 5th IconSWM 2015, Bangalore, India (2015)

Characterization of Market Solid Waste at Source in Kolhapur City, Maharashtra, India

Manjusha Sarnobat, Girish Kulkarni and Sandip Mali

Abstract Increasing quantity of municipal solid waste is a problem for many nations to manage it properly. There is a need for environmentally sound management practices of the wastes. Kolhapur is one of the rapidly growing cities in the state of Maharashtra, in India. Kolhapur city is famous for very ancient Mahalaxmi temple. Many pilgrims visit the temple every year. Kolhapur is having main central market area around Mahalaxmi temple. Four major vegetable markets are placed in 1.5 km periphery of the temple. These markets are mainly of vegetable, fruit, and flower markets. The aim of this study is to find out the characterization as well as composition of municipal solid waste generated in these marketplaces. Focus of the work is to treat the waste in the same area by sustainable manner. Daily analysis made over these four market waste for one week for physical and chemical parameters. After analysis the results show that the organic content and ph of Kapil Tirth Market were 88 and 7.9%, that of Runmukteshwar Market were 72 and 7.0%, that of Shingoshi Market were 78 and 7.8%, and that of Mahalaxmi Mandir flower market were 81.2 and 8.0%. Chemical analysis show the moisture content 30% and volatile solids for Kapil Tirth Market 35%, for Runmukteshwar Market were 30 and 33%, for Shingoshi Market were 35 and 26.7%, and for Mahalaxmi Mandir flower market were 40 and 32%. The average density of solid waste of four marketplaces was 790 kg/m^3. A large amount of organic wastes are found in these market waste bins. If the organic waste is to be treated in that premises only, then the cost of collection, transportation, and treatment on dumping site would get reduced directly and the final by-product would get after processing the waste is of good quality manure.

M. Sarnobat (✉)
Kit's College of Engineering, Kolhapur, India
e-mail: manjusha3475@gmail.com

G. Kulkarni
Shivaji University, Kolhapur, India

S. Mali
Pimpri Chinchwad College of Engineering, Pune, Maharashtra, India

A. S. Kalamdhad et al. (eds.), *Advances in Waste Management*,
https://doi.org/10.1007/978-981-13-0215-2_33

Keywords Municipal solid waste · Characterization of MSW · Kolhapur
Composting

1 Introduction

Growing amount of municipal solid waste due to high human population has paid attention to manage it wisely. Solid waste management is chief confront for many nations. It is observed that there is a significant increase in municipal solid waste generation in India in last few decades. Per capita municipal solid waste generated daily in India ranges from 100 to 500 g [1]. Proper management of municipal solid waste is a neglected issue in India. Today, there is need to change old and unsustainable practices of municipal waste management by a major concern for urban local bodies (ULBs) with more of these authorities creating new infrastructure [2].

In our daily life activities, waste products arise and they are generated at every stage of process of production and development. When the source and type of waste are known precisely, then the design and operation of proper solid waste management can be possible [3]. Municipal solid waste is mixed in nature and consists of many diverse materials which are derived from various types of activities. Waste is any unwanted material intentionally thrown away for disposal. So the waste disposal by systematic manner is the important present need.

At a standstill so far, more than 90% of municipal solid waste generated in India is directly disposed on dump land in a poor manner. Nowadays, solid waste management has turned into significant issue, in addition to other environmental problems. Therefore, expansion of solid waste management facilities and their operation and maintenance in a sustainable manner by urban local bodies is essential.

It has been observed that in India, the amount of waste generated per capita is increasing at a rate of 1–1.33% annually. The increasing amount of municipal solid waste demands for high amount of land for waste disposal. So looking to this scenario, introduction of proper waste management practices is the present needs to prevent and control environmental pollution and health problems. One of the best practices to dispose municipal solid waste is by decentralized system. By knowing different pockets of the city having same kind of waste generation, we can give proper waste treatment process for that waste. This would decrease the demand of land for waste disposal, and also good quality by-product can be achieved. For this study, the central part of the city was selected around Mahalaxmi temple where four major vegetable, fruit, and flower markets are situated. The objectives of this study were (1) to know characteristics, composition, and quantity of solid waste generated at study area, (2) to suggest decentralized management plan for the municipal solid waste for study area.

1.1 Description of Kolhapur City

Kolhapur is situated at southwest of Maharashtra at 160° 40′ 31.81″N and 740° 15′ 12.10″E and is at altitude of 607 m above mean sea level. It covers an area of about 853 ha. Kolhapur is situated on the banks of the river Panchgang and is having historical Mahalaxmi temple at the center of the city. Chatrapti Shahu Maharaja, the great emperor of Kolhapur, was having the vision of modern Kolhapur. He developed the entire region of Kolhapur district in all respects. Kolhapur district is having plentiful natural resources like water, soil, natural vegetation, animal wealth, and minerals. So Kolhapur became one of the most agriculturally advanced districts of Maharashtra. One of the negative impacts of the city's rapid development is the increase in municipal solid waste generation, resulting in environmental degradation. Considering all the factors, this study focused on to suggest decentralized waste management plan for the waste generation in marketplaces of the city. Total nine vegetable markets are there in Kolhapur city shown in Fig. 1, which generate high organic wastes daily. Out of those nine markets, four markets are selected which are closely located around in 1.5 km periphery of this city. The names of that markets chosen for the study are Kapil Tirth Market, Runmukteshwar Market (Padalkar Market), Shingoshi Market, and flower market at Mahalaxmi temple. These marketplaces are selected for the study because they are closely located, and decentralized solid waste management system can be applied here after studying the waste characteristics. Marketplaces of Kolhapur city are shown in following two figures. Figure 1 shows the study area of closely located marketplaces, and Fig. 2 shows all the nine markets in the city.

1.2 Current Municipal Solid Waste Management Practice of Kolhapur City

Kolhapur is rapidly developing city in state of Maharashtra. India is having population around 549,283 residents. Daily municipal solid waste generation in Kolhapur city is around 180–220 ton/day. Kolhapur Municipal Corporation handles this waste by sweeping, collection, Transportation, and disposal. The city is divided into 81 wards and subdivided into 11 health divisions. Daily municipal solid waste collection system is run by Kolhapur Municipal Corporation. Municipal waste (residential and commercial establishments) is collected by door-to-door collection system and deposited by residents in small and large community bins. Addition of waste to MSW is from street sweeping and drain cleaning. The large collection bins are made up of steel material having capacity of one ton.

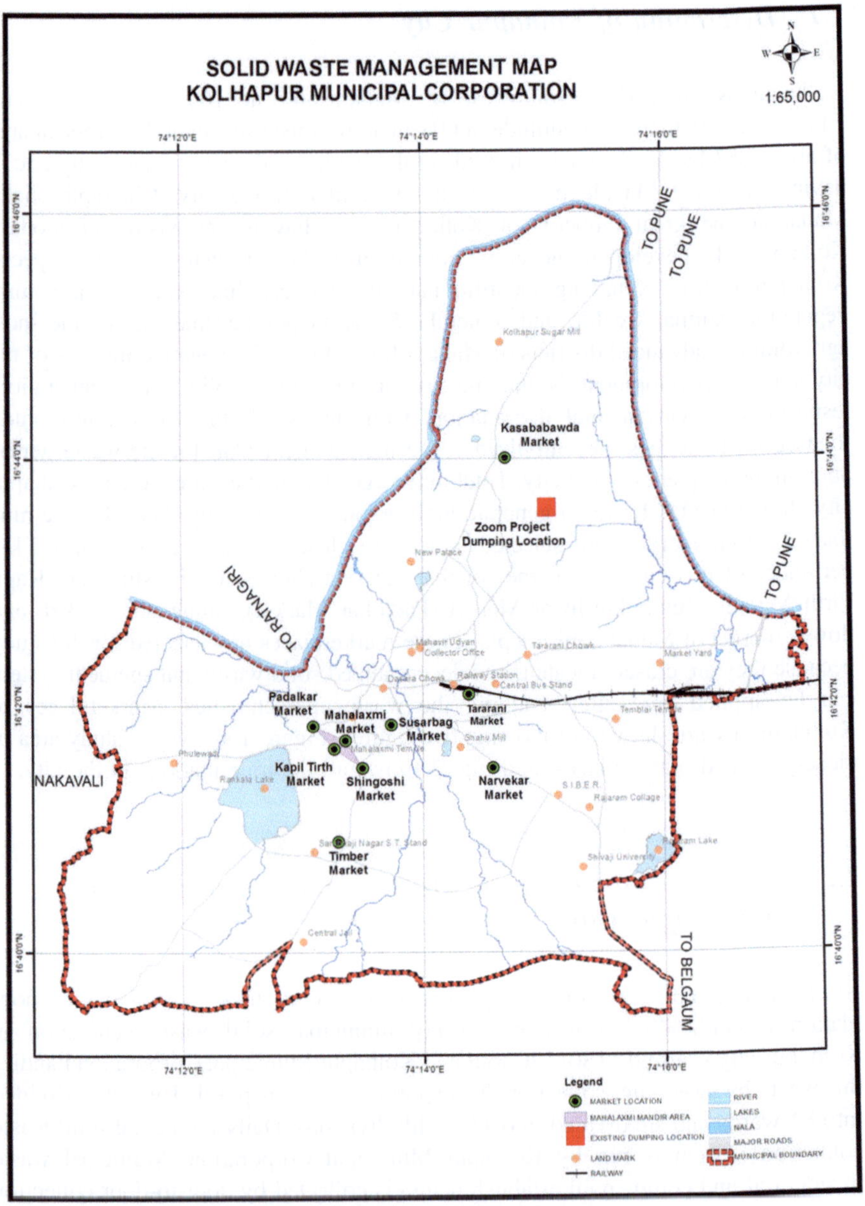

Fig. 1 Details of study area of vegetable markets of Kolhapur city

Total waste containers in Kolhapur city are 200. Collected waste from the containers mechanically transfers into refuse compactors and then transferred to the disposal site at Kasaba Bawada. Kolhapur Municipal Corporation is having three refuse compactors each of capacity 7–8 tones and also two dumpers having

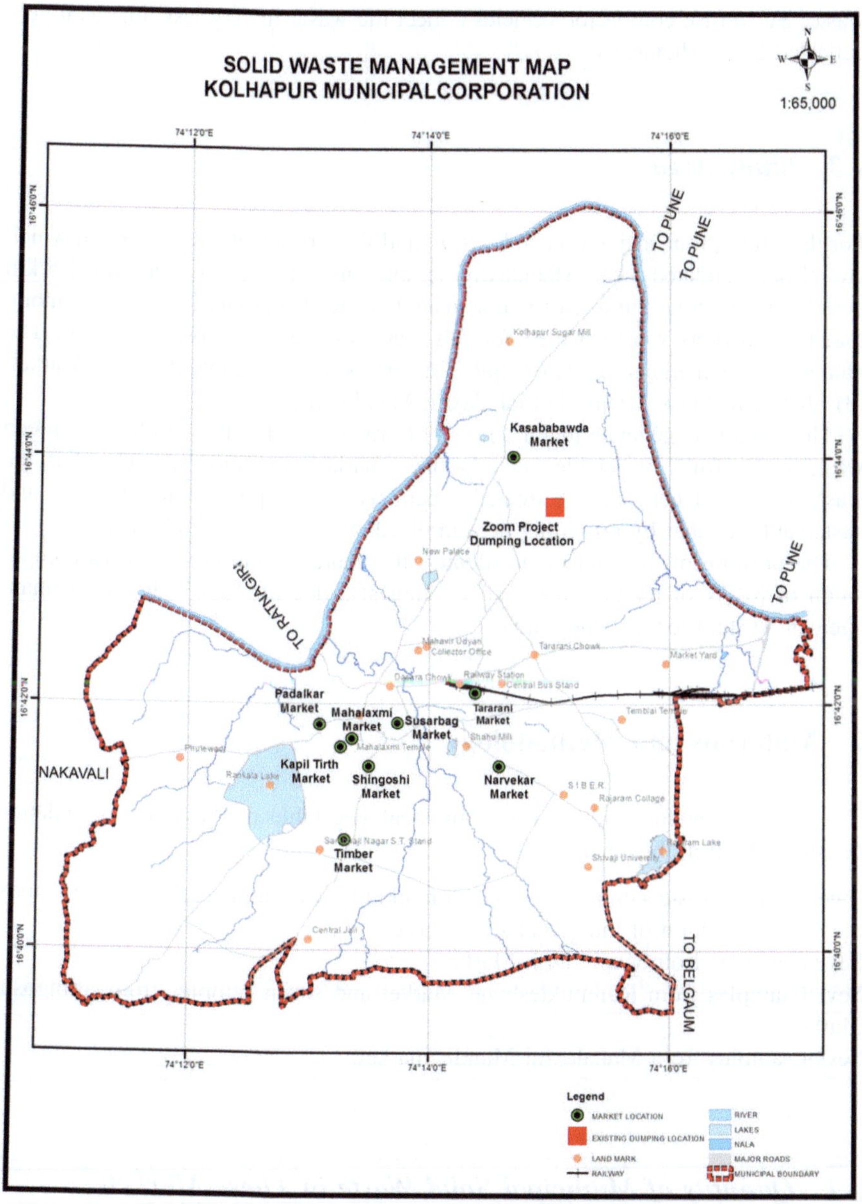

Fig. 2 Details of all vegetable markets and dumping site of Kolhapur city

capacity of 2 metric tons for transportation of solid waste. For each vehicle, one driver and two collection laborers are provided which makes two trips per day (8-h work day). To fill the dumper, it takes 2 h time for loading and unloading of the waste. But refuse compactor vehicles collect the waste in very less time as it gets collected by mechanically.

1.3 Study Area

For this study, four major vegetable, fruit, and flower markets were selected which are closely situated near Mahalaxmi temple in center of the city in 1.5 km periphery. The total number of major markets in Kolhapur city is nine, but among that four markets were selected for this study as they are closely situated. The names of the markets are (1) Kapil Tirth Market, (2) Runmukteshwar Market, (3) Shingoshi Market, and (4) Mahalaxmi Mandir market.

These markets generate major amount of organic wastes daily, in form of green waste, rotten fruits, vegetables, etc. As these markets are closer to each other, the waste generated here can be treated separately and dispose of in scientific and sustainable manner by providing decentralized system for the study area.

The purpose of the study is to identify the characterization as well as composition of MSW of market areas and to suggest a decentralized solid waste management system for the study area.

2 Materials and Methodology

The area chosen for study covers four local vegetable markets. The population density is high in this area.

Twenty-eight waste samples were taken randomly from four market areas to study the characterization of municipal solid waste.
Seven samples from Kapil Tirth Market.
Seven samples from Runmukteshwar Market and seven samples from Shingoshi Market.
Seven samples from Mahalaxmi Mandir market.

2.1 Quantity of Municipal Solid Waste in These Markets

It was found that daily solid waste generated and collected in Kapil Tirth Market is around 1600 kg/day which is collected twice in a day. Waste generated in Runmukteshwar Market is around 1200 kg/day collected twice in a day, in

Shingoshi Market, the waste generation is around 400 kg/day, and in Mahalaxmi temple area, the waste generated is around 250 kg/day.

2.2 *Quality of Waste Generated*

It was found that total quantity of waste generated in this area is around 3500 kg/day.

The waste is of heterogeneous in nature consisting of waste from residential area, waste from hotels, restaurants, and waste from vegetable markets. As the quantity of waste from vegetable markets is higher in these waste bins, so it can be treated separately without getting it mixed with total municipal solid waste of the city.

The total quantity of waste generated in the study area is around 3 stones which is very less amount compared to total quantity of solid waste of the city, so it can be easily manageable decentrally. In Kapil Tirth Market area and Runmukteshwar Market area, it was observed that the plastic waste is sorted at the time of collection of solid waste by the waste collector. So very less amount of plastic waste is found in waste bins here.

2.3 *Method of Waste Characterization*

Collection of sample and segregation of sample:

A number of municipal solid waste samples were collected for one week from February 1 to February 18, 2016, from each four market waste containers. Their locations are shown in Fig. 2. Totally 28 samples were segregated on-site. The sampling was done by quartering method, totally 100 kg sample was collected and mixed thoroughly, and 12.5 kg sample was ready by using quartering method. Manual process of sorting, was carried out for classifying and weighing all items in each sampling units for the collected 12.5 kg sample. This sample was then segregated manually into different physical components like paper, cloth, plastics, organic material, and glass. Sorting was based on visual inspections. Each category of waste was weighed separately by an electronic weighing balance. Standard personal safety procedures were followed during the sorting process such as wearing gloves, apron, and shoes. Each of these components is weighed to determine their fractions in the waste sample collected. Two kilograms of sample was collected in plastic bags and then brought to the laboratory and analyzed for physical characteristics, like moisture content. The remaining samples were kept stored and analyzed for remaining parameters like pH, volatile solids.

Characterization of solid waste:

Following characteristics were measured for the collected waste samples.

1. **Specific Weight (Density)**:
   It is mass per unit volume of solid waste. It varies from region to region. It is significant in storage, collection, and transportation of solid waste. Also, it is useful parameter while designing transport vehicle.
   The density of solid waste was found out by adding the known quantity of solid waste in 1 m^3 volume cardboard box. The waste was added into the box by noting the weight every time, and then, density was calculated from the weight added.
2. **Moisture Content**:
   It is expressed as mass of moisture per unit mass of wet or dry material. It has significant effect on transport facility.

$$\text{Moisture Content } = \frac{\text{Wet weight} - \text{Dry weight} * 100}{\text{Wet weight}}$$

3. **pH**:
   The pH is important parameter when treating the waste by composting method. The dried organic waste sample from municipal solid waste was analyzed for pH. The pH was analyzed by shaking 50 gm of powdered waste material in 250 mL of distilled water for 24 h and analyzed by pH meter (USEPA 2001).
4. **Organic Matter**:
   To find organic matter in waste sample, 25 g of dry waste is ignited at 360 °C for 24 h (USEPA 2001).
5. **Total Solids, Volatile Solids, Fixed Solids**:
   As there are no specific standard methods for analyzing solids in municipal solid waste samples, the solid waste samples were analyzed based on the same methods recommended for river and lake sediment samples. The methods used for solid waste samples and semisolid waste samples from water and wastewater treatment method.

Total solids are defined as the solids left in the sample after it has been dried to constant temperature of 105 °C. After ignition of total solids in a muffle furnace at 550 °C, the fraction of solids remaining is defined as fixed solids and the fraction lost during ignition is termed as volatile solids.

3 Results and the Discussion

The final results are from average result values of seven samples. The work was done for one-week duration.

3.1 Characteristics of Solid Waste Generation

Density:

Average density of municipal solid waste of these four markets was found high, 790 kg m^3. The waste found was having high quantity of organic waste. All selected four market waste bin samples show high density of municipal solid waste. It was observed in these areas that the waste is coming from commercial sectors like shops, offices, hotels, and residential areas also. The workers in all these market places strictly sorts out the recyclables at the source and colds it out it to scrap markets. So very less amount of recyclable waste enters into the waste bins. Major amount of waste observed was of organic in nature in all these market bins (Figs. 3 and 4; Tables 1 and 2).

3.2 Analysis of Results

From the graphs, it is observed that the solid waste in these four markets is having enough amount of organic wastes. It was found that the waste is more dense and humid, due to common consumption of fruits and vegetables and unpackaged food. The other fractions are considerably in lower side. The waste collected in the sample bins is from residential zone and vegetable market area. Commercial wastes like plastic and cardboards are sorted out separately and sold out to scrap market by the waste collector workers.

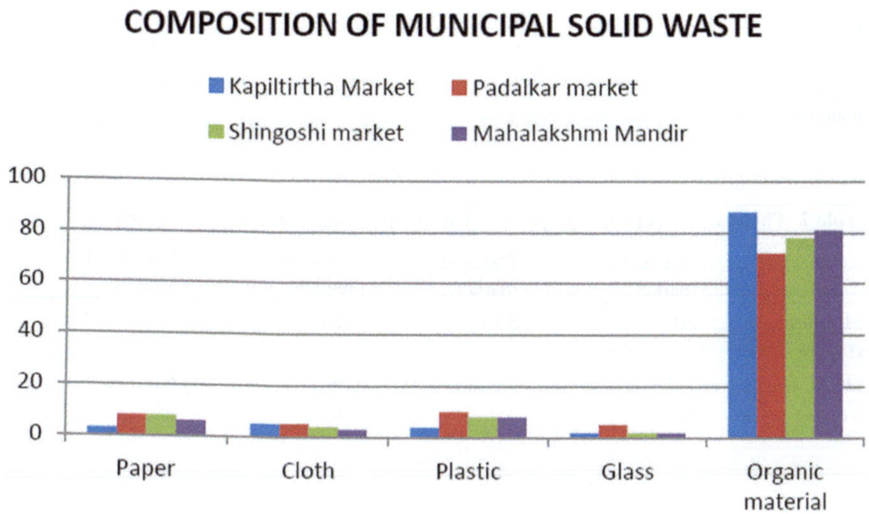

Fig. 3 Graph showing composition of municipal solid waste of four markets

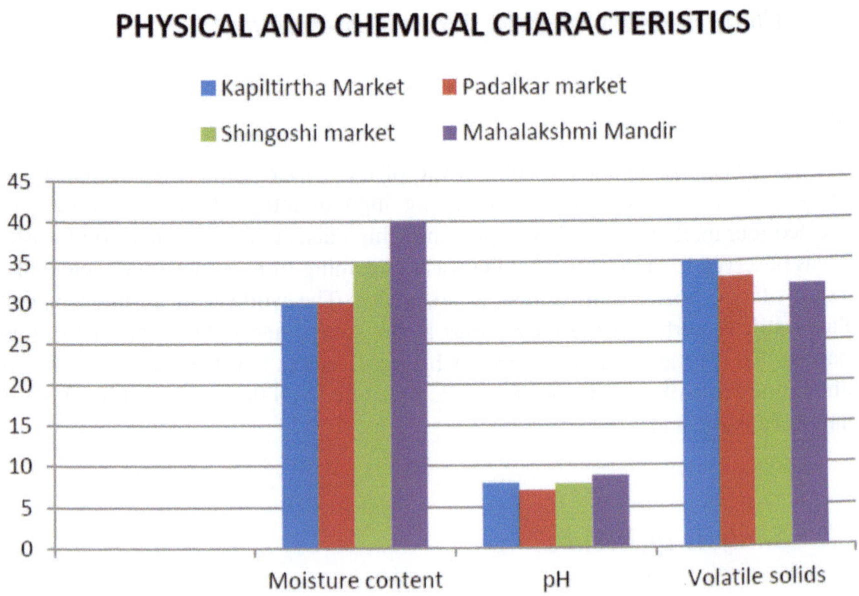

Fig. 4 Graph showing physical and chemical properties of municipal solid waste of four markets

Table 1 Composition of municipal solid waste at four markets

	Kapil Tirth Market in (%)	Padalkar Market in (%)	Shingoshi Market in (%)	Mahalaxmi Mandir in (%)
Paper	3	8	8	6
Cloth	5	5	4	3
Plastic	4	10	8	8
Glass	2	5	2	2
Organic material	88	72	78	81

Table 2 Physical and chemical properties of municipal solid waste of four markets

	Kapiltirtha market	Padalkar market	Shingoshi market	Mahalakshmi Mandir
Moisture content	30%	30%	35%	40%
pH	7.9	7	7.8	8.8
Volatile solids	35	33	26.7	32

High moisture content gives rise for higher density of the said solid wastes. So it is concluded that daily collection of solid waste is essential due to high moisture content of solid waste and considerable amounts of biodegradable materials in the study area. As the major portion of non-biodegradable wastes is separated at source, a good quality compost can be prepared from this type of waste. As the quantity of organic waste generates here is considerable, large-scale composting plants can be set up that would allow huge amounts of organic wastes to be converted to useful product like compost. This compost can then be used as a soil conditioner and fertilizer. Plant like this can be set up in nearby open area. The suggested compost plant can be run without a profit constraint if funded by ULB. An example of a vermi-composting plant is the one in Bhadreshwar, West Bengal. Currently, urban local bodies (ULBs) are running about ten large-scale composting plants in different cities of India and five vermi-composting plants. Other treatment options that are more difficult to operate and maintain are biogas generation. Pelletization of waste for energy generation is feasible when the waste has low moisture content and high organic content (having high calorific value) [2].

4 Conclusion

For proper waste management, first step is to understand (waste types being generated in order to design suitable collection and disposal systems. The maximum proportion of waste in the studied area can be composted by proper technical methods instead of disposed of. Sorted recyclables, like plastics and glass, can be sent to scrap merchants. Paper and textiles may be allowed to remain in the organic waste as it may not interfere the suggested composting process. The valuable organic material should not be sent to open dump site, but from that high-quality organic waste good quality manure can be prepared. The waste in the studied area can be treated by decentralize system, where the waste can be collected separately and sent to the gardens of municipal corporation where windrow composting can be done. Two gardens of municipal corporations are in the nearby area of the markets.

The total quantity of waste generated in study areas was found 3.5 tons/day, whereas the total amount of waste generated in Kolhapur city limits is approximately 180 tons/day. If such pockets can be made of the city depending on the type and nature of waste, proper treatment can be given to it.

Solid wastes in study area are characterized physically as well as chemically. The results give average values of various parameters

Moisture content 33.75% volatile solids 31.674, average density 790 kg/m^3, average pH as 7.8 and average organic, content 79.75%. All these results show that the waste can be treated by composting process. The by-product compost could be utilized in municipal-owned gardens nearby these market areas.

As all these markets are placed close to each other in 1.5 km periphery area, so it is possible to collect the waste from these markets and then to treat it in the near place. If the waste can be treated in the same area, then the transportation cost of solid waste to dumping site which is 8 km distance away from the city up to Rs. 1920/day for 15 tons/day quantity of solid waste can be saved. If such decentralized processing technologies would be adopted in the different parts of the cities, then the ultimate load of solid waste on dumping site would get reduced. Kolhapur Municipal Corporation should follow such decentralized plan which will reduce the ultimate load on the transportation and dumping site.

Acknowledgements We would like to express thanks to the staff of Kolhapur Municipal Corporation, for providing the necessary data used to undertake the research showed in this article.

References

1. Vivek S. Agrawal.: Sustainable waste management. Case study of Nagpur, India
2. Naresh Kumar, K., Goel, S.: Characterization of municipal solid waste (MSW) and a proposed management plan for Kharagpur, West Bengal, India. Res. Conserv. Recycl. J.**53**(3), 166–174 (2009)
3. Tchobanoglous, G., Theisen, H., Vigil, S.: Integrated solid waste management. McGraw-Hill Inc, New York (1993)
4. Oyelola, O.T., Babatunde, A.I.: Characterization of domestic and market solid wastes at source in Lagos metropolis, Lagos. Nigeria. Afr. J. Environ. Sci. Technol. **3**(12), 430–437 (2008)

Utilization of Thermal Power Plant By-Product Fly Ash in Waste Management

Chandra Bhanu Gupt, M. Mohamed Younus and S. Sreedeep

Abstract Landfill liners are integral part of waste management for isolating the harmful wastes from surrounding environment. In the absence of impermeable natural soils, compacted mixture of expansive soil and sand have found wide applications as landfill liners. In this study, attempt was made to replace sand with thermal power plant by-product fly ash. Annual production of fly ash in India is approximately 170 million tons per annum out of which only 61% is being used in construction of road and embankments, production of cement, reclamation of low lying areas, agriculture, making bricks, tiles and others. There is limited literature on the utilization of waste product (fly ash) for waste management applications. The major objective of this research work is to explore the utility of waste fly ash in waste management. The study clearly demonstrates the utility of fly ash as a substitute of sand in the construction of liners.

Keywords Landfill liners · Expansive soil · Sand · Fly ash · Waste containment

1 Introduction

Safe disposal of municipal and harmful wastes is a serious concern for environmentalist and planners. Landfill liners are mandatory for safe containment of these wastes. Landfills are usually lined with an impermeable material to prevent contamination of the surrounding soil and underlying groundwater by waste leachate.

C. B. Gupt (✉) · M. Mohamed Younus · S. Sreedeep
Department of Civil Engineering, IIT Guwahati, Guwahati 781039, Assam, India
e-mail: c.gupt@iitg.ernet.in

M. Mohamed Younus
e-mail: mohamed@iitg.ernet.in

S. Sreedeep
e-mail: srees@iitg.ernet.in

A. S. Kalamdhad et al. (eds.), *Advances in Waste Management*,
https://doi.org/10.1007/978-981-13-0215-2_34

In the absence of impermeable natural soils, bentonite and its mixture with specific geomaterials are the most widely used materials for the construction of hydraulic barriers such as liners due to its low hydraulic conductivity (<10^{-7} cm/s) [1]. In the past, compacted sand–bentonite mixture was one of the most commonly adopted hydraulic barriers investigated for the hazardous and nuclear waste containment system [2–6]. However, sand is an important useful construction material and its availability is limited due to the environmental issues associated with its dredging. Therefore, it is important to explore alternate material as a substitute for sand in the construction of facilities like landfill liners.

Fly ash is a by-product material being generated by combustion of pulverized coal from thermal power plants. In India, major source of power generation is coal-based where 75% of the total power is obtained from the thermal power plants and results in generation of large amount of fly ash. Such a huge quantity does pose challenging problems, in the form of land usage, health hazards and environmental dangers. Fly ash is disposed off in dry or wet state. But studies show that wet disposal of this waste does not protect the environment from the migration of metal into the soil through leaching process. Environmental Protection Agency, Japan, observed that the coal ash satisfies the criteria for landfill disposal. According to the hazardous waste management and handling rule of 1989, fly ash is considered as non-hazardous.

In this study, effort has been made to evaluate the usefulness of fly ash for liner construction and thereby substituting sand. A large amount of the produced fly ash is disposed in mono-fills [7]. The disposal of fly ash is becoming expensive each year due to the large area of land needed for its disposal. One of the amicable solutions to the problem is reuse of fly ash for some meaningful applications. The pozzolanic and self-hardening properties of fly ash justify its use in cement, concrete and pavements [7]. However, the use of fly ash for the construction of landfill liners has not been explored in detail.

With this in view, the present study purports to examine the suitability of fly ash as a landfill liner material. The major objective of this study is to maximize the use of fly ash for the said application. Therefore, different fly ash–expansive soil mixes were subjected to hydraulic conductivity and sorption test to verify the potential use of fly ash in waste management projects. These two criteria for evaluating the suitability of material for landfill liner have been proposed in this study. Based on the obtained results, it is noted that a fly ash–expansive soil mix with less than or equal to 70% of fly ash is found appropriate for landfill liner application.

2 Materials and Methods

The fly ash (designated as FA) used in this study is a waste by-product from a coal-fired power generating station at Farakka, West Bengal, India. The bentonite (B) used in this study is commercially available and designated as B. All the basic characteristics of B and FA such as particle size distribution, Atterberg limits,

Table 1 Physical properties and classification of the materials used in this study

Property	Fly ash (FA)	Bentonite (B)
Specific gravity	2.07	2.82
Grain size characteristics		
Sand (4.75–0.075 mm)	0	0
Coarse sand (4.75–2 mm)	0	0
Medium sand (2–0.425 mm)	0	0
Fine Sand (0.425–0.075 mm)	25	7
Silt size (0.075–0.002 mm)	75	44
Clay size (<0.002 mm)	0	49
Atterberg limits (%)		
Liquid limit	Non-plastic	224
Plastic limit	–	31
Plasticity index	–	193
Classification	Class F	CH
SSA (m^2/g)	1.4	219
Mineral present	Quartz	Quartz, montmorillonite, kaolinite

specific gravity have been determined by following the guidelines provided in the literature [8–11] during experimental investigation and presented in Table 1. The total specific surface area (SSA) was determined using ethylene glycol monoethyl ether (EGME) method [12], and the cation exchange capacity (CEC) was measured by the ammonium replacement method [13]. The mineralogical compositions of the materials were investigated by X-ray diffraction analysis (Bruker AXS D8 model, Germany) by using a graphite monochromatic and Cu-Kα radiation. A summary of these characterizations along with mineralogy is presented in Table 1. The chemical characteristics of B and FA are listed in Table 2. The FA used in this study belongs to class F (ASTM C 618) and B belongs to CH as per USCS classification.

2.1 Compaction Test

The non-plastic FA was uniformly dry mixed with B in different proportion as listed in Table 3, along with its designation. Variation of weight percentage of FA was taken from 0 to 100 for the present study. Such a wide range is required to explore the possibility of maximizing the use of FA as discussed above. These mixes were subjected to standard Proctor compaction test [14] to investigate MDD and OMC of all mixes. The standard Proctor compaction curves are depicted in Fig. 1. The highest MDD value of B as depicted in Fig. 1 might be due to its higher specific gravity value in comparison with FA as presented in Table 1. The highest value of OMC of B is due to its high water retention capacity. Also, it can be noted that

Table 2 Chemical characteristics of materials used in this study

Compounds (%)	FA	B
SiO_2	47.5	58.2
Al_2O_3	26.1	15.4
Fe_2O_3	8.4	3.3
CaO	0.9	0.2
MnO	0.2	1.3
MgO	0.3	5.6
Loss on ignition	2.3	–
pH	7.1	7.6
CEC (meq/100 g)	1.23	56.90

Note CEC—Cation exchange capacity

Table 3 Mix designation of bentonite–fly ash

Mix	FA (%)	B (%)
B 100-FA0	0	100
B 80-FA 20	20	80
B 60-FA40	40	60
B 50-FA50	50	50
B 30-FA70	70	30
B 0-FA100	100	0

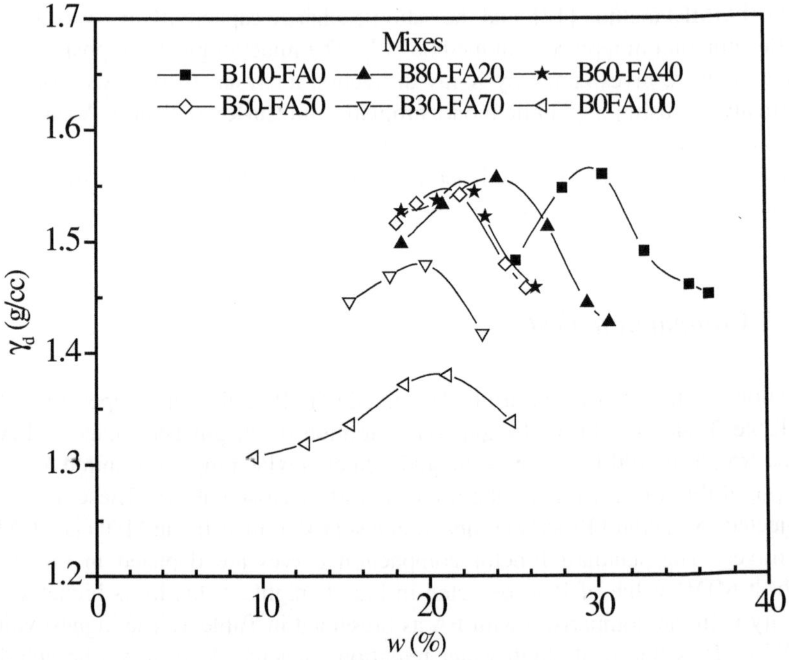

Fig. 1 Standard Proctor compaction curves for different B-FA mixes

MDD and OMC of the mixes gradually decrease as the percentage of FA increases, which may be attributed to its low value of specific gravity and water retention capacity of FA.

2.2 *Hydraulic Conductivity Test*

Measurement of hydraulic conductivity in porous media is a complicated process, and many laboratory apparatus exist for different soil and condition. Measuring hydraulic conductivity of fine-grained swelling materials provides additional challenges since large change in volume is anticipated during the testing. Therefore, the specimen volume should be controlled during the tests. Bentonite has a low hydraulic conductivity *k*, in the order of 10^{-8}–10^{-12} cm/s depending on its density [15]. This would considerably increase the saturation and the test duration for samples compacted at OMC. A new permeability test set-up has been developed for minimizing the saturation duration. The purpose of new set-up is to determine hydraulic conductivity under no swell condition and also to achieve quick saturation of the sample.

A schematic diagram of the new upward flow hydraulic conductivity apparatus is shown in Fig. 2. The thickness and diameter of the soil specimen used in this set-up are 30 and 60 mm, respectively. As explained earlier, the expansive soil has a tendency to swell when it comes in contact with water. To prevent this, a hollow movable screw provided would secure the water collection reservoir in position by pushing it downwards and thereby prevents swelling during flow of water through the soil mass. It maintains constant volume set-up which does not allow changes in total volume during testing.

Another feature of this set-up is that there is provision for applying vacuum for quick saturation as shown in Fig. 2. During the application of vacuum, the plug has been placed in the position to make the system airtight. During the flow test, this plug is removed and an extension tube is placed to facilitate the flow of water from the collection reservoir. In this set-up, experiment can be performed with varying height of soil. In the present study, the height of the soil specimen is 3 cm.

Permeability (*k*) from the falling head hydraulic conductivity test was performed at different compaction state of these mixes. Different compaction mass of sample was taken on the mass volume basis for the sample preparation and statically compacted in the perspex cylindrical mould. The results were used to evaluate the appropriate compaction states of mixes which satisfies the permeability criteria for designing liner ($\leq 10^{-7}$ cm/s). To ensure the repeatability of the observation, the value of *k* in this study has been obtained as the average of three test results with negligible variation.

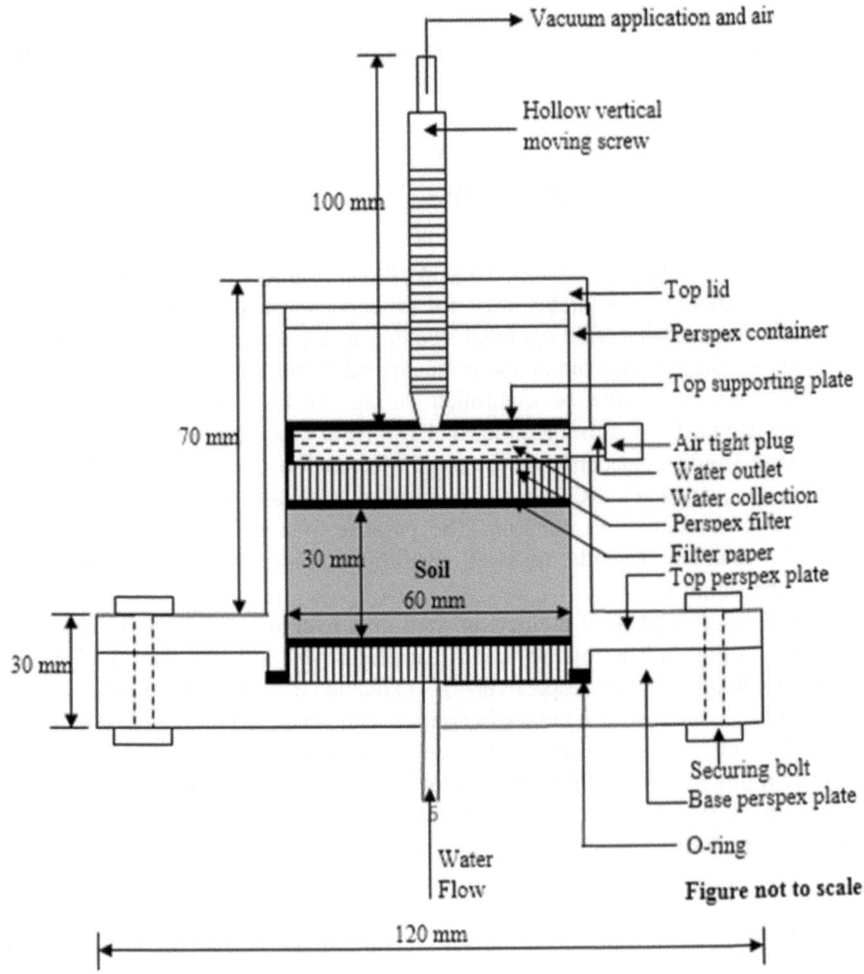

Fig. 2 Schematic diagram of permeability set-up used in this study

2.3 *Contaminant Retention Study*

The contaminant retention study of the mixes was performed by considering lead (Pb^{+2}) as the model contaminant. Lead nitrate stock solution was prepared by dissolving $Pb(NO_3)$ in deionized water. Solutions at different concentration were obtained by adding appropriate amount of stock solution to deionized water to obtain ion range of concentration varying from 20 to 1000 mg/L the solution pH was adjusted using dilute hydrochloric acid or sodium hydroxide solution. For minimizing the precipitation effect, the initial pH of the solution was adjusted to 5 before the start of the batch experiment. 5.5 g of air-dried soil sample sieved through 2 mm sieve is weighed and transferred into a 125 mL glass container. It is then mixed with 50 mL of contaminated solution to maintain a liquid to solid (L/S) ratio of 10. The container is then closed and placed on a rotary shaker and subjected to shaking for 24 h [16]. After shaking, the soil solution slurry was centrifuged at 3000 rpm for 20 min and then supernatant was filtered through 0.45 μm pore size filter paper. The resulting solution is subjected to chemical analysis for determining the concentration of Pb^{+2} by using an ion chromatograph (Metrohm Switzerland). Each analysis was done three times to ensure repeatability, and average of the test results is used for plotting retention isotherm.

3 Results and Discussion

3.1 *Hydraulic Conductivity Characteristics of Bentonite–Fly Ash Mixes*

Falling head test was used to determine the hydraulic conductivity using Eq. 1 for all bentonite–fly ash mixes at different compaction state with varying water content. Three points: optimum moisture content (OMC), dry of optimum and wet of optimum on standard Proctor curves presented in Table 4 have been selected for *k* determination.

$$k = \frac{aL}{At}\ln\left(\frac{h_1}{h_2}\right) \quad (1)$$

where a is the area of the standpipe, L is the length of the soil sample, A is the cross-sectional area of the soil sample, t is the time to drop head from h_1 to h_2 and h_1 and h_2 are initial and final heads, respectively, as measured at the start of the experiment and after time t. The results obtained from the test are depicted in Fig. 4 corresponding to different compaction states.

The variation of k as a function of fly ash percentage is shown in Fig. 3. It can be noted that for all the three compaction points investigated in this study, k increases

Table 4 Compaction states of samples used for hydraulic conductivity determination

Mix	Dry of OMC		OMC		Wet of OMC	
	γ_d (g/cc)	w (%)	$\gamma_{d\ max}$ (g/cc)	w (%)	γ_d (g/cc)	w (%)
B100-F0	1.55	27.97	1.57	29.66	1.49	32.83
B80-F20	1.53	20.80	1.56	24.22	1.44	29.42
B60-F40	1.54	20.48	1.55	22.04	1.46	26.35
B50-F50	1.52	18.03	1.54	21.01	1.48	24.55
B30-F70	1.46	16.22	1.48	19.66	1.42	23.16
B0-F100	1.32	14.27	1.38	19.66	1.34	24.89

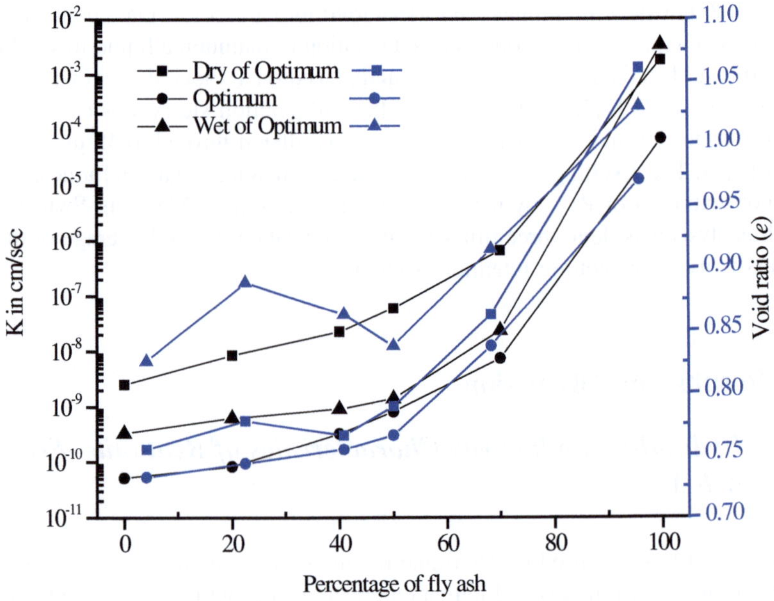

Fig. 3 Variation of hydraulic conductivity and void ratio with fly ash percentage

with an increases in fly ash percentage. This is mainly due to the inclusion of bigger cohesionless particles of fly ash as compared to B. The hydraulic conductivity decreases approximately four orders of magnitude when 30% expansive soil is used. Further, it can be noted that compaction state corresponding to OMC has minimum *k* value for all the fly ash–expansive soil mixes, followed by wet of optimum and dry of optimum state. The difference among all these states is more for samples with lower percentage of fly ash.

It can be observed from Fig. 4 that *k* value for all the mixes decreases from dry of optimum point to OMC and further increases when wet of optimum point is considered. Except fly ash, the minimum *k* value for fly ash–bentonite mix is obtained at water content 0.5–1.6% more than optimum *k* moisture content. Several investigators also reported that generally the lowest hydraulic conductivity of clayey soil is achieved when the soil is compacted at water content slightly higher than the optimum water content [17]. This is due to the development of a water film around the particles, which causes swelling of the clay exerting pressure against the surrounding particles. Consequently, the effective pore size reduces resulting in low hydraulic conductivity. For utility of a material as landfill liners, the compacted layer should have *k* value $\leq 1 \times 10^{-7}$ cm/s [18, 19]. Figure 4 shows that except fly ash, all the fly ash–bentonite mixes compacted corresponding to OMC yield *k* less than acceptable value.

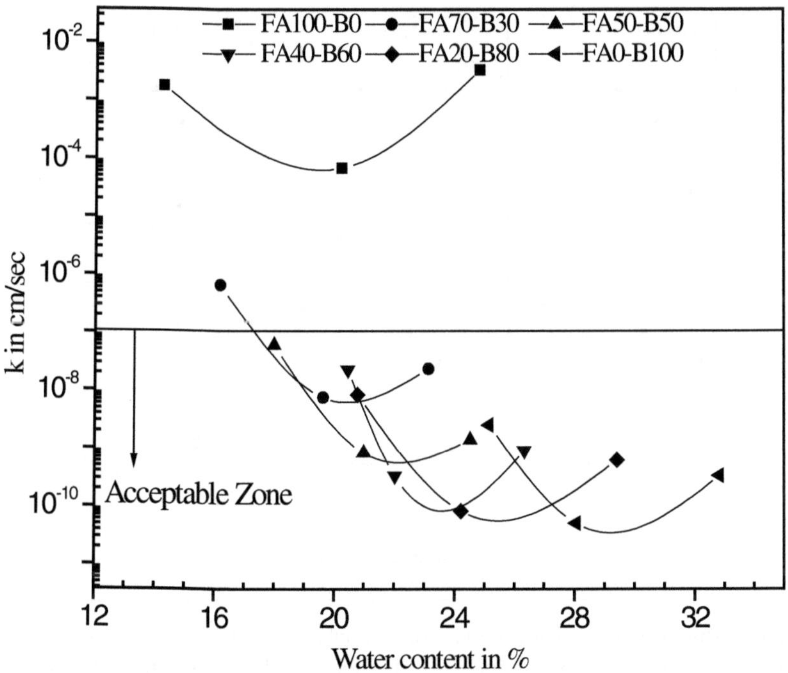

Fig. 4 Variation of hydraulic conductivity of mixes with varying water content

3.2 *Retention Characteristics of Bentonite–Fly Ash Mix*

The data presented in Fig. 5 shows that the amount of Pb^{+2} retained by mixes increased with the increase in Pb^{+2} concentration in the pore water solution used in the experiment, especially in lower concentration zone. The linear variation of sorption of metals at lower concentration was also reported by [20]. This is mainly because of an increase in concentration gradient with an increase in Pb^{+2} ions in the solution, which might act as a driving force for increased retention. As the concentration increased, the increase in the retention is less pronounced. The process of retention of Pb^{+2} possibly occurred in two stages with the saturation of different retention sites. With the increase in metal concentration, more sites were filled and Pb^{+2} retention became more difficult [21]. Similar result was reported by the [22], who demonstrated the occurrence of two phases in the adsorption curve. The initial phase of adsorption curve corresponds to high binding energy, although the soil had a low capacity for the adsorption. In second phase, higher quantities of Pb^{+2} are retained, although the retention energy is lower. When the retention sites become saturated with Pb^{+2} the retention decreases with further increasing of Pb^{+2} in the solution.

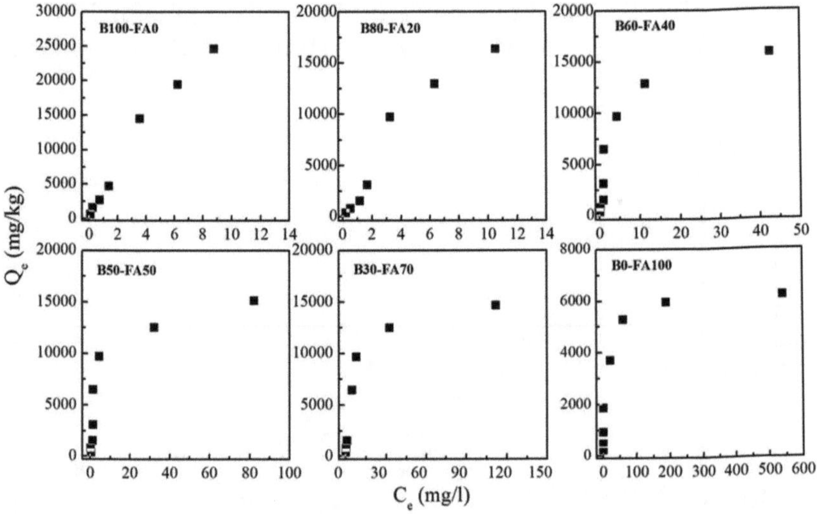

Fig. 5 Retention characteristics of Pb^{+2} in different B-FA mixes

The results obtained from the chemical analyses of the bentonite–fly ash mixes were utilized to develop nonlinear Freundlich and Langmuir retention isotherms using Eqs. 2 and 3 [23]. The isotherms represent the relation between adsorbed concentration (Q_e in mg/kg) and soil pore water concentration (C_e in mg/L) at equilibrium of solute species.

$$Q_e = K_F.C_e^N \tag{2}$$

$$Q_e = \frac{QK_LC_e}{1 + K_LC_e} \tag{3}$$

where K_F and N are Freundlich constants, K_L is the contaminant retention constant related to the binding energy (L/mg) and Q_e is the maximum contaminant adsorbed by solid (mg/kg). The retention characteristic curves for lead contaminant solution are shown in Fig. 5. It can be noted from the figures that Q_e increases as the percentage of bentonite increases in the mix. As compared to bentonite, Q_e of fly ash is quite less, and this is mainly due to larger specific surface area of B in comparison to FA, which provides large number of site for retention of the metal ions. This indicates that there is a need of clearly optimizing bentonite–fly ash mix considering the retention parameter also for reducing the thickness of liner, maximizing the use of fly ash and thereby economizing the project.

Figures 6 and 7 show the fitting of Freundlich and Langmuir isotherms model, and the model parameters are listed in Table 5. It was observed from Table 5 that both the isotherms gave satisfactory fitting even though the best fit was observed for Langmuir isotherm as is evident from R^2 values. The results also show that an

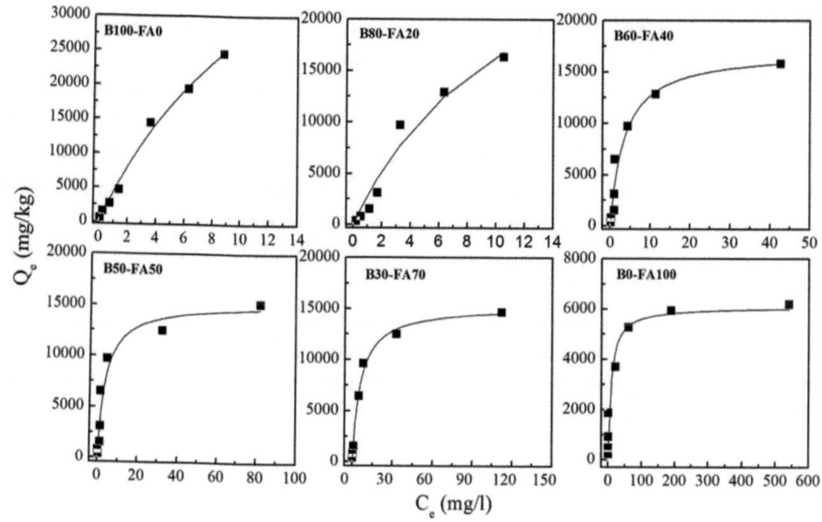

Fig. 6 Langmuir sorption isotherm of Pb^{+2} for different fly ash–expansive soil mixes

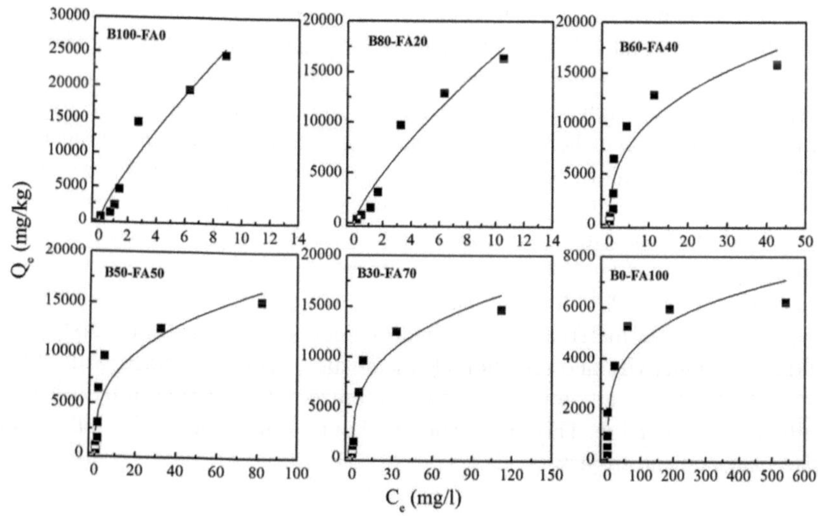

Fig. 7 Freundlich sorption isotherm of Pb^{+2} for different fly ash–expansive soil mixes

increase in the initial Pb^{+2} concentration leads to a reduction in percentage removal efficiency of Pb^{+2} ion, showing the process to be highly dependent on the initial concentration.

On this basis of the hydraulic conductivity criteria, bentonite mixed with 70% fly ash satisfies the requirement of $<10^{-7}$ cm/s for the landfill liner. From the retention test, it can be noted that bentonite mixed with 70% fly ash shows approximately

Table 5 Sorption isotherms parameter for Pb^{+2} as model contaminant

Mix	Langmuir		Freundlich		
	K_L (L/mg)	R^2	K_F	N	R^2
B100-FA0	0.080	0.9481	4427	0.804	0.9357
B80-FA20	0.085	0.9626	2914	0.759	0.9428
B60-FA40	0.291	0.9888	4336	0.369	0.8696
B50-FA50	0.266	0.9821	3667	0.336	0.8407
B30-FA70	0.172	0.9932	3694	0.312	0.9015
B0-FA100	0.114	0.9716	1465	0.251	0.8798

90% removal of Pb^{+2} up to the concentration range of 1000 mg/L of stock solution, which signify the optimum use of fly ash for the contaminant retention in landfill liners.

4 Conclusions

This study deals with laboratory investigation and evaluation of bentonite–fly ash mixes for its application in waste containment system as compacted liner material using hydraulic conductivity and retention as qualifying parameter. The study highlights that fly ash alone should not be used as landfill liner material. The objective of the research work is to maximize the utilization of class F fly ash (a waste by-product from thermal power plant) in waste containment projects. The study indicates that all the bentonite–fly ash mixes compacted at optimum moisture content satisfy the regulatory requirements of landfill liners. Based on the study, it is proposed that bentonite can be mixed up to 70% dry weight of class F fly ash for its use as compacted landfill liner. It was observed that bentonite mixed with 70% fly ash remove 90% Pb^{+2} for the concentration range of 1000 mg/L, which may be thought very good material for the landfill liner application. For satisfying the hydraulic conductivity and retention characteristics, bentonite mixed with 70% fly ash can be floated as an optimal mix for maximize the utilization of fly ash for the landfill liner application. This observation will quite handy and useful in the field of waste management projects.

References

1. Chalermyanont, T., Arrykul, S.: Compacted sand—bentonite mixtures for hydraulic containment liners. Songklanakarin J. Sci. Technol. **27** (2005)
2. Heineck, K.S., Lemos, R.G., Flores, J.A., Consoli, N.C.: Influence of particle morphology on hydraulic behavior of coal fly ash and sand. Geotech. Geol. Eng. **28**, 325–335 (2010)
3. Gleason, M.H., David, E.D., Gerald, R.E.: Calcium and sodium bentonite for hydraulic containment applications. J. Geotech. Geoenviron. Eng. **123**(5), 438–445 (1997)

4. Ogata, N., Komine, H.: Permeability changes of bentonite—sand mixture before and after swelling. SMIRT **12**, 357–362 (1993)
5. Cowland, J.W., Leung, B.N.: A field trial of a bentonite landfill liner. Waste Manage. Res. **9**, 277–291 (1991)
6. Abele, W.V.: The influence of bentonite on the permeability of sandy silts. J. Nuc. Chem. Waste Manag. **6**, 8l–88 (1986)
7. Nhan, C.T., Graydon, J.W., Kirk, D.W.: Utilizing coal fly ash as landfill barrier material. Waste Manage. **16**(7), 587–595 (1996)
8. ASTM D 698.: Test method for laboratory compaction characteristics of soils using standard effort. West Conshohocken, PA (2000)
9. ASTM D 422.: Standard Test Method for Particle Size Analysis of Soils. West Conshohocken, PA, (1998)
10. ASTM D 4318.: Standard test method for liquid limit, plastic limit, and plasticity index of soils. West Conshohocken, PA (2000)
11. ASTM D854.: Standard test method for specific gravity of soils. West Conshohocken, PA (2000)
12. Cerato, A.B., Lutenegger, A.J.: Determination of surface area of fine-grained soils by the ethylene glycol monoethyl ether (EGME) method. Geotech. Test. J. **25**(3), 314–320 (2002)
13. Horneck, D.A., Hart, J.M., Topperand, K., Koespell, B.: Methods of soil analysis used in the soil testing laboratory at Oregon State University. Ag. Expt. Station SM 89, Oregon State University, Corvallis (1989)
14. ASTM C 618.: Standard specification for fly ash and raw or calcined natural pozzolan for use as a mineral admixture in Portland cement concrete. West Conshohocken, PA (2008)
15. Dixon, D.A., Graham, J., Gray, M.N.: Hydraulic conductivity of clays in confined tests under low hydraulic gradients. Can. Geotech. J. **36**, 815–825 (1999)
16. ASTM D 4646–03.: Standard Test Method for 24-h Batch-Type Measurement of Contaminant Sorption by Soils and Sediments. West Conshohocken, PA, (2003)
17. Daniel D, Benson C.: Water content–density criteria for compacted soil liners. J. Geotech. Eng. **116**(12) (1990)
18. Benson, C.H., Trast, J.M.: Hydraulic conductivity of thirteen compacted clays. Journal of clay mineralogy **43**(6), 669–681 (1995)
19. Daniel, D.E.: Predicting hydraulic conductivity of clay liners. J. Geotech. Eng. **116**(12) (1984)
20. McLaren, R.G., Swift, R.S., Williams, J.G.: The adsorption of copper by soil materials at low equilibrium solution concentrations. J. Soil Sci. **32**, 247–256 (1981)
21. Petruzzelli, G., Guidi, G., Lubrano, L.: Ionic strength effect on heavy metal adsorption by soil. Commun. Soil Sci. Plant Anal. **16**, 971–986 (1985)
22. Jordão, C.P., Alves, N.M., Pereira, J.L., Bellato, C.R.: Adsorção de íons Cu^{+2} em Latossolo Vermelho-Amarelo húmico. Quím. Nova. **23**, 5–11 (2000)
23. Sharma, H.D., Reddy, K.R.: Geoenvironmental engineering: site remediation, waste containment, and emerging waste management technologies. Wiley, New York (2004)

Effectiveness Study of Dolochar as a Filter Media in Wastewater Filtration

Purushottam Das Vairagi and Rajesh Roshan Dash

Abstract The study was carried out to evaluate the feasibility of slow sand filter by using dolochar, a solid waste generated from sponge iron industry, as a filter media and compared with the sand media filter. Effectiveness of slow sand filters assessed by two laboratory-scale filter columns (10 cm diameter and 0.40 m media depth): one consisting sand media and other with dolochar. The effective size of both sand and dolochar media used were 0.42 mm to study the performance at filtration rate of 0.06 m/h. For 160 h of run period, the dolochar media showed higher removal efficiencies, especially chemical oxygen demand (COD) removal up to 89% compared to sand media filter of 69%, turbidity removal of 96% compared to 90% of sand media filter, and phosphate removal also observed more in case of dolochar media filter—especially, the phosphate removal from the dolochar media filter was up to 92.71% compared to sand filter of 85.24%. Since the dolochar media filter exhibited to possess higher removal efficiencies, it supports the utility of dolochar as a low cost and highly efficient filter media for slow sand filtration.

Keywords Wastewater filtration · Effective size · Dolochar · Chemical oxygen demand (COD)

1 Introduction

Slow sand filters (SSFs) are probably the most effective, simplest and least expensive water treatment process for developing countries. They are very efficient in removal of bacteria, organics, cysts, ova viruses, etc. Today, the need of

---

P. D. Vairagi (✉) · R. R. Dash
School of Infrastructure, Indian Institute of Technology Bhubaneswar, Bhubaneswar 752050, India
e-mail: pv11@iitbbs.ac.in

A. S. Kalamdhad et al. (eds.), *Advances in Waste Management*,
https://doi.org/10.1007/978-981-13-0215-2_35

sustainable techniques increased due to the more water demand and resulting proper use of wastewater, and also to treat this making use of simple techniques like using the suitable filter can provide a good option to improve the efficiency of filter. Using waste material as a filter media is advantageous as it reduces environmental impact and provides the beneficial use of waste material; same is tried in this study where use of dolochar, a waste material, was used as a media in filter and aimed to check its suitability. Dolochar produced as a by-product of direct reduction of iron (DRI) process for the production of sponge iron. Development of suitable processes and techniques for utilisation of the dolochar is important for bulk utilisation, value addition and/or moving towards zero waste technology [1]. For the production of 100 tonnes of sponge iron, the amount of iron ore and coal requirement is 154 and 120 tonnes, respectively, in which case the solid waste generated is around 45 tonnes and out of which 25 tonnes are char and widely known as dolochar [2]. Till date, the application of various adept solid materials such as sand, involving synthetic materials, natural materials, agricultural by-products, industrial by-products and other material as filter media like zeolite, etc., had proved either expensive or having low removal efficiency. Thus, this study aimed to determine the efficiency of removal of impurities from the wastewater by using low-cost industrial waste material dolochar as a filter media and make comparison with the conventional filter media sand.

2 Materials and Methods

2.1 Set-up of the Experimental Column

The principal apparatus employed in this investigation consists of two 100 mm (10 cm) internal diameter vertical Perspex tube of 120 cm (1.2 m) in height as shown in Fig. 1. First filter is packed with single media of sand to a height of 20 cm. The column was filled with locally available natural sand. The quality of the treated water and maintenance requirements for the system also depend on the selected variables like sand size, flow rates and sand bed depth. The sand used is characterised by its effective size (ES or d10) and uniformity coefficient (UC or d60/d10). The recommendations for ES vary between 0.15 and 0.40 mm [3]. 1975 g of sand was sieved mechanically on a set of sieves, and sieve fractions were weighed. From sieve analysis, the effective size and uniformity coefficient of sand obtained were of 0.42 and 3.25 mm, respectively. 1100 g of dolochar sieved from sieve analysis effective grain size of 0.42 mm and uniformity coefficient of 3.09 were obtained. To evaluate the effectiveness, it is decided to keep the filtration rate lower than the optimum so that for the more retention time there are more chances of removal of the impurities. Before filling into the column, the sand and dolochar were thoroughly washed with tap water to remove clay and other mineral contaminants from the sand particles and dried in an oven at 120 °C overnight.

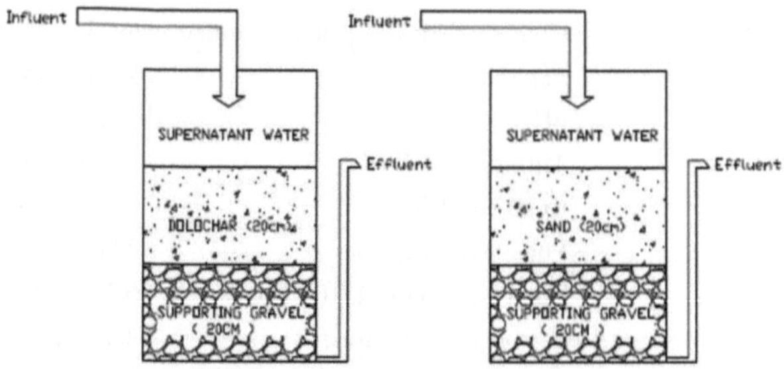

Fig. 1 Schematic diagram of dolochar and sand filter

The lower section of the columns contains 12 cm depth of coarse gravel (4.75–10 mm), overlaid by 8 cm layer of 2–4.75 mm fine gravel. For sand and dolochar media filters, 10 cm of 1.18–4.75 mm coarse media was overlaid with 10 cm of 0.15–1.18 mm fine media at the top. As the depth of the filter media increases, more impurity removal can be achieved [4, 5]; this study mainly aims to evaluate the efficiency of the filter media, so the depth of filter media was kept small compared to the standard size [6].

The depth of the sand bed was so arranged that its top surface was at a level with the flanged joint in the filter column. A small amount of gravel, 4.75–10 mm in size was provided on the top of the filter beds in order to homogeneous supply of wastewater to filter beds. A constant inflow of primary settled effluent at hydraulic loading rate of 0.06 m/h was maintained by a peristaltic pump.

2.2 Sample Collection

A primary settled effluent from sewage treatment plant having capacity of 100 KLD at the city of Bhubaneswar was selected for regular monitoring over a period of 3 months (January 2016–March 2016). Grab samples of wastewater were obtained at the outlet of primary settled effluent. Effluent samples were collected from the effluent pipe provided to each filter column.

2.3 Sample Analysis

Samples were collected from the outlet of filter column at the different intervals for various parameters. Samples were assayed for microbiological (Total Bacteria and *Escherichia coli*) and physico-chemical parameters (COD, Phosphate and turbidity) as per the standard methods [7].

Table 1 Characteristics of primary settled wastewater effluent

Characteristics of primary settled effluent	
Parameters	Values (average)
pH	6.82–7.96 (7.24)
Turbidity (NTU)	17–96 (45.2)
COD (mg/L)	180.8–446.8 (302.04)
Phosphate (mg/L)	27.85–4.35 (12.31)
Total bacteria (cfu/100 mL)	6.2×10^7–8.7×10^7
E. coli (cfu/100 mL)	2.2×10^6–3.7×10^6

3 Results and Discussion

The range and average values of the physico-chemical characteristics of the primary settled wastewater effluent used as the influent for slow sand filter are reported in Table 1. Average values and standard deviation of laboratory measurements, regarding the main parameters of sand and dolochar filter, are given in Table 2.

3.1 Turbidity Removal

Turbidity is one of the most important parameters for monitoring the performance of a filter. It is believed that turbidity serves as a carrier for nutrients and pathogens which can result in biological activity. Figure 2 shows turbidity removal was observed to be significantly high and stable since the start of the experiment in case of dolochar filter. The turbidity level in the filter influent ranged from 17 to 96 NTU (avg. 45.2 NTU), whereas at a particular operational period, the effluent turbidity levels lie in the range of 3–1 NTU (avg. 1.4 NTU) and for sand media filter it is 6–1 NTU (avg. 3.2 NTU) during the entire study. The average turbidity removals were 96.25 and 90.45% from dolochar filter and sand filter, respectively.

Table 2 Average values and standard deviation of laboratory measurements, regarding the main parameters for the operation of filter beds

Laboratory measurements	Average value	Standard deviation
Turbidity at the inlet, NTU	45.2	22.61
Turbidity at the outlet of sand filter, NTU	3.2	1.60
Turbidity at the outlet of dolochar filter, NTU	1.4	0.66
COD at the inlet, (mg/L)	302.04	76.96
COD at the outlet of sand filter, (mg/L)	83.29	18.38
COD at the outlet of dolochar filter, (mg/L)	31.37	9.19
Phosphate at the inlet, (mg/L)	12.31	7.15
Phosphate at the outlet of sand filter, (mg/L)	3.55	0.70
Phosphate at the outlet of dolochar filter, (mg/L)	1.84	0.59

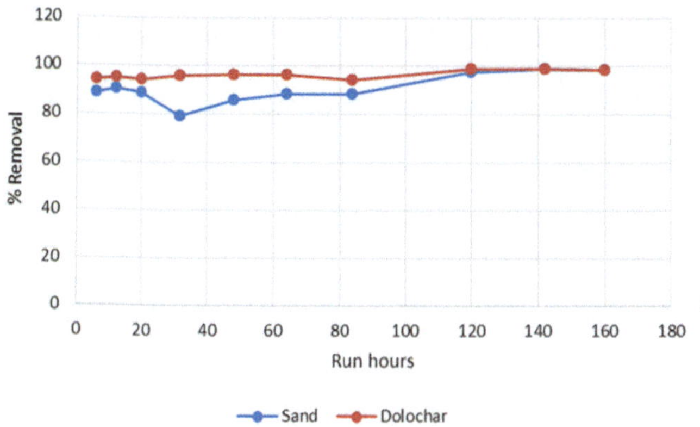

Fig. 2 Results of turbidity removal

This higher removal for dolochar filter may be due to the adsorption on the dolochar media, which is a carbonaceous material having active surface and provides more surface area for adsorption [8]. Although for the intermittent flow, the development of the biological layer was low, resulting in lower removal of the impurities. The development of this layer makes purifying bacteria well established and plays an important part in the treatment process, which makes further possibility of turbidity removal.

3.2 *Chemical Oxygen Demand (COD) Removal*

COD removal profile of both filter columns is shown in Fig. 3. It was observed that the COD removal efficiency of the dolochar filter was more than the sand filter. The average COD removal from dolochar filter for the run period of 160 h achieved about 88.96%, which is far more than the sand media filter of 69.28%. Although the results of the experiments are lower than the results obtained in studies having nearly similar operating conditions [9, 10], but the comparison will not be appropriate because of intermittent flow. Due to off period the removal efficiency also decreased as there was widen of the pore spaces of surface layer resulting easy flow of water.

3.3 *Phosphate Removal*

Phosphate removal from dolochar filter obtained was higher compared to sand media filter, which can be seen in Fig. 4. It has been confirmed that dolochar is highly effective media for the phosphate removal from the sample of wastewater, due to its high adsorbent capacity [11]. The influent phosphate concentration was

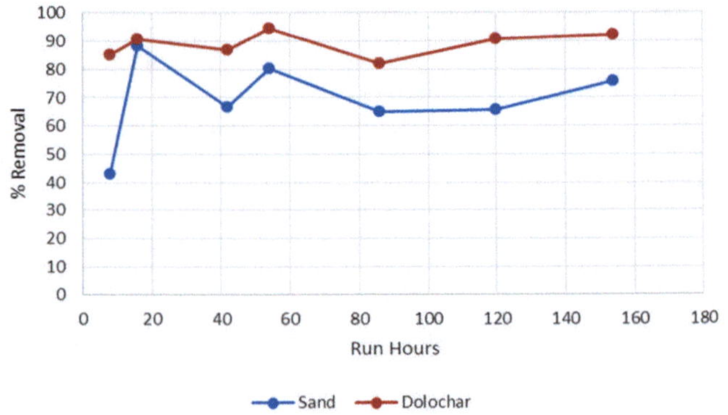

Fig. 3 Results of COD removal

19.57–41.46 mg/L. The average effluent concentrations of sand and dolochar filter were 3.55–1.84 mg/L. Previous study [11] suggest that, phosphate ions react with iron and aluminium oxides by ligand exchange, forms inner-sphere complexes and the presence of magnesium and calcium ion facilitates phosphate removal via precipitation. Phosphate is one of the sources of nutrients in water bodies, which causes eutrophication in water bodies. Other than eutrophication, the increasing loads of phosphates in water bodies stimulate the activity of a damaging microbe known as Pfiesteria and speed up the production of microcystin, a toxin that poisons aquatic animals and can cause hepatocellular carcinoma in humans [11, 12]. This study suggests that dolochar media filter can be effective in removing the phosphate from wastewater and lessen the potency of eutrophication. Also, it has been showed that the dolochar media favours in regard to the nitrate removal [12] which further proves the dolochar can be effective media for filtration of the wastewater where nutrient removal is desired.

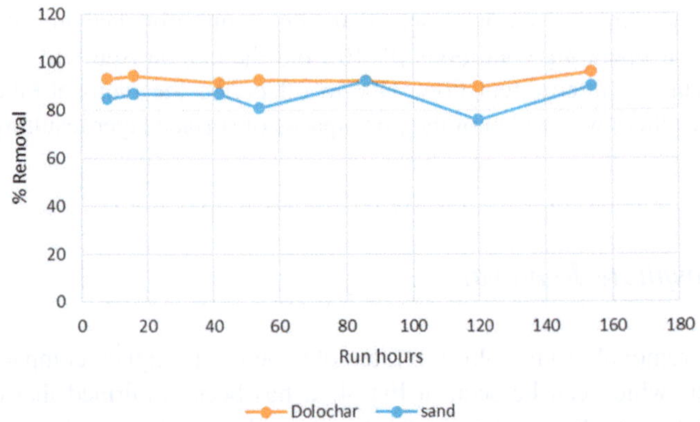

Fig. 4 Results of phosphate removal

3.4 Microbial Removal

Some preliminary and qualitative results in regard to check the microbial removal efficiency were carried out for the total bacterial and *E. coli*; the influent total bacteria and *E. coli* count were of 8.7 × 107 cfu/100 mL and 3.7 × 106 cfu/100 mL, respectively. The effluent concentration of total bacteria from sand filter and dolochar filter was of 5.3 × 106 cfu/100 mL and 6.56 × 106 cfu/100 mL, respectively. The effluent concentration of *E. coli* from sand filter and dolochar filter was of 1.3 × 105 cfu/100 mL and 1.56 × 105 cfu/100 mL, respectively. The removal efficiency of both the filters observed was nearly same. The dolochar filter removing microbial impurities favours by the study of [13], which showed that the bacterial removal can be due to the adsorption on the material, mainly on the carbon materials, adding to the straining action. [8, 14] explained adsorption also plays important role in the bacterial removal. From the experiment, it was determined that the pHzpc was 8.56 of dolochar in present study, due to most of the bacteria having negative charge and solution pH lower than the pHzpc; dolochar develops positive charges on its surface which makes possible the cause of bacterial removal from dolochar filter.

4 Conclusion

In this study, two filter media, sand and other cheap media, a sponge iron industry by-product, dolochar, were used in two laboratory-scale filters. Filters were operated and their removal efficiencies were compared. The pH, turbidity, COD, phosphate, *E. coli* and total bacteria removal in these filter set-ups were measured. The higher removal efficiencies were related to dolochar filter, other than the ambiguous results on the removal of the microbial impurities. It can be concluded that the dolochar as a filter media can be a very good option for removal of impurities from wastewater in communities that have some problems with these parameters. For the run period of 160 h, the average removal of turbidity was above 96% with the dolochar filter and with sand filter it was 90%, and other impurity removals were also higher in case of dolochar filter. For the filter run of 160 h, the head loss developed was low due to the intermittent flow to both the filters. Although more study is required in this aspect to prove the effectiveness of the dolochar, tests such as X-Ray detection (XRD) and scanning electron microscopy (SEM) can explain the more removal efficiency in case of dolochar filter. The results of this study support the utility of dolochar as a low cost and highly efficient filter media.

Acknowledgements The authors are thankful to the School of Infrastructure, Indian Institute of Technology Bhubaneswar, India, for providing facilities to carry out the research work in the concerned area.

References

1. Dwari, R.K., Rao, D.S., Swar, A.K., Reddy, P.S.R., Mishra, B.K.: Characterization of Filter having dolochar wastes generated by the sponge iron industry. Int. J. Miner. Metall. Mater. **19** (11), 992 (2012)
2. Panda, H., Sahoo, S., Tiadi, N., Dash, R.R., Mohanty, C.R.: Sorption of hexavalent chromium from synthetic waste water using Dolochar. Recent Res. Sci. Technol. **6**(1), 106–110 (2014)
3. Ellis, K.V., Wood, W.E.: Slow sand filtration. Crit. Rev. Environ. Control **15**(4), 315–354 (1985)
4. Nancy, A.B., Josephine, M., Lizzy, M.A.: Slow sand filtration of secondary sewage, effluent: effect of sand bed depth on filter performance. Int. J. Innov. Research Science, Engineering and Technology **3**(8), 2319–8753 (2014)
5. Langenbach, K., Kuschk, P., Horn, H., Kastner, M.: Slow sand filtration of secondary clarifier effluent for wastewater reuse. Environ. Sci. Technol. **43**, 5896–5901 (2009)
6. Ellis, K.V.: Slow sand filtration as technique for the tertiary treatment of municipal sewages. Water Res. **21**(4), 403–410 (1987)
7. Association, American Public Health: APHA. Standard methods for the examination of water and wastewater. APHA, Washington, D.C. (1998)
8. Hatt, J.W., Germain, E., Judd, S.J.: Granular activated carbon for removal of organic matter and turbidity from secondary wastewater. Water Sci. Technol. **67**(4), 846–853 (2013)
9. Tyagi, V.K., Khan, A.A., Kazmi, A.A., Mehrotra, I., Chopra, I.: Slow sand filtration of UASB reactor effluent: a promising post treatment technique. Desalination **249**, 571–576 (2009)
10. Farooq, S., Al-Yousef, A.K.: Slow sand filtration of secondary effluent. J. Environ. Eng. Div. **119**(4), 615–630 (1993)
11. Rout, P., Bhunia, P., Dash, R.R.: Effective utilization of a sponge iron industry by-product for phosphate removal from aqueous solution: a statistical and kinetic modelling approach. J. Taiwan Inst. Chem. Eng. **46**, 98–108 (2015)
12. Rout, P.R., Dash, R. R., Bhunia, P.: Nutrient removal from binary aqueous phase by dolochar: highlighting optimization, single and binary adsorption isotherms and nutrient release. Process Safety Environ. Protect. http://dx.doi.org/10.1016/j.psep.2016.01.001 (2016)
13. Busscher, H.J., Dijkstra, J.B., Engels, E., Langworthy, D.E., Collias, D.E., Bjorkquist, D.W., Mitchell, M.D., Vander, H.C.: Removal of two waterborne pathogenic bacterial strains by activated carbon particles prior to and after charge modification. Environ. Sci. Technol. **40**, 6799–6804 (2006)
14. Stevik, T.K., Kari, A., Ausland, G., Hanssen, J.F.: Retention and removal of pathogenic bacteria in wastewater percolating through porous media: a review. Water Res. **38**, 1355–1367 (2004)

Assessment of Bio-medical Waste Management in Different Hospitals in Aligarh City

Izhar Alam, Gulfaraz Alam, Sohail Ayub and A. A. Siddiqui

Abstract This study reveals that most of the hospitals are lacking in bio-medical waste management system. These hospitals do not properly segregate the waste dispose of the bio-medical waste without any treatment. Hospital should have vigilance cells especially to monitor the handling and treatment of the waste but not working well. Waste should be segregated and treated as a bio-medical waste management and with handling rules, 1998. A series of programs should be organized for general public awareness and to make the employee of the healthcare facilities aware in the handling and management of bio-medical waste. It is observed that most of the hospitals have no management system in place. The hospital does not give priority to the proper disposal of the waste. Several methods of disposal of medical waste are proper dumping, burning, burial, selling, reuse, and removal by municipal trucks without any treatment. On the basis of detailed study and data collection from working hospitals, it can be assessed that total quantity of bio-medical waste generation in Jawaharlal Nehru Medical College and other hospitals in Aligarh is about 750–850 kg/day.

Keywords Bio-medical waste · Hospitals · Segregated · Treatment
Waste management

1 Introduction

The establishment of human health care is a basic requirement of each and every civilized society. Food, medicines, chemicals, equipment, and instruments are used in hospitals while treating out-patients and patients admitted into different wards of hospital, producing a variety of wastes and bio-medical wastes. The wastes generated from the treatment of patients suffering from infectious diseases may spread

---

I. Alam (✉) · G. Alam · S. Ayub · A. A. Siddiqui
Department of Civil Engineering, ZHCET, A.M.U., Aligarh, Uttar Pradesh, India
e-mail: izhar.alam090@gmail.com

A. S. Kalamdhad et al. (eds.), *Advances in Waste Management*,
https://doi.org/10.1007/978-981-13-0215-2_36

infection either directly or indirectly through the environment. Globally, this issue has been seriously considered and appropriate waste management systems are being developed and installed. In India, this problem has been recognized by the concerned agencies, and government regulations have been framed to systematize implementation. Accordingly, authorities at healthcare units (HCUs) are trying to install systems to meet legal requirements. A number of difficulties are being faced at many places for implementation of these requirements in practice. In order to help resolve these problems, further, an improvement plan to develop appropriate waste management systems is suggested [1]. Bio-medical waste has been considered as one of the major health and environmental management issues over the many decades. Poor waste management, lack of handling knowledge of bio-medical waste, and unscientific disposal of various healthcare wastes pose serious problems and direct and indirect public health threats to healthcare personnel, nurses, technicians, workers, hospital visitors, patients, surrounding communities, and the environment as well [2]. Though medical care is vital for our life, health and well-being, the bio-medical waste generation from the medical activities is hazardous, toxic, and even lethal owing to their high potential for different types of diseases transmission. Bio-medical waste has recently emerged as an issue of major concern not only for hospitals but also for nursing homes and surrounding to the environment, according to the law conforcement agencies, media and general public. Bio-medical waste is generated during the diagnosis, treatment, immunization of human being and animals and also from research activities—production or testing of biological methods or using human body parts, blood samples, etc. It may include waste like sharp materials (injections, blades), soiled wastes, disposables, anatomical wastes, cultures, chemical waste, discarded medicines. The illegal and unethical reuse of bio-medical waste can be extremely dangerous and even fatal. Diseases such as cholera, plague, TB, hepatitis (especially HVB), AIDS (HIV), diphtheria in either epidemic or even endemic form pose grave public health risks (Fig. 1).

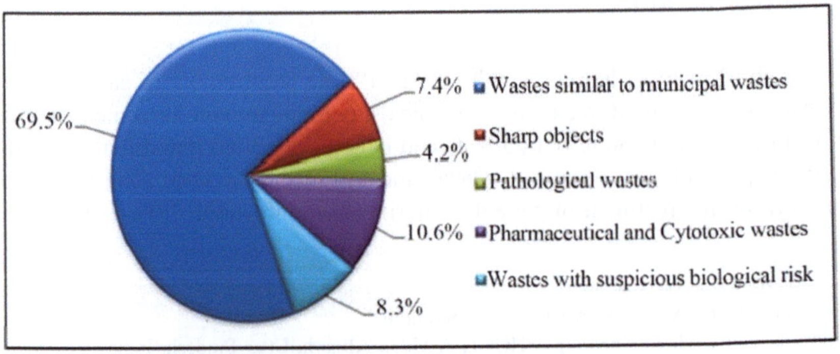

Fig. 1 Percentage of composition of bio-medical waste [4]

The importance of management in the healthcare sector is that positive correlations have been found between clinical and economic performance. Through the analysis of 37 studies, they found that the performance of healthcare systems and organizations seems to be correlated with management practices, leadership, manager characteristics, and cultural attributes that are associated with managerial values and approaches [3]. Medical waste management is of major concern due to the potentially high risks to both human health and the environment caused by inadequate waste management practices [4]. The assessment of Bio-Medical waste has a big influence on infrastructure development and use undertaken and executed without fully recognizing the complexity, diversity, and social and technological evolution of the system almost inevitably squander economic, environmental, social, and cultural resources [5]. Bio-medical waste management (BMWM) has become crucial issue as it poses new health risks and harm to the environment. Bio-medical waste (BMW), if not handled in a proper way, is a potent source of diseases like HIV, hepatitis B and C, and other bacterial diseases causing serious threat to human health. This aspect and step-by-step approach is to establish BMWM system in tertiary-level hospital, and planning to make this center a model to provide hands-on training to all state medical colleges and other hospitals is to help them establish BMWM system [6]. A single dumping practice is not a solution to the problems of waste management of hospital. In most of the cases, a number of practices including landfills, burning, autoclaving, and recycling used in combination are used. Every practice has its own flaws and strength [7].

2 Materials and Methods

Essentially bio-medical wastes are those solid wastes, which have fluids or liquid wastes (including its in-between products) which are generated through diagnosis processes, monitoring, and preventive, curative, or palliative activities, in vast field of veterinary and human medicine, in research pertaining thereto, or in the production or testing of biological and animal wastes from slaughterhouses or any others like establishments.

In Jawaharlal Nehru Medical College and other hospitals, 80–85% of bio-medical wastes are non-hazardous wastes; about 10% are infectious, and 5% are non-hazardous waste but infectious. It is very hard to identify and segregate each and every article of bio-medical waste from the solid waste. Therefore, the bio-medical waste is divided into many groups, which are amenable to specific treatments of different disposal methods. According to the BMW management and handling rules, BMW has been grouped into the following categories (Fig. 2).

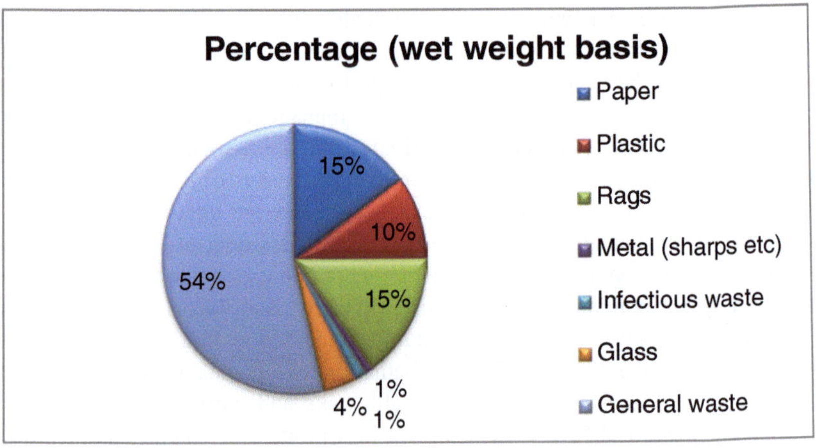

Fig. 2 Average composition of bio-medical waste in JNMC and other hospitals

- **Human Anatomical Waste**: Different types of body parts, human tissues, organs, bones, etc.
- **Animal Waste**: Animal tissues, different types of body parts, carcasses, bleeding parts, fluid, blood, and experimental animals used in research purposes, waste generated by veterinaries hospitals, colleges, discharge from hospitals, animal houses, etc.
- **Microbiology and Biotechnology Waste**: Waste from different laboratories of various cultures, stocks of specimens of microorganisms, live or attenuated vaccines, various human parts and animal cell culture used in research, and infectious agent from research laboratories and industrial laboratories, waste from production of biological, toxins, etc.
- **Waste Sharps**: Needles, syringes, scalpels, type of blades, glassware, etc. This includes both used and unused sharps wastes.
- **Discarded Medicines and Cytotoxic Drugs**: Waste comprising the out-of-dated, contaminated, and discarded medicines and many unused drugs.
- **Soiled Waste**: It contains bloods and body fluids including cotton items, dressing rolls, soiled plaster casts, linens, beddings, other material contaminated with bloods.
- **Solid Waste**: Waste generated from disposable items other than the wastes sharp such as tubings, catheters, intravenous sets, bottles, hand gloves.
- **Liquid Wastes**: Waste generated from testing laboratories and washings, cleaning, housekeeping, and disinfecting activities.
- **Incinerator Waste**: Almost ash from incinerator.
- **Chemical Waste**: Chemicals used in biological process, disinfection, insecticides, laboratory testing, etc.

2.1 Segregation and Storage

Segregation and storage both are important aspects of bio-medical waste management process, where waste are segregated as infectious and non-infectious, thus minimizing the risk of staff and public as well as resource used for treatment purpose.

For the purpose of segregation, the bio-medical waste can be broadly divided into the following categories.

(1) General or non-infectious/non-hazardous medical wastes, which are not contaminated with blood of living being, other body fluids, or infectious agents.
(2) Infectious/hazardous medical waste.

The container for segregated waste should be clearly identifiable. The best system is to use colored plastic bags/containers. The color coding, and type of containers, shall be allowed as per schedule II of Bio-Medical Waste Management Rules. Sharp materials need special prior while segregating and storing because needle and other sharp materials can act as a reservoir of pathogens. It may survive for a long time because of the presence of soiled waste.

2.2 Transportation

The medical wastes, except the general and non-hazardous wastes, are transported in container which is completely closed and stored in secured place near the treatment unit. Transported and segregated bio-medical waste inside the medical premises go for further treatment and disposal facility.

2.3 Waste Minimization

Waste minimization practice usually benefits the waste producer. It helps in reducing the cost for both the purchaser of goods and cost of waste treatment, and disposal is reduced and the liabilities associated with the disposal of hazardous waste are less. All health facilities for employees have a role to play in this process and should be trained in waste minimization and the management of hazardous materials. This process is most important particularly for the staff of department.

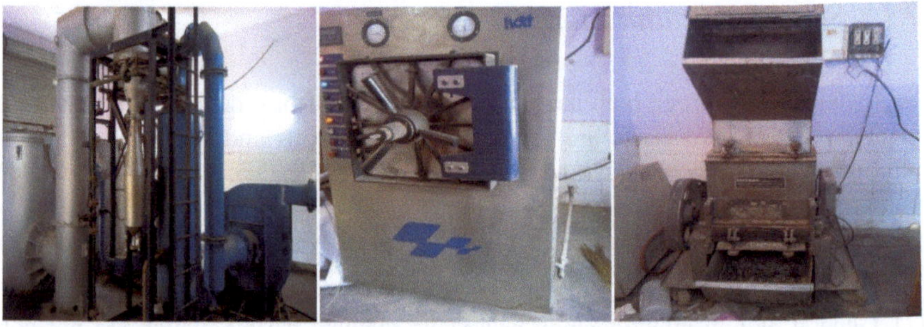

Incinerator Autoclave Shredder

Fig. 3 Waste treatment systems

2.4 *Available Waste Treatment System*

For treatment of bio-medical wastes, various methods, techniques, and processes are designed to change the biological characters and composition of any bio-medical waste as to reduce or eliminate its potential for causing disease (Fig. 3).

In Jawaharlal Nehru Medical College (JMNC) these waste treatment equipments are available:

- Incinerator (capacity of incinerator 35–40 kg/h);
- Autoclave;
- Shredder.

3 Result and Discussion

3.1 *Jawaharlal Nehru Medical College and Hospital (JNMC)*

In order to develop different types of proper waste management strategies in the hospital, it is important to have accurate information of the bio-medical waste generation rate in hospital. In JNMC, it is about 1315 beds for patients and is divided into different departments. Results from the surgery department are that the average generation rate of bio-medical waste generation at the Jawaharlal Nehru Medical College and hospital is between 0.45 and 0.7 kg/bed/day and infectious waste is 0.03–0.04 kg/bed/day.

The use of disposable instruments in hospitals and packaging materials rather than the use of reusable items in the healthcare centers has made India the largest producer of waste generation. A sound understanding of the contents of the waste stream is helpful in the development of realistic waste reduction and recycling programs. According to the survey, it was found that around 82.5% of bio-medical

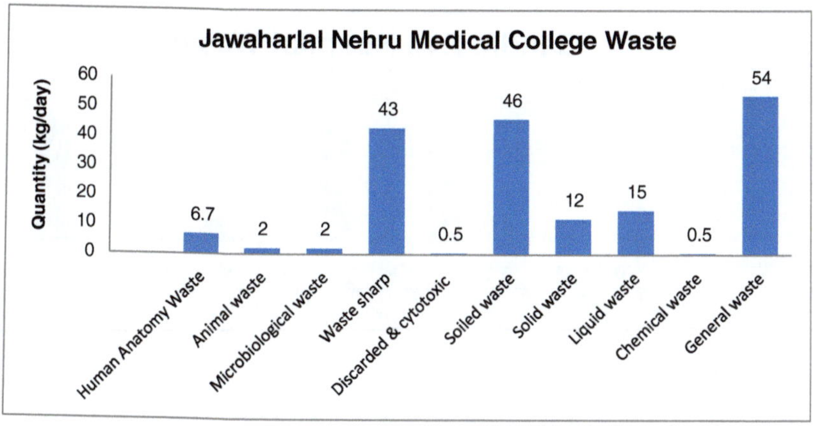

Fig. 4 Jawaharlal Nehru Medical College waste

wastes were general wastes similar in properties to municipal wastes. The remaining 17.5% was infectious and hazardous wastes. Jawaharlal Nehru Medical College and Hospital produced high quantity of waste generation, and this may be because of more number of patients as it is less expensive and has got the opportunity to have multiple departments for treatment for mostly all types of diseases (Fig. 4).

3.2 *Gandhi Eye Hospital*

Gandhi Eye Hospital is an eye special hospital. As shown in the figure, a large amount of soiled waste, waste sharp, and general waste is produced. Capacity of hospital is 750 beds, and waste produced is 0.127 kg/bed/day (Fig. 5).

3.3 *University Health Centre*

University Health Centre serves only university staff and student of Aligarh Muslim University. About 50 beds are available in this dispensary. Emergency facility is not available at this centre, and serious patients are referred to the JNMC hospital. So the bio-medical waste generation rate is small. There is no human anatomy waste and no animal waste, and infectious waste is also negligible. The main waste in the University Health Centre is general waste, soiled waste, some of discarded medicine and waste sharp (Fig. 6).

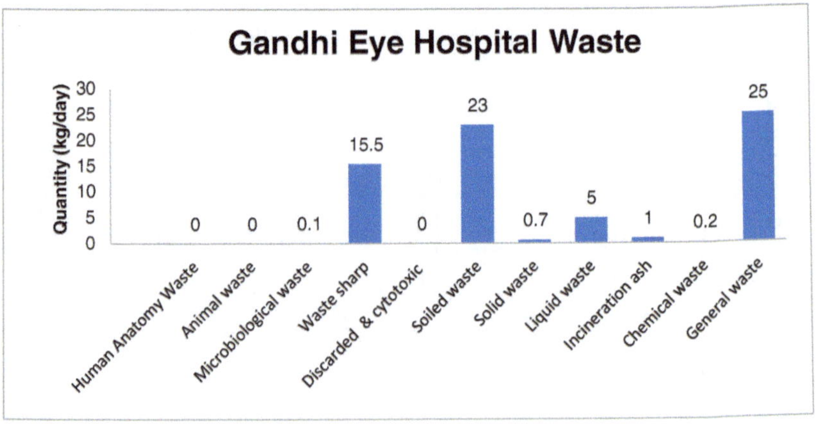

Fig. 5 Gandhi Eye Hospital waste

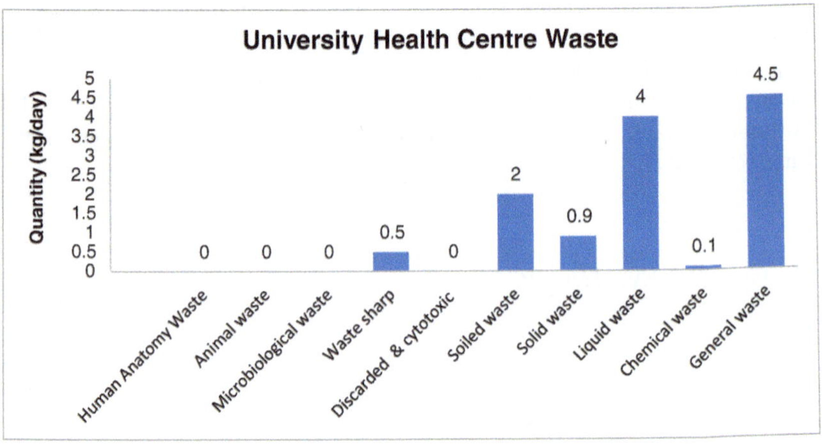

Fig. 6 University Health Centre waste

3.4 Sir Ziauddin Dental College AMU

Sir Ziauddin Dental College serving as an Outpatient Department (OPD) contains mainly liquid waste, general waste, and soiled waste. As shown in figure, there is not any human anatomy waste, animal waste, and microbiological waste (Fig. 7).

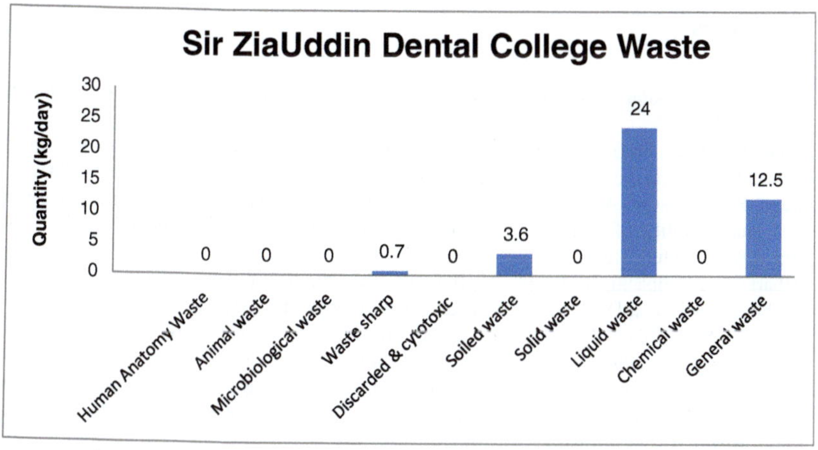

Fig. 7 Sir Ziauddin Dental College waste

4 Conclusion

The study reveals that most hospitals are lacking waste management system but some hospitals have proper management of waste. Most of the hospitals do not segregate and treat the bio-medical waste before disposal. Waste picker or cleaners sort out this waste and separate items that can be sold again, and this cycling spreads deadly diseases through the reuse of discarded materials. There are no proper storage facilities, and all the waste is dumped in open by some hospitals. Wastes are dumped in place like public garbage disposal bins in the side of roads surrounding hospitals or municipal corporations collection site. There is no hospital where color-coded bag/container is used. Most of the hospitals are not financially sound to install incinerator or different type of treatment facilities. A variety of methods are used for disposals of waste. These include burning of wastes, burial, selling, dumping, reuse, and removal by different methods of municipal corporations. There is no competent authority in any hospital to quantify the total bio-medical waste generation. On the basis of present study, it can be assessed that the total quantity of bio-medical waste generation in Jawaharlal Nehru Medical College and other hospitals in Aligarh is 750–850 kg/day. In most of the hospitals, the lower staffs are the responsible for collection and disposal of these wastes and are ignorant about the hazards associated with the waste.

Acknowledgements It gives me immense pleasure to owe my humble gratefulness to my supervisors and their generosity and interest to help me in all my endeavors. I am also thankful to **Chairman** of the Department of Civil Engineering, Faculty of Engineering and Technology, AMU, Aligarh, for the help and guidance. I am also thankful to all the hospitals which helped in collecting data and authors and publishers who gave us constant source of inspiration and support. I would also like to extend my thanks to all for helping me throughout the work.

References

1. Patil, A.D., Shekdar, A.V.: Health-care waste management in India. J. Env. Manag. **63**, 211–220 (2001)
2. Tamplin, S.A., Davidson, D., Powis, B., O'Leary, Z.: Issues and options for the safe destruction and disposal of used injection materials. Waste Manag. **25**, 655–665 (2005)
3. Lega, F., Prenestini, A., Spurgeon, P.: Is management essential to improving the performance and sustainability of health care systems and organizations? A systematic review and a roadmap for future studies. Value Health S46–S51 (2013)
4. Mbarki, A., Kabbachi, B., Ezaidi, A., Benssaou, M.: Medical waste management: a case study of the Souss-Massa-Drâa region. Morocco. J. Env. Prot. **4**(9), 914–919 (2013)
5. Michele, S., Daniela, L.: Decision-support tools for municipal infrastructure maintenance management. Procedia Comput. Sci. 36–41 (2011)
6. Kumari, R., Srivastava, K., Wakhlu, A., Singh, A.: Establishing biomedical waste management system in Medical University of India—A successful practical approach. Clin. Epidemiol. Glob. Health 131–136 (2013)
7. Nemathaga, F., Maringa, S., Chimuka, L.: Hospital solid waste management practices in Limpopo Province, South Africa: a case study of two hospitals. Waste Manag. **28**, 1236–1245 (2008)

Extraction of Cu and Zn from High-Grade Printed Circuit Board Scraps by Conventional and Hybrid Bioleaching

Anshu Priya and Subrata Hait

Abstract The objectives of the present study were twofold: physicochemical characterization of PCBs from e-waste, viz. mobile phone, digital video disc (DVD) player, inverter, and subsequent assessment of bioleaching of selected metals from high-grade PCB. Cu was the most abundant metal among base metals, while content of Zn was high in mobile phone PCB. Thus, mobile phone PCB was selected as high-grade material and was subjected to bioleaching using pure culture of *Acidiphilium acidophilum*, with and without lemon juice with citric acid having chelating activity to simulate hybrid and conventional conditions, respectively. Results indicated increased recovery of both the metals in hybrid approach. Under hybrid condition, the maximum concentrations of Cu and Zn leached were 2.419 mg/L and 1.272 mg/L, respectively, while it was 0.424 and 0.132 mg/L in conventional approach at 15 days. These findings highlight potential of hybrid bioleaching in efficient metal recovery from e-waste.

Keywords E-waste · Printed circuit board · Metals · Conventional bioleaching Hybrid bioleaching · *Acidiphilium acidophilum*

1 Introduction

The constantly growing demand for products such as electrical and electronic equipments (EEE) in association with shorter obsolesce age have built up e-waste worldwide. E-waste generation rate in the urban environment globally is determined to be twice to thrice rapid as compared to other waste types [1]. Additionally,

A. Priya · S. Hait (✉)
Department of Civil and Environmental Engineering,
Indian Institute of Technology Patna, Bihta 801103, Bihar, India
e-mail: shait@iitp.ac.in

A. S. Kalamdhad et al. (eds.), *Advances in Waste Management*,
https://doi.org/10.1007/978-981-13-0215-2_37

the trans-boundary movements of e-waste from the developed and industrialized economies to the developed and developing nations like India and China in considerable quantities (50–80%) have amplified the e-waste problem [2, 3]. Among e-waste, printed circuit board (PCB) is the basic unit of EEE, making up 3–6% of the total equipment weight [4–6]. PCBs are composite and heterogeneous with typical content approximately 28–30% metals and 70–72% non-metals covering items such as polymers, wood, plastics and ceramics [6, 7]. The metal content in PCBs includes base metals such as Cu, Zn, Pb, Ni, Fe with contents of 10–20%, 1–5%, 1–3%, 1–4%, respectively, and precious metals, viz. Pt, Au, Ag with content of 0.3–0.4% [3, 8]. The diversity and abundance of metals make PCB a prospective secondary repository of metal. Further, the presence of toxic metals such as As, Cr, Hg, Pb, Ni and non-metals including flame retardants like polychlorinated biphenyls, polybrominated diphenyl ethers in concentrations beyond permissible limit turns PCBs hazardous [6, 9]. Inappropriate and indiscriminate dumping of waste PCBs can cause degradation in human and environmental health along with loss of vital metal resources [3, 10]. Therefore, metals recycling from e-waste assume its vitality not only to avert environmental pollution but also from the aspect of metal recovery for economic development [10, 11]. However, e-waste management is difficult task because the traditional metallurgical processes, both pyrometallurgy and hydrometallurgy, are exhaustive processes in terms of energy and economy along with being linked to severe secondary environmental pollution through emission of toxic gasses like furans, dioxins, release of metal dusts, hazardous chemical lixiviants and generation of toxic wastewater [12–15]. The amalgamation of biotechnology with the hydrometallurgy by incorporation of bioagents such as microbes, enzymes has contributed to the development of environmentally sustainable and economical alternative to the traditional metal recovery process called biometallurgy or biohydrometallurgy for metal recovery and e-waste recycling [16–18]. Microbial leaching or bioleaching technique exploits the phenomenon of microbial solubilization of metals and metallic compounds to the recoverable state [19–22]. The conventional bioleaching process involves role of microbial diversity, majority of which belong to the chemolithotrophic group especially *Thiobacillus* spp., like *Acidithiobacillus thiooxidans and Acidithiobacillus ferrooxidans* which can fix carbon and energy from CO_2 and inorganic compounds for solubilization of metallic compounds [20, 23]. *Acidithiobacillus* spp., a gammaproteobacteria, is extensively exploited for bioleaching studies, while metal leaching ability of alphaproteobacteria like *Acidiphilium* spp. from e-waste is yet to be explored. *Acidiphilium* spp. are mesophilic, Gram-negative rods that are capable of growing under acidic and aerobic environment on mixotrophic nutrition utilizing variety of substrates and consequently oxidizing and bioleaching metals from solid substrate [24]. In the last few decades, the studies related to microbiological leaching have shown to have an essential role in metal solubilization from e-waste; however, major disadvantage linked to the process is its slow rate as compared to chemical leaching [20–22, 25]. Various metal chelating agents have been applied for metal recovery from mineral deposits, but majority of the chemicals have disadvantage of environmental persistence due to their low biodegradability and higher residual effect [26]. However, scarce information is available on use of natural as well as synthetic chelating agent for metal extraction from e-waste. Citric acid, abundantly present in citrus fruits such as lemon, has tetradentate chelating activity [27] which has not been applied in metal dissolution from e-waste. In comparison with other chelating agents such as ethylene diamine disuccinic acid (EDDS), ethylene diamine tetraacetic acid (EDTA), citric acid from lemon juice are biodegradable, absolutely safe, naturally occurring and non-toxic chelator with a good level of complexing capacity with metals [27].

Although bioleaching of metals is efficient, cost-effective and eco-friendly approach, its slow recovery rate is the main constraint in metal recovery. An integrated approach combining biological with chemical method can potentially enhance metal extraction efficiency. In this context, the present research aims at characterization of PCBs from electronic scraps, viz. mobile phone, digital video disc (DVD) player, inverter for their metallic contents and subsequent assessment of conventional bioleaching using pure culture of *A. acidophilum* as well as hybrid bioleaching using combination of lemon juice (natural chelator) along with the bacteria, *A. acidophilum,* in bioleaching of metals, viz. Cu and Zn from high-grade PCB only.

2 Materials and Methods

2.1 E-waste

PCBs, the core constituent of e-waste, from mobile phone, DVD player and inverter (Fig. 1), were acquired from the electrical and electronic shops of Patna, India. The PCBs collected were processed by manual dismantling followed by removal of mounted items such as capacitors, batteries, resistors. For experimental use, PCB scarps were pulverized and divided into particle size fractions of <1 and <600 μm using standard sieves. Comminuted PCBs of <1 mm particle size were subjected to

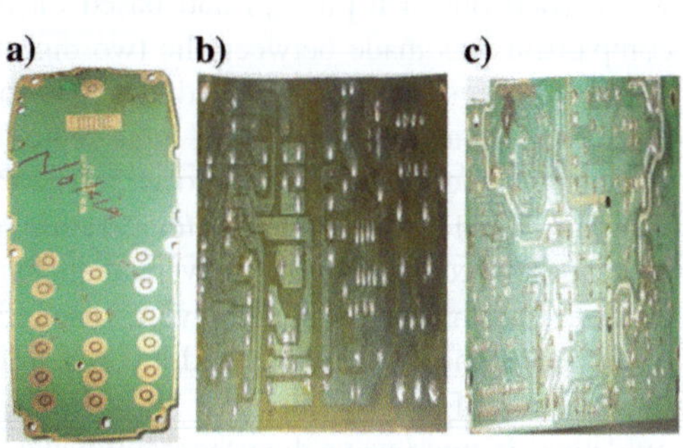

Fig. 1 Photographs of PCB scraps of **a** mobile phone, **b** DVD player, **c** inverter used in the study

physicochemical characterization, while with particle size range <600 μm were used for bioleaching studies. The pulverized PCBs were morphologically determined using scanning electron microscope (SEM).

2.2 Determination of Moisture and Volatile Matter Content of PCBs

The content of moisture and volatile matter in waste PCBs were quantified using methods prescribed in ASTM-D2216-98 [28] and standard methods [29], respectively. PCB powder weighing to about 100 mg was dried in crucibles for 24 h at 110 ± 5 °C in drying oven. The samples upon being cooled to room temperature in desiccator were quantified for moisture content by determining the loss in weight of the samples. The volatile matter fraction was quantified by heating the dried samples in muffle furnace at temperature of 550 ± 5 °C for 20 min. On cooling, the weight difference of dried and muffled samples was calculated to obtain volatile matter content. All experiments were conducted in triplicates for each of PCB scraps.

2.3 Analysis of Metal Content of PCB Scraps and a Comparative Assessment of Digestion Methods

For elemental analysis, the pulverized PCBs of mobile phone, DVD player and inverter with sieve size of <1 mm were subjected to two different chemical digestion procedures: aqua regia digestion procedure, ISO 1995 [30], and microwave-assisted digestion, USEPA 3052 [31] for solubilization of selected metal constituents', viz. Cu, Zn, Mn, Ni, Co and Cr. The concentration of each element in digestate was then analysed by atomic absorption spectrophotometer (AAS) (iCE3500, Thermo Scientific, USA). Metals' concentration quantification was carried out in triplicates and based on metal solubilization efficiency, and a comparison was made between the two digestion techniques.

In aqua regia digestion procedure, well-homogenized 200 mg pulverized PCB samples were added to 12 ml of aqua regia, i.e. a mix of nitric acid (HNO_3) and hydrochloric acid (HCl) in ratio of 1:3. The PCB comminution fines in aqua regia were incubated for 16 h at room temperature, prior to heating on hotplate for 2 h at 130 °C. The cooled digestates were then filtered and diluted to appropriate concentration. Similarly, in microwave-assisted digestion, 200 mg of pulverized PCBs of mobile phone, DVD player and inverter were digested in closed digestion vessels under action of acid mixture of 9 ml of HNO_3 and 3 ml of HF as per USEPA 3052 guidelines in microwave digester (Multiwave Pro, Anton Paar GmbH, Austria). The digested solutions obtained were filtered using filter paper of pore size 0.22 μm, and deionized water was used for volume make-up. The digestates obtained from the two acid digestion procedures were then analysed for Cu, Zn, Mn, Ni, Co and Cr content using AAS.

2.4 *Microorganism*

A. acidophilum (NCIM 5344; ATCC 27807) bacteria were used in the present study for metal bioleaching from e-waste. The bacteria were purchased from National Collection of Industrial Microorganism (NCIM), a unit of National Chemical Laboratory (NCL), Pune, India. The procured bacteria were initially revived and maintained on NCIMB 206 medium having composition $(NH_4)_2SO_4$: 3 g/l, $MgSO_4 \cdot 7H_2O$: 1 g/l, KH_2PO_4: 0.5 g/l, KCl: 0.1 g/l, $Ca(NO_3)_2$: 0.018 g/l, $FeSO_4 \cdot 7H_2O$: 0.01 g/l, $FeSO_4 \cdot 7H_2O$: 0.01 g/l at pH of 3.5 and 30 °C growth temperature with shaking speed of 150 rpm for period of 6–8 days. The strains were then cultured on solid media plates, and single bacterial isolates were observed under SEM.

2.5 *Bioleaching Experiments*

Owing to the rich metallic content of Cu and Zn, mobile phone PCB was considered as high-grade material and was subjected to bioleaching studies for selected metal recovery. Bioleaching of Cu and Zn from pulverized waste mobile phone PCB with pulp density of 1 g/l and size <600 μm was assessed at 170 rpm using pure acidophilic culture of *A. acidophilum*, an alphaproteobacteria with and without the addition of lemon juice containing citric acid as active constituent to simulate hybrid and conventional conditions, respectively. All bioleaching experiments were conducted at temperature of 30 °C owing to its suitability for optimum bacterial growth.

2.5.1 Conventional Bioleaching

Bioleaching experiments with conventional approach were conducted in 500 ml NCIMB 206 culture media in 1 l conical flasks at acidic pH of 3.5. Pulverized mobile phone PCB weighing to 500 mg of was added into the culture media and sterilized by autoclaving at 121 °C at 15 lb/in.^{2} for 20 min. The culture flasks were then inoculated with 5 ml of *A. acidophilum* and incubated at temperature of 30 °C and 170 rpm in shaker incubator (SIF 5000R, Jeio Tech, South Korea). Cell-free incubations were performed under the same conditions as the control experiments.

All experiments were conducted in triplicates. The progression of bioleaching was monitored at an interval of 3 days by measuring the fall in pH using pH meter (Hach, India) and determination of metals, i.e. Cu and Zn concentration in the liquid phase using AAS.

2.5.2 Hybrid Bioleaching

This approach involved a mix of chemical and biological apparatus. Lemon juice with citric acid as active ingredient in combination with bacterial culture was applied as hybrid bioleaching approach. 500 mg of pulverized mobile phone PCB of particle size <600 μm was added to 500 ml of NCIMB 206 culture media with 5 ml of lemon juice in 1 l Erlenmeyer flasks. The pH was maintained to 3.5, and flasks were autoclaved for 20 min at 121 °C at pressure of 15 $lb/in.^2$. Pure inoculum of 5 ml of *A. acidophilum* was added to the bioleaching flasks and incubated at 30 °C and 170 rpm in shaker incubator. Control experiments were conducted in the absence of bacterial cells under same conditions. The leached solution of all flasks was periodically analysed for pH as well as Cu and Zn contents using AAS standard procedure.

2.6 Analytical Determination

During bioleaching experiments, the leached solutions were analysed for pH and metals, viz. Cu and Zn contents at regular intervals. The pH of leached liquor drawn from each of the flasks was determined using pH meter. The samples were filtered through 0.22-μm Millipore filters for removal of cells and suspended particles followed by volume make-up using deionized water prior to analysis of Cu and Zn contents using AAS. All analyses were carried out in triplicates. Conventional and hybrid bioleaching approaches were compared for Cu and Zn extraction efficiency.

3 Results and Discussion

3.1 Characterization of PCB Scraps

The SEM images depicted heterogeneous matrix of PCBs with particle of different texture, shape and sizes. Figure 2 presents scanning electron micrographs of pulverized PCBs of mobile phone, DVD player and inverter. PCB particles shown in the SEM micrographs constitute mainly of rods and chips of variable size and shape which may be are of cables, insulating platform material, metal sheets, etc.

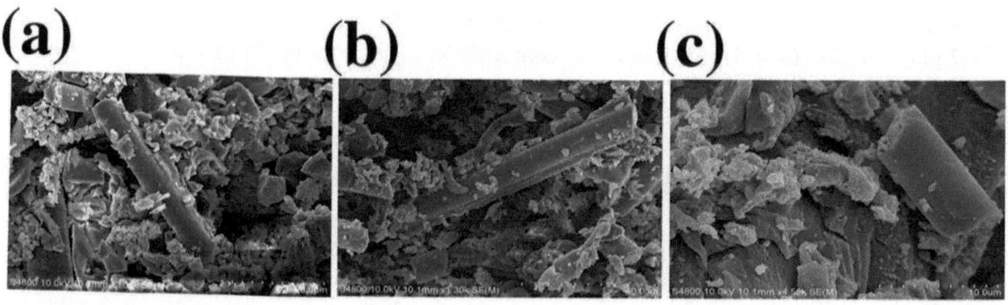

Fig. 2 SEM images of pulverized PCBs from **a** mobile phone, **b** DVD player, **c** inverter

3.2 *Moisture and Volatile Matter Content in PCBs*

The results of moisture and volatile matter content revealed that DVD player PCB has maximum moisture content of 1.140 ± 0.002% and inverter PCB has highest content of volatile matter of 0.64 ± 0.04 mg/g dw (Table 1). The amount of water lost during moisture content analysis was considered to be freely bound water which escaped on drying. The volatile matter content of PCBs indicates the organic fractions which are lost on the ignition. Weight loss during volatile matter content analysis is due to the release of volatile pyrolytic components of organic nature present in PCB scraps. It has been reported that PCBs contain polymers such as polyethylene, polypropylene, polystyrene [15]. These chemical species from PCBs easily volatilize at high temperature, leading to the release of hazardous gases like dioxins, furans and poly aromatic hydrocarbons causing environmental pollution [10].

3.3 *Analysis of Metal Content of PCB Scraps and a Comparative Assessment of Digestion Methods*

Metal analysis of acid digestate of PCBs revealed that microwave-assisted digestion, USEPA 3052, efficiently digested the PCB scraps as compared to aqua regia digestion, ISO 1995. This might be because of hydrofluoric acid (HF) use in USEPA 3052 method. HF efficiently dissolved the metals embedded in the silicon platform of PCBs, because of its extremely corrosive nature, leading to greater metal dissolution. Cu content was the most plentiful among the base metals with more than 80 mg/g dw, while other metals like Mn, Ni, Co were present in low

Table 1 Moisture and volatile matter content in PCBs (n = 3)

E-waste	Moisture content (%)	Volatile content (mg/g dw)
Mobile phone	0.121 ± 0.005	0.14 ± 0.02
DVD player	1.140 ± 0.002	0.54 ± 0.33
Inverter	0.00214 ± 0.00034	0.64 ± 0.04

Table 2 (a) Elemental composition of mobile phone PCBs ($n = 3$), (b) Elemental composition of DVD player PCBs ($n = 3$), (c) Elemental composition of inverter PCBs ($n = 3$)

E-waste	Metal	Concentration (mg/g dw)	
		Aqua regia digestion (ISO, 1995)	Microwave-assisted digestion (USEPA 3052)
(a)			
Mobile phone	Cu	343.98 ± 4.02	366.07 ± 1.77
	Zn	29.51 ± 0.14	32.62 ± 0.03
	Mn	0.10 ± 0.02	0.19 ± 0.07
	Ni	8.72 ± 3.15	5.59 ± 1.05
	Co	0.07 ± 0.05	0.865 ± 0.08
	Cr	ND	ND
(b)			
DVD player	Cu	79.19 ± 2.34	81.72 ± 0.47
	Zn	0.65 ± 0.33	0.70 ± 0.60
	Mn	0.07 ± 0.01	0.12 ± 0.03
	Ni	0.13 ± 0.05	6.70 ± 0.58
	Co	0.03 ± 0.01	0.27 ± 0.07
	Cr	ND	ND
(c)			
Inverter	Cu	86.48 ± 2.22	103.52 ± 4.40
	Zn	0.75 ± 0.02	0.82 ± 0.30
	Mn	ND	0.006 ± 0.003
	Ni	0.1 ± 0.02	0.52 ± 0.16
	Co	0.15 ± 0.09	0.45 ± 0.08
	Cr	ND	ND

concentrations in all the three PCBs (Table 2a–c). However, Cr was not at all detected in any of the PCB scraps. Among the three PCB scraps, waste mobile phone PCB contained rich metal quantity with content of 366 mg/g dw of Cu and 33 mg/g dw of Zn.

3.4 Bioleaching Studies

Owing to high metallic content of Cu and Zn, mobile phone PCB was considered as high-grade material and was subjected to bioleaching studies under conventional and hybrid bioleaching approaches exploiting bacteria *A. acidophilum* with and without lemon juice.

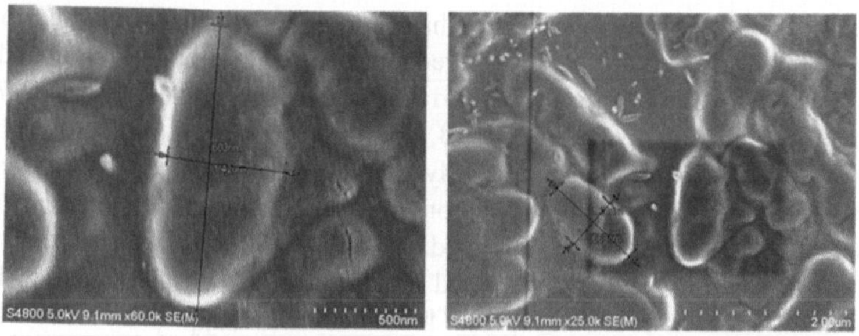

Fig. 3 *A. acidophilum* cells under SEM

3.4.1 Bioleaching Bacteria

A. acidophilum employed for metal bioleaching study observed under SEM appeared as rods of size 0.5–0.8 μm × 1.0–1.5 μm (Fig. 3). The bacteria occur as single cells, often as pairs and occasionally as chains. Growth on solid medium forms smooth, cream coloured round, small, convex and translucent colonies.

3.4.2 Metal Bioleaching

The pH of the bioleaching solution is an essential factor dictating metal dissolution. The optimized pH not only supports microbial growth but also dictates metal solubilization [32]. However, a very high and low pH has an inhibitory effect on bacterial growth and thus low metal solubilization [33]. Figure 4 shows the pH variation in bioleaching reactors with the progression of solubilization of metals, Cu and Zn. The pH of the bioleaching reactors increased from 3.5 to 5.46 during the start of bioleaching with slow leachability but consequently concluded with rapid leaching rates with pH drop to 4.43 and 4.12 in conventional as well as hybrid bioleaching experiments, respectively. Leaching was observed to increase with decrease in pH to optimum value. In the literature, it has been reported that electronic scrapes are alkaline which contribute to elevation of pH of the leaching solution during initial stage of bioleaching which later decreased with bacterial growth.

Fig. 4 pH variation of bioleaching medium with leaching time for **a** conventional bioleaching and **b** hybrid bioleaching

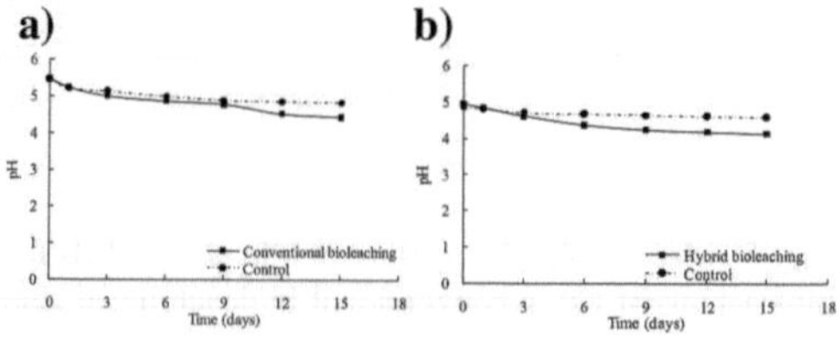

Metal bioleaching results indicated that dissolution of both the metals from the crushed PCB varied greatly under conventional and hybrid bioleaching conditions showing better solubilization in hybrid approach. At sample size of 1 g/l and comminution fines of <600 μm of PCB, dissolution of 0.416 mg/L was achieved for Cu at 12 days which asymptotically increased to 0.424 mg/L at 15 days of conventional bioleaching. Similar asymptotic bioleaching trend was obtained for Zn with an increase to 0.132 mg/L at 15 days from 0.130 mg/L at 12 days in conventional approach. However, increased bioleaching of both the metals in hybrid approach was observed as compared to conventional approach. Under hybrid condition, the maximum concentrations of Cu and Zn leached were 2.419 and 1.272 mg/L, respectively, at 15 days. Similar study conducted by Choi et al. [25] also reported use of commercial chelator in the form of citric acid along with bioleaching bacteria *A. ferrooxidans* to enhance metal recovery. However, no study related to hybrid bioleaching employing combination of *A. acidophilum* and chelating agents for metal recovery from e-wastes exists in the literature. Increased dissolution of the selected metals from the PCB in hybrid bioleaching approach can be attributed to the presence of citric acid as metal chelating agent in lemon juice along with the action of *A. acidophilum*. Insignificant leaching of metals was observed in case of uninoculated controls (Fig. 5).

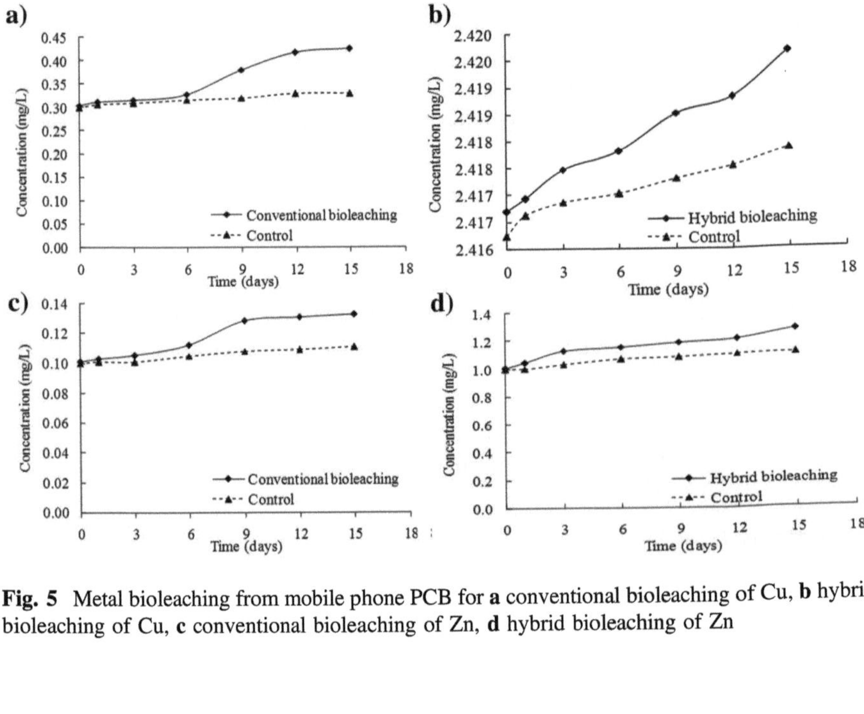

Fig. 5 Metal bioleaching from mobile phone PCB for **a** conventional bioleaching of Cu, **b** hybrid bioleaching of Cu, **c** conventional bioleaching of Zn, **d** hybrid bioleaching of Zn

4 Conclusions

This research on extraction of metals, Cu and Zn, from high-grade PCB scraps has demonstrated the practical applicability of bioleaching technique and successful development of fast recovery of metals by hybrid bioleaching as compared to conventional bioleaching. The results revealed that hybrid bioleaching was notably more efficient compared to conventional bioleaching. The maximum concentrations of Cu and Zn leached under hybrid approach were 2.419 and 1.272 mg/L, respectively, while under conventional condition recovery was 0.424 and 0.132 mg/L, respectively, at 15 days. Maximum mobilizations of major metals like Cu and Zn from PCB by mix of *A. acidophilum* and lemon juice containing citric acid proved effectiveness of hybrid bioleaching approach in metal recovery from e-waste. The key findings of bioleaching assessment highlight the practical applicability of hybrid bioleaching for the development of green technique for metal recycling from e-waste. However, metal recovery rate and the leaching time are required to be improved by methodical exploration, optimizing the various abiotic and biotic process parameters, governing the process of bioleaching for commercial applications.

Acknowledgements The authors are grateful to the Department of Science and Technology, Government of India for fellowship grant (IF130860) for the research work.

References

1. Grossman, E.: High Tech Trash: Digital Devices, Hidden Toxics, and Human Health, pp. 336–352. Island Press, Washington, DC (2006)
2. Hicks, C., Dietmar, R., Eugster, M.: The recycling and disposal of electrical and electronic waste in China—legislative and market responses. Environ. Impact Assess. Rev. **25**, 459–471 (2005)
3. Huang, K., Jie Guo, J., Xu, Z.: Recycling of waste printed circuit boards: a review of current technologies and treatment status in China. J. Hazard. Mater. **164**, 399–408 (2009)
4. Basdere, B., Seliger, G.: Disassembly factories for electrical and electronic products to recover resources in product and material cycles. J. Environ. Sci. Tech. **37**, 5354–5362 (2003)
5. Das, A., Vidyadhar, A., Mehrotra, S.P.: A novel flowsheet for the recovery of metal values from waste printed circuit boards. Resour. Conserv. Recycl. **53**, 464–469 (2009)
6. Li, J., Shrivastava, P., Gao, Z., Zhang, H.C.: Printed circuit board recycling: a state-of-the-art survey. IEEE Trans. Electron. Packag. Manuf. **27**, 147–222 (2004)
7. Zhou, Y., Qiu, K.: A new technology for recycling materials from waste printed circuit boards. J. Hazard. Mater. **175**, 823–828 (2010)
8. Li, J., Lu, H., Guo, J.: Recycle technology for recovering resources and products from waste printed circuit boards. J. Environ. Sci. Tech. **41**, 1995–2000 (2007)
9. Hageluken, C.: Recycling of electronic scrap at Umicore's integrated metals smelter and refinery. World Metall. Erzmetall **59**(3), 152–161 (2006)

10. Deng, W.J., Zheng, J.S., Bi, X.H., Fu, J.M., Wong, M.H.: Distribution of PBDEs in air particles from an electronic waste recycling site compared with Guangzhou and Hong Kong, South China. Environ. Int. **33**, 1063–1069 (2007)
11. Niu, X., Li, Y.: Treatment of waste printed wire circuit boards in electronic waste for safe disposal. J. Hazard. Mater. **145**, 410–416 (2007)
12. Cui, J., Zhang, L.: Metallurgical recovery of metals from electronic waste: a review. J. Hazard. Mater. **158**, 228–256 (2008)
13. Dalrymple, I., Wright, N., Kellner, R.: An integrated approach to electronic waste (WEEE) recycling. Circuit World **33**, 52–58 (2007)
14. Ilyas, S., Ruan, C., Bhatti, H.N., Ghauri, M.A., Anwar, M.A.: Column bioleaching of metals from electronic scrap. Hydrometallurgy **101**, 135–140 (2010)
15. Owens, C.V., Lambright, C., Bobseine, K.: Identification of estrogenic compounds emitted from the combustion of computer printed circuit boards in electronic waste. J. Environ. Sci. Tech. **41**, 8506–8511 (2007)
16. Atlas, R.M., Bartha, R.: Microbial ecology: Fundamentals and Applications, 2nd edn., pp. 148–151. Benjamin Cummings, US (1997)
17. Wang, J., Bai, J., Xu, J., Liang, B.: Bioleaching of metals from printed wire boards by *Acidithiobacillus ferrooxidans* and *Acidithiobacillus thiooxidans* and their mixture. J. Hazard. Mater. **172**, 1100–1105 (2009)
18. Xie, F., Cai, T., Ma, Y., Li, H., Li, C., Huang, Z., Yuan, G.: Recovery of Cu and Fe from printed circuit board waste sludge by ultrasound: evaluation of industrial application. J. Clean. Prod. **17**, 1494–1498 (2009)
19. Beolchini, F., Fonti, V., Dell'Anno, A., Rocchetti, L., Veglio, F.: Assessment of biotechnological strategies for the valorization of metal bearing wastes. Waste Manag. **32**, 949–956 (2012)
20. Brandl, H., Bosshard, R., Wegmann, M.: Computer-munching microbes: metal leaching from electronic scrap by bacteria and fungi. Hydrometallurgy **59**, 319–326 (2001)
21. Brierley, J.A., Brierley, C.L.: Present and future commercial applications of Biohydrometallurgy. Hydrometallurgy **59**, 233–239 (2001)
22. Bosecker, K.: Bioleaching: metal solubilisation by microorganisms. FEMS Microbio. Rev. **20**, 591–604 (1997)
23. Ilyas, S., Anwar, M.A., Niazi, S.B., Afzal, G.M.: Bioleaching of metals from electronic scrap by moderately thermophilic acidophilic bacteria. Hydrometallurgy **88**, 180–188 (2007)
24. Zhang, Y., Peng, A., Yang, Y., Liu, J., Qiu, G.: Isolation, characterization of *Acidiphilium* sp. DX1-1 and ore bioleaching by this acidophilic mixotrophic organism. Trans. Nonferr. Met. Soc. China **23**, 1774–1782 (2013)
25. Choi, M.S., Cho, K.S., Kim, D.S., Kim, D.J.: Microbial recovery of copper from printed circuit boards of waste computer by *Acidithiobacillus ferrooxidans*. J. Environ. Sci. Health A **39**(11–12), 2973–2982 (2004)
26. Tandy, S., Bossart, K., Mueller, R., Ritschel, J., Hauser, L., Schulin, R., Nowack, B.: Extraction of heavy metals from soils using biodegradable chelating agents. J. Environ. Sci. Tech. **38**, 937–944 (2004)
27. Muller, B.: Citric acid as corrosion inhibitor for aluminium pigment. Corros. Sci. **46**, 159–167 (2004)
28. American Society for Testing and Materials (ASTM): Standard Test Methods for Laboratory Determination of Water (Moisture) Content of Soil and Rock by Mass. ASTM D2216-98, ASTM International (1998)
29. American Public Health Association (APHA): Standard Methods for the Examination of Water and Wastewater, 22nd edn., pp. 61–68. American Public Health Association-American Water Works Association-Water Environment Federation, Washington, DC (2012)

30. International Organization for Standardisation (ISO): Soil Quality—Extraction of Trace Elements Soluble in Aqua Regia. ISO 11466 (1995)
31. US Environmental Protection agency, Method 3052: Microwave Assisted Acid Digestion of Siliceous and Organically Based Matrices SW-846. Washington, DC (1996)
32. Sand, W., Gehrke, T., Jozsa, P.G., Schippers, A.: (Bio)chemistry of bacterial leaching—direct vs. indirect bioleaching. Hydrometallurgy **59**, 159–175 (2001)
33. Rohwerder, T., Gehrke, T., Kinzler, K., Sand, W.: Bioleaching review part A. Appl. Microbiol. Biotechnol. **63**(3), 239–248 (2003)

Bio-inspired Facile and Green Synthesis of Au@Ag@AgCl Nanoparticles Using *Benincasa Hispida* Peel Extract and Their Photocatalytic Activity for the Removal of Toxic Dye Under Solar Irradiation

Th. Babita Devi and M. Ahmaruzzaman

Abstract The present study reports a simple environmental-friendly process for the synthesis Gold@silver@silver chloride nanoparticles (Au@Ag@AgCl NPs) by using peel extract of *Benincasa hispida* (ash gourd) without using any toxic external reagents and external halide source. The synthesized nanoparticles were well characterized by using XRD, FT-IR, UV, TEM, and EDX analyses. The phytochemicals present in the peel extract were responsible for the formation of the nanoparticles. The synthesized Au@Ag@AgCl NPs showed an excellent photocatalytic activity for the degradation of toxic dye under solar irradiation. Approximately, 98% degradation of toxic dye was observed using Au@Ag@AgCl NPs as a photocatalyst under solar irradiation.

Keywords Au@Ag@AgCl · Biomaterials · Nanostructures · Toxic dye Photodegradation

1 Introduction

Nowadays, noble metals NPs like Pd, Au, Ag, and Pt have attracted considerable interest because of their unique property, surface plasmon resonance (SPR), showing excellent optical, electrical, and catalytic properties [1, 2]. Recently, many studies have been reported for the synthesis of alloy, core–shell, core–double shell, metal–metal halide nanoparticles (NPs) using toxic chemicals and solvents [3–8]. The utilization of these highly toxic chemicals and solvents makes the environment polluted, and therefore, the development of totally green synthesis is very important

Th. Babita Devi (✉) · M. Ahmaruzzaman
Department of Chemistry, National Institute of Technology Silchar, Silchar 788010, Assam, India
e-mail: babisana89@gmail.com

A. S. Kalamdhad et al. (eds.), *Advances in Waste Management*,
https://doi.org/10.1007/978-981-13-0215-2_38

to reduce environmental pollution. Therefore, biosynthesis using plant-based materials has been exploited for the green fabrication of various NPs.

Keeping in view the shortcoming of previous studies, we design a green method for the synthesis of Au@Ag@AgCl NPs using biomaterial (*Benincasa hispida* peel extract). It was for the first time a green method for the synthesis of Au@Ag@AgCl NPs was reported using *B. hispida* peel extract. The phytochemical presents in the peel extract helped in the formation of NPs. The synthesized nanoparticles were further confirmed by EDX, XRD, TEM, and UV–visible spectroscopy.

2 Materials and Methods

2.1 Materials

The starting materials were of AR grade silver nitrate ($AgNO_3$), hydrogen tetrachloroaurate monohydrate ($HAuCl_4 \cdot 3H_2O$), and malachite green oxalate (MGO) was purchased from Sigma-Aldrich, Silchar.

2.2 Preparation of Benincasa Hispida *Peel Extract*

A total of 10 g of fresh peel of *B. hispida* was boiled for 30 min with 250 mL of distilled water in a round-bottom flask. The peel extract was filtered using Whatman No. 41 filter paper to obtain the pure extract.

2.3 Synthesis of Au@Ag@AgCl Nanoparticles

The biosynthesis of Au@Ag@AgCl nanoparticles was synthesized by using 20 mL of 10% *B. hispida* peel extract with $HAuCl_4 \cdot 3H_2O$:Ag NO_3 (1:1 molar ratio) by heating at 70 °C for 20 min.

2.4 Characterization of Au@Ag@AgCl Nanoparticles

Absorption spectra were recorded on Cary 100 BIO UV–visible spectrophotometer. The prepared nanoparticles were characterized by P-XRD method using Phillips X'Pert PRO diffractometer with CuK radiation of wavelength 1.5418. For the morphology study, FT-IR spectroscopy and elemental study, JEM-2100 transmission electron microscope, Bruker Hyperion 3000, FEG-SEM, Model: JSM-7600F, Magnification: ×25 to 1,000,000a have been used.

2.5 Photocatalytic Study of Au@Ag@AgCl Nanoparticles

The photocatalytic activity of the synthesized Au@Ag@AgCl nanoparticles was evaluated by degradation of toxic dye like malachite green oxalate (MGO). A total of 0.02 g of the catalyst was dispersed in 250 ml of MGO (6×10^{-4} M) aqueous solution. The experiment was carried out on a sunny day at Silchar City, Assam, between 10 a.m. to 2 p.m. (outside temperature 35–40 °C).

3 Results and Discussion

3.1 UV–Visible Analysis

Ag and Au NPs have unique phenomena known as surface plasmon resonance (SPR). Figure 1 represents the UV–visible spectrum of Au@Ag@AgCl nanoparticles. Figure 1 shows two absorption bands, one at 400 nm and another at 530 nm. The absorption band at around 400 nm is due to SPR of silver nanoparticles presence on the surface of AgCl nanoparticles, and the absorption band at 530 nm is due to the SPR of gold.

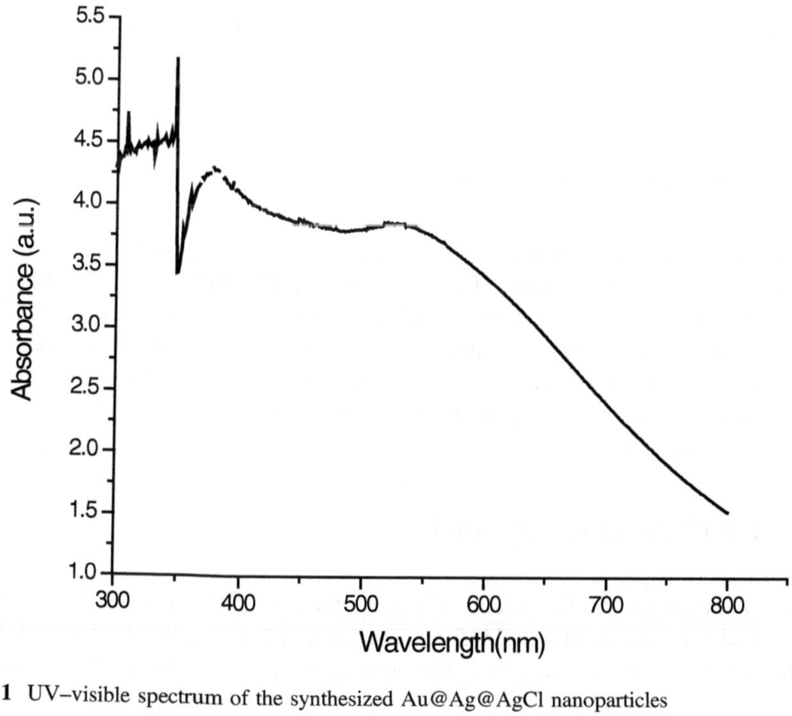

Fig. 1 UV–visible spectrum of the synthesized Au@Ag@AgCl nanoparticles

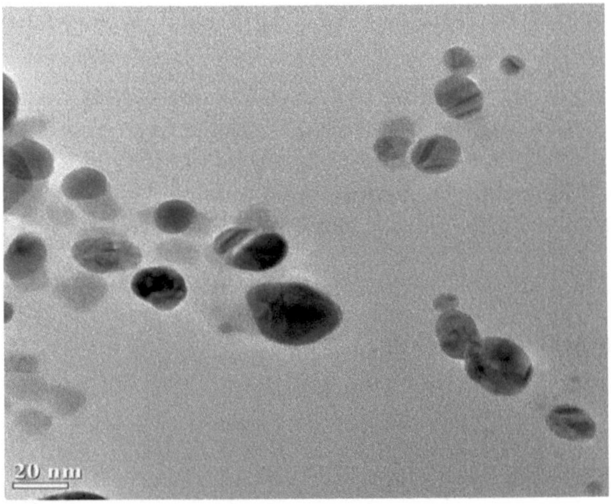

Fig. 2 TEM image of the synthesized Au@Ag@AgCl nanoparticles

3.2 *Morphology Study*

Figure 2 corresponds to the TEM images of the synthesized Au@Ag@AgCl NPs. The TEM images revealed that the particles are spherical in morphology. The average particles sizes of the synthesized Au@Ag@AgCl nanoparticles were found to be (15–25) nm.

3.3 *X-Ray Diffraction (XRD) Studies*

The characteristic XRD peaks at 27.8°, 32.2°, 46.2°, 54.9°, and 57.6° were corresponded to the lattice planes of (111), (200), (221), (311), and (222) and also reflected the face-centered cubic (FCC) structure of AgCl crystal (JCPDS No. 31-1238) (Fig. 3). The remaining peaks at 38.2°, 66.4°, and 77.2° were corresponded to the lattice planes of (111), (220), and (311) reflected the face-centered cubic (fcc) structure of Au–Ag (JCPDS file: 65-2871).

3.4 *FT-IR Spectroscopy Study*

Figure 4 represents FT-IR spectra of the synthesized NPs. The FT-IR spectrum showed absorption bands at 3445 cm^{-1} correspond to hydrogen bonded O–H stretching vibrations of alcohols. The absorption peak at 1731 cm^{-1} could be

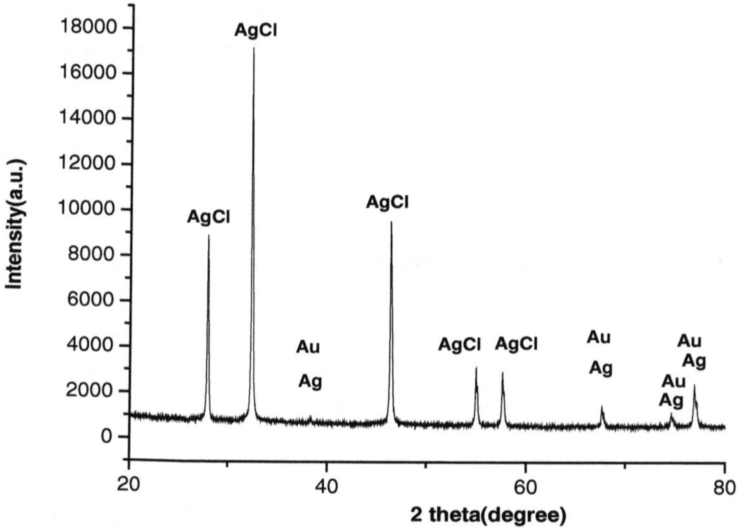

Fig. 3 XRD spectrum of the synthesized Au@Ag@AgCl nanoparticles

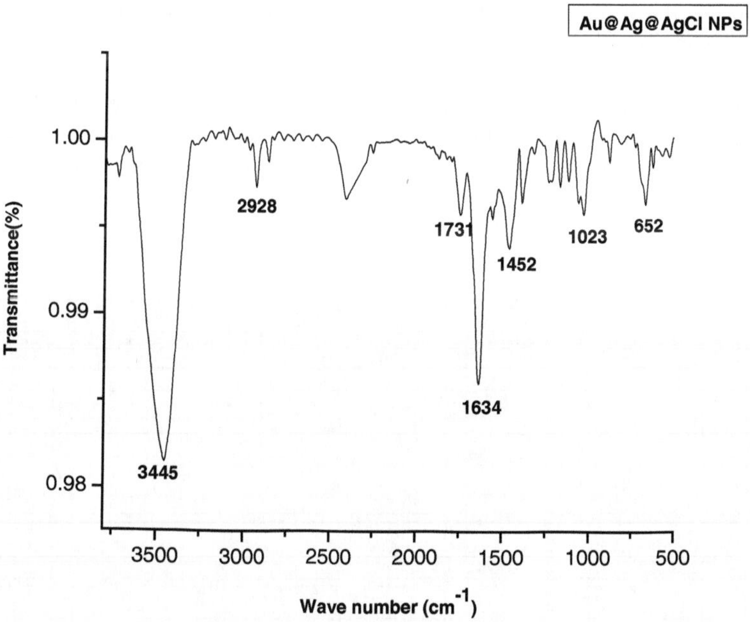

Fig. 4 FT-IR spectrum of the synthesized Au@Ag@AgCl nanoparticles

attributed to C=C stretching vibrations of about C=O amide conjugated C=O of the proteins that are responsible for capping and stabilizing of Au@Ag@AgCl NPs. The remaining absorption peaks at 1634, 1452, and 1023 cm^{-1} were due to the presence of –C=O asymmetric stretching vibration, aromatic stretching of –C–N, C–H group (aromatic), respectively.

3.5 Energy Dispersive X-Ray Spectroscopy

The elemental analysis of the synthesized nanoparticles was performed using EDX. In the spectrum, the peaks around 2.2–2.5 keV correspond to Au, 3 keV correspond to Ag, and 2.7 keV correspond to binding energies of AgCl (Fig. 5).

3.6 Evaluation of Photocatalytic Activity of Au@Ag@AgCl Nanoparticles

The photocatalytic activity of the synthesized nanoparticles was carried out by monitoring the changes in optical absorption spectra of malachite green oxalate (MGO) solution under solar irradiation. The UV–visible spectra of the dye (MGO) showed a strong absorption band at 615 nm (Fig. 6a). The absorption band at 615 nm gradually decreases with increase in irradiation time and disappears completely within 110 min. The degradation of dye followed the pseudo-first-order reaction, and its kinetics may be expressed by the following reaction

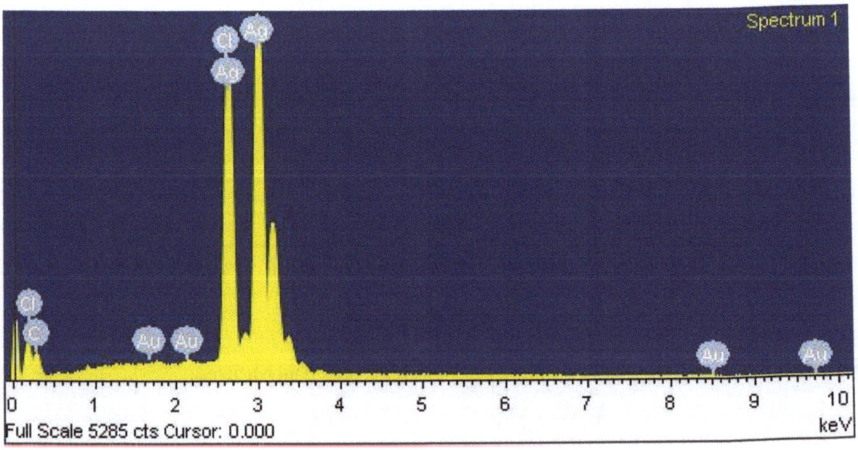

Fig. 5 EDX spectrum of the synthesized Au@Ag@AgCl nanoparticles

(a)

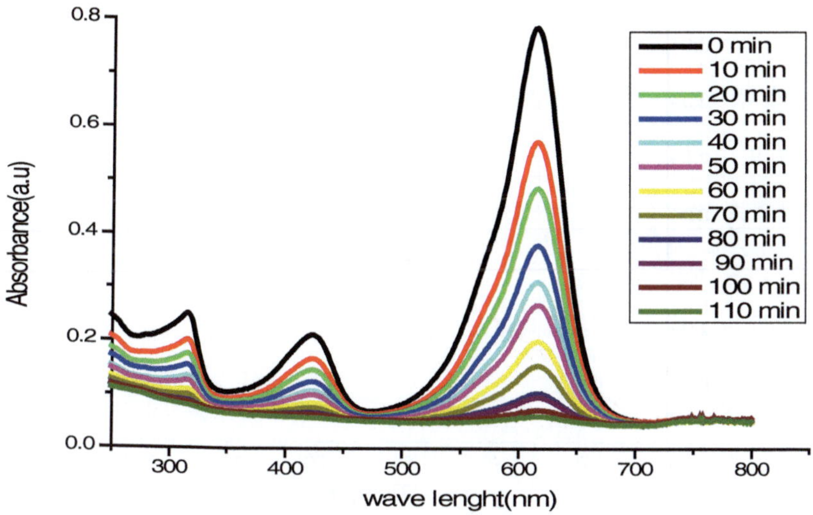

Fig. 6 **a** Photodegradation of MGO dye under solar irradiation using Au@Ag@AgCl NPs as a photocatalyst. **b–c** Plot of ln(Co/C_t) with irradiation time of degradation of MGO dye and the efficiency of the photodegradation of MGO

$$\ln(C_0/C_t) = kt \tag{1}$$

where k is the rate constant, C_0 and C_t are the absorbance or concentration before and after degradation of MGO dye, respectively. The rate constant for the photodegradation of MGO dye was calculated using Eq. (1). Figure 6b represents the plot of ln (C_0/C) versus irradiation time (t), and it gives a linear relationship. Hence, the slope of the line represents the rate constant (k) for the photodegradation of MGO dye and the value of k is found to be 4.66×10^{-2} min^{-1}.

The percentage efficiency of photodegradation of MGO dye was determined using the following equation

$$X = [(C_0 - C)/C] \times 100 \tag{2}$$

where C_0 and C are the absorbance or concentration before and after degradation of MGO dye, respectively. Figure 6c shows the percentage efficiency of photodegradation of MGO dye with time and observed that 98% of the dye was photochemically degraded within 110 min using Au@Ag@AgCl nanoparticles.

(b)

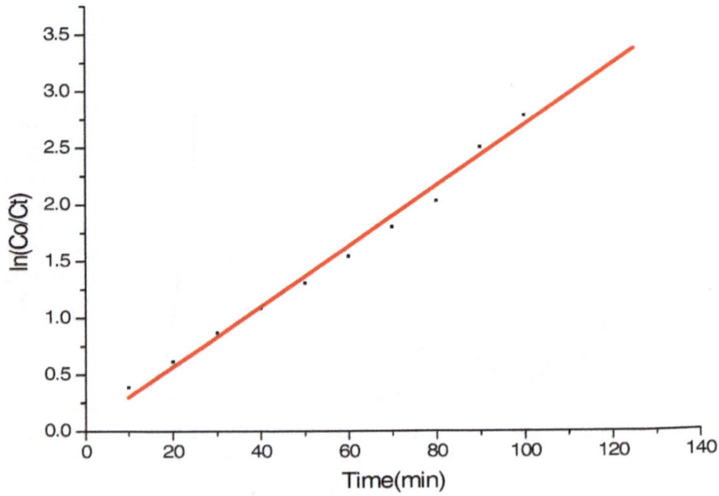

(c)

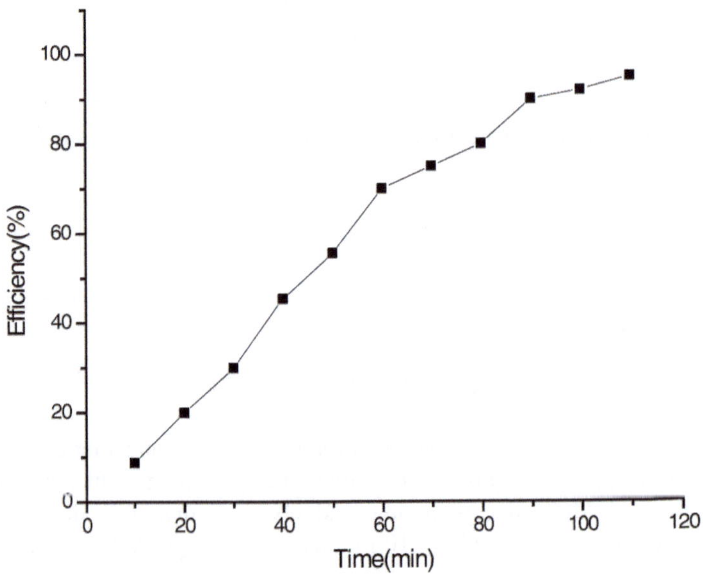

Fig. 6 (continued)

3.7 Mechanism of Photodegradation of Malachite Green Oxalate Dye Using Au@Ag@AgCl NPs

Due to the SPR effect of Ag metal, NPs present on the surface of AgCl nanoparticles, and an electron–hole pairs on the surface of Ag NPs were generated under visible irradiation. The electrons will be trapped by adsorbed O_2 to form $^{\bullet}O_2^-$. In the meantime, the holes were transferred to the surface of AgCl because of the high oxidation ability of the chloride ion present in AgCl and lead to the oxidation of Cl^- ions to form Cl^o atoms (radicals) [9, 10]. In the photodegradation of MGO dye, Au NPs do not show any photocatalytic activity because of trapped by Ag and AgCl NPs that makes it unable to absorb the light.

4 Conclusions

In summary, we developed a green method for the production of Au@Ag@AgCl nanoparticles using *B. hispida* peel extract. The phytochemical presents in the peel extract acts as reducing and stabilizing agent during the synthesis of Au@Ag@AgCl NPs. The UV–visible spectroscopy, XRD, EDX, and FT-IR spectroscopic studies confirmed the formation of Au@Ag@AgCl NPs. The synthesized Au@Ag@AgCl NPs were successfully utilized as a photocatalyst for the degradation of MGO dye under solar light. The strong SPR effect of Ag NPs can generate electron–hole pairs and leading to the enhancement of the photocatalytic activities of Au@Ag@AgCl NPs. The reactive species, such as $^{\bullet}O_2^-$ and Cl^0, are mainly responsible for the degradation of malachite green oxalate dye.

Acknowledgements The authors would like to express their heartfelt thanks to the Director, NIT Silchar for providing laboratory facilities and financial assistance. Our special thanks are extended to SAIF-NEHU Shillong, SAIF IIT Bombay, and CSMCRI Gujarat for providing the TEM, EDX, and XRD facilities.

References

1. Devi, T.B., Ahmaruzzaman, Md., Begum, S.: A rapid, facile and green synthesis of Ag@AgCl nanoparticles for the effective reduction of 2,4-dinitrophenyl hydrazine. New J. Chem. **40**, 1497–1506 (2016)
2. Devi, T.B., Begum, S., Ahmaruzzaman, M.: Photo-catalytic of plasmonic Ag@AgCl nanoparticles (synthesize via a green route) for the effective degradation of Victoria green B from aqueous phase. J. Photochem. Photobiol., B **160**, 260–270 (2016)
3. Devi, T.B., Ahmaruzzaman, M.: Bio-inspired sustainable and green synthesis of plasmonic Ag/AgCl nanoparticles for enhanced degradation of organic compound from aqueous phase. Environ. Sci. Pollut. Res. Int. **23**, 17702–17714 (2016)

4. Chaudhuri, R.G., Paria, S.: Au and Ag/Au double-shells hollow nanoparticles with improved near infrared surface plasmon and photoluminescence properties. J. Colloid Interf. Sci. **461**, 15–19 (2016)
5. Ghosh, T., Satpati, B., Senapati, D.: Characterization of bimetallic core-shell nanoring synthesized via ascorbic acid controlled galvanic displacement followed by epitaxial growth. J. Mater. Chem. C. **2**, 2439–2447 (2014)
6. Knauer, A., Eisenhardt, A., Krischok, S., Koehler, J.M.: Nanometer precise adjustment of the silver shell thickness during automated Au–Ag core–shell nanoparticle synthesis in micro fluid segment sequences. Nanoscale **6**, 5230–5238 (2014)
7. Kulkarni, A.A., Bhanage, B.M.: Ag@AgCl nanomaterial synthesis using sugar cane juice and its application in degradation of Azo dyes. ACS Sustain. Chem. Eng. **2**, 1007–1013 (2014)
8. Haldar, K.K., Kundu, S., Patra, A.: Core-size-dependent catalytic properties of bimetallic Au/Ag core-shell nanoparticles. ACS Appl. Mater. Interfaces. **6**, 21946–21953 (2014)
9. Wang, P., Huang, B.B., Qin, X.Y., Zhang, X.Y., Dai, Y., Wei, J.Y., Whangbo, M.H.: Ag@AgCl: a highly efficient and stable photo catalyst active under visible light Angew. Chem. Int. Ed. **47**, 7931–7933 (2008)
10. Wang, P., Huang, B.B., Zhang, X.Y., Qin, X.Y., Dai, Y., Wang, Z.Y., Lou, Z.Z.: Highly efficiently visible light plasmonic photocatalyst Ag@Ag(Cl, Br) Ag@AgCl-AgI. Chem cat Chem. **3**, 360–364 (2011)

High Energy Absorbent Sustainable Concrete Element From Used Tin Cans

Rahul Kumar Jaiswal, Namrata Singh Solanki, Siddharth Satish and K. S. Satyanarayanan

Abstract Large quantities of solid wastes are being generated worldwide which lead to environmental concerns. One such major component is tin cans. Utilization of tin cans wastes in making building construction materials can reduce the magnitude of many associated problems. "Energy Block" is a type of block which has high energy absorption when kept under load. It is mainly composed of cement, sand, fibers, and tin cans. This paper shows the energy block is an environment friendly and lightweight building block from cement mortar, waste tin cans and fibers, without the use of coarse aggregate providing high energy absorption to the concrete. The paper concludes the results of laboratory tests and the actual behavior and performance of such building materials in practical application providing lightweight and the usage of tin cans can be economical for conservation of environment.

Keywords Energy block · Tin cans · Economical · High energy absorption

1 Introduction

'Energy Block' is a type of block which can resist impact load to a great extent. It is mainly composed of cement, sand, and tin cans (of specified volume of 250 mL). The energy absorption test was conducted using universal testing machine to get ultimate strength. While conducting the experiments, the results clearly show that the energy absorbed by combination of plain mortar, fiber, and tin cans is

R. K. Jaiswal (✉) · N. S. Solanki · S. Satish · K. S. Satyanarayanan
Civil Engineering, SRM University, Chennai 603203, India
e-mail: rahuljais50@gmail.com

N. S. Solanki
e-mail: simmidstar15@gmail.com

A. S. Kalamdhad et al. (eds.), *Advances in Waste Management*,
https://doi.org/10.1007/978-981-13-0215-2_39

maximum; hence, it can be concluded that there is a wide difference on comparing energy absorption among all type of block.

2 Review of Literature

Fanella and Naaman [1] performed an experimental study on "Stress–Strain Properties of Fiber-Reinforced Mortar in Compression." An experimental program was conducted in various volume fractions to show the comparison of stress–strain curves on the addition of steel, monofilament polypropylene, and twisted polypropylene fibers [1].

Abrishami and Mitchell [2] performed an experimental study on "Influence of Steel Fibers on Tension Stiffening." The main objective is to showcase the structural response of reinforced concrete tension members. The influence of steel fibers on the control of splitting cracks and on tension stiffening was investigated [2].

Kukreja et al. [3] performed an experimental study on "Ultimate Strength of Fiber-Reinforced Concrete Slabs." An experimental test is conducted, and it is found that the concrete slab with reinforcement provides the higher strength than the unreinforced concrete slab [3].

Zollo [4] performed an experimental study on "Fiber-Reinforced Concrete—An Overview after 30 Years of Development." This paper presents a rhetorical discussion on the subject of fiber-reinforced concrete. It discusses commonly applied terminology and models of mechanical behavior that form a basis for understanding material performance without presenting mathematical details [4].

Ravinder [5] performed an experimental study on "Strength Characteristics of Coca-Cola Tin Waste as Fibers in Concrete." The use of recycled fibers from post-consumer waste offers additional advantages of waste reduction and resources conservation. The addition of fiber increases the properties of concrete, i.e., flexural strength, impact strength, and shrinkage properties. The addition of fiber increases the properties of concrete [5].

3 Initial Methods Adopted and Material Used

A base mix of cement–sand mortar (1:3 mix) is made up to a thickness of 12.7 mm (½ in.). A chicken mesh wire is placed on the raw base instantly. Then mortar is placed above the chicken mesh up to a height of 12.7 mm. The block is of the length 370 mm, breadth 180 mm, and height 70 mm. Mesh wire is of the thickness 0.5 mm. Then, the total surface area is divided into two parts. Three tin cans (of volume 250 mL) are placed in each of the two rows such that there is a total of six cans in two rows. A minimum clearance gap of 12.7 mm is kept between the cans in shorter direction and 71.12 mm (2.8 in.) gap in longitudinal direction. This clearance gap is used for surrounding the cans with chicken mesh in a definite pattern. After mesh wire is placed, cement mortar of same mix 1:3 is filled in the clearance gap to allow the mesh wire to set. Mortar layer having thickness of 12.7 mm is placed on top of the cans; then again chicken mesh is placed. Over the mesh wire, again a layer of 12.7 mm thick mortar is placed. This freshly prepared block is allowed to dry for one day. Then testing is done.

4 Testing of "Energy Block"

The block is kept for one day to attain strength. Then, it is plastered with cement mortar of mix 1:4. Weight of the block recorded before curing is 18.25 kg. It is kept for 7 days curing.

5 Test Result on Energy Block

The block failed in the test as it is not made according to the standard dimensions allowed for testing. It is also very heavy and costly. A new concept of making cylindrical block of standard dimensions is presented, taking into account the weight factor and cost involved.

5.1 Construction of Cylindrical Block (Modification of Energy Block)

A new concept which is the modified form of earlier cuboidal "Energy Block" was accepted that led to the construction of cylindrical block. Instead of making a single block, multiple blocks with different properties and parameters were constructed. About four primary cases or parameters were kept in mind before construction, whether block should be plain, reinforced with tin or fibers or a combination of all the three. The work of constructing 12 cylindrical blocks was divided into 4 batches of 3 cylinders in each batch. In a single attempt, 6 blocks were casted. Then later, another 6 blocks were casted. All blocks were cured for a period of 28 days to attain characteristics compressive strength. The energy absorption test was conducted using universal testing machine to get ultimate strength. While conducting the experiments, the results clearly show that the energy absorbed by combination of plain mortar, fiber, and tin cans is maximum; hence, it can be concluded that there is a wide difference on comparing energy absorption among all type of block (Fig. 1; Table 1).

Fig. 1 Mold along with tin can

Table 1 Details of tin cans

Specifications of tin cans	Dimensions of tin cans
Length (*L*)	120 mm
Diameter (*d*)	52 mm
Volume	250 mL

6 Construction

For the construction, a PVC pipe mold of 300 mm height and 150 mm diameter is required. With the help of mold, it is easy to cast the blocks in given standard dimensions. Proper measures are taken so that mortar should not stick on inside the mold. Oil is applied to the inner surface of the mold to prevent sticking of mortar (Tables 2 and 3).

A total of 12 blocks need to be constructed [batch1-plain mortar only, batch2-plain mortar with tin only, batch3-plain mortar with fiber only, batch4-plain mortar with both tin and fiber]. Each of them should be constructed three in number and the average value is taken for writing the final result.

During casting, special care is given in placing the tin cans: one on top of the other with a minimum clearance gap of 20 mm. Surrounding space is filled with plane mortar. The tins have a volume of 250 mL and placed in center of the block such that longitudinal axis of tin is parallel to longitudinal axis of the constructed block. After casting, they are label accordingly for further identification (Fig. 2).

Table 2 Details of cylindrical block

Dimensions of cylindrical block	
Diameter	150 mm
Height	300 mm
Volume	5301.4×10^3 mm^3

Table 3 Details of fiber

Specification of fiber	Dimensions of fiber
Length (*L*)	50 mm
Diameter (*d*)	1 mm
Aspect ratio (*L*/*d*)	50

Fig. 2 Galvanized iron fiber

7 Mix Design of the Cylindrical Blocks

For the mix design, raw materials used are cement, sand, and water. Coarse aggregate is not used. Cement and sand are taken in the ratio of 1:3 to form cement–sand mortar with water–cement ratio is taken as 0.50. For the plain mortar block, the mix is taken directly and constructed. But for other cases, certain additions are done in such a way that the chemical properties of the mix remain same for all the cases (Table 4).

Table 4 Mix ratio

Mix ratio	Proportions
Cement:Sand	1:3
W/*c* ratio	0.50

Fig. 3 Casting of cylinders

For blocks reinforced with tin, a clearance as specified of 20 mm is given while placing the tin cans one on top the above.

For the block reinforced with fiber, steel (binding wires) fibers having length of 50 mm and diameter 1 mm is taken and added uniformly throughout the mix evenly. The amount of fibers added to a concrete mix is expressed as a percentage of the total volume of the composite (concrete and fibers), termed “volume fraction” (V_f). V_f typically ranges from 0.1 to 3%. The optimum value of 0.75% is considered. The aspect ratio (L/d) is calculated by dividing fiber length (L = 50 mm) by its diameter (d = 1 mm). In our case, the aspect ratio is 50.

For blocks involving tin and fiber, both the points mentioned above is considered. Initially, plain mortar (3) blocks and plain mortar and tin (3) blocks were constructed. Later on another 3–3 blocks were made in which onset consisting of mortar and fiber and another set consisting of both tin and fiber with mortar. So in this way, all 12 blocks were made and cured for a minimum period of 28 days (Fig. 3).

8 Testing of the Cylindrical Blocks

Special care is taken while transporting the blocks from place of manufacture to place of testing. The blocks were tested for compression. The test was done by “Universal Testing Machine” having capacity of 10,000 kg. UTM is a dynamic machine which gives multiple results in a single test with graph. The UTM machine is operated by software that provide the data accurately at every interval in variation of loading (Fig. 4).

The testing is done after 28 days and following figures are obtained showing graph between load and crosshead travel for each block (Figs. 5, 6, 7, 8, and 9).

Fig. 4 Failure occur in cylinder

This graph is very crucial for finding the ultimate strength. While conducting the experiments, the results clearly showed that the energy absorbed using plain mortar is 450 J, plain mortar with fibers is 520 J, plain mortar with tin cans is 1200 J, and plain mortar, fiber, and tin cans are 2550 J. Hence, it can be concluded that there is a wide difference in the energy absorption between the plain mortar (fibers and tins cans) and the other three mixes (Figs. 10, 11, 12, and 13).

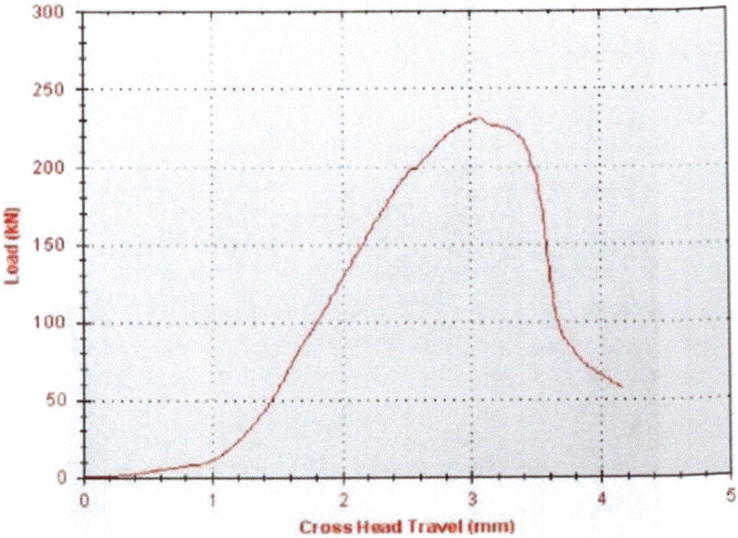

Fig. 5 Load versus deflection curve for type 1-plain mortar

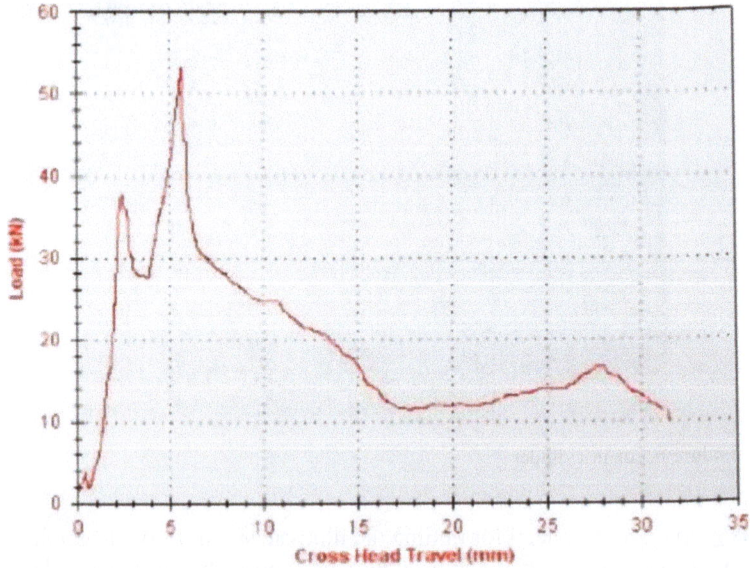

Fig. 6 Load versus deflection curve for type 2-mortar + fiber

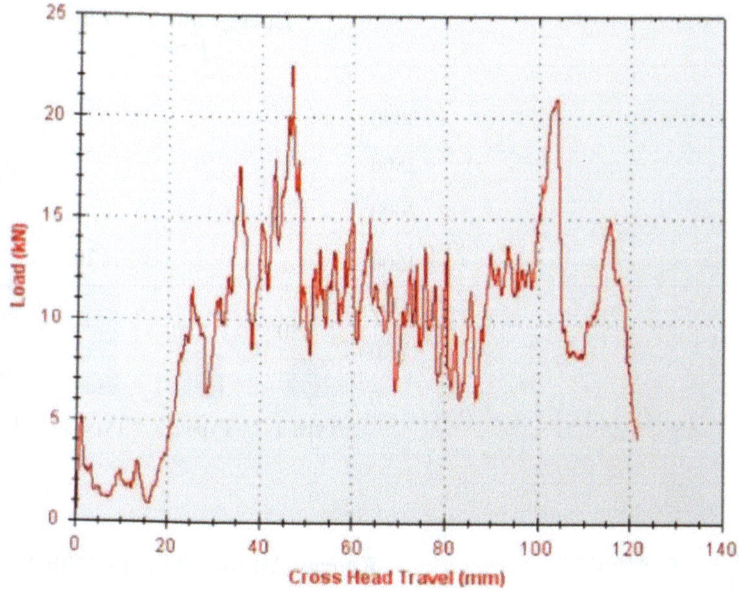

Fig. 7 Load versus deflection curve for type 3-mortar + tin cans

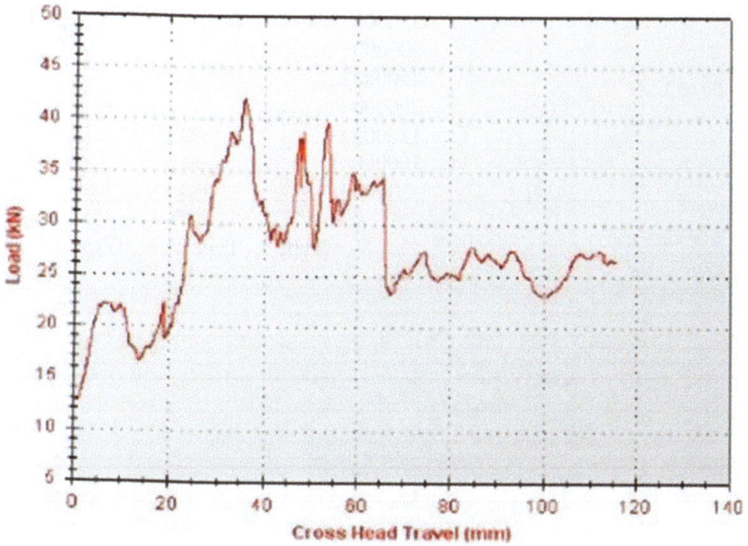

Fig. 8 Load versus deflection curve for type 4-mortar + fiber + tin cans

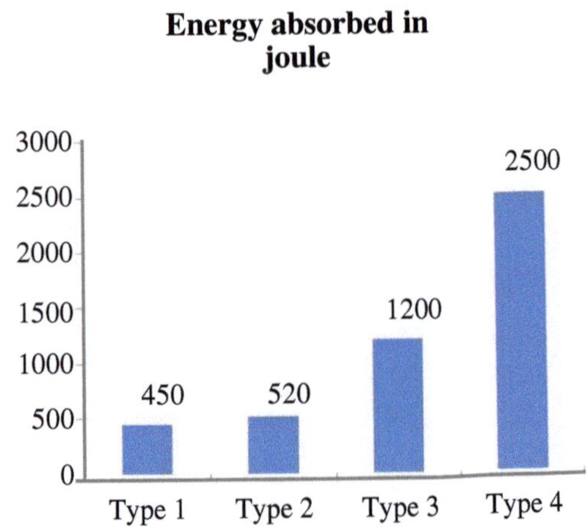

Fig. 9 Energy absorption graph (J)

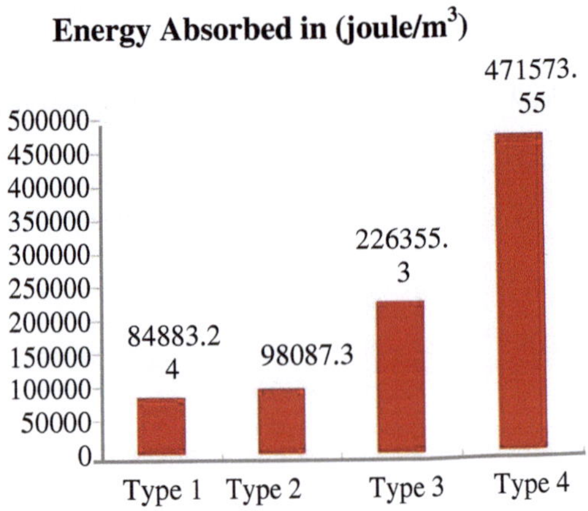

Fig. 10 Energy absorption graph (J/m^3)

Fig. 11 Quantity of mortar (m^3)

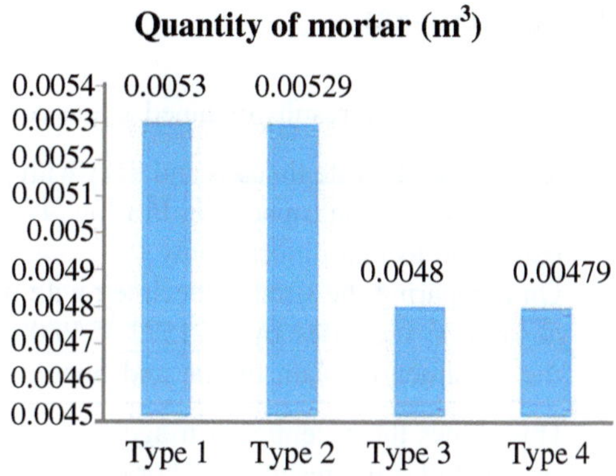

Fig. 12 Cost of cylinder (Rs.)

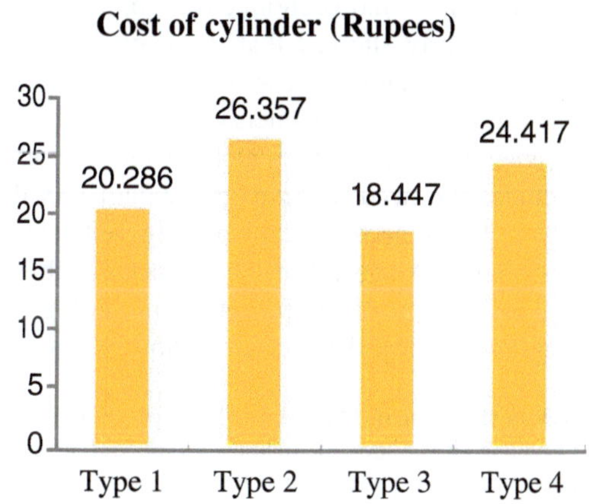

Fig. 13 Cost of cylinder per m^3 (Rs.)

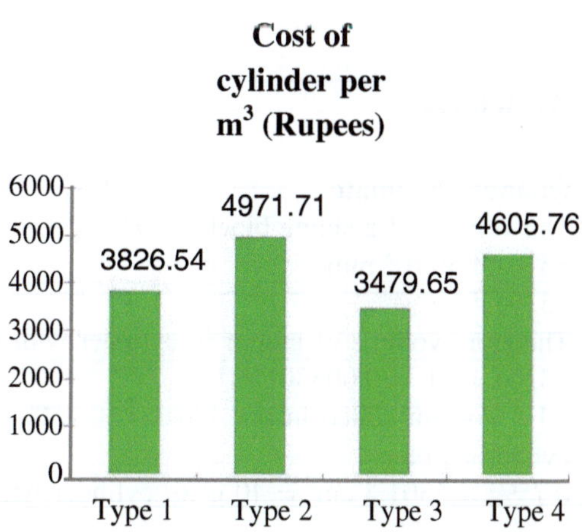

9 Conclusion

Followings are the results obtained while conducting the experiments-

1. The usage of waste tin cans and fiber with plain mortar increases the volume and energy absorption capacity in J/m^3 by (471573.55/84883.24) = 5.56 times when compared to plain mortar only.
2. On comparing the similar specimen with plain mortar, tin, and fiber, the energy absorption increases by (471573.55/98087.3) = 4.81 times when compared to the specimen of plain mortar and fiber.

This shows the potential of usage of waste tin cans for energy absorption. This paper concludes the results of laboratory tests and important research findings and the potential of using tin cans wastes in building construction materials which focus on sustainable and economic development. While conducting the experiments, the results clearly showed that the energy absorbed using the combination of plain mortar, fiber, and tin cans is very high. Hence, it can be concluded that there is a wide difference in the energy absorption between the plain mortar (fibers and tins cans) and the other three mixes. This paper also showcase the research gaps, which includes the need to develop standard mix design for solid waste-based building materials, the need to develop energy-efficient method of processing solid waste use in concrete, the need to study the actual behavior and performance of such building materials in practical application providing lightweight, and the usage of tin cans can be economical for conservation of environment.

Acknowledgements We would like to place on record, our grateful thanks to SRM University for providing all facilities and help in carrying out this project. We are extremely grateful to Dr. K. S. Satyanarayanan, Professor and Head, Department of Civil Engineering for the encouragement and support provided during the project work. Also, we would like to thank Mr. S. Srinivasa Senthil, our sub-guide for the valuable guidance from the beginning to the end.

Appendix

Quantity Estimate
The volume of a single block is $\pi r^2 h = \pi \times 75^2 \times 300$
$=5301.4 \times 10^3$ mm^3
=5301.4 mL
Therefore volume of mortar in cylinder with plain mortar only
=5301.4 mL = 0.005301 m^3
Fiber content added in one block is 0.75% of the volume of the mortar filled in cylindrical block.
0.75% $\times$ 5301.4 cm^3 = 40 g for cylinder with plain mortar and fiber only

Therefore volume of fiber used
=40/8.05 (since density of steel = 8.05 g/cm^3) = 4.97 cm^3
=4.97 mL
Volume of mortar in cylinder with mortar + fiber
=5301.4 − 4.97
=5296.43 mL = 0.005296 m^3
Volume of mortar in cylinder with mortar + tin
=5301.4 − 500
=4801.4 mL = 0.004801 m^3
For the cylinder having plain mortar, tin and fiber, mass of fiber used is calculated in following steps
Volume of plain mortar
=[volume of cylinder − volume of tin cans]
=5301.4 − 500
=4801.4 mL (reduced volume)
Mass of fiber used
=0.75% of 4801.4
=(0.75/100) × 4801.4
=36 g
Volume of fiber
=36/8.05 = 4.47 cm^3 = 4.47 mL
Therefore volume of total mortar is used in cylinder
=(volume of cylinder − volume of tin cans − volume of fiber)
=5301.4 − 500 − 4.47
=4796.92 mL = 0.004797 m^3

Details of Cost Estimation
Total volume of cylinder = 0.0053 m^3
Volume of mortar in plain mortar cylinder
=0.0053 m^3
As cement:sand = 1:3
Therefore quantity of cement
=[0.0053014/(1 + 3)] = 0.001325 m^3
Density of cement = 1440 kg/m^3
1 bag of cement = 50 kg of cement
Volume of cement = 50/1440 = 0.0347 m^3
Therefore for volume of 0.001325 m^3
=[(50/0.0347) * 0.001325]
=1.9 kg of cement for 1 cylinder
Therefore for the plain mortar cylinder the volume of mortar used is 1.9 kg
1 bag = 50 kg of cement = Rs. 375
1.9 kg = [(375/50) * 1.9]
=Rs. 14.25 (cement price)
Volume of sand = 3 * 0.001325 = 0.003975 m^3
0.0283 m^3 = Rs. 43

Therefore for 0.14037 m^3 = [(43/1) * 0.14037]
=Rs. 6.036 (sand price)
Total price = 14.25 + 6.036
=Rs. 20.286 (for plain mortar cylinder)
For 1 m^3 = [20.286/0.0053014]
=Rs. 3826.54
Similarly,
Volume of mortar used in cylinder filled with plain mortar + fiber only
=0.005296 m^3
Therefore volume of cement used
=[0.005296/(1 + 3)]
=0.001324 m^3
Quantity of cement used in cylinder
=0.001324/0.0347
=0.0382 bags
1 bag = 50 kg
Therefore 0.0382 bag = 1.91 kg for 1 cylinder
1 bag = 50 kg of cement = Rs. 375
Therefore for 1.91 bag = Rs. 14.325
Volume of sand = 3 * 0.001324
=0.00397 m^3
0.0283 m^3 = Rs. 43
Therefore for 0.00397 m^3 = Rs. 6.032
Quantity of fiber used = 4.97 mL = 0.04 kg
1 kg of fiber = Rs. 150
Therefore for 0.04 kg = Rs. 6
Total cost of 1 cylinder
=[14.325 + 6.032 + 6]
=Rs. 26.357
For 1 m^3 = [26.357/0.0053014]
=Rs. 4971.71
Now estimation for cylinder with plain mortar + tin cans only
Volume mortar used in cylinder
=0.0048 m^3
Quantity of cement used
=[0.0048/(1 + 3)]
=0.0012 m^3
Quantity of ccmcnt in bag
=[0.0012/0.0347]
=0.0346 bags
1 bag of cement = 1.73 kg
For 0.0346 bags = 1.73 kg
1 bag = 50 kg = Rs. 375
For 1.73 kg = Rs. 12.975
Quantity of sand used

=3 * 0.0012
=0.0036 m^3
0.0283 m^3 = Rs. 43
For 0.0036 m^3 = Rs. 5.4667
Total cost of a cylinder
=[5.4667 + 12.975]
=Rs. 18.4417
For 1 m^3 = [18.447/0.0053014]
Now estimation for cylinder with all mortar, fiber and tin cans
Volume of mortar used
=0.004797 m^3
Quantity of cement used
=[0.004797/(1 + 3)]
=0.001199 m^3
Quantity of cement in term of bag
=0.001199/0.0347
=0.03455 bags
1 bag = 50 kg
Therefore in 0.03455 bags = 1.7277 kg
1 bag = 50 kg = Rs. 375
For 1.7277 kg = Rs. 12.957
Quantity of sand used in cylinder = 0.001199 * 3
=0.003597 m^3
0.0283 m^3 = Rs. 43
Therefore in 0.003597 m^3 = Rs. 5.46
Now quantity of fiber used = 4.97 mL = 0.04 kg
1 kg = Rs. 150
Therefore in 0.04 kg = Rs. 6
Total cost of a cylinder = [12.957 + 5.46 + 6] = Rs. 24.417
For 1 m^3 = [24.417/0.0053014]
=Rs. 4605.76

References

1. Fanella, D.A., Naaman, A.: Stress strain properties of fiber reinforced mortar in compression. ACI J **82**(4), 475–483 (1985)
2. Abrishami, H.H., Mitchell, D.: Influence of steel fibers on tension stiffening. ACI Struct J **94**, 769–776 (1997)
3. Kukreja et al.: Ultimate strength of fiber reinforced concrete slabs. In: Proceedings International Symposium of Fiber Reinforced Concrete, pp. 237–255. Madras, 1 (1987)
4. Zollo, R.F.: Fiber reinforced concrete: an overview after 30 years of development. Cem. Concr. Compos. **19**, 107–122 (1997)
5. Ravinder, K.: Strength characteristics of coca-cola tin waste as fibres in concrete. Int. J. Adv. Res. Found. **3**(2), 9–12 (2016)

MIX
Papier aus verantwortungsvollen Quellen
Paper from responsible sources
FSC® C105338

If you have any concerns about our products,
you can contact us on
ProductSafety@springernature.com

In case Publisher is established outside the EU,
the EU authorized representative is:
Springer Nature Customer Service Center GmbH
Europaplatz 3, 69115 Heidelberg, Germany

Printed by Libri Plureos GmbH
in Hamburg, Germany